BIRD COLORATION

Bird Coloration

VOLUME I

Mechanisms and Measurements

EDITED BY

Geoffrey E. Hill
and Kevin J. McGraw

HARVARD UNIVERSITY PRESS
Cambridge, Massachusetts
London, England · 2006

Copyright © 2006 by the President and Fellows of Harvard College
All rights reserved
Printed in the United States of America

Library of Congress Cataloging-in-Publication Data

Bird coloration / edited by Geoffrey E. Hill and Kevin J. McGraw.
    p.   cm.
  Includes bibliographical references (p.) and indexes.
  ISBN 0-674-01893-1 (alk. paper) (volume 1)
  ISBN 0-674-02176-2 (alk. paper) (volume 2)
    1. Birds—Color. I. Hill, Geoffrey E. (Geoffrey Edward) II. McGraw, Kevin J.
QL673.B555   2006
598.147'2—dc22   2005046126

Contents

Preface   *vii*

I. Perception and Measurements

1. Color Perception   *3*
   INNES C. CUTHILL

2. Quantifying Colors   *41*
   STAFFAN ANDERSSON AND MARIA PRAGER

3. Analyzing Colors   *90*
   ROBERT MONTGOMERIE

4. Effects of Light Environment on Color Communication   *148*
   MARC THÉRY

II. Mechanisms of Production

5. Mechanics of Carotenoid-Based Coloration   *177*
   KEVIN J. MCGRAW

6. Mechanics of Melanin-Based Coloration   *243*
   KEVIN J. MCGRAW

7. Anatomy, Physics, and Evolution of Structural Colors   *295*
   RICHARD O. PRUM

8. Mechanics of Uncommon Colors:
Pterins, Porphyrins, and Psittacofulvins　　　　*354*
KEVIN J. MCGRAW

9. Cosmetic and Adventitious Colors　　　　*399*
ROBERT MONTGOMERIE

III. Controls and Regulation of Expression

10. Hormonal Control of Coloration　　　　*431*
REBECCA T. KIMBALL

11. Genetic Basis of Color Variation in Wild Birds　　　　*469*
NICHOLAS I. MUNDY

12. Environmental Regulation of Ornamental Coloration　　　　*507*
GEOFFREY E. HILL

Acknowledgments　　　　*561*

Contributors　　　　*565*

Species Index　　　　*569*

Subject Index　　　　*581*

Color illustrations follow p. 340

Preface

Interest in the bold and brilliant colors of birds comes naturally. Like birds, but unlike most other mammals, humans are visual animals. We think visually; we express ourselves visually; we respond strongly to visual stimuli. Much more than the family dog or cat, we are struck by the vivid scarlet wings of a Ross's Lourie or the iridescent blue chest of a Fork-tailed Woodnymph. To wonder how and why birds should display such pleasing coloration is inescapable.

Interest in the colorful displays of birds has followed an interesting cycle over the past 150 years, from an initial focus on evolution and function by Darwin and Wallace, to an almost exclusive focus on mechanistic control through the first three-quarters of the twentieth century, and back to a focus on function and evolution with relatively little interest in mechanisms through the last quarter of the century. At the close of the twentieth century and into first few years of the new millennium, a more balanced approach to the study of bird color displays is emerging. It has become clear to scientists interested in bird coloration that a firm understanding of function and evolution can only be achieved through an understanding of the mechanistic basis for coloration, and likewise that the mechanisms of avian coloration can only be fully understood in the context of their function and evolution. An approach combining the studies of proximate control, function, and evolution is becoming the paradigm in research on bird coloration. Thus it is fitting that the first of our two volumes on bird coloration is a thorough review of the measurement, analysis, and mechanisms of avian color displays.

The mechanistic basis of the color displays of animals was nicely summarized in several books in the mid-twentieth century (Fox and Vevers 1960; Needham 1974; Fox 1976), but breakthroughs in melanin and carotenoid analytical techniques and a new theoretical and microscopic approach to the study of the nanostructural basis of nonpigment colors have left these reviews of color mechanisms long out of date. Light environment was rarely considered before John Endler took up the topic in detail in the late 1980s, so this line of research represents an entirely new topic since previous reviews of the proximate control of bird coloration were written. Likewise, avian perception, and particularly the importance of the ultraviolet vision in birds, is a new topic, not previously incorporated into comprehensive reviews of avian coloration.

Not only do we know more about the pigments and the structural mechanisms involved in color displays, but the tools for studying bird colors have improved tremendously in the past two decades. Quantification of color displays with subjective rankings or by visual comparison to color plates has given way to accurate spectrometric techniques that now routinely include the ultraviolet portion of the spectrum that is part of the perceptual world of birds. New and better means to analyze spectral data are also now available. New mathematical models have been developed to account for the light environment and the visual sensitivity of intended receivers when quantifying color. Since the topic was last comprehensively reviewed, many experiments have been conducted exploring the role of genes and the environment in shaping color expression, and bird colors have emerged as a model system in behavioral ecology for understanding how different types of sexual signals may communicate different sets of information about an individual.

The need for a review and synthesis of this tremendous and growing literature on avian coloration was clear, but the task was too big to be undertaken by a single author or pair of authors. So we solicited the aid of the world's leading experts on avian coloration to summarize the current state of knowledge. From the outset, we intended this to be an edited volume that was tightly structured. We began with a table of contents of the topics that we thought were most important to be covered in such a review and then searched for leading experts on each topic to write the chapters. The authors were instructed to create a thorough synthetic review of their topic, rather than a showcase of their own research programs. The result is a comprehensive review of the literature on avian coloration. Originally we intended one volume to include measures, mechanisms, function, and evolution, but with comprehensive treatment of topics came large chapters and a manuscript too long for one volume.

The topics fell naturally into two volumes, however, and nothing of our original goal was sacrificed when we divided one long manuscript into two more manageable parts.

We divided this volume into three sections. The first section, on perception and measurement, provides the necessary background for the studies of avian coloration. The second section of the book focuses on the mechanisms of production. Chapters in this section review the pigment literature and approaches to studying the mechanisms by which nanostructures produce coloration. The third section of the book considers factors that control color expression. Here the effects of genes, environment, and hormonal control, which connects genetic and environmental effects, are discussed as a prelude to considerations of trait function. We are now at the inception of studies that map color traits onto avian genomes, and the potential for such gene mapping on the study of ornamental coloration is enormous.

It is truly an exciting time to be studying avian coloration. We hope that this book serves as a basic reference for a broad range of topics related to bird coloration, and that it will act to stimulate new studies to fill gaps in our current understanding of the proximate controls of the brilliant, striking, and varied color displays of birds.

References

Fox, D. L. 1976. Animal Biochromes and Structural Colours. Berkeley: University of California Press.

Fox, H. M., and G. Vevers. 1960. The Nature of Animal Colors. New York: Macmillan.

Needham, A. E. 1974. The Significance of Zoochromes. New York: Springer-Verlag.

I

Perception and Measurements

1

Color Perception

INNES C. CUTHILL

Objects reflect, transmit, or emit varying amounts of electromagnetic radiation of different wavelengths due to differences in their chemical and structural composition. Color vision has evolved many times to exploit this fact, so that animals can detect and discriminate between objects. Indeed, it is the capacity for discriminating between objects based on differences in the relative amounts of different wavelengths of light, rather than the absolute amount of light, that defines color vision. However, different taxa have come up with different solutions to the problem of extracting this information from the light entering their eyes, dependent on their ecology, the biological significance of different visual tasks, and design constraints imposed by such factors as body size and evolutionary history. Even in a small taxonomic group, such as the vertebrates, visual pigments and other key determinants of visual perception have been lost and gained over relatively short evolutionary periods. For this reason, it is unwise to extrapolate from human color experience to that of other animals. In this chapter, I start by reviewing the general principles of color vision, mainly with reference to the species we understand best (humans), and then explore what is to us the rather exotic visual system of birds.

General Principles of Color Vision

Although there is serious debate among philosophers and cognitive psychologists about the subjective experience we call "seeing color," neuroscientists

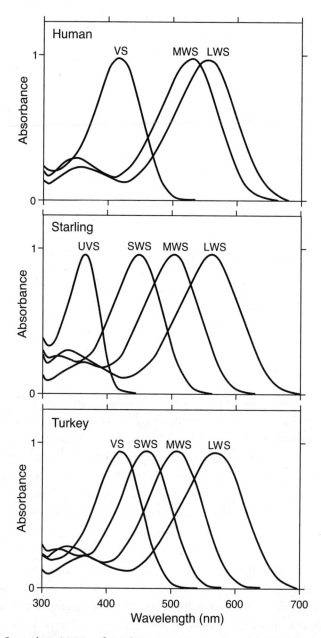

Figure 1.1. Spectral sensitivities of visual pigments, measured using MSP, in the single cones of humans, European Starlings, and Domestic Turkeys. Curves are fitted rhodopsin templates that have been standardized to a maximum of 1 for ease of comparison. Mean values for the wavelengths of maximum absorbance (λ_{max}) for the visual pigments are: human VS = 419 nm, MWS = 531 nm, LWS = 558 nm (Dartnall et al. 1983); starling UVS = 362 nm, SWS = 449 nm, MWS = 504, LWS = 563 nm (Hart et al. 1998); turkey VS = 420 nm, SWS = 460 nm, MWS = 505 nm, LWS = 564 nm (Hart et al. 1999).

studying other species adopt more pragmatic definitions, the most common being the capacity to discriminate between different wavelengths of light independent of their intensity (Jacobs et al. 1981; Thompson et al. 1992; Kelber et al. 2003). If a physicist wants to determine the amount of light of different wavelengths coming from an object (the spectral radiance), he or she would use a spectrometer. A typical spectrometer uses a diffraction grating to split the incoming light into its constituent wavelengths and measures the intensity of light at each wavelength or fraction of a wavelength (Chapters 2 and 3). The measurement, whether in light quanta (photons) or energy per unit area, time, and wavelength, is an objective physical quantity independent of the observer. The waveband measured and the degree of precision depends only on one's interests and on how much one is willing to spend on the machine. In principle, one could design an eye that worked something like a spectrometer, but in practice, evolution has found different solutions. The eyes of humans, birds, and most other animals do not split light into component wavelengths. Instead the same light enters all photoreceptive cells, but different photoreceptors respond selectively to different, and fairly broad, wavebands. The waveband to which different photoreceptor cells respond depends on the light-absorbing pigments that they contain. In humans, color vision is based on just three photoreceptor types, the so-called "red," "green," and "blue" cones, each respectively more likely to absorb light in the long-, medium-, and short-wave portions of the human-visible spectrum (Figure 1.1). This design would seem to entail a dramatic loss of information compared with what the spectrometer can offer: three crude samples compared to the thousand-plus measurements that a budget machine can offer over the same wavelength range. However, whereas the spectrometer gives great precision about the spectrum of light at a single point, the eye gains by providing spatial information: the vertebrate retina is a two-dimensional array of spectral samples. In fact, the trade-off is not huge in terms of loss of information about spectral radiance, because the spectra from natural objects are usually themselves broad waveband. Therefore three broad waveband samples are sufficient to discriminate between the vast array of objects that humans naturally encounter (Buchsbaum and Gottschalk 1983; Chiao et al. 2000). Three wavebands would not seem to be enough for birds, however, because they have color vision based around four cone types: birds are tetrachromatic, whereas humans are only trichromatic.

Color vision depends on having different photoreceptor types tuned to different wavebands of light, but the perception of color is not the product of how much a given photoreceptor type is stimulated in isolation, but how much it is stimulated relative to other photoreceptor types. Any one photoreceptor is color-blind, in that although it is more likely to absorb photons of one wavelength than another, once absorbed, the information on the photon's wavelength is lost. Thus light of a high intensity away from a receptor's peak sensitivity can produce the same response from that cell as lower intensity light at the wavelength of maximum sensitivity. Information on the wavelength distribution of the light entering the eye is instead captured by comparing the stimulations of different photoreceptor types. This is the principle of "opponent processing," for which specialized nerve cells in the retina (ganglion, horizontal, and amacrine cells) have inhibitory and excitatory responses when different photoreceptor types are stimulated and thus "compare" the relative stimulation of photoreceptors (for accessible introductions, see Jacobs 1981; Thompson et al. 1992; Bradbury and Vehrencamp 1998).

In an important sense, the color information is encoded by the upstream nerve cells, not the photoreceptors themselves, but the actual color perceived depends on further higher-level neural interactions and, some have argued, a true subjective experience of color requires consciousness (see Kelber et al. 2003: 88). The latter issue is at the boundaries of neuroscience and philosophy, but the important message is that color is not a physical quantity, but a psychophysical one that is entirely dependent on the photoreceptors and neural processing of the organism concerned. Human color experience can be divided into quasi-orthogonal dimensions of brightness or value (how light/dark an object is; an achromatic dimension), hue (e.g., red, green, purple; the term most commonly equated with color), and saturation or chroma (the amount of white added to a hue; e.g., pink is a less saturated form of red). These psychological constructs have neurobiological correlates (e.g. ganglion cells in opponent-processing networks), but currently we cannot be certain whether other animals with color vision have subjective experiences equivalent to the human percept of, for example, saturation. What we can be certain of, is that if an animal has different photoreceptors in terms of spectral sensitivity or number of types, and/or combines their outputs in different ways from humans, then colors will be perceived differently (Endler 1990; Thompson et al. 1992; Bennett et al. 1994). When a bird and a human look at the sky, they really do not see the same color, even though the light reaching them is the same. This conclusion is not a philosophical point about the nature of subjective experience; it is a consequence of basic neurobiology.

Dimensionality of Color Vision

Normally sighted humans are termed "trichromatic," not because they have three functional cone types, but because color discrimination is based on neural comparison of all three photoreceptor types, and, as a corollary, all human-perceived colors can be simulated by the appropriate mixing of monochromatic red, green, and blue lights. All colors perceptible by humans can thus be represented in a three-dimensional color space. Depending on the question at hand, the axes of the color space can be high-level cognitive constructs measured through psychophysical experiments (e.g., brightness, hue, and saturation, as utilized in the Munsell color system; Munsell Color Company 1976; Plate 4), known neural channels (luminance plus red-green and blue-yellow opponency; e.g., Sumner and Mollon 2000b), or the photon catches of the photoreceptors (for further discussion of color spaces, see Goldsmith 1990; Thompson et al. 1992; Kelber et al. 2003; Chapter 3). Photon capture spaces are the form most commonly used if, as with birds, the opponent channels have not been fully characterized. Often it is convenient to rescale the photon captures of each cone to a proportion of the photon captures when viewing a reference stimulus, such as a perfect white reflector (e.g., Fleishman and Endler 2000; Maddocks et al. 2001) or the background (e.g., Vorobyev et al. 1998). The former is conceptually easier for representation, but the latter is more biologically realistic, as it captures the phenomenon of sensory adaptation, whereby receptors respond to differences relative to their adapted state rather than to the absolute stimulus. Also, modeling the cone photon captures as relative to the adapting background provides a simple means of implementing color constancy, the phenomenon by which the colors of objects are relatively invariant when the illuminating light changes (e.g., see Kelber et al. 2003).

If color (i.e., hue) is the variable of primary interest, then elimination of the brightness component reduces the number of dimensions by one (Figure 1.2a) and the resulting color space is called a "chromaticity diagram" (detailed descriptions of color spaces and the equations used to plot them can be found in Kelber et al. 2003; Endler and Mielke, in press). For a dichromat, if the short- and long-wave cone captures are Q_S and Q_L, respectively, then any hue can be represented as a projection from the two-dimensional color space onto the line $Q_S + Q_L = 1$. Such a chromaticity diagram therefore disregards information on absolute stimulus intensity and represents only relative cone

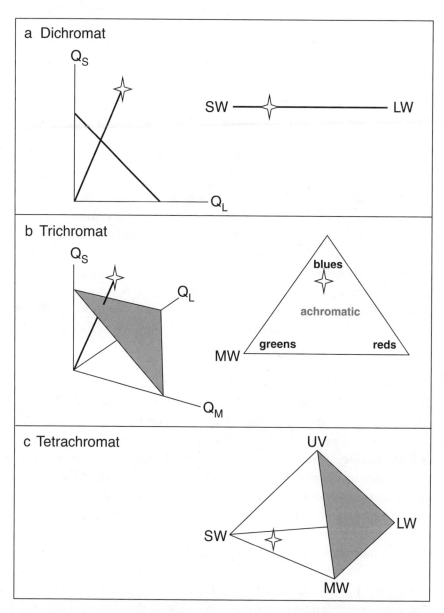

Figure 1.2. Color spaces for photoreceptor (cone) photon catches (left) and corresponding chromaticity diagrams (right) for (a) a dichromat, such as a typical mammal; (b) a trichromat, such as a human; and (c) a tetrachromat, such as a bird. Depiction of the cone-capture color space of a bird is not possible, as it is four-dimensional. As an example, in each of the chromaticity diagrams, the star represents the location of a short-wave radiance spectrum in the three cone capture spaces and how it projects onto the line, triangle, or tetrahedron, respectively, of the corresponding chromaticity diagrams. The adapting background used for the calculations (e.g., white or gray) would lie at the center of each chromaticity diagram.

captures; a dichromat's chromaticity color space is therefore one-dimensional. For trichromats, such as humans, the color space can be represented as a triangle with the apices representing the proportionate cone captures of the short-wave sensitive (SWS), medium-wave sensitive (MWS), and long-wave sensitive (LWS) cones (Figure 1.2b). The chromaticity color space is a projection onto the plane $Q_S + Q_M + Q_L = 1$ in the volume represented by the absolute short-wave, medium-wave, and long-wave photon captures. It is important to realize that this approach makes an implicit assumption that hue perception is independent of the absolute brightness of the stimuli. The reduction of dimensions is designed to capture the fact that hue perception relies on relative photon captures of the photoreceptors; thus it makes sense to disregard the information on absolute photon capture when constructing the color space. Avian (or turtle or goldfish) chromaticity color space, being based on the relative photon captures of four cones, is the volume $Q_{UV} + Q_S + Q_M + Q_L = 1$ within a four-dimensional space, with Q_{UV} the photon capture of an ultraviolet (UV)-sensitive (or violet-sensitive) cone type. Analogous with the triangular color space of humans, such a tetrachromatic color space is a tetrahedron with the proportionate captures of each of the four cone types at each apex (Figure 1.2c).

How Animals See Color

The main problem facing scientists when assessing whether another species can see color is that, unless one is careful, achromatic (brightness) cues can often also be used to discriminate between objects. Those readers old enough to have seen black-and-white television may remember that, in fact, they did not have that great difficulty distinguishing between their favorite football teams on the basis of the "grayness" of their shirts. For example, reds often look darker than greens because the mammalian luminance mechanism is most sensitive at medium wavelengths; on this basis, your pet dog can be trained to pick up red, but not green, balls. Ideally, in a discrimination or associative learning paradigm, one would equalize brightness differences between the objects to be discriminated, so that only hue differences could be used (Jacobs 1981). In practice, this requires knowledge of exactly how the animal perceives brightness, which might be a function of the simple sum of cone sensitivities, a weighted sum, or just one photoreceptor type (Thompson et al. 1992). To circumvent this problem, the usual approach is to vary the brightness of the stimuli randomly between presentations (Jacobs 1981) or introduce random spatial variation in brightness within each stimulus (e.g., Osorio et al. 1999b). Even so, with a finite number of stimuli, there may be an unintended average difference in the luminance of the test colors; the solution is to introduce "probe" trails in which test stimuli are deliberately varied in either brightness or hue to assess which component of color is being used (e.g., Smith et al. 2002).

Having assessed that an animal has color vision (i.e., can discriminate objects on the basis of wavelength distribution alone), three methods have been most widely used to determine the mechanism behind this. Procedurally, the most simple is electroretinography (Jacobs et al. 1996a). This method involves placing a surface electrode on the eye and measuring the electrical potential changes when monochromatic (single wavelength) lights are shone onto it. Typically, the neural response is highest where a given photoreceptor has its peak sensitivity, and so the number of peaks can be used to infer how many photoreceptors are involved in color vision (e.g., there would be three peaks in humans). This method is most powerful when the data are compared to a particular model of how photoreceptors interact (e.g., Vorobyev and Osorio 1998). The drawbacks of this method are that: (1) different photoreceptor combinations may contribute to sensitivity under different lighting conditions and so some receptor mechanisms (e.g., luminance perception) may swamp the signal from chromatic mechanisms; (2) it does not provide detailed information on the spectral sensitivities of the different photoreceptors; and (3) even if several photoreceptors are present and detectable via electroretinography, there is no guarantee that they are all involved in opponent processing mechanisms. For example, mice have two opsin gene types, a UV and a long-wave, but they are usually co-expressed in the same cells (Applebury et al. 2000). These two visual pigments contribute two peaks to the electroretinogram (in the UV and green; Jacobs et al. 2004), but because each cone possesses both pigment types, this response might represent achromatic, not color, vision (Neitz and Neitz 2001). (Note that, probably because the relative expression of the two pigment types varies across the retina, the mouse is still capable of some color vision [Jacobs et al. 2004]; my point here is simply that demonstrating multiple peaks in an electroretinogram does not, by itself, prove that an animal has color vision.) The first and second problems can be solved using microspectrophotometry (MSP), which is the only method that can directly measure visual pigment absorption properties. MSP is simply spectrometry at the microscopic scale, such that individual cells or parts of cells can be measured (Liebman 1972; Hawryshyn et al. 2001). However, because

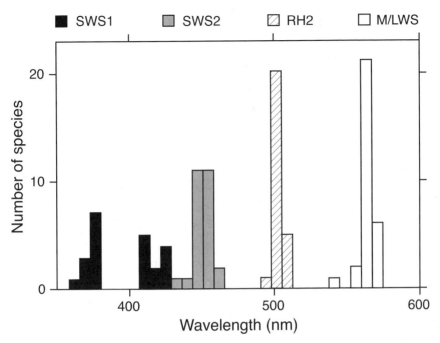

Figure 1.3. Frequency distribution of wavelengths of maximum sensitivity (λ_{max}) of avian cone visual pigments. Only species measured using microspectrophotometry have been included (using data from Hart 2001, 2004). Visual pigments are classified according to the opsin type they contain (see Box 1.1): SWS1 in UVS or VS cones, SWS2 in SWS cones, RH2 in MWS cones and M/LWS in LWS cones. See Box 1.1 for an explanation of the abbreviations.

the necessary combined spectrometer-microscopes have to be custom built, and the objects to be measured are small and delicate (the outer segments of cones, where the visual pigments lie, can be as small as 1.5 μm in diameter in birds), visual pigments have been measured in relatively few species of bird (Figure 1.3).

MSP characterizes the visual pigments, but proof that the photoreceptors with different visual pigments are actually involved in color vision (problem 3 above) requires behavioral experiments. The stringent requirements for wavelength-based discrimination experiments have already been described. By extending the number of colors to be discriminated, one can determine not merely whether an animal can see color, but the precision and extent of its color discrimination and, in principle, one can isolate particular opponent mechanisms. If carefully designed with respect to known properties of the photoreceptors (e.g., from MSP) and compared to predictions from specific models, this method is very powerful: a good example is Goldsmith and Butler's (2003) test of Vorobyev and Osorio's (1998) prediction that color discrimination is limited by noise in the photoreceptors. Such experiments can involve projection of narrow waveband colored lights onto response keys in a "Skinner Box" (as in Goldsmith and Butler 2003) or printed color patterns on paper targets containing food (as in Osorio et al. 1999b). Although the design of appropriate stimuli is more complicated, the latter approach requires less training and also lends itself to the investigation of other questions about color cognition, such as the role of memory (Osorio et al. 1999a) and ontogenetic experience (Miklosi et al. 2002).

Rather than establishing the limits of color discrimination, an alternative approach is that of color mixture experiments (e.g., Neumeyer 1992). These experiments determine whether any color can be simulated by mixing together two or more monochromatic lights. For example, all the colors perceived by a dichromat, such as a dog, can be simulated by mixing together two monochromatic light sources corresponding to the sensitivities of the two cone types. Thus a monochromatic 500 nm light could not be discriminated from an appropriately weighted sum of, say, monochromatic 420 nm and 570 nm lights. A human, being trichromatic and thus with the extra information available from a third cone, could successfully discriminate between the 500 nm light and the 420/570 nm mixture (the former would look green and the latter purple). However, all human-perceived colors can be simulated by mixtures of three monochromatic "red," "green," and "blue" lights; this is, of course, the basis of color television. Such experiments have shown that Goldfish (*Carassius auratus*), for example, are tetrachromatic, with four monochromatic lights, appropriate to goldfish's four cone types, being necessary to reproduce all goldfish-perceived colors (Neumeyer 1992). However, this method has only been attempted for one species of bird, the Rock Pigeon (*Columba livia*; Palacios et al. 1990; Palacios and Varela 1992), probably because of the technical sophistication and extensive animal training required.

Behavioral experiments are the litmus test for color vision and can establish what opponent channels must contribute to it, but they do not identify the mechanism at the neural level. To do this, one needs single-cell recordings from the various types of nerve cell upstream from the photoreceptors (i.e., ganglion, horizontal, and amacrine cells) when the retina is illuminated with different monochromatic lights. This experiment has not been done with birds, but data exist for turtles, which have broadly similar retinas to birds, and these data are discussed below.

Avian Color Vision

Three important differences are immediately apparent between avian and human color vision (reviewed by Cuthill et al. 2000; Hart 2001): birds have (1) a wider spectral range, (2) more cone types in the retina, and (3) colored oil droplets that filter the light entering the cones (Plate 1). I review these differences in turn, before considering the evidence that birds utilize interactions between their greater diversity of cone types to enjoy a color space of higher dimension than that experienced by humans. First, however, it is worth discussing the possibility of polarization sensitivity in birds, because, if present, this sensitivity would considerably increase the complexity of investigating both avian color vision and the color patterns relevant to birds (e.g., sexual and social signals, cryptic and aposematic colors in prey). Light coming from the sun has wave energy in all planes orthogonal to the direction of transmission, but on striking or passing through certain media (including dust and water in the atmosphere), the energy in some planes is absorbed more than in others—the light has become (partially) plane polarized. Because different objects polarize light to varying degrees, polarization can provide information about objects independent of their hue or brightness. Bees are the best-known examples of animals that exploit such information, using the polarization patterns in the sky to orient with respect to the sun even when the sun's disk is not visible (e.g., see Brines and Gould 1982). However, polarization vision has also been implicated in the detection of water and oviposition surfaces by insects (Horvath et al. 1997), prey detection by fish and squid (Shashar et al. 1998; Flamarique and Browman 2001), and social signaling in cuttlefish (Shashar et al. 1996). Indeed, cephalopods appear to use polarization vision as an alternative to color vision; most are color-blind, but can discriminate objects on the basis of their surface polarization patterns (Shashar and Cronin 1996). For animals with both color and polarization vision, one cannot assume that the two systems provide independent information in the way that hue and brightness are perceived largely independently in humans. If the same photoreceptors are used, then a change in the angle of polarization (e.g., through the object or viewer moving) provides an equivalent variation in photoreceptor stimulation to that of a change in light intensity. This situation would lead to a change in polarization being perceived as a change in color, and vice versa. In some butterflies, such perception of polarization as coloration seems to be the case (Kelber et al. 2001), and one can only assume that, for the objects that are biologically important to them, this co-variation of color and polarization is not a problem. Other organisms, however, have designed solutions to decouple polarization and color information. Honeybees (*Apis mellifera*), for example, have spirally twisted photoreceptors in most of the eye, to eliminate the inherent sensitivity of insect photoreceptors to plane polarized light; but in the sky-facing portion of the eye, where polarization sensitivity is important for orientation, the photoreceptors are untwisted (Wehner and Bernard 1993). Do birds have polarization vision? This issue is still controversial. In the orientation and navigation literature, experiments using depolarizers or rotation of the angle of polarization show significant effects on orientation (e.g. Kreithen and Keeton 1974; Phillips and Moore 1992; Able and Able 1995; Munro and Wiltschko 1995), as they would with bees. However, such manipulations can also unintentionally alter brightness and, when tested in paradigms that exclude this possibility, pigeons, European Starlings (*Sturnus vulgaris*; Plate 23), and Common Quail (*Coturnix coturnix*) do not show evidence for polarization sensitivity (Coemans et al. 1994a; Vos Hzn et al. 1995; Greenwood et al. 2003). Certainly mechanisms for polarization vision exist in other vertebrates, notably fish (Hawryshyn and McFarland 1987; Cameron and Pugh 1991; Flamarique et al. 1998), but for birds, the jury is still out (Martin 1991).

Spectral Range

The upper limit of vision in birds and humans is about 700 nm (the figure is approximate because very high-intensity infrared radiation can still stimulate the long-wave red cones, although this is a situation not normally encountered in nature). The lower limit of vision in humans is about 400 nm (although short-wave sensitivity is progressively reduced as one ages), whereas in most birds it is closer to 315 nm in the near UV (Cuthill et al. 2000; Hart 2001; near UV wavelengths from 315 to 400 nm are defined as "UV-A"). Humans normally cannot see UV light, not because they lack an appropriate visual pigment (the blue cone of humans contains a SW1 pigment with moderate UV sensitivity; Figure 1.1a and Box 1.1), but because the ocular media preceding the retina absorb wavelengths below 400 nm (Douglas and Marshall 1999). Birds have lenses, corneas, and aqueous and vitreous humors that are transparent to UV-A (i.e., from 400 nm down to ~315 nm), so for birds, UV light is visually useful. The first unambiguous evidence for UV vision in birds came from discrimination experiments with a White-vented Violetear (*Colibri serrirostris*; Huth and Burkhardt 1972) and pigeons (Wright 1972). Subsequent experiments using electroretinography suggested that UV vision was wide-

> **Box 1.1. Visual Pigment Terminology and Evolution**
>
> All vertebrate visual pigments consist of a chromophore molecule bound to a protein called an "opsin." The chromophore of most terrestrial vertebrates, and all birds, is 11-cis retinal, derived from vitamin A_1. Visual pigments containing 11-cis retinal are called "rhodopsins," and their spectral sensitivities depend on the amino acid sequence of their constituent opsin. The nomenclature of opsins is deeply confusing, largely because the most commonly used terms were coined with reference to humans and prior to a broad picture of vertebrate opsin evolution. It is now clear that five opsin families arose early in vertebrate evolution. There are two shortwave-sensitive groups: SWS1, which includes the ultraviolet- and violet-sensitive (UVS and VS) pigments of birds, reptiles, and teleosts, and also the "blue" (or, confusingly, S-) cone of humans and other mammals. An SWS2 group includes the SWS or "blue" cone pigment of birds and nonmammalian vertebrates, but has no homolog among eutherian mammals. There are two medium-wavelength-sensitive opsin families: RH1 is found in the rods, and seems to have evolved from the pigment in the medium wavelength-sensitive cone, termed "RH2." This is the pigment found in the MWS or "green" cone opsin of birds and non-mammalian vertebrates, but again has no homolog among eutherian mammals. Finally, there is an opsin family tuned to medium-to-long wavelengths (M/LWS) that includes the avian "red" cone pigments and both the primate "red" and "green" cone pigments (Okano et al. 1992; Bowmaker and Hunt 1999; Yokoyama and Shi 2000; Shi et al. 2001; Yokoyama 2002).

spread (Chen et al. 1984; Chen and Goldsmith 1986), but it was not until 1993 that an avian UV-sensitive visual pigment was directly measured using MSP (Maier and Bowmaker 1993). At present, positive behavioral, electrophysiological or MSP evidence exists for more than 35 species of birds (Bennett and Cuthill 1994; Cuthill et al. 2000; Hart 2001), with circumstantial evidence from visual pigment gene sequences for another 40 or so (Ödeen and Håstad 2003). The only birds for which there is evidence for lack of UV vision are owls, based on MSP of visual pigments in the Tawny Owl (*Strix aluco*; Bowmaker and Martin 1978) and a behavioral test in Tengmalm's Owl (*Aegolius funereus*; Koivula et al. 1997). Ödeen and Håstad (2003) also failed to amplify the appropriate gene sequence for a UV-sensitive pigment in the Tawny Owl. These results are not themselves conclusive, as UV visual pigments are small and sparsely distributed, and so hard to find using MSP, and sequence amplification may fail for technical reasons (Ödeen and Håstad also failed to amplify DNA in some species where there is no reason to doubt UV sensitivity). However, given their nocturnal habits and thus reduced need for color vision, owls are some of the few birds where one might anticipate the loss of some visual pigments. All other birds appear to have broad similarities in their visual systems, but there are important interspecific differences in visual pigment tuning, oil droplet characteristics, and thus predicted visual performance.

Avian Cone Types and Visual Pigments

Although behavioral or electrophysiological measurements can be used to infer the number and types of photoreceptors an animal possesses (e.g., Goldsmith et al. 1981; Chen and Goldsmith 1986; Goldsmith 1986), measurement of the absorption properties of the visual pigments themselves requires MSP. The typical avian pattern is four types of single cone and one type of double cone (Cuthill et al. 2000; Hart 2001). The double type is not found in mammals and consists of a larger and smaller cell in close physical and electrical contact (Smith et al. 1985; Bowmaker et al. 1997). This diversity of cone types is not unusual among vertebrates, and, among other taxa, turtles have a similar retinal complement (Bowmaker 1991a). Of the cone types, double cones are usually the most abundant (e.g., ~50% of the cone in a typical passerine, the starling; Hart et al. 1998; Figure 1.4). Both members of the pair contain the same visual pigment as the long-wave-sensitive single cone, but the larger member has a less pigmented oil droplet (see the section below on oil droplets), and the smaller member has a reduced, or no, oil droplet (Bowmaker et al. 1997; Hart 2001). For this reason, the double cones of birds have a broad spectral sensitivity that overlaps both long- and medium-wave single cones (Figure 1.5c); a priori, this overlap would suggest that they add nothing to color vision beyond that possible with the single cones. Indeed, measurement of overall spectral sensitivity in the Pekin Robin (*Leiothrix lutea*; Plate 1) showed peaks in sensitivity corresponding to the sensitivities of the four single cone types, but with no apparent contribution from the double cones (Maier and Bowmaker 1993). Subsequent modeling of color discrimination thresholds are consistent with color vision involving only single cones (Vorobyev and Osorio 1998), and behavioral tests on Domestic Chicken (*Gallus gallus*;

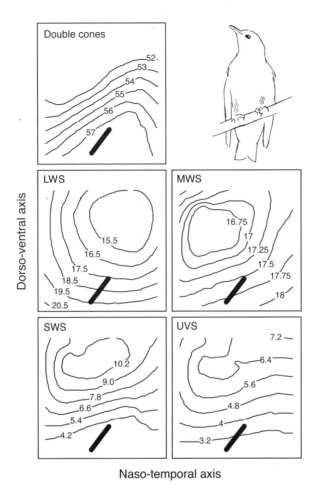

Figure 1.4. Contour plots of the percentages of different cone types in the eye of the European Starling (adapted from Hart et al. 1998). The view is of a notional left eye, seen from the corneal aspect, but values are means across males and females, which do not differ statistically, and left and right eyes, which do (see Figure 1.7). The thick line on each graph indicates the approximate position of the pecten. LWS = long-wave sensitive cones; MWS = medium-wave sensitive cones; SWS = short-wave sensitive cones; UVS = ultraviolet sensitive cones.

Plate 26) suggest that double cones are involved in achromatic (noncolor-based) tasks (Jones and Osorio 2004). The latter research showed that texture discrimination was possible when the patterns differed in (calculated) intensity for the double cones, but not when intensity or color cues were only available to single-cone mechanisms. This result gives double cones the same role

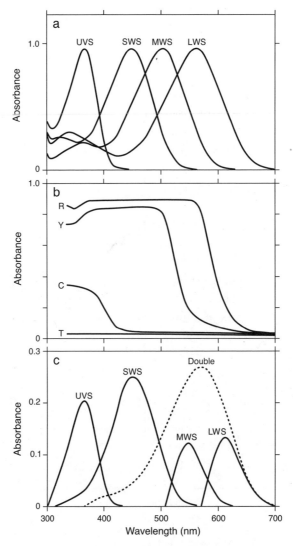

Figure 1.5. Effects of oil droplets on the effective spectral sensitivities of a starling's cones. (a) Visual pigment sensitivities, as shown for the starling in Figure 1.1. (b) Absorbance spectra of oil droplets: T (transparent), C (colorless), Y (yellow), and R (red), as found in the UVS, SWS, MWS, and LWS single cones, respectively. Curves are based on the data in Hart et al. (1998), as calculated for Cuthill et al. (2000), but may underestimate the efficiency of short-wave absorption by the droplets (see Hart 1998). (c) Predicted effective spectral sensitivities of the single cones, after taking into account absorption by the oil droplets and other ocular media. Wavelengths of peak sensitivity for the UVS, SWS, MWS, and LWS single cones shift to 371 nm, 453 nm, 543 nm, and 605 nm, respectively. Also, plotted (dotted curve) is the average spectral sensitivity of the double cones after taking into account droplet and media absorption (they contain the same visual pigment as the LWS single cones). In fact, the spectral sensitivity of a starling's double cones varies across the retina, because cones in the ventral retina (upward looking) have more densely pigmented droplets and so absorb more UV and short-wave radiation.

as the achromatic luminance signal from mammalian single cones. Indeed, motion detection in the pigeon shows a similar broad spectral sensitivity to that of the double cones (Campenhausen and Kirschfeld 1998), and motion detection in humans is also a task that is primarily color-blind. Other functions, such as detection of polarization patterns, have been suggested for double cones (Young and Martin 1984; Cameron and Pugh 1991), and a contribution to color perception cannot be ruled out (see Palacios and Varela 1992), but the current consensus is that avian color vision is primarily subserved by the single cones.

The four types of avian single cone are differentiated by virtue of the visual pigment they contain, one from each of the four main classes of opsin that diverged early in vertebrate evolution. Opsins are the protein component of visual pigments and are members of the large class of G-protein–linked cell membrane receptors. The opsin is bound to a chromophore that in birds is always 11-cis-retinal, and so in birds, it is the opsin type that determines the spectral absorbance properties of the visual pigment. All visual pigments have bell-shaped absorbance spectra, and in birds, the wavelength of peak absorbance (λ_{max}) is determined only by the amino acid sequence of the opsins. The LWS cone of birds contains a M/LWS opsin type (Box 1.1) with a λ_{max} of about 565 nm. The MWS cone contains an RH2 opsin with a λ_{max} of about 505 nm (similar to that of the pigment in the rods). The SWS cone contains a SWS2 opsin of λ_{max} between 430 and 460 nm. Finally the pigment conferring UV sensitivity is a SWS1 opsin with a λ_{max} between about 355 and 420 nm. The sensitivities of the SWS1 and SWS2 opsins appear to co-vary, with species that have a longer wavelength λ_{max} for the SWS1 opsin also having a longer wavelength λ_{max} for their SWS2 opsin (Hart 2001). In fact, spectral sensitivities of the avian SWS1 opsins fall broadly into two classes: types with maximal sensitivity around 405–420 nm (species with these pigments are said to have "violet-sensitive" [VS] cones), and opsins with maximal sensitivity firmly in the UV (λ_{max} between 355 and 370 nm; species with these pigments are said to have "ultraviolet-sensitive" [UVS] cones; Bowmaker et al. 1997; Hart 2001; Ödeen and Håstad 2003). Owls aside, there therefore seem to be two main types of color vision system in birds (Figure 1.1): species with VS, SWS, MWS, and LWS cones; and those with UVS, SWS, MWS, and LWS cones.

Until recently, it seemed that the distribution of VS and UVS type color vision followed simple phylogenetic lines (Bowmaker et al. 1997; Cuthill et al. 2000): passerines, including all the songbirds, and psittacines (parrots) having

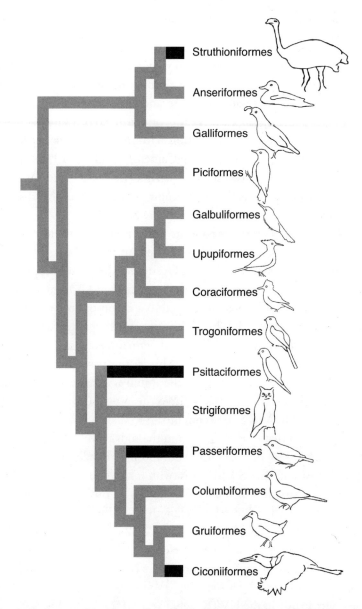

Figure 1.6. Phylogeny of the main orders of birds (following Sibley and Ahlquist 1990), with gray bars representing taxa whose members have exclusively VS cone types (SWS1 pigments with maximum absorbance $\lambda_{max} > 400$ nm) and black bars taxa with some or all members possessing UVS cones (SWS1 pigments with $\lambda_{max} < 370$ nm). In the Struthioniformes, the rhea has UVS pigments, the ostrich VS pigments; in the Passeriformes, most have UVS pigments, but crows and sub-oscines have VS; in Ciconiiformes, only gulls have UVS pigments; among Psittaciformes, only UVS pigments have been found. Data are from Ödeen and Håstad (2003), based on the predicted effects of amino acid substitutions at known spectral tuning sites in the opsin molecules, supplemented by measured values from MSP tabulated in Cuthill et al. (2000) and Hart (2001).

MSP or in vitro expression of these sequences is required. However, if upheld, the work does suggest that the distribution of the UVS and VS variants of avian color vision is not as phylogenetically conservative as previously thought.

All the above leaves humans looking rather impoverished when it comes to photoreceptor hardware. The assumed superiority of human color vision is largely an effect of a comparison with other mammals, so it is worthwhile asking why mammalian color vision is so poor and why primates have secondarily regained some, but not all, of the richness of the birds' color world. The three cone types of humans are shared, with minor differences, with all Old World (catarrhine) primates, but is unusual among eutherian mammals (trichromacy has been discovered in some marsupials; Arrese et al. 2002). Most other eutherian mammals are dichromats, possessing only a single SWS cone type, maximally sensitive in either the UV or violet waveband, and a LWS cone peaking in the green-yellow waveband (Jacobs 1993). Such mammals are color-blind in the sense that red-green color-blind humans are; they can discriminate between some colors, but not to the same extent as trichromats. Most obviously, blues can be discriminated from greens, yellows, and reds, but discrimination among greens, yellows, and reds is compromised. Trichromacy evolved in Old World primates, and independently in some New World (platyrrhine) primates (Jacobs et al. 1996b), through the duplication and subsequent mutation of the gene determining spectral sensitivity of the visual pigment in the long-wave cone (reviewed by Surridge et al. 2003). The "extra" cone type in the medium-to-long-wave part of the spectrum has been suggested to aid the detection and discrimination of fruits of varying ripeness against green foliage (Osorio and Vorobyev 1996; Sumner and Mollon 2000a,b; Regan et al. 2001), or young from old leaves (Dominy and Lucas 2001), and there is some experimental evidence for these advantages to primate trichromacy (Caine and Mundy 2000; Lucas et al. 2003; Smith et al. 2003). However, as superior as primate trichromacy is compared to the color vision of most mammals, it is impoverished by comparison to many other vertebrates. Early in eutherian mammal evolution, two classes of visual pigments were lost, presumably because of a nocturnal niche that did not favor color vision (Jacobs 1993). The four classes of cone visual pigment seen in many teleost fish, reptiles, and birds, arose early in vertebrate evolution (Bowmaker 1998; Yokoyama and Shi 2000). This early origin can create problems of nomenclature (Box 1.1), because gene-sequencing indicates that the blue cone of humans actually contains a visual pigment homologous to the UV/violet class the UVS system, with all other nonpasserines having the VS system. Sequencing of the SWS1 opsins across vertebrates suggests that the VS system is ancestral for birds, and therefore the UVS system of passerines and parrots could be explained by perhaps a single mutational event (Yokoyama et al. 2000). However, recent work by Ödeen and Håstad (2003) and Håstad (2003) has uncovered a far more complicated pattern. The spectral tuning of the avian SWS1 opsin is determined largely by the amino acids present at relatively few sites, of which, site 90 (labeled as per bovine rhodopsin) has the greatest effect (Wilkie et al. 2000; Yokoyama et al. 2000; Shi and Yokoyama 2003). Serine at position 90 appears to be the ancestral state for birds and results in a VS pigment, whereas cysteine at position 90 shifts the λ_{max} downward by about 35 nm to produce a UVS pigment (Wilkie et al. 2000; Yokoyama et al. 2000; Shi and Yokoyama 2003). This seemingly simple correspondence between spectral tuning of the visual pigment and sequence for a small portion of the opsin gene allowed Ödeen and Håstad (2003) to use fast, noninvasive molecular methods to screen 45 species from 35 families of birds, something that would have taken years using traditional approaches. Of the 45 species, eight had been measured using MSP, and the correspondence between the sequence data and the directly measured λ_{max} was good, giving confidence in the molecular approach. From the data on the 37 new species, it appears that UVS pigments have evolved independently at least four times (Figure 1.6). As expected from previous work using MSP, most passerines (measured species came from the Fringillidae, Muscicapidae, Paridae, Passeridae, Sturnidae, and Sylvidae) had opsin sequences indicative of UVS pigments, as did the two species of Psittacidae measured (the Budgerigar [*Melopsittacus undulatus*] and the African Grey Parrot [*Psittacus erithacus*]). Conversely, among the Passeriformes, members of the Corvidae and Tyrannidae had VS pigments. UVS pigments were found in gulls (but not in the closely related Razorbill [*Alca torda*], Common Murre/Guillemot [*Uria aalge*], or any other Ciconiiform). Surprisingly, in a subsequent study that focused on sister taxa to gulls in the order Ciconiiformes, the closely related and ecologically similar terns (Sternidae) were found to have VS pigments (Håstad 2003), as have all other seabirds measured (Hart 2004). Also unexpected on grounds of phylogeny and ecology, was that, whereas the Ostrich (*Struthio camelus*) had the expected VS pigment opsin, the Common Rhea (*Rhea americana*) had a UVS-type sequence. Ödeen and Håstad (2003) and Håstad (2003) also found some novel mutations of unknown effect, so it is possible that these modify the λ_{max} predicted solely on the basis of cysteine or serine at position 90. For this reason, confirmation by of birds and other vertebrates. Likewise, the green and red cone pigments of humans are both members of the LWS class of vertebrate photopigments. Mammals lack homologs of the blue and green cone pigments found in birds and many other vertebrates.

Oil Droplets and Ocular Filtering

Even visual pigments with a λ_{max} at long wavelengths, such as in the LWS cones, have significant absorption in the UV due to a secondary, or β, peak in the absorption spectrum (Jacobs 1992). The color-blind nature of individual photoreceptors means that such secondary peaks are deleterious for color vision, because photoreceptor output would be ambiguous with respect to the spectral composition of the incoming light. For this reason, the effective sensitivities of the photoreceptors are often modified by filters in the light path (Douglas and Marshall 1999). In humans and most (but not all) mammals, UV light is absorbed by the lens and macular pigments before it ever reaches the photoreceptors (Stark 1987; Bone and Landrum 1992). In all birds for which the transmission of the ocular media has been measured, considerable UV-A is transmitted, and in those species with UVS cones, the pre-cone light path is essentially transparent to UV-A (e.g., see Hart et al. 1998, 1999, 2000a,c; Hart 2002). Instead, birds fine-tune their effective spectral sensitivity of their photoreceptors at the level of the cones themselves.

Avian cones possess carotenoid-containing oil droplets in the vitreal part of the cell inner segments. Cone oil droplets occur in many vertebrates, including some fish, amphibians, reptiles, and marsupial (but not eutherian) mammals (Walls 1963; Hailman 1976; Armengol et al. 1981; Ohtsuka 1985; Bowmaker 1991b; Ahnelt et al. 1995), but it is the birds and turtles that are notable for droplets that are densely pigmented with carotenoids. If one looks at a light microscope preparation of a bird's retina, a mosaic of differently colored droplets are apparent (see the figures in Hart et al. 2000b; Hart 2004; Plate 1), with a tight match between the cone type and the droplet color it possesses (Goldsmith et al. 1984; Bowmaker 1991b; Hart 2001). On account of their carotenoid content, the droplets act as long-pass cut-off filters (Figure 1.5b), blocking light below a certain wavelength from reaching the visual pigment in the cell, and the spectral location of the cut-off (and so the observed color of the droplet) is highly correlated with the λ_{max} of the cone's visual pigment (Bowmaker et al. 1997; Hart 2001). Thus UVS and VS cones possess transparent (T-type) droplets that allow all light, including UV-A, through to their UV-sensitive visual pigment. SWS cones have clear (C-type) droplets that block UV, so the resulting cone is only readily stimulated by short-wave blue light. MWS cones have yellow (Y-type) droplets, the yellow color being the result of absorption of short-wave and UV light, and so these droplets fine-tune the MWS cone to only medium-wave green light. Finally, the LWS cones have red (R-type) droplets that block UV, short-, and medium-wave light. In the MWS and LWS cones particularly, there is a notable cost in terms of reduced overall sensitivity of the cone to light (Figure 1.5), because the oil droplet's cut-off wavelength is often similar to the λ_{max} of the cone's visual pigment, and so a significant portion of the available light never reaches the visual pigment. The effective λ_{max} of the entire cone is also shifted toward longer wavelengths than the λ_{max} of the cone's visual pigment alone (Figure 1.5). The cost of reduced overall sensitivity to light, which must reduce the effectiveness of color vision at low light levels (Vorobyev 2003), is offset by the droplets' effect of narrowing the waveband to which each cone type responds and so reducing overlap in the spectral sensitivity of the different cone types (Figure 1.5c versus a). This narrowing should theoretically increase the discriminability of colors, provided that light levels are not limiting (Govardovskii 1983; Vorobyev et al. 1998; Vorobyev 2003) and may improve color constancy (Osorio et al. 1997; Vorobyev et al. 1998). In this regard, it is notable that in a central strip of the retina of the Wedge-tailed Shearwater (*Puffinus pacificus*; Plate 1), where the cells are smaller and densely packed to increase acuity, these cones also have less pigmented oil droplets (Hart 2004). The reduced pigmentation would partially offset the reduced photon capture due to small size, albeit at the presumed cost of reduced color discrimination for images on this part of the retina (Hart 2004).

Intra-retinal variation in both the density of oil droplet pigmentation and the abundance of different cone types is common in birds. The pigeon is well known for having a distinct "red field" in the posterior-dorsal retina, so-called because of the high numbers of LWS cones with red oil droplets, and longer wavelength cut-offs for both the R- and Y-type oil droplets in this region (Bowmaker 1977). As this part of the eye looks downward and forward, presumably this concentration of LWS cones is an adaptation for foraging. However, the European Starling has more UVS and less LWS cones in this part of the retina, so the pigeon pattern is far from universal. Both starlings and Eurasian Blackbirds (*Turdus merula*) have more double cones in the ventral retina, perhaps advantageous for detection of aerial predators via movement

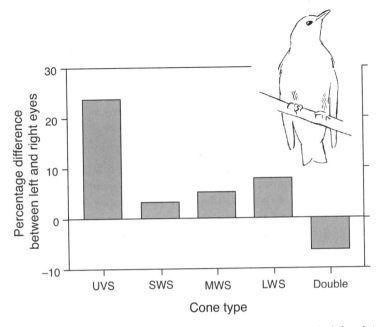

Figure 1.7. Differences in relative abundance of each cone class between the left and right eyes of the European Starling (from Hart et al. 2000b). Values are percentage differences for each cone class. Double = double cones; LWS = long-wave sensitive cones; MWS = medium-wave sensitive cones; SWS = short-wave sensitive cones; UVS = ultraviolet sensitive cones. In absolute terms, double cones are more abundant than any other cone class.

(Hart et al. 2000c). Starlings also show a dorsoventral gradient of increasing oil droplet pigmentation in the double cones, perhaps acting as "graduated sunglasses" to protect against UV, or to balance the radiance received by the upper and lower halves of the retina (Hart et al. 1998); this pattern has been seen in other species as well (Goldsmith et al. 1984; Hart et al. 2000a). Starlings are, in fact, the only species for which enough individuals have been sampled in the same experimental protocol to attach narrow confidence intervals to statements about intra-retinal and inter-individual differences. Hart et al. (1998) found no evidence of sex differences, but did establish a small but highly significant asymmetry between the left and right eyes (Hart et al. 2000b; Figure 1.7), a pattern also seen in a smaller sample of the Blue Tit (*Parus caeruleus*; Plate 18, Volume 2; Hart et al. 2000c). The right eye has relatively more double cones, which presumably relates to asymmetries in development (Rogers and Bolden 1991; Rogers and Krebs 1996) and/or eye use (Güntürkün 1997), but the functional significance has not been investigated.

All diurnal birds that have been investigated have greater numbers of double cones than any other photoreceptor class and, among the single cones, LWS and MWS cones outnumber SWS and UVS/VS cones (Bowmaker et al. 1997; Hart et al. 1998, 2000c). For example, in the Blue Tit, the percentages of double, LWS, MWS, SWS, and UVS cones are 37, 20, 20, 15, and 8%, respectively (Hart et al. 2000c), whereas in the blackbird and starling, the percentage of double cones exceeds 50% (Hart et al. 1998; Hart et al. 2000c). The low abundance of SWS and UVS/VS cones does not imply that they are unimportant in color discrimination; the central human fovea lacks SWS cones and is surrounded by a ring in which the density of SWS cones is only 8% (Bumsted and Hendrickson 1999), and yet blue is an important part of our color world. Nevertheless the low abundance of UVS/VS and SWS cones suggests lower acuity at short wavelengths. Broad ecological patterns seem to exist in the relative abundance of the cone types and/or the pigmentation of their oil droplets (Muntz 1972; Partridge 1989; Hart 2001); as most studies have estimated cone type abundance using light microscopy to count oil droplet types, the two effects are often hard to separate. Nevertheless, as oil droplets change the effective spectral sensitivity of cones and the relative abundance of single cone types affects the thresholds for color discrimination (Vorobyev and Osorio 1998), these differences will create differences in color perception among species, even when they possess the same visual pigments. Predictably, nocturnal birds have higher numbers of rods and fewer pigmented oil droplets in their cones, the premium being on light capture per se rather than on spectral discrimination. Seabirds often have higher numbers of LWS cones and/or more pigmented oil droplets; the earlier suggestion that this was an antiglare adaptation for seeing through the water surface now seems unlikely, as short-to-medium wavelengths in fact pass through the surface more readily than do long wavelengths (Hart 2001). There is some evidence that plumage coloration might co-vary with cone type distribution and behaviorally measured color discrimination, at least on a broad taxonomic scale (Peiponen 1992), but other comparisons between more closely related groups have revealed no obvious pattern (Hart et al. 2000a). The fact is that good data, quantifying intra-retinal variation, gross cone type distribution, and oil droplet pigment density, with sufficient numbers of replicate individuals, are too scarce to draw strong conclusions about inter-specific variation.

Dimensionality of Avian Color Space

The possession of UV or violet cones in one or more opponent mechanisms undoubtedly means that birds see colors that humans cannot experience (Bennett and Cuthill 1994; Bennett et al. 1994). This statement has been tested directly (Osorio et al. 1999b; Smith et al. 2002), and any of the numerous experiments that show that UV affects behavioral decisions in birds (see Cuthill et al. 2000; Chapter 4, Volume 2) is an indirect demonstration of it. However, this is insufficient evidence to conclude that the avian color world is richer than ours. There remains the possibility that the UVS/VS cone's output is only compared with that of one other cone class (e.g., VS-SWS opponency, as shown for domestic chicks by Osorio et al. 1999b), and that parts of the theoretically tetrahedral color space are not filled. In humans, there are two primary opponent channels (Wyszecki and Stiles 1982): a red-green system that compares the outputs of the red and green cones, and a yellow-blue system that compares both long-wave cone classes to the short-wave cone class. There is also an achromatic "luminance" channel that sums outputs from (mainly the red and green) cones. This channel has higher spatial resolution than the chromatic channels and is important in movement detection and resolving surface-texture information. Of the chromatic channels, the yellow-blue system is the ancestral one for mammals (because most mammals have only two cone types: a short versus medium/long-wave comparison is the only one available); the red-green system is an innovation of some primates. Given the particular circumstances of primate evolution—the loss of two visual pigment types in mammalian ancestry and then the re-evolution of a third cone pigment—one should not expect other vertebrates, such as birds, to process cone output in exactly the same way as humans. For example, birds may not have a yellow-blue (i.e., MWS plus LWS versus SWS) channel because, unlike humans, the MWS and LWS have not recently diverged from a single LWS opsin that was involved in LWS-SWS opponency, as it is in most mammals. Furthermore, birds might not experience nonspectral colors resulting from the stimulation of nonadjacent cone classes. In humans, purple, which is not part of the rainbow, as it does not correspond to any single region of monochromatic light, is experienced when the SWS and LWS cones are highly stimulated relative to the MWS cones. Do birds see purple, or additional nonspectral colors, such as UV-green, UV-yellow, or UV-red? The necessary data do not exist for birds, and acquisition of such data is an important direction for future neurobiological research. Confidence that birds do have a rich array of possible opponent channels is raised by electrophysiological measurements from turtles (Ammermüller et al. 1998; Ventura et al. 1999, 2001), which have similar retinal structures to those of birds: four single-cone types plus colored oil droplets. Ventura et al. (2001) have found retinal cells with opponent processing of USM/L, US/ML, U/SML, UL/SM and UML/S light (where U = UV, S = SW, M = MW, L = LW). Particularly interesting are the UL/SM and UML/S opponencies as these correspond to nonspectral colors.

A further complication (or exciting prospect, depending on one's perspective) is that the dimensionality of color vision may alter with changes in illumination. It does in goldfish (Neumeyer and Arnold 1989), where tetrachromacy switches to trichromacy as the available light drops. This change makes adaptive sense, because by pooling instead of comparing cone outputs, spectral discrimination ability is reduced in favor of light-capture capacity when photons become limiting. Until extensive color-mixing experiments on birds are conducted, the dimensionality of avian vision will remain unknown, as will the capacity for adaptive modulation of color vision with changing light levels.

Functions of Avian Color Vision

Three important functional questions come to mind with regard to bird color vision: (1) What are the advantages of tetrachromacy? (2) Does UV sensitivity have a special function? (3) Why do some groups have a UVS cones and others VS cones? Considering the first question, possession of several narrowband photoreceptors rather than a few, overlapping, broadband photoreceptors has advantages in terms of color discrimination ability and color constancy, and oil droplets enhance this advantage of tetrachromacy (Osorio et al. 1997; Vorobyev et al. 1998; Vorobyev 2003). All things being equal, these advantages are bought at the price of lower spatial acuity (because more receptors require more space in the retina). Birds offset this reduced acuity somewhat by having very small cones, but the latter itself incurs a price of reduced photon capture per cone. Taken together, the typical avian eye is well designed for high color-discrimination performance at acuities comparable to those of humans (Hodos 1993), but only at high light intensities. The typical mammalian eye is instead designed for high photon capture but poor wavelength discrimination; it has far higher numbers of rods than found in birds, and only two cone types. Trichromatic primates have regained moderate wavelength discrimination, but the system still appears to be built around a compromise between luminance and color discrimination (Osorio and Vorobyev 1996; Osorio et al. 1998). In short, phylogenetic constraint (the loss of SWS and MWS visual pigments in eutherian ancestry, and the corresponding neural wiring) may play a significant role in explaining the lack of tetrachromacy in primates.

Because humans lack UV vision, it is tempting to assume that this ability has some specific adaptive significance. It may, but the UVS/VS cone may just be part of a general-purpose color vision system, and the relevant question instead is: why do humans lack UV vision? Phylogenetic constraint might explain the lack of tetrachomacy, but it is unlikely to explain the absence of UV vision. The human blue cone has a SWS1 opsin homologous to the UVS/VS class of birds (Box 1.1), and, in mammals as in birds, a single mutation can switch its sensitivity maximum to the UV (Yokoyama et al. 2000). Instead the possession or absence of UV sensitivity is likely to reflect the trade-off between the costs and benefits of allowing UV to enter the retina. The costs are several, most obviously that several important biomolecules, not least DNA, absorb UV, leading to damage and mutagenesis. Second, short wavelengths are absorbed more than are long ones, so that UV may be less visually useful to animals with larger eyes and thus with greater intra-ocular absorption (Hart 2001). Third, chromatic aberration (i.e., focal length varies with wavelength, leading to a problem bringing all wavelengths of light to a focus at the same point) is more severe for animals with larger eyes seeking high spatial resolution (Kröger et al. 1999). All of these costs tend to weigh against use of UV by larger, long-lived organisms, unless there are counterbalancing benefits. Those benefits may apply to a wide range of visual tasks, as many objects vary in UV reflectance (Bennett and Cuthill 1994), so that it may not be appropriate to single out a specific function for the UV component of avian color vision. Certainly, there is now widespread evidence that UV has a significant role in sexual signaling (reviewed by Cuthill et al. 1999, 2000; Chapter 4, Volume 2) and foraging (reviewed by Cuthill et al. 2000; Honkavaara et al. 2002; Church et al. 2004), and may be involved in orientation (Bennett and Cuthill 1994; Coemans et al. 1994b). If there are special benefits, they may involve the ability to signal using a "private channel" that mammalian predators cannot see (Guilford and Harvey 1998) or, because of short-wave scattering by the atmosphere, are less visible to predators at a distance than to conspecifics close-up (Bennett and Cuthill 1994). Alternatively, as UV signals are produced by nanostructures in the feathers (Andersson 1998; Prum et al. 1998, 1999; Chapter 7), UV discrimination may facilitate honest signaling of condition

(Andersson 1998; Keyser and Hill 1999, 2000). There is some evidence that UV-based colors are more common in plumage regions used in signaling (Hausmann et al. 2003), and, because backgrounds such as leaves and bark reflect little UV, UV (like red) should provide a good contrast (Burkhardt 1982). It seems more likely that avian plumage evolved to exploit a pre-existing sensitivity to UV than vice versa (Cuthill et al. 2000), but, that said, the role of UV in signaling could be a powerful force maintaining the UVS/VS cone type.

Why should some birds have a UVS cone, whereas others possess a VS cone, the latter probably being the ancestral state for birds (Shi et al. 2001; Shi and Yokoyama 2003)? The general cost-benefit arguments discussed above should apply; one would predict that the UVS cone type should only be found in smaller, less long-lived species (as it appears to be in mammals, with only some rodents having UVS cones; Jacobs et al. 1991, 2001). Unfortunately, the variation documented by Ödeen and Håstad (2003) does not readily fit this picture: gulls, parrots, and rheas are large and fairly long-lived birds, yet have UVS opsins. To test the hypothesis that, in gulls at least, high UV sensitivity aids detection of fish against upwelling UV light, Håstad (2003) investigated opsins in other plunge-diving seabirds. However, terns and gannets possess VS, not UVS, opsins. Likewise, although one might predict that the UVS system should be more common in UV-rich habitats, and there is some evidence of such a correlation in fish (Losey et al. 2003), it is hard to see what the habitat differences are between gulls and terns, or rheas and ostriches. Thus the adaptive significance of UVS versus VS vision in birds remains elusive and a challenge for future research.

Summary

Birds, and indeed vertebrates as a whole, are ancestrally tetrachromatic. Their color vision is based on four single cone types, maximally sensitive to the UV/VS, SWS, MWS, and LWS sections of a visible spectrum that extends from the UV-A (~315 nm) to far red (~700 nm). Birds also have double cones, which eutherian mammals lack, and these may provide the sensory basis for achromatic (color-blind) tasks, such as texture and motion detection, which in humans are performed by the luminance pathway from the LWS and MWS cones. The spectral tuning of avian cones is a function of the amino acid sequence in the opsin pigment protein and, importantly, of the colored oil droplets that act as cut-off filters to modify the light reaching the visual pigments. Tetrachromacy and the narrowing of overlap in spectral sensitivity of the four single-cone types is predicted to give birds better color discrimination than humans, as long as light intensities are high, and birds undoubtedly can see colors to which humans are blind. The majority of experiments that have investigated whether UV information is utilized in, for example, foraging or sexual signaling, shows that UV detection has a significant role. Whether UV vision has a special function is unclear—wavelength-dependent properties of light suggest that it could—but it may just be part of a general-purpose tetrachromatic vision system. Instead, the relevant evolutionary question is: why do mammals lack tetrachromacy? The answer probably lies in an ancestral nocturnal niche in which color vision was unimportant and so two of the ancestral vertebrate opsin genes were lost. Phylogenetic constraints regarding neural organization may explain why diurnal mammals have not re-evolved tetrachomacy (only primates are trichromatic; most mammals are dichromats), but constraints are unlikely to explain why UV vision is not more common among mammals, as a single mutation can produce a shift from a VS to a UVS opsin. A VS opsin was probably the ancestral state for birds, but UVS opsins have evolved at least four times: in parrots, in most passerines (but not crows or flycatchers), gulls (but not terns or other Ciconiiformes), and rheas (but not Ostriches). The UVS/VS dichotomy in birds (and other groups; e.g., mammals) presumably reflects the trade-off between the value of very-short-wave visual information and the costs, notably retinal damage and mutagenesis, chromatic aberration, and the higher scattering of short wavelengths. However, the complex distribution of UVS pigments among birds does not readily fit the hypotheses that have been proposed. That said, the ever-increasing understanding of the mechanisms of avian visual perception are revolutionizing behavioral and evolutionary studies of color in birds and bird-relevant objects, because we can now realistically model the avian perception of color.

References

Able, K. P., and M. A. Able. 1995. Manipulations of polarized skylight calibrate magnetic orientation in a migratory bird. J Comp Physiol A 177: 351–356.

Ahnelt, P. K., J. N. Hokoc, and P. Rohlich. 1995. Photoreceptors in a primitive mammal, the South American opossum, *Didelphis marsupialis aurita*: Characterization with anti-opsin immunolabeling. Visual Neurosci 12: 793–804.

Ammermüller, J., A. Itzhaki, R. Weiler, and I. Perlman. 1998. UV-sensitive input to horizontal cells in the turtle retina. Euro J Neurosci 10: 1544–1552.

Andersson, S. 1998. Morphology of UV reflectance in a Whistling-thrush: Implications for the study of structural colour signalling in birds. J Avian Biol 30: 193–204.

Applebury, M. L., M. P. Antoch, L. C. Baxter, L. L. Y. Chun, J. D. Falk, F. Farhangfar, K. Kage, M. G. Krzystolik, L. A. Lyass, and J. T. Robbins. 2000. The murine cone photoreceptor: A single cone type expresses both S and M opsins with retinal spatial patterning. Neuron 27: 513–523.

Armengol, J. A., F. Prada, and J. M. Genis-Galvez. 1981. Oil droplets in the chameleon (*Chamaleo chamaleo*) retina. Acta Anatomica 110: 35–39.

Arrese, C. A., N. S. Hart, N. Thomas, L. D. Beazley, and J. Shand. 2002. Trichromacy in Australian marsupials. Curr Biol 12: 657–660.

Bennett, A. T. D., and I. C. Cuthill. 1994. Ultraviolet vision in birds: What is its function? Vision Res 34: 1471–1478.

Bennett, A. T. D., I. C. Cuthill, and K. J. Norris. 1994. Sexual selection and the mismeasure of color. Am Nat 144: 848–860.

Bone, R. A., and J. T. Landrum. 1992. Distribution of macular pigment components, zeaxanthin and lutein, in human retina. Meth Enzymol 213: 360–366.

Bowmaker, J. K. 1977. The visual pigments, oil droplets and spectral sensitivity of the pigeon. Vision Res 17: 1129–1138.

Bowmaker, J. K. 1991a. The evolution of vertebrate visual pigments and photoreceptors. In J. R. Cronly-Dillon and R. L. Gregory, ed., Vision and Visual Dysfunction, Volume 2, 63–81. Boston: CRC Press.

Bowmaker, J. K. 1991b. Visual pigments, oil droplets and photoreceptors. In P. Gouras, ed., Vision and Visual Dysfunction, Volume 6, 108–127. Boston: CRC Press.

Bowmaker, J. K. 1998. Evolution of colour vision in vertebrates. Eye 12: 541–547.

Bowmaker, J. K., and D. M. Hunt. 1999. Molecular biology of photoreceptor spectral sensitivity. In S. N. Archer, M. B. A. Djamgoz, E. R. Loew, J. C. Partridge, and S. Valerga, ed., Adaptive Mechanisms in the Ecology of Vision, 439–464. London: Chapman & Hall.

Bowmaker, J. K., and G. R. Martin. 1978. Visual pigments and colour vision in a nocturnal bird, *Strix aluco* (Tawny Owl). Vision Res 18: 1125–1130.

Bowmaker, J. K., L. A. Heath, S. E. Wilkie, and D. M. Hunt. 1997. Visual pigments and oil droplets from six classes of photoreceptor in the retinas of birds. Vision Res 37: 2183–2194.

Bradbury, J. W., and S. L. Vehrencamp. 1998. Principles of Animal Communication. Sunderland, MA: Sinauer.

Brines, M. L., and J. L. Gould. 1982. Skylight polarization patterns and animal orientation. J Exp Biol 96: 69–91.

Buchsbaum, G., and A. Gottschalk. 1983. Trichromacy, opponent colour coding and optimum colour information transmission in the retina. Proc R Soc Lond B 220: 89–113.

Bumsted, K., and A. Hendrickson. 1999. Distribution and development of short-wavelength cones differ between *Macaca* monkey and human fovea. J Comp Neurol 403: 502–516.

Burkhardt, D. 1982. Birds, berries and UV. Naturwissenschaften 69: 153–157.

Caine, N. G., and N. I. Mundy. 2000. Demonstration of a foraging advantage for trichromatic marmosets (*Callithrix geoffroyi*) dependent on food colour. Proc R Soc Lond B 267: 439–444.

Cameron, D. A., and E. N. J. Pugh. 1991. Double cones as a basis for a new type of polarization vision in vertebrates. Nature 353: 161–164.

Campenhausen, M. V., and K. Kirschfeld. 1998. Spectral sensitivity of the accessory optic system of the pigeon. J Comp Physiol A 183: 1–6.

Chen, D. M., and T. H. Goldsmith. 1986. Four spectral classes of cone in the retinas of birds. J Comp Physiol A 159: 473–479.

Chen, D. M., J. S. Collins, and T. H. Goldsmith. 1984. The ultraviolet receptor of bird retinas. Science 225: 337–340.

Chiao, C. C., T. W. Cronin, and D. Osorio. 2000. Color signals in natural scenes: Characteristics of reflectance spectra and effects of natural illuminants. J Opt Soc Am A 17: 218–224.

Church, S. C., A. T. D. Bennett, I. C. Cuthill, and J. C. Partridge. 2004. Avian ultraviolet vision and its implications for insect protective coloration. In H. van Emden and M. Rothschild, ed., Insect and Bird Interactions, 165–184. Andover, UK: Intercept Press.

Coemans, M. A. J. M., J. J. Vos Hzn, and J. F. W. Nuboer. 1994a. The orientation of the e-vector of linearly polarized light does not affect the behaviour of the pigeon *Columba livia*. J Exp Biol 191: 107–123.

Coemans, M. A. J. M., J. J. Vos Hzn, and J. F. W. Nuboer. 1994b. The relation between celestial colour gradients and the position of the sun, with regard to the sun compass. Vision Res 34: 1461–1470.

Cuthill, I. C., J. C. Partridge, and A. T. D. Bennett. 1999. Avian UV vision and sexual selection. In Y. Espmark, T. Admundsen, and G. Rosenqvist, ed., Animal Signals, 61–82. Trondheim: Royal Norwegian Society of Sciences and Letters, The Foundation, Tapir Publishers.

Cuthill, I. C., J. C. Partridge, A. T. D. Bennett, S. C. Church, N. S. Hart, and S. Hunt. 2000. Ultraviolet vision in birds. Adv Stud Behav 29: 159–214.

Dartnall, H. J. A., J. K. Bowmaker, and J. D. Mollon. 1983. Human visual pigments: Microspectrophotometric results from the eyes of seven persons. Proc R Soc Lond B 220: 115–130.

Dominy, N. J., and P. W. Lucas. 2001. Ecological importance of trichromatic vision to primates. Nature 410: 363–366.

Douglas, R. H., and N. J. Marshall. 1999. A review of vertebrate and invertebrate ocular filters. In S. N. Archer, M. B. A. Djamgoz, E. R. Loew, J. C. Partridge, and S. Valerga, ed., Adaptive Mechanisms in the Ecology of Vision, 95–162. London: Chapman & Hall.

Endler, J. A. 1990. On the measurement and classification of colour in studies of animal colour patterns. Biol J Linn Soc 41: 315–352.

Endler, J. A., and P. W. J. Mielke. In press. Comparing entire colour patterns as birds see them. Biol J Linn Soc.

Flamarique, I. N., and H. I. Browman. 2001. Foraging and prey-search behaviour of small juvenile rainbow trout (*Oncorhynchus mykiss*) under polarized light. J Exp Biol 204: 2415–2422.

Flamarique, I. N., C. W. Hawryshyn, and F. I. Harosi. 1998. Double-cone internal reflection as a basis for polarization detection in fish. J Opt Soc Am A 15: 349–358.

Fleishman, L. J., and J. A. Endler. 2000. Some comments on visual perception and the use of video playback in animal behavior studies. Acta Ethol 3: 15–27.

Goldsmith, T. H. 1986. Interpreting trans-retinal recordings of spectral sensitivity. J Comp Physiol A 159: 481–487.

Goldsmith, T. H. 1990. Optimization, constraint, and history in the evolution of eyes. Quart Rev Biol 65: 281–322.

Goldsmith, T. H., and B. K. Butler. 2003. The roles of receptor noise and cone oil droplets in the photopic spectral sensitivity of the budgerigar, *Melopsittacus undulatus*. J Comp Physiol A 189: 135–142.

Goldsmith, T. H., J. S. Collins, and D. L. Perlman. 1981. A wavelength discrimination function for the hummingbird *Archilochus alexandri*. J Comp Physiol A 143: 103–110.

Goldsmith, T. H., J. S. Collins, and S. Licht. 1984. The cone oil droplets of avian retinas. Vision Res 24: 1661–1671.

Govardovskii, V. I. 1983. On the role of oil drops in colour vision. Vision Res 23: 1739–1740.

Greenwood, V. J., E. L. Smith, S. C. Church, and J. C. Partridge. 2003. Behavioural investigation of polarisation sensitivity in the Japanese Quail (*Coturnix coturnix japonica*) and the European Starling (*Sturnus vulgaris*). J Exp Biol 206: 3201–3210.

Guilford, T., and P. H. Harvey. 1998. The purple patch. Nature 392: 867–868.

Güntürkün, O. 1997. Avian visual lateralization: A review. NeuroReport 8: R3–R11.

Hailman, J. P. 1976. Oil droplets in the eyes of adult anuran amphibians: A comparative survey. J Morphol 148: 453–468.

Hart, N. S. 1998. Avian Photoreceptors. Ph.D. diss., University of Bristol, Bristol, UK.

Hart, N. S. 2001. The visual ecology of avian photoreceptors. Prog Ret Eye Res 20: 675–703.

Hart, N. S. 2002. Vision in the peafowl (Aves: *Pavo cristatus*). J Exp Biol 205: 3925–3935.

Hart, N. S. 2004. Microspectrophotometry of visual pigments and oil droplets in a marine bird, the Wedge-tailed Shearwater *Puffinus pacificus:* Topographic variations in photoreceptor spectral characteristics. J Exp Biol 207: 1229–1240.

Hart, N. S., J. C. Partridge, and I. C. Cuthill. 1998. Visual pigments, oil droplets and cone photoreceptor distribution in the European Starling (*Sturnus vulgaris*). J Exp Biol 201: 1433–1446.

Hart, N. S., J. C. Partridge, and I. C. Cuthill. 1999. Visual pigments, cone oil droplets, ocular media and predicted spectral sensitivity in the Domestic Turkey (*Meleagris gallopavo*). Vision Res 39: 3321–3328.

Hart, N. S., J. C. Partridge, A. T. D. Bennett, and I. C. Cuthill. 2000a. Visual pigments, cone oil droplets and ocular media in four species of estrildid finch. J Comp Physiol A 186: 681–694.

Hart, N. S., J. C. Partridge, and I. C. Cuthill. 2000b. Retinal asymmetry in birds. Curr Biol 10: 115–117.

Hart, N. S., J. C. Partridge, I. C. Cuthill, and A. T. D. Bennett. 2000c. Visual pigments, oil droplets, ocular media and cone photoreceptor distribution in two species of passerine: The Blue Tit (*Parus caeruleus* L.) and the Blackbird (*Turdus merula* L.). J Comp Physiol A 186: 375–387.

Håstad, O. 2003. Plumage Colours and the Eye of the Beholder. The Ecology of Colour and Its Perception in Birds. Ph.D. diss., Uppsala University, Uppsala, Sweden.

Hausmann, F., K. E. Arnold, N. J. Marshall, and I. P. F. Owens. 2003. Ultraviolet signals in birds are special. Proc R Soc Lond B 270: 61–67.

Hawryshyn, C. W., and W. N. McFarland. 1987. Cone photoreceptor mechanisms and the detection of polarized light in fish. J Comp Physiol A 160: 459–465.

Hawryshyn, C. W., T. J. Haimberger, and M. E. Deutschlander. 2001. Microspectrophotometric measurements of vertebrate photoreceptors using CCD-based detection technology. J Exp Biol 204: 2431–2438.

Hodos, W. 1993. The visual capabilities of birds. In H. P. Zeigler and H. J. Bischof, ed., Vision, Brain and Behaviour in Birds, 63–76. Cambridge, MA: MIT Press.

Honkavaara, J., M. Koivula, E. Korpimaki, H. Siitari, and J. Viitala. 2002. Ultraviolet vision and foraging in terrestrial vertebrates. Oikos 98: 505–511.

Horvath, G., J. Gal, and R. Wehner. 1997. Why are water-seeking insects not attracted by mirages? The polarization pattern of mirages. Naturwissenschaften 84: 300–303.

Huth, H. H., and D. Burkhardt. 1972. Der spektrale Sehbereich eines Violetta Kolibris. Naturwissenschaften 59: 650.

Jacobs, G. H. 1981. Comparative Color Vision. New York: Academic Press.

Jacobs, G. H. 1992. Ultraviolet vision in vertebrates. Am Zool 32: 544–554.

Jacobs, G. H. 1993. The distribution and nature of colour vision among the mammals. Biol Rev 68: 413–471.

Jacobs, G. H., J. K. Bowmaker, and J. D. Mollon. 1981. Behavioral and microspectrophotometric measurements of color vision in monkeys. Nature 292: 541–543.

Jacobs, G. H., J. Neitz, and J. F. Deegan II. 1991. Retinal receptors in rodents maximally sensitive to ultraviolet light. Nature 353: 655–656.

Jacobs, G. H., J. Neitz, and K. Krogh. 1996a. Electroretinogram flicker photometry and its applications. J Opt Soc Am A 13: 641–648.

Jacobs, G. H., M. Neitz, J. F. Deegan, and J. Neitz. 1996b. Trichromatic color vision in New World monkeys. Nature 382: 156–158.

Jacobs, G. H., J. A. Fenwick, and G. A. Williams. 2001. Cone-based vision of rats for ultraviolet and visible lights. J Exp Biol 204: 2439–2446.

Jacobs, G. H., G. A. Williams, and J. A. Fenwick. 2004. Influence of cone pigment coexpression on spectral sensitivity and color vision in the mouse. Vision Res 44: 1615–1622.

Jones, C. D., and D. Osorio. 2004. Discrimination of oriented visual textures by poultry chicks. Vision Res 44: 83–89.

Kelber, A., C. Thunell, and K. Arikawa. 2001. Polarisation-dependent colour vision in Papilio butterflies. J Exp Biol 204: 2469–2480.

Kelber, A., M. Vorobyev, and D. Osorio. 2003. Animal colour vision—Behavioural tests and physiological concepts. Biol Rev 78: 81–118.

Keyser, A. J., and G. E. Hill. 1999. Condition-dependent variation in the blue-ultraviolet coloration of a structurally based plumage ornament. Proc R Soc Lond B 266: 771–777.

Keyser, A. J., and G. E. Hill. 2000. Structurally based plumage coloration is an honest signal of quality in male Blue Grosbeaks. Behav Ecol 11: 202–209.

Koivula, M., E. Korpimaki, and J. Viitala. 1997. Do Tengmalm's Owls see vole scent marks visible in ultraviolet light? Anim Behav 54: 873–877.

Kreithen, M. L., and W. T. Keeton. 1974. Detection of polarized light by the Homing Pigeon, *Columba livia*. J Comp Physiol 89: 83–92.

Kröger, R. H. H., M. C. W. Campbell, R. D. Fernald, and H.-J. Wagner. 1999. Multifocal lenses compensate for chromatic defocus in vertebrate eyes. J Comp Physiol A 184: 361–369.

Liebman, P. A. 1972. Microspectrophotometry of photoreceptors. In H. J. A. Dartnall, ed., Photochemistry of Vision, Volume VII/1, 481–528. Berlin: Springer-Verlag.

Losey, G. S., W. N. McFarland, E. R. Loew, J. P. Zamzow, P. A. Nelson, and N. J. Marshall. 2003. Visual biology of Hawaiian coral reef fishes. I. Ocular transmission and visual pigments. Copeia: 433–454.

Lucas, P. W., N. J. Dominy, P. Riba-Hernandez, K. E. Stoner, N. Yamashita, E. Loria-Calderon, W. Petersen-Pereira, Y. Rojas-Duran, R. Salas-Pena, S. Solis-Madrigal, D. Osorio, and B. W. Darvell. 2003. Evolution and function of routine trichromatic vision in primates. Evolution 57: 2636–2643.

Maddocks, S. A., S. C. Church, and I. C. Cuthill. 2001. The effects of the light environment on prey choice by zebra finches. J Exp Biol 204: 2509–2515.

Maier, E. J., and J. K. Bowmaker. 1993. Colour vision in the passeriform bird, *Leiothrix lutea:* Correlation of visual pigment absorbency and oil droplet transmission with spectral sensitivity. J Comp Physiol A 172: 295–301.

Martin, G. R. 1991. Ornithology—The question of polarization. Nature 350: 194.

Miklosi, A., Z. Gonda, D. Osorio, and A. Farzin. 2002. The effects of the visual environment on responses to colour by domestic chicks. J Comp Physiol A 188: 135–140.

Munro, U., and R. Wiltschko. 1995. The role of skylight polarization in the orientation of a day-migrating bird species. J Comp Physiol A 177: 357–362.

Munsell Color Company. 1976. Munsell Book of Color. Glossy Finish Collection, 2 vols. Baltimore: Munsell/Macbeth/Kollmorgen.

Muntz, W. R. A. 1972. Inert absorbing and reflecting pigments. In H. J. A. Dartnall, ed., Photochemistry of Vision, Volume VII/1, 529–565. Berlin: Springer-Verlag.

Neitz, M., and J. Neitz. 2001. The uncommon retina of the common house mouse. Trends Neurosci 24: 248–249.

Neumeyer, C. 1992. Tetrachromatic color vision in goldfish: Evidence from color mixture experiments. J Comp Physiol A 171: 639–649.

Neumeyer, C., and K. Arnold. 1989. Tetrachromatic colour vision in the goldfish becomes trichromatic under white adaptation light of moderate intensity. Vision Res 29: 1719–1727.

Ödeen, A., and O. Håstad. 2003. Complex distribution of avian color vision systems revealed by sequencing the SWS1 opsin from total DNA. Mol Biol Evol 20: 855–861.

Ohtsuka, T. 1985. Relation of spectral types to oil droplets in cones of turtle retina. Science 229: 874–877.

Okano, T., D. Kojima, Y. Fukada, Y. Shichida, and T. Yoshizawa. 1992. Primary structures of chicken cone visual pigments: Vertebrate rhodopsins have evolved out of cone visual pigments. Proc Natl Acad Sci USA 89: 5932–5936.

Osorio, D., and M. Vorobyev. 1996. Colour vision as an adaptation to frugivory in primates. Proc R Soc Lond B 263: 593–599.

Osorio, D., N. J. Marshall, and T. W. Cronin. 1997. Stomatopod photoreceptor tuning as an adaptation to colour constancy underwater. Vision Res 37: 3299–3309.

Osorio, D., D. L. Ruderman, and T. W. Cronin. 1998. Estimation of errors in luminance signals encoded by primate retina resulting from sampling of natural images with red and green cones. J Opt Soc Am A 15: 16–22.

Osorio, D., C. D. Jones, and M. Vorobyev. 1999a. Accurate memory for colour but not pattern contrast in chicks. Curr Biol 9: 199–202.

Osorio, D., M. Vorobyev, and C. D. Jones. 1999b. Colour vision of domestic chicks. J Exp Biol 202: 2951–2959.

Palacios, A. G., and F. J. Varela. 1992. Color mixing in the pigeon (*Columba livia*). 2. A psychophysical determination in the middle, short and near-UV wavelength range. Vision Res 32: 1947–1953.

Palacios, A., C. Martinoya, S. Bloch, and F. J. Varela. 1990. Color mixing in the pigeon. A psychophysical determination in the longwave spectral range. Vision Res 30: 587–596.

Partridge, J. C. 1989. The visual ecology of avian cone oil droplets. J Comp Physiol A 165: 415–426.

Peiponen, V. A. 1992. Colour discrimination of two passerine bird species in the Munsell system. Ornis Scand 23: 143–151.

Phillips, J. B., and F. R. Moore. 1992. Calibration of the sun compass by sunset polarized light patterns in a migratory bird. Behav Ecol Sociobiol 31: 189–193.

Prum, R. O., R. H. Torres, S. Williamson, and J. Dyck. 1998. Coherent light scattering by blue feather barbs. Nature 396: 28–29.

Prum, R. O., R. Torres, C. Kovach, S. Williamson, and S. M. Goodman. 1999. Coherent light scattering by nanostructured collagen arrays in the caruncles of the Malagasy asities (Eurlyaimidae: Aves). J Exp Biol 202: 3507–3522.

Regan, B. C., C. Julliot, B. Simmen, F. Vienot, P. Charles-Dominique, and J. D. Mollon. 2001. Fruits, foliage and the evolution of primate colour vision. Phil Trans R Soc B 356: 229–283.

Rogers, L. J., and S. W. Bolden. 1991. Light-dependent development and asymmetry of visual projections. Neurosci Lett 121: 63–67.

Rogers, L. J., and G. A. Krebs. 1996. Exposure to different wavelengths of light and the development of structural and functional asymmetries in the chicken. Behav Brain Res 80: 65–73.

Shashar, N., and T. W. Cronin. 1996. Polarization contrast vision in octopus. J Exp Biol 199: 999–1004.

Shashar, N., P. S. Rutledge, and T. W. Cronin. 1996. Polarization vision in cuttlefish —A concealed communication channel. J Exp Biol 199: 2077–2084.

Shashar, N., R. T. Hanlon, and A. de M. Petz. 1998. Polarization vision helps detect transparent prey. Nature 393: 222–223.

Shi, Y. S., and S. Yokoyama. 2003. Molecular analysis of the evolutionary significance of ultraviolet vision in vertebrates. Proc Natl Acad Sci USA 100: 8308–8313.

Shi, Y. S., F. B. Radlwimmer, and S. Yokoyama. 2001. Molecular genetics and the evolution of ultraviolet vision in vertebrates. Proc Natl Acad Sci USA 98: 11731–11736.

Sibley, C. G., and J. E. Ahlquist. 1990. Phylogeny and Classification of Birds: A Study in Molecular Evolution. New Haven, CT: Yale University Press.

Smith, A. C., H. M. Buchanan-Smith, A. K. Surridge, D. Osorio, and N. I. Mundy. 2003. The effect of colour vision status on the detection and selection of fruits by tamarins (*Saguinus* spp.). J Exp Biol 206: 3159–3165.

Smith, E. L., V. J. Greenwood, and A. T. D. Bennett. 2002. Ultraviolet colour perception in European Starlings and Japanese Quail. J Exp Biol 205: 3299–3306.

Smith, R. L., Y. Nishimura, and G. Raviola. 1985. Interreceptor junction in the double cone of the chicken retina. J Submicrosc Cytol Pathol 17: 183–186.

Stark, W. S. 1987. Photopic sensitivities to ultraviolet and visible wavelengths and the effects of the macular pigments in human aphakic observers. Curr Eye Res 6: 631–638.

Sumner, P., and J. D. Mollon. 2000a. Catarrhine photopigments are optimized for detecting targets against a foliage background. J Exp Biol 203: 1963–1986.

Sumner, P., and J. D. Mollon. 2000b. Chromaticity as a signal of ripeness in fruits taken by primates. J Exp Biol 203: 1987–2000.

Surridge, A. K., D. Osorio, and N. I. Mundy. 2003. Evolution and selection of trichromatic vision in primates. Trends Ecol Evol 18: 198–205.

Thompson, E., A. Palacios, and F. J. Varela. 1992. Ways of coloring: Comparative color vision as a case study for cognitive science. Behav Brain Sci 15: 1–74.

Ventura, D. F., J. M. de Souza, R. D. Devoe, and Y. Zana. 1999. UV responses in the retina of the turtle. Visual Neurosci 16: 191–204.

Ventura, D. F., Y. Zana, J. M. de Souza, and R. D. Devoe. 2001. Ultraviolet colour opponency in the turtle retina. J Exp Biol 204: 2527–2534.

Vorobyev, M. 2003. Coloured oil droplets enhance colour discrimination. Proc R Soc Lond B 270: 1255–1261.

Vorobyev, M., and D. Osorio. 1998. Receptor noise as a determinant of colour thresholds. Proc R Soc Lond B 265: 351–358.

Vorobyev, M., D. Osorio, A. T. D. Bennett, N. J. Marshall, and I. C. Cuthill. 1998. Tetrachromacy, oil droplets and bird plumage colours. J Comp Physiol A 183: 621–633.

Vos Hzn, J. J., M. A. J. M. Coemans, and J. F. W. Nuboer. 1995. No evidence for polarization sensitivity in the pigeon electroretinogram. J Exp Biol 198: 325–335.

Walls, G. L. 1963. The Vertebrate Eye and its Adaptive Radiation. New York: Hafner.

Wehner, R., and G. D. Bernard. 1993. Photoreceptor twist: A solution to the false color problem. Proc Natl Acad Sci USA 90: 4132–4135.

Wilkie, S. E., P. R. Robinson, T. W. Cronin, S. Poopalasundaram, J. K. Bowmaker, and D. M. Hunt. 2000. Spectral tuning of avian violet- and ultraviolet-sensitive visual pigments. Biochemistry 39: 7895–7901.

Wright, A. A. 1972. The influence of ultraviolet radiation on the pigeon's color discrimination. J Exp Anal Behav 17: 325–337.

Wyszecki, G., and W. S. Stiles. 1982. Color Science: Concepts and Methods, Quantitative Data and Formulae. New York: John Wiley.

Yokoyama, S. 2002. Molecular evolution of color vision in vertebrates. Gene 300: 69–78.

Yokoyama, S., and Y. S. Shi. 2000. Genetics and evolution of ultraviolet vision in vertebrates. FEBS Lett 486: 167–172.

Yokoyama, S., F. B. Radlwimmer, and N. S. Blow. 2000. Ultraviolet pigments in birds evolved from violet pigments by a single amino acid change. Proc Natl Acad Sci USA 97: 7366–7371.

Young, S. R., and G. R. Martin. 1984. Optics of retinal oil droplets: A model of light collection and polarisation detection in the avian retina. Vision Res 24: 129–137.

2

Quantifying Colors

STAFFAN ANDERSSON AND MARIA PRAGER

As the Danish ornithologist Jan Dyck predicted and pioneered 40 years ago, reflection curves have become "of great value in the investigation of the biological functions of colours" (Dyck 1966). Analyses of spectral reflectance and radiance data have indeed revived the central roles of animal coloration in evolutionary and behavioral ecology. It is now widely appreciated that color is not a property of objects or organisms, but a highly dynamic and context-dependent product of light conditions, reflectance, and receiver psychology. Whereas most of the theoretical framework for the objective analysis of optical signaling in animals, including the physics and psychophysics of color, was formulated long ago (e.g., Hailman 1977; Endler 1978; Lythgoe 1979; Jacobs 1981), it is only during the past few years that spectrometric color quantification has become a standard tool in behavioral ecology, replacing the human-subjective color charts and photometric or photographic devices.

This development was influenced by several coincident events in the early 1990s. First, Endler (1990) published a seminal paper on the quantification of animal color patterns. It may have been prohibitively technical for some readers, but it widely introduced the insight that color communication is a function of light conditions and observer sensitivity and that it should be objectively measured with a *spectro*-something. Second, ultraviolet (UV) vision and tetrachromacy was "rediscovered" in birds (Chapter 1), which suggested the exciting possibilities of (to the human eye) invisible plumage colors and patterns, originally proposed by Burkhardt (1982, 1989) and strongly emphasized by Bennett et al. (1994). Together with demonstrations of avian mate choice based on ultraviolet color variation (Maier 1994; Bennett et al. 1996; Andersson and Amundsen 1997), these papers inspired many researchers to look for UV signaling in their study animals, which, of course, called for methods of color quantification that were not dependent on human vision. Last, but not least important, was the advent of miniature diode-array spectroradiometer systems, pioneered by Ocean Optics (Dunedin, Florida). These devices drastically cut down both the costs and weight of bringing spectrometry into the field, using convenient fiber-optic probes and sensors to measure plumage reflectance of birds in the hand and light conditions directly at the display grounds. In less than a second, a spectrum could be captured over the entire 300- to 700-nm range of avian vision, allowing explorations of the neglected UV channel. Not surprisingly, these soon became popular instruments and today a spectrometer system is either in the hands or on the grant proposals of most projects studying color communication in the wild. Although still rather expensive (U.S. $4000–5000 for a complete ultraviolet/visible (UV/VIS) system, excluding the interfaced computer), it is money well invested because diode-array spectroradiometers have become indispensable tools for color researchers. From reading this volume, it will be obvious that spectral reflectance is the most objective and context-independent measure of those properties of birds that we perceive as their dazzling color variation.

After a brief recapitulation of the history of the study of plumage reflectance, this chapter starts off somewhat philosophically by trying to put reflectance and color in the perspective of the general problem of objective analyses of signals and signal cognition. The second and main section provides some basics and principles of measuring light and reflectance, building on several excellent recent texts (e.g., Endler 1990; Bradbury and Vehrencamp 1998). Throughout, the focus is on the acquisition of plumage reflectance, with only brief notes on human-subjective photometric and photographic methods. Although such techniques can be useful if they are carefully applied and interpreted, there is no space to describe them here.

In the final section, we discuss some general reflectance characteristics and consequences for quantification of the two major types of plumage coloration (pigmentary and structural), with emphasis on the subtractive effects of pigment absorptance and the additive effects of spectrally selective scattering. Although the focus of this chapter is on the acquisition of spectral data, the physics of these mechanisms also has immediate implications for how such variation should be interpreted and analyzed with regard to color signaling

(Guilford and Dawkins 1991; Endler 1993b; Andersson 1999a). Preceding the comprehensive reviews and practical guides to statistical reflectance analysis and color space models in the following chapters (Chapters 3 and 4), we therefore conclude this chapter with a set of "objective colorimetrics." Based directly on reflectance shape and general features of color production and cognition (i.e., bypassing specific visual models), these colorimetrics capture some important proximate and ultimate aspects of avian plumage coloration.

Brief History of Plumage Reflectance

As in many other areas of quantitative biology, much of the early work on physical measurements of animal coloration and color vision is in German (e.g., Peiponen 1963; Lubnow and Niethammer 1964). The very first measures of avian plumage reflectance, however, seem to have been carried out by Bowers (1956) in a study of color variation in the Wrentit (*Chamaea fasciata*). A few years later, in a study of hummingbird iridescence, Greenewalt et al. (1960) measured the reflectance of microscopic feather parts. Juhn (1964) similarly measured reflectance of feather parts to characterize the melanin pigmentation of chickens. Shortly thereafter, spectrometry of feathers and its potential was described and illustrated by Dyck (1966), who later went on to use spectrometry to give the first evidence that the so-called "Tyndall colors" are caused by interference phenomena (Dyck 1976; Chapter 7). It was not until 20 years ago, however, that reflectance was first used to describe avian color variation in an ecological study, in this case addressing the adaptive significance of coloration in American wood-warblers (Parulinae) (Burtt 1986).

A striking aspect of these pioneering applications is the view of reflectance as simply a more accurate way of measuring the human-visible color variation. The thought of different color perception in birds, let alone a wider spectral range, does not seem to have occurred. Both Dyck (1966) and Burtt (1986), for example, calculated the industrial standard CIE (Commission Internationale d'Eclairage) tristimulus values and chromaticity coordinates for the "standard observer" (CIE 1971), without discussing the human subjectivity inherent in these measures. The bird's-eye view of plumage reflectance, including the UV, was instead introduced by Burkhardt (1989), providing one of the main inspirations behind the accelerating spread of reflectance spectrometry in ornithology and behavioral ecology. A recent search (September 2004) using the keywords "reflectance " and "plumage" in a major science citation database produced 55 records, 44 of which were published in the past 5 years (1999–2004)!

At such speed, there is always a risk that an application spreads faster than its technical understanding and theoretical underpinnings. If the present chapter provides some of this understanding and contributes to some unification of techniques and terminology, then it has served its purpose.

The Human Beholder on Trial

Before focusing entirely on techniques for reflectance spectrometry, some discussion is needed on the subjectivity and mismeasure of methods based on, or adapted to, human color vision. These include the classic color chart systems (e.g., Munsell, NCS, Methuen, Ridgway; see Zuk and Decruyenaere 1994; de Repentigny et al. 1997; Hill 2002), but also digital image analysis (e.g., Villafuerte and Negro 1998) and psychometrically based color spaces and colorimetrics (e.g., chromaticity, CIELAB, RGB, CMYK; CIE 1978). To get the message across a decade ago, it was perhaps necessary to reject such methods and instruments based on human vision as "entirely inappropriate for studies of animal color patterns" (Endler 1990: 319) and "seriously distorting our judgment of most evolutionary hypotheses involving color" (Bennett et al. 1994: 848). Today, however, we can afford to be a little less drastic, for a number of reasons.

First, that human perception can capture biologically relevant color signal variation is obvious from much of the older literature that, despite human colorimetry, has established practically all the central mechanisms and adaptive functions of avian coloration reviewed in this volume. For the same reason, it would be unfortunate, even arrogant, to discourage students of bird coloration in countries and circumstances where a spectrometer system is not affordable, but where a digital camera might be.

Second, with explicit considerations of the limitations of human vision (such as UV-blindness and yellow-biased sensitivity), many aspects of human color technology and computational color vision (Billmeyer and Saltzman 1981; Wyszecki and Stiles 1982) can be quite useful. Based on "standard observers" (CIE 1971, 1978), human color psychometrics are spectrally and mathematically formulated in such detail that in some respects, our color vision can be used as a physically calibrated instrument (i.e., controlling for its biases and constraints). In other words, if we can project a signal spectrum accurately in our human color space, we should also be able to use that position (determined photometrically or, less reliably, by color charts) to derive objective features of the signal, such as the relative amount of long- and short-wavelength radiation. We cannot (and probably never will) make that projection with "beauty," or "vigor," or other more deeply subjective sensations, but we can, to some extent, with color.

Subjectivity and the Mismeasure of Signals

Now let us scrutinize the main criticism of using our eyes or human colorimetrics to assess signals intended for other receivers: that it is "subjective" in neglecting both signal conditions and signal cognition (Guilford and Dawkins 1991; Endler 1992; Real 1992; Enquist 1998). Upon closer inspection, this criticism applies to measurement of many kinds of signals, not just color. To illustrate, we will be color-blind for a while and consider a simpler signal trait, an elongated bird tail, the length of which is measured with a ruler to the nearest 0.5 mm. When studying this tail variation in relation to sexual selection, we make several assumptions about how length maps onto the perceived variation in the eyes and mind of a choosy female or an approaching rival, and how viewing conditions (e.g., distance, angle, background) affect perceived length. The perceived resolution may be lower or higher than that of our ruler, and it is not likely to be uniform across tail lengths; for example, a 1-cm difference may be difficult to detect between two long-tailed (say, 50- and 51-cm) males, but obvious between two relatively short-tailed (5- and 6-cm) males. Furthermore, under some minimum length (perhaps 1 mm), there is no sensation of length, and shorter still, there is no visual sensation at all. There is thus a limited range of perceivable lengths and a nonuniform distribution of resolution within that range—parameters that may vary between species and be adaptively tuned to the relevant length discrimination tasks and viewing conditions. Sound familiar? Yes—replacing "length" (of tails) with "wavelength" (of light) in the preceding sentence would be a spot-on definition of species-specific color vision.

The point of this analogy is that consideration of sensitivity ranges and discrimination functions are not more (and not less!) important for color than it is for other signals like tail length, badge size, or display rate, for which receiver psychology is rarely discussed or investigated, at least not among behavioral ecologists. A notable exception is fluctuating asymmetry, in which the debated signal function has triggered research and important insights into its perception (e.g., Swaddle and Ruff 2004). Operant-conditioning experiments is perhaps not what most students want to do for a doctoral degree, but it is the way forward to get a handle on the discrimination functions, preference generalizations (Jansson et al. 2002), and other central aspects of signal cognition that are rarely demonstrated empirically. Only then, with behaviorally supported color spaces of our study animals (or an ecologically and phylogenetically close relative), can we move completely from more or less "human subjective" to "species-subjective" signal classifications and measurements. Ideally the latter will be the end-point of research into the function of avian coloration, but it will take decades until even the most popular "beholders" have been investigated.

In summary, projecting human color discrimination onto birds, with the limitations taken into account, may not be the only, nor even the worst, misrepresentation of the avian receiver. In fact, when a variety of natural plumage reflectances were mapped into the simple but behaviorally supported "cone space" models by Vorobyev and Osorio (1998), there were strong positive correlations between avian and human color discrimination (Vorobyev et al. 1998). Not surprisingly, the poorest correspondence was found for spectra differing primarily in the UV, but even so, the results suggest that human color discrimination may be a quite reasonable indication of signal variation perceived by birds. Conversely, for human-invisible or hardly detectable color variation in the center of the human range (450–650 nm), where we seem to have substantially higher resolution than have birds (Vorobyev et al. 1998), we should not be too quick to equate reflectance variation with avian signal variation until this has been experimentally confirmed.

Subjectivity and the Unmeasure of UV

The most obvious human "mismeasure" of avian coloration is of course our complete blindness to ultraviolet wavelengths (UV-A, 320–400 nm), which birds not only can see but do so with a dedicated fourth cone type in the retina, most likely contributing a new "dimension" of color vision (i.e., a tetra- versus trichromatic color space; Chapter 1). With a striking analogy, Bennett et al. (1994) compared the neglected UV channel to watching a television broadcast from which the blue (B) channel of the human-adapted RGB system was filtered away. This would of course be a dramatic change in color sensations, but in terms of perceivable variation, it is a loss of chromatic resolution only to the extent that the lost spectral channel (blue) varies independently from the visible channels (green and red). For most natural reflectance curves, such independence between spectral channels seems rare, allowing a great deal of discrimination of variation in one spectral region based on cues in another.

Likewise, the problem of not including UV in avian color quantification should depend on how unpredictable it is from measurements in the human-visible spectral range (e.g., VIS reflectance or digital photography). It is, however, confusing that there are so many different views on how to interpret such covariation between spectral regions. For example, whereas ubiquity and variability of UV reflectance within visible plumage colors have been reported as evidence that UV signaling is widespread (Eaton and Lanyon 2003), we think the same data show that there is a striking absence of UV-specific reflectance and absorptance mechanisms, and that most UV components are highly predictable from the type and intensity of human-visible plumage coloration. It is important in this respect to realize that absence as well as presence of reflectance may indicate a signaling function: what matters is the variation in spectral shape in the region of interest, and the extent to which is autocorrelated to neighboring parts of the spectrum.

Thus, although UV reflectance certainly has been confirmed to be an integral part of avian color communication (Chapter 1; Chapter 4, Volume 2), and its contribution to plumage colors always should be considered (Eaton and Lanyon 2003), the message here is simply that it is quite predictable from human-visible color variation and the current knowledge of avian color mechanisms. As a result, for plumage color quantification, human-based methods can be acceptable if used with care and combined with some objective confirmation of the UV contribution, and high quality VIS reflectance or photometrics extrapolated to UV may provide as good (or even better) resolution than would noisy measurements including UV.

With these observations, we conclude our attempts to balance the sometimes too-polarized views on human-subjective methods and turn our attention to reflectance spectrometry, without doubt the ultimately objective and most powerful method to quantify coloration.

Let There Be Reflectance

Whereas "beauty is in the eye of the beholder," the beauty of reflectance is that it is in the surface of the "beheld"! It may be in skin, fur, scales, or feathers, but the important thing is that reflectance is a physical property of senders or objects, independent of receivers (hence, objective) and of light conditions (hence, context-independent). In other words, without a beholder, there is no visual communication; and without light, there are no optical signals in the form of reflection or reflected radiance (see below), but there is reflectance!

Just like a guitar is tuned when no one is playing it, reflectance is a property of a surface even in the absence of light.

Where then is color? Is it in objects or subjects, in senders or receivers, or in between? If color is only in the beholder, is it in the retina, visual cortex, or, like the proverbial beauty, at a higher cognitive level? Endler (1990) talks about reflectance as the inherent color of a surface, which implies an objective aspect of color after all. Although some of this discussion can be left to philosophers (e.g., Thompson et al. 1992; Hardin 1993), students of animal coloration should at least stop to identify which particular aspect of color is to be measured and, even more important, on which aspect conclusions are going to be drawn. This clarification will simplify the choice and use of quantification methods considerably. In addition, much of the confusion and debate concerning the human-subjective mismeasure of color (Bennett et al. 1994) most certainly derives from vaguely formulated questions.

As regards animal color signaling, it is important to make clear whether the focus is on describing the present or historical (phylogenetic) variation among senders and signals (the beheld), such as ornamental coloration, or on exposing the signal selection pressures exerted by a receiver (the beholder), such as mate choice or predation. For example, when addressing the proximate mechanisms or fitness consequences of color variation within or between species, color (or rather, its physical origin) is in the beheld: Any consistent and repeatable quantification of that variation is, in principle, valid. In short, there is no philosophical flaw in concluding that, for example, red finches fledge more young than do yellow finches.

To help distinguish between these different aspects, what follows is a "light version" of the research program on animal color communication laid out by Hailman (1977), Lythgoe (1979), and Endler (1990, 1993a). Figure 2.1 provides a simplified account of the equations and figure in Endler (1990), and the mnemonic "ARTS" as a reminder that a color sensation (in the beholder) is a product of four main components—ambient light (A), reflectance (R), transmittance (T), and sensitivity (S)—that are spectral functions of wavelength over the visual range of the study species. For the process of color communication to take place, the product ARTS must be sufficiently large (or, more correctly, sufficiently different from the corresponding product for the background; Chapter 1). And because all components potentially are significant sources of variation in a communicated color signal, all of ARTS should ideally be quantified in the pursuit of its adaptive explanations.

Quantifying Colors

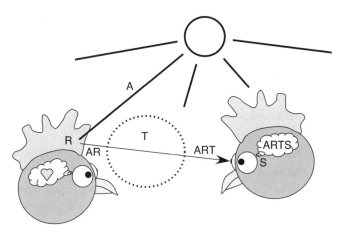

Figure 2.1. The fine ARTS of color communication. The color perceived by the observer is a product of ambient light (A), reflectance (R), transmittance (T), and receiver sensitivity (S). Whereas reflectance (R) is a property of the sender, the reflected radiance (the product AR) is the signal emitted from the surface of the sender. The fraction of this radiance that is carried to the receiver (ART) depends on distance and the transmittance (T) of the medium. Adapted from Endler (1990).

In most avian coloration projects, however, even when targeting a single color signal in a particular context (receiver and environmental signal conditions), it is practically impossible to gather all the relevant data, especially on the color perception (S) of the receiver. This difficulty means that one must make a decision about which components to measure and which to either disregard or generalize from other studies. For example, and as described in Chapters 3, one may resort to measure only reflectance (R), and then multiply this value with a general ambient (A) and transmittance (T) spectra and the sensitivity functions (S) from a color vision model, such as the tetrachromatic "cone space" of Vorobyev and Osorio (1998).

Measuring Light and Reflectance

Light

According to a recent science dictionary, light is "the form of electromagnetic radiation to which the human eye is sensitive and on which our visual awareness of the universe and its contents relies" (Isaacs et al. 1999: 394). It is obvious that the human eye sets the standards for the many kinds of photometric devices developed for commercial lighting and printing applications—photography, TV, computer monitors, and the like. These sensors are restricted to the human waveband (400–700 nm) and tuned to the human sensitivity within it. A simple ground rule for objective color quantification is thus to avoid instruments and methods with names containing "photo" or "light." However, the concept of light is so strongly tied to biological vision in general (man or beast), that it would be pointless to argue that it should be replaced by "radiance" or some other term. To talk about light in the context of animal vision, we must redefine light as the spectrum available to any seeing creature. This spectrum would typically be the 300- to 800-nm waveband (Bradbury and Vehrencamp 1998), which generously covers the approximately 320- to 700-nm range of avian color vision (Chapter 1).

Radiometry

The general name for an instrument that measures electromagnetic radiation, including light, is "radiometer." Practically all radiometers (and photometers) are based on photoelectric detectors that measure radiant power (joules per second = watts), also called "energy flux." (There are no quantum sensors that measure photon flux, but some operating software can be set to make the conversion—see below). The modern silicon photodiodes respond to a wide range of wavelengths (100–1100 nm) and can be made very small. If, in analogy with color vision, a radiometer also measures the spectral composition of the radiation as a function of wavelength, it is a spectroradiometer. The separation into wavelengths is accomplished by either a single sensor measuring a few wavelengths at a time as the light beam passes through a filter wheel (e.g., the model LI-1800 from Li-Cor, www.licor.com, used by Endler), or by diffracting the wavelengths with a prism or grating to a rainbow that falls on the array of photodiodes on a charge-coupled device (CCD) from which the accumulated (analog) voltage in each element is directly converted to digital counts displayed in real-time by the measuring software. The principle of such a diode-array spectroradiometer is illustrated in Figure 2.2 by the optical bench of the USB2000 from Ocean Optics (www.oceanoptics.com). With fast scan rates, no moving parts, and the trend toward reduced size and price (as for other CCD-based products), these miniature diode-array spectrometers have become by far the most popular instruments for field spectrometry.

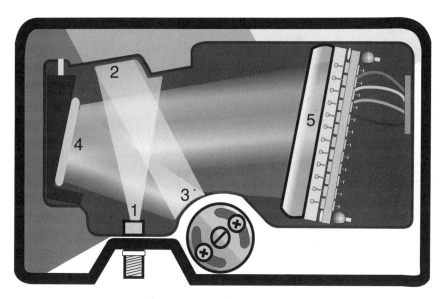

Figure 2.2. Diode-array spectroradiometer, illustrated by the USB2000 from Ocean Optics. As light enters the spectrometer, undesired wavelengths are removed by a filter (1). Filtered light then encounters a collimating mirror (2) that focuses the beam on a diffraction grating (3). Like a prism, this grating scatters the different wavelengths onto a focusing mirror (4) that directs the light via a collector lens onto the CCD detector (5). The CCD converts the optical signal to digital format. Redrawn from Ocean Optics (2003).

Quantities and Units

Endler (1990) gives an exhaustive review and discussion of radiometric quantities and units, which need not be spelled out here. However, some confusion seems to have arisen from his strong emphasis on using quantum units or "photon flux" (moles per second) as opposed to the spectral power (i.e., watts or counts of the photovoltaic effect) recorded by the photoelectric instruments. Although the distinction between a retina and a CCD is important (because biological photoreceptors indeed respond to the number rather than the summed energy of photons; Chapter 1), this emphasis may (unintentionally) have given the impression that the popular diode-array spectroradiometers are not measuring the "right thing" after all, not even in reflectance mode. This would be an unfortunate misunderstanding and needs some clarification: First, spectral reflectance (R) (as well as transmittance, T) are ratios (percentages), wavelength by wavelength, between sample and illuminant spectra, which thus can be measured in any unit, as long as both quantities are measured in the same units. Second, when radiance or irradiance spectra are needed (e.g., visual modeling; see Chapters 3 and 4), the conversion from power (or the corresponding photovoltaic counts) to photon flux is straightforward: in short, spectral energy is converted to spectral photon flux by using a transfer function based on Planck's constant; $Q(\lambda) = 0.00835919 \cdot \lambda \cdot E(\lambda)$ (Endler 1990), in which Q is the photon flux, λ is wavelength, and $E(\lambda)$ is the spectral power output of the spectroradiometer. In simple terms, this conversion corrects the spectral distribution for the fact that an energy unit of red light represents more photons than the same unit of blue or UV light. Note, however, that even in visual modeling (ARTS), although not the norm among visual ecologists, unconverted spectral power distributions can be used as long as this currency is used for both illuminant (A) and sensitivity (S) spectra. Use of the unconverted distributions is typically the tradition in human color technology (Billmeyer and Saltzman 1981) and probably reflects the problem we started out with; namely, that light is measured as energy. Note also that all the simple interconversions used above only concern spectral distribution functions. If total power (or total photon flux, for that matter) has been measured, there is no way to derive a spectral distribution and hence color.

Finally, it should be mentioned that there are a number of human-based quantities and units associated with photometry, such as luminous flux (lumens) and luminous intensity (candelas), and older ones (e.g., lux, lambert, footcandle). These are of course not directly applicable to bird vision and are mentioned here primarily so they can be avoided. However, if you are restricted to photometric measurements (e.g., RGB readings of a digital image, or spectral luminosity curves), the "best of a bad job" may be to use human color technology (CIE 1978; Billmeyer and Saltzman 1981) "backwards" to convert data to crude but less subjective estimates of the corresponding reflectance or ambient spectra.

Principles of Reflection and Measuring Geometry

To set the stage for some of the terms used in conjunction with reflectance measurements, we take a brief look at the general principles of light reflection. When a beam of incident light encounters the surface of an optically denser (slower) medium, such as the keratin cortex of a feather, it is partially refracted into the medium (i.e., bent toward) the surface normal (the line perpendicular to the surface) and partially reflected at the same but opposite angle to the normal. From an overall smooth surface, such as polished metal, or a surface

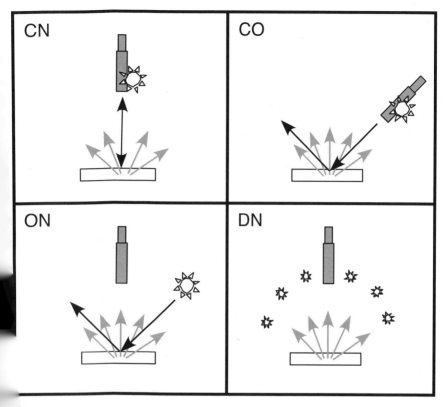

Figure 2.3. Measuring geometries categorized according to angles of illumination and observation, with respect to the surface normal. Solid arrows represent central axes of incident light and specular reflectance; gray arrows indicate diffuse reflectance. Positions of light source and radiance detector are marked by the large sun and probe symbols, respectively. Smaller sun symbols represent diffuse illumination. CN = coincident normal (0°/0°); CO = coincident oblique (45°/45°); DN = diffuse normal (d/0°); ON = oblique normal (45°/0°).

White-Standards and Dark Currents

In spectral-acquisition software, reflectance (R) is derived from the ratio between the reflected radiance of the sample surface (AR) and that of a reference surface with known reflectance (AR_r). If the reference reflectance is high (close to 100%) and spectrally uniform over the waveband of interest, the reference surface is usually called a "white-standard," and its reflected radiance will thus be a good estimate of the incident light (i.e., $AR_r \approx A$). Hence, with measuring geometry and incident light (A) held constant between measurements of

composed of a mosaic of many small parallel surfaces, most of
beams coincide to produce a strong specular reflection or gloss
the object a shiny, highlighted appearance. Rough surfaces,
microscopic components at many different angles to the incid
the barb ridges that make up a typical feather vane), give pri
reflection and a duller or matte appearance. Because both specu
surface glosses are usually spectrally uniform (achromatic or "c
tend to desaturate or even completely conceal (at angles of
lights") any spectrally nonuniform (chromatic) reflections comi
lying structures and pigment filtering. A common objectiv
geometry (i.e., the angles of illumination and observation) is t
imize the contribution of specular surface reflection.

Although reflection is the optical phenomenon as such, re
2.1) refers to the ratio of reflected to incident light for a give
and measuring geometry. It can be derived from any measur
ation intensity (photon flux or energy), as long as the incid
reflected radiance (AR) are measured in the same units. Tl
radiance" has sometimes been used as synonymous with "ref
usage is wrong, because the former is a measure of the radi
ject (hence, a property of the transmitted signal), wherea
conversion factor by which reflected radiance derives from
ance (hence, a property of the object). Likewise, light tha
either absorbed or transmitted by the material, according t
proportions absorptance (commonly, but erroneously, c
and transmittance.

Another goal of the selection of measuring geometry, sc
with the first, is to mimic as closely as possible the natura
of a color signal. In terrestrial light environments, excludi
surfaces are more directionally than diffusely illuminate
most chromatic signals are diffuse and desaturated in the
face reflections. Measuring geometries with directional lig
angled sensor (15–30°; Endler 1990) and a small specul
is therefore reasonably similar to the natural situation
common measuring geometries for recording surface re
tions of terminology to use in method descriptions. N
nonuniform surface topography (e.g., a feather vane),
rotation should also be considered. In the discussions b
coincident normal (CN) is the most practical and r
measuring geometry for both pigmentary and structu

standard and sample, reflectance can be directly estimated as $R \approx AR/AR_r$. Note, however, that this approximation primarily concerns the spectral shape of the reflectance, whereas the intensity ("brightness") has a more complicated dependence on the specific white-standard and measuring geometry used. The close-to-100% reflectance specified for some white-standards refers to their total reflection over a 180° solid angle, of which only a fraction is collected by the narrow probe acceptance angle (~30°). Estimation of absolute reflectance intensity therefore requires that the reference and sample reflect equal proportions of their respective total reflectance toward the sensor or that such differences are adjusted for (Endler 1990).

In practice, this point is rarely considered, and various diffuse white-standards are used without adjustments, in principle assuming that feathers or bare parts are also perfect diffusers, although they are not. As long as the standard reflectance is spectrally uniform, however, and spectral shape ("color") is the primary signal variation under study, the assumption of perfect diffusion is not a major problem, especially not for relative reflectance variation within a population. But for interspecific differences or comparisons among populations measured with different methodologies, it should be kept in mind that plumage or bare-part brightness will always be inflated or deflated, depending on tissue type and measuring geometry (see below).

In addition to using a reference surface for indirectly estimating patch reflectance, another modification of the reflectance formula is required to control for any noise detected by the spectrometer during practical measurements. Noise is caused both by electromagnetic disturbance from the external environment and by the spectrometer system itself. This so-called "dark signal" or "dark current" needs to be subtracted from both sample and reference spectra before calculation of the reflectance ratio. In other words, sample reflectance is estimated as reflected radiance minus dark signal, divided by reflected radiance from reference surface minus dark signal:

$$R = (AR - D)/(AR_r - D).$$

Equipment and Techniques for Field Reflectance Spectrometry

Spectrometer Systems

The Ocean Optics model USB2000 (Figure 2.2), its predecessor S2000, and the AvaSpec from Avantes (www.avantes.com) are probably the most common spectrometers used in behavioral ecology research today, but there are other interesting and similarly priced alternatives (e.g., the EPP2000, www.stellarnet-inc.com). All fulfill (or can be tailored to meet) the basic requirements for field research on avian coloration: 300 to 800 nm sensitivity, fiber-optic input, small size, ruggedness, computer-powered USB or serial (EPP2000) interface, and a range of spectrometric accessories and software. In addition to these small and relatively inexpensive instruments, there is a number of spectroradiometer platforms and accessories available from the large optoelectronics companies (e.g., Zeiss, Oriel, EG&G), some of which have been used for plumage or feather reflectance, primarily in the laboratory (e.g., Burkhardt 1989; Endler and Théry 1996; Vorobyev et al. 1998; Siitari et al. 2002).

Although we have not attempted a complete survey of diode-array spectrometers on the market, or even a direct comparison between the mentioned models, there are some less obvious details to consider when choosing a spectrometer. First, a spectrally uniform sensitivity of the detector (CCD) can be more important than its absolute sensitivity in allowing a sufficient, but not saturating, signal intensity across the entire spectral range (320–700). If the peak spectral sensitivity of the detector largely coincides with the peak spectral intensity of the light source, the short integration time needed to avoid detector saturation (~4,000 counts) at the peak wavelengths may result in a weak and noisy signal elsewhere in the spectrum. Low signal-to-noise ratios can, in principle, be avoided by measuring the ranges separately with different settings (or different detectors) and subsequently combining them, but this means twice as many scans and additional spectral processing. A better alternative might be to chromatically filter the light source (at the lamp or at the entrance to the spectrometer), but there are still good reasons to shop around for uniform detectors (ask the vendor to produce the detector response curve).

In addition, look for uniform noise without spikes or glitches from second-order effects or "cross-talk" in the CCD when choosing a spectrometer. If present, such spikes or distortions are typically strongest at medium wavelengths (500–600 nm) when using powerful UV light sources (e.g., deuterium lamps) to measure bright surfaces. They vary in amplitude and are difficult to remove even with a new dark current and white standard before each scan. And even if these spikes can be identified as obvious instrument artifacts in otherwise smooth reflectance curves, they are a nuisance to remove before spectral analysis.

Operating Software

After the first versions of Ocean Optics SpectraScope in the early 1990s, spectral acquisition software has evolved and speciated rapidly. Since 1993, we have mostly used CSpec™ from Omni Spec (www.omnispec2000.com), initially designed for Ocean Optics spectrometers. At the time, it had several advantages, most of which now seem to be present in the latest versions of OOIBase™ (Ocean Optics), AvaSoft™ (Avantes), and SpectraWiz® (StellarNet, www.stellarnet-inc.com).

Although all currently available software packages in principle are adequate for reflectance spectrometry, we still argue that software quality should be among the most important considerations when choosing a spectrometer system. The best advice is to first try out the complete system in the expected field situation. When working alone outdoors or in a car with a poorly shaded screen and trying to hold the probe against an uncooperative wild bird, reflectance measuring can be quite different from the demonstration at the lab bench. Small details become important for the quality and speed of measurements: How many commands does it take to create a new directory/folder and how awkward are such commands? How many commands to capture, name, and save dark spectra, reference spectra, and a series of reflectance spectra? Because the handling time per bird in many projects is already forbiddingly long (e.g., due to blood sampling and experimental manipulations), there is no spare time and often no spare fingers to navigate complicated dialog boxes or sub-subsubmenus. Important features and comparisons between software include:

1. *Convenient keyboard shortcuts.* CSpec, but not OOIBase, stops scanning and "freezes" the spectrum when the space bar or a mouse button is pressed. The frozen spectrum can then be named and saved by pressing Ctrl+A. Both programs have keyboard shortcuts for storage of dark (Ctrl+D) and reference (Ctrl+R) files in the current directory.

2. *Automatic saving of dark and reference.* Automatic saving for each reflectance spectrum can be specified in the general settings of OOIBase. However, one may want to consider how automated saving of two ancillary files for each data scan will affect requirements for storage space.

3. *Immediate (autoscaled) inspection of dark and reference files.* If the software, despite being in reflectance mode with fixed Y-axis (0–1), automatically autoscales dark and reference curves, inspections of these curves are easy to do. An earlier version of CSpec had this feature, but in current CSpec and OOIBase versions, one needs to change the Y-axis to autoscale and then change

the axis back after the inspections. The need for manual rescaling might seem like a minor problem, but in many situations, it means that one skips the inspections, acquiring the dark in the dark, so to speak.

4. *Autoincrementation of filenames.* By specifying a basename and starting index, one can use the autoincrement feature of OOIBase for assignment of consecutive filenames to subsequent scans, saved by pressing Ctrl+S.

5. *Text output.* In both CSpec and OOIBase, reflectance data are saved in text format, which means that no conversion or renaming is needed before import to a spreadsheet or database program. Likewise, AvaSoft offers export to ASCII format.

6. *Synchronization of spectrometer and xenon strobe.* Although easily accomplished in OOIBase (see the section on light sources below), synchronization is not possible in our version of CSpec, but it has evidently been added to recent versions of the software (M. Wood, pers. comm.).

7. *Automated calibration.* In combination with a mercury-argon light source, AvaSoft provides automatic peak detection and calculation of wavelength calibration coefficients.

8. *Scripting possibilities.* In both SpectraWiz and OOIBase32 (the Platinum version), you can customize software functions by writing your own scripts. Notably, SpectraWiz offers compatibility with a whole range of programming languages (e.g., Visual Basic, C/C++, LabView).

9. *Photon flux conversion.* Spectral power can be expressed in a multitude of units, and converted to photon flux (i.e., moles/m^2/nm/s) in SpectraWiz.

10. *Software documentation.* Finally, the level and usefulness of software documentation and help facilities differ substantially among the different software packages. For instance, whereas CSpec supplies little more than one-line explanations of menu items, OOIBase ships with a comprehensive help document, complemented by online facilities.

Light Sources

The small DC-powered tungsten-halogen light sources (e.g., Ocean Optics HL2000 and Avantes AvaLight-HAL) provide strong, stable, and spectrally smooth light from approximately 380 to 700 nm, with a weak short-wave tail that allows noisy but usually informative reflectance down to about 365 nm. Tungsten-halogen light sources may take a few minutes to stabilize, and should therefore be switched on a few minutes ahead of measuring and then left running. Although it is claimed that bulbs may last thousands of hour (even 10,000

Quantifying Colors

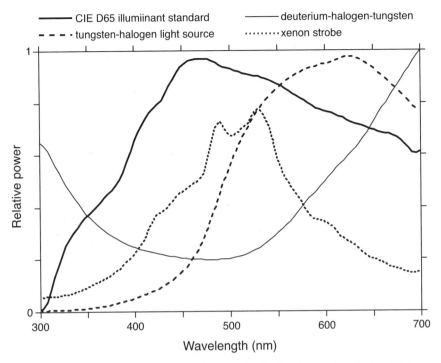

Figure 2.4. Comparison of relative spectral power distribution from a clear sky (the CIE D65 illuminant standard, solid thick line), a tungsten-halogen light source (Ocean Optics HL2000, dashed line), a deuterium-tungsten-halogen combination lamp (Ocean Optics DT-MINI, solid thin line) and a xenon strobe (Ocean Optics PX-2, dotted line). Note that these are relative outputs, whereas the absolute intensities differ substantially: in particular, the DT-MINI output is less than 20% of the HL2000, giving rise to noisier but UV-inclusive measurements.

for the more expensive long-life bulbs), it is wise to carry a spare bulb at all times. Using these lamps instead of the UV/VIS alternatives mentioned below in effect means trading the first bit of UV (320–365 nm) for excellent reflectance in the human-visible range and a reliable indication of any significant UV reflectance (Figure 2.4). Combined with confirmations (using full-spectrum lamps) that UV reflectance is absent or weak, these light sources are useful primarily for pigmentary color variation; for example, the spectral location of slopes ("hue") in saturated carotenoid pigmentation. When a distinct secondary UV peak is present, as in many unsaturated yellow carotenoid colors (Chapters 5 and 9), this is usually sufficient reason to switch to a UV/VIS light source.

The requirement of a portable, sufficiently strong and stable 300- to 700-nm light remains the major limitation for battery-operated UV/VIS reflectance work. The clear sky irradiance spectrum is a frequently used standard (CIE D65; Figure 2.4) in color technology, but there is unfortunately no artificial light source with a similar spectral output. For constant (steady) light, deuterium-halogen combination lamps are the main options (e.g., DH-2000), but these are heavy and require 110/220 V AC. Using a 12-V DC to 110/220-V AC inverter (at least 1200 W), they can be run from the battery of a car, if available; otherwise, a heavy battery or a gasoline-driven generator (plus gasoline!) must be carried along. A promising improvement is the lightweight, DC-operated miniature deuterium-halogen lamp. An early version that we tried was disappointingly weak (actually giving noisier UV reflectance than the 360- to 400-nm tail of the HL2000). However, the 1.2-W output of the present DT-MINI from Ocean Optics, which is ~20% of that of HL2000, may provide a sufficient signal for plumage reflectance. The spectral distribution (Figure 2.4) also seems smoother than that from DH-2000, in which a pronounced 660-nm deuterium spike can cause glitches in the reflectance curves.

Another attractive and portable solution is a xenon strobe (flash), such as Ocean Optics PX-2 or Avantes XE-2000. Via a serial 15-pin cable from the spectrometer to the PX-2 strobe, pulse rate can be automatically synchronized with the scan to either provide a single flash per scan, and thus the same signal intensity independent of integration time (single trigger mode), or a variable number of flashes per scan (multiple trigger mode). In the latter situation, the pulse rate is fixed, and signal adjustment is made by changing the integration time. For a stable signal, this interval must be in multiples of some integration time (in ms) determined by the spectrometer model and the analog/digital interface (see PX-2 section in OOIBase help files). We have limited experience with the xenon strobe, but have heard mostly positive judgments (except on its annoying, high-pitched noise!). Just like nonuniform detector sensitivities (see above), however, the two strong peaks around 500 nm are a problem, because they saturate the detector if the integration time is increased to obtain a stronger UV signal.

Reflection Probes

Fiber-optic reflection probes are used to guide the light from the light source to the measured surface and then direct the reflected radiance back to the spectrometer. The most commonly used is a bifurcated (Y-shaped) 7×200 μm

Quantifying Colors 61

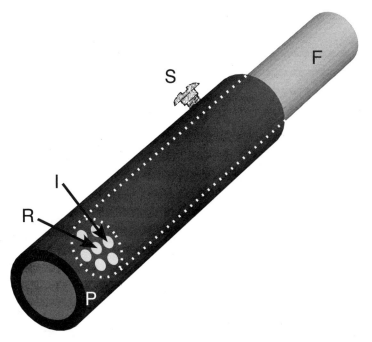

Figure 2.5. Probe pointer used for coincident normal (CN) reflectance measurements. The black PVC tube (P) fits tightly on the probe ferrule (F), but can be moved to adjust the diameter of the measuring spot, and can be locked in position by tightening the screw (S). Ferrule ending shows (not to scale) the exits of six illumination fibers (I) surrounding the reading fiber (R).

probe (e.g., Ocean Optics R200-7) in which a bundle of six 200-μm-wide optic fibers in the illumination leg (attached to the light source) join with and surround a single fiber from the reading leg attached to the light entrance of the spectrometer. The probe ends in a roughly 30 × 6 mm steel cylinder called a "ferrule," at the end of which the ring of illumination fibers surrounds the read fiber (Figure 2.5). From a single 200-μm fiber, the light exits in a 25° solid angle, and the combined radiation angle from the circle of six fibers is thus slightly larger. For plumage reflectance on live birds or museum specimens using the CN measuring geometry (see Figure 2.3), the 7 × 200 probes (e.g., Ocean Optics R200-7) are the most suitable, allowing the right light intensities from the portable light sources. There are also long, thin ferrules that can be used for aiming at small or narrow patches (<3 mm), such as nestling gape flanges (e.g., Hunt et al. 2003) or single feathers measured in the lab.

Probes and ferrules can be custom made for particular applications. Furthermore, Ocean Optics offers a variety of do-it-yourself kits providing necessary tools and fiber material for probe assemblage.

The fiber probes can be damaged from intensive use, excessive bending, being squeezed by the lid of the instrument case, and the like. Avantes' fiber probes have a metal cover instead of the standard plastic cladding, which makes them less sensitive, but certainly not immune, to abuse. Fibers and probes should therefore be handled and transported with care and checked regularly. To check that a fiber probe is working correctly, point one end toward the sky or a lamp and make sure that all fiber windows shine at the other end. It should also be noted that all fibers and probes from the companies mentioned here use the same standard connectors (SMA905), which means that fibers, light sources, and other accessories can be combined from the different suppliers.

Probe Aiming and Alignment Devices

Variation between scans in measuring geometry (distances and angles of illumination and reading) can be a major source of error in reflectance studies, particularly for structural colors, which is why some kind of aiming device is essential to achieve and standardize the alignment of light guides with the sample surface. For measuring feathers or other samples in the lab or under a microscope, illumination and measuring fibers, as well as the sample surface itself, can be aligned very precisely in any conceivable measuring geometry. Given a fixed viewing axis, Osorio and Ham (2002) identify five additional dimensions of angular variation: elevation (angle between illumination and reading axes), azimuth (angle; i.e., height, above the surface plane), plus the three axes about which the sample surface can rotate. As they point out, to deal with all these parameters is a formidable task, but they describe a simple but ingenious apparatus for adjusting three of them—elevation, azimuth, and one axis of surface rotation.

For measuring live birds with a handheld reflection probe, elevation is fixed to 0° (i.e., coincident), while azimuth and the three surface rotations can vary over the entire hemisphere of angles from which the probe can be aimed. The problem is thus to standardize the measuring geometry to one of these angles, in addition to a fixed measuring distance. In color technology (e.g., Wyszecki and Stiles 1982), the two commonly used angles (from the surface normal) are 0° and 45°, the latter being used for glossy surfaces. The angle can be fixed by using a block holder in which the ferrule can be inserted at either 90° (for CN geometry) or 45° (for Coincident Oblique [CO] geometry), and at a fixed distance from the measuring opening on the opposite side. The main problem with this technique is the difficulty of aiming at a small plumage area. To solve both the aiming problem and fixation of distance, we use a probe pointer (Figure 2.5), which is a simple but efficient solution for the CN measuring geometry. The reasons for using CN measurements whenever possible, even when some desaturating gloss is picked up, are several: First and foremost, it avoids the variation from rotation of the measuring axis around the surface normal (i.e., the often large effect of how the barbs are aligned and tilted in relation to the viewing axis; Osorio and Ham 2002). Second, even if rotation around the surface normal is kept reasonably constant (e.g., by always aiming perpendicular to the length axes of feathers), the oblique angle chosen (e.g., 45°) is more difficult to standardize with a handheld probe. If necessary, for example because of too much gloss at CN measurements, the probe pointer can be cut at the desired angle.

Integrating Spheres

Although often impractical in the field and for live birds, the principle of integrating spheres (Figure 2.6) should be mentioned, as it is the standard way to achieve perfectly diffuse illumination of the sample. In the typical configuration, the illumination enters from the side (in relation to the sample port) and is uniformly scattered around the highly reflective white interior of the sphere. To exclude nondiffuse components of reflection, the reading fiber attached at the top of the sphere is shielded from first reflections by a baffle. Switching fibers between ports transforms the sphere into a reflection-integrating device, measuring total light reflected from a sample in all directions. Operating the device in this mode is not so useful for color signals, which typically are received at a narrow acceptance angle, but it can be used to obtain a "cosine-corrected" (Endler 1990) measure of ambient irradiance. Integrating spheres yielding near-perfect diffuse illumination over a wide wavelength range (250 nm to 2500 nm) are available from several manufacturers (e.g., Avantes AvaSpheres, Ancal IC09).

White-Standards and Dark Currents

The white-standards in common use today are made from stable, white materials (e.g., Spectralon® in Ocean Optics WS-1 and PTFE® [polytetrafluoroethylene], in Avantes WS-2), that give strong ($\approx$100%, but see above) diffuse

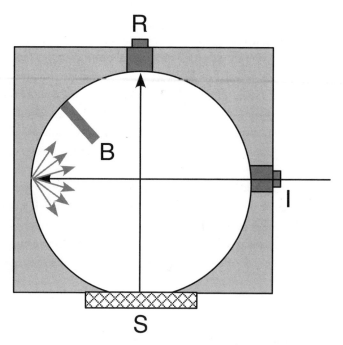

Figure 2.6. Integrating sphere for diffuse illumination and/or diffuse reflectance. As configured here, illuminating light (I) enters from the side and is scattered multiple times by the white reference material lining the inside of the sphere. This creates diffuse illumination of the sample (S), the reflected radiance of which is measured normal to its surface. The direct (first) interior reflections (gray arrows) are prevented from reaching the reading port (R) by a baffle (B).

UV and VIS reflectances. Alternatives include barium sulphate, polished aluminium, or other materials with sufficiently high and uniform reflectance in the desired wavelength interval (Endler 1990). Being considerably cheaper than manufactured reference tiles, Teflon™ tape has been used as white-standard by some researchers.

Keeping the reference surface immaculate is of vital importance, and for this reason, it is generally supplied with a protective casing of some kind. Furthermore, the use of a probe socket (Figure 2.7) ensures that readings are taken at a fixed short distance (1–2 mm) above the surface, avoiding contact with the probe and minimizing exposure in general. However, it also entails a decrease of reference reflectance in comparison with sample reflectances, as these are acquired with the probe held directly at the feather surface. To compensate for this decrease, sample spectra need to be multiplied with a correction

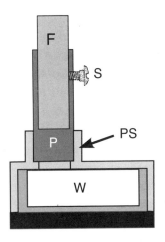

Figure 2.7. White-standard with probe socket. Drawing and photo of the WS-2 reference tile (from Avantes, The Netherlands) enclosing the white-standard surface (W). The tile has been fitted with a home-made probe socket (PS) to hold the probe pointer during reference measurements. See Figure 2.4 for details of the probe pointer.

factor equivalent to the ratio of standard reflectance taken with and without the probe socket.

If the standard surface has been soiled (e.g., by a fingerprint), it can usually be cleaned by using the appropriate practices recommended by the manufacturer. Cleaning recommendations for the WS series of white-standards include airbrushing with clean air or nitrogen gas for light impurities, and ultrasonic baths in deionized water (2 h at 40–60 °C) followed by vacuum drying (12 h at 60 °C, ≤1 torr) when dealing with heavier soiling. Alternatively, applying emery paper or packing tape to the surface may be a quick solution for removing dust (R. Montgomerie, pers. comm.). Unless explicitly recommended by the manufacturer, soapy water, alcohol, or organic cleaners must not be used, however, as these may substantially reduce reflectance, especially in the UV. In general, it cannot be overemphasized how sensitive the UV band is to corruption of the reference surface. Fats, proteins, and just about any kind of dirt have their main absorptances in the UV, and even the slightest soiling or contact with skin of the standard can reduce reflectance and thereby boost measurements of UV reflectance, even creating artefactual peaks that may be interpreted as UV signals.

Obtaining the dark spectrum (i.e., the detector signal when no light enters the probe) requires that ambient light is completely excluded from the acceptance angle of the probe and that the illumination beam is not reflected back into the probe. Such conditions are most efficiently arranged by a cardboard "dark-box" with a tight-fitting opening for insertion of the probe. The box should, of course, be impenetrable to light and sealed up with tape if necessary, and large enough (or lined with, e.g., black velvet) so that there is no detectable reflectance off the inner walls (also remove loose flaps).

Calibrating the Spectrometer

Before putting your spectrometer to work for the first time, it must be configured to use the correct wavelength calibration coefficients, which most likely have been determined by the manufacturer and shipped along with the instrument. The calibration coefficients are extracted from the third-order polynomial function that describes how the wavelength of incident light varies with pixel number in the CCD detector. This relationship is subject to drift from, for example, small displacements of mirrors or grating, and will therefore require recalibration at some point. The adjustment is done with a mercury-argon lamp (HG-1, OceanOptics), associating its output of known monochromatic peaks with detector pixels and computing a new set of polynomial regression coefficients (see detailed information in OOIBase help files). Although drift seems to be very slight in the recent models, it should be checked regularly (either with the mercury-argon lamp, if available, or by some other reference spectra), at least at the beginning and end of each field season, and of course after any kind of rough handling or transportation.

Acquiring Plumage Reflectance

We now move to a consideration of some practicalities of acquiring plumage reflectance data. A step-by-step guide to the measuring process is given in Box 2.1, and below is a more detailed discussion on some important elements of reflectance acquisition.

Inspecting Dark and Reference Spectra

Just before taking a continuous series of scans (typically a new individual or a set of samples measured in quick succession), one must first obtain a dark spectrum and then a reference (white-standard) spectrum, both of which will be incorporated in subsequent reflectance measurements, and on which the quality of the reflectance data critically depends. It is important that both dark and reference data (dark.txt and referenc.txt in CSpec; *.Dark and *.Reference in OOIBase) are saved together with the scans for which they are used. By saving them together, reference curves can quickly be found and inspected as potential sources of odd-looking reflectance curves or outlier colorimetrics, which typically affect brightness but can also affect spectral shape, especially in UV, if the problem is a soiled white-standard. When corrupted references indeed are exposed, the reflected radiance raw data can be recovered by, for example, multiplying a reflectance spectrum with its associated reference spectrum. The recovered radiance can then be divided by another, uncorrupted reference spectrum; if possible, one obtained at the same occasion.

Because the dark signal is there to remove external disturbance and detector noise (in the absence of light), it should be measured as recently and similarly as possible to the reflectance scans (e.g., with the lamp running, fibers hooked up), except that no light is allowed to enter the probe. Aiming the probe into a dark corner may be sufficient, but a dark-box (see above) is a simple guarantee that neither ambient light nor reflections off a surface are recorded.

Acquired darks should be fairly flat and less than ten units in amplitude (Figure 2.8b). Provided that dark subtraction has been configured in the software (see Box 2.1), repeating measurement in the box after saving a dark should give a baseline at zero. When starting up the system, the first one or two dark scans may for some reason be abnormal (typically, in our experience, a steep slope above 600 nm), but usually stabilizes after a few scans in quick

Box 2.1. Reflectance Spectrometry

The following instructions provide a guide from starting up the system (in the field or in the lab) to saving the first reflectance scan. The equipment assumed, and illustrated in Figure B2.1, is a fiber-optic spectrometer with either a stable halogen lamp or a xenon strobe, a bifurcated reflection probe fitted with a probe pointer (Figure 2.5), and a white standard with probe socket (Figure 2.7). Software commands refer to OOIBase32® (O) and Cspec® (C), respectively, but can be translated to other measuring software. Also note that clickable icons are not mentioned.

System Setup and Configuration

1. Connect light source to AC with adaptor or directly to DC (12 V). Connect pulsed light source to spectrometer with synchronizing 15-pin cable. If using a halogen lamp, switch it on to allow 5 minutes for stabilization.

2. Connect computer to AC or DC, if needed, and start it. After startup, connect spectrometer to computer using USB cable or PCMCIA card. Attach illumination fiber leg (large "window") to light source, and measuring leg (small "window") to spectrometer.

3. Start measuring software. For strobe, select flash mode via toggle switch and configure software to operate light source (O: Spectrum | Configure Data Acquisition | Strobe Enable). Consult manual on how to adjust integration time to ensure constant number of flashes per integration cycle.

4. Make sure software is in scope mode (O: Spectrum | Scope Mode; C: Acquire | Acquire setup | Auto divide off). Set scale of X axis to 300–750 nm; Y axis to auto scale (O: View | Spectrum scale; C: Setup | Scale). In CSpec, turn on dark correction option (Setup | Operational | Dark correction).

5. Create sample directory, and direct subsequent dark and reference files to this directory by saving a dummy file in it (C: Ctrl+A), or enable auto saving of reference and/or dark with samples (O: Edit | Settings | File Saving).

Dark Current and White-Standard

1. Aim probe into dark box or dark corner, and take Dark (O: Spectrum | Store Dark + File | Save Dark; C: Ctrl+D;). The noisy dark spectrum should be fairly horizontal and <10 units in amplitude.

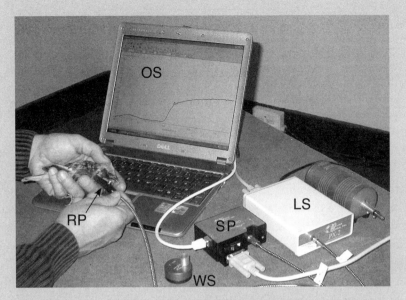

Figure B2.1. Setup of equipment for reflectance spectrometry: LS = light sources; OS = operating software; RP = reflection probe; SP = spectrometer; WS = white-standard.

2. Check reference signal: Place probe in reference probe socket and resume scanning. Freeze acquisition and inspect spectrum (O: Spectrum | Snapshot Mode; C: any key). Peak intensity should be about 3500 counts. If not, adjust integration time or measuring distance accordingly (see text) and, when adequate, take new Dark and reference.

3. Save reference (with probe still in socket), (O: Spectrum | Store Reference + File | Save Reference; C: Ctrl+R).

Acquire Reflectance Scan

1. Set software in reflectance/transmission mode (O: Spectrum | Transmission Mode; C: Acquire | Acquire setup | Auto divide on). Set Y axis scale to 0–125 (O) or 0–1.25 (C).

2. Check reference reflectance. If not a flat line at 100% or 1.0 (deviations most likely in the UV), take new reference and replace the former.

3. Scan plumage reflectance by holding (not pressing) probe pointer perpendicularly against plumage until a stable signal. Stop scanning and save and name reflectance file (O: Ctrl+S; C: Any key, then Ctrl+A).

succession. If not, high variability in successive darks may indicate bad connections or nearby sources of electromagnetic interference (e.g., power lines, radar stations). PCMCIA cables are particularly sensitive and may need to be shielded with aluminum foil. Some temporal fluctuations in the dark signal are to be expected. Therefore, dark spectra need to be checked and updated at regular intervals during reflectance measurements.

A typical white-standard spectrum is shown in Figure 2.8a. Note that maximum radiance should not exceed 3,500 counts, safely below saturation of the detector. Dividing the reference spectrum with some previously obtained reference should always yield a flat line close to 1. The exact value will vary if the ferrule has been repositioned inside the probe sheath or the integration time has been adjusted since the last measurements. Any deviations from a flat line, however, reveal changes in either lamp output or fiber and detector attenuation (not critical) or a corrupted reference tile (critical), which must be dealt with immediately. Because most kinds of dirt and grease absorb strongly in the UV, problems with a white-standard are virtually always in the UV, appearing as a dip below 400 nm where the reference is less reflective than it was before. The use of a soiled and UV-deprived reference may substantially inflate sample reflectance in the affected regions of the spectra.

Measuring Geometry

A detailed account of the effects of measuring geometry on plumage reflectance is beyond the scope of this chapter, but it was recently treated in great detail by Osorio and Ham (2002). The aim here is only to make some general remarks and suggestions on simple and easily standardized methods for measuring plumage reflectance in the field.

Despite the complex surface structure of avian plumage, feathers have a number of advantages for reflectance work. At the completion of molt, feathers are dry, dead tissue without the blood vessels or chromophores that can produce rapid color change in living integument. Although feather wear (e.g., Ornborg et al. 2002), bleaching, soiling, and cosmetics (Chapter 9) may change plumage reflectance over time, these factors can easily be controlled for, and plumage color can be considered comparatively constant and insensitive to handling and measuring procedures. Furthermore, the ridged vane surfaces, from parallel and vertically tilted barbs (Figure 2.9), produce largely diffuse achromatic surface reflectance (from barb keratin cortexes; Figure 2.9). Except in classic interference colors, the internally scattered and pigment-filtered light

Quantifying Colors 71

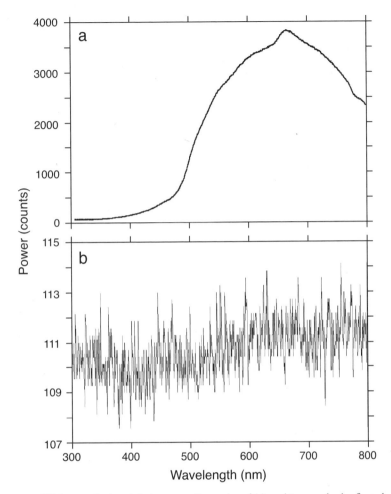

Figure 2.8. White-standard and dark spectra. Examples of (a) a white-standard reflected radiance, in CSpec software, using a tungsten-halogen light source (Ocean Optics HL2000); (b) a dark spectrum.

is also rather diffusely emitted. This is true even when the emission is chromatic (UV, violet, blue) from coherent scattering (Chapter 7), which is why such coloration for long was termed "non-iridescent structural coloration" (Auber 1957b). This also means that the color signal is not strongly desaturated by specular reflections (gloss). Furthermore, because most barb morphologies (Auber 1957a) are not apically flattened in the plane of the vane (but see Andersson 1999b), CN need not produce the strongest specular reflectance.

Figure 2.9. Scanning electron microscopy image of cross-sectioned part of apical vane of a blue tit crown feather. The parallel aligned and primarily vertically tilted barbs create a ridged and largely diffusely reflecting vane surface, usually allowing coincident normal (CN) measuring without desaturating specular reflectance (gloss). White scale bar is 20 μm.

Instead, maximum reflectance should appear at some elevational and rotational angle normal to the largest achievable surface of parallel barb sides. (Think of it as shining a flashlight at a steel washboard.)

The main practical reason to use CN measuring whenever possible (using the simple probe pointer; Figure 2.5) is that it is most easily standardized. First, compared to using a sheath cut at 45° degrees (CO), we find it easier to apply consistent pressure against the plumage between measurements, which may affect how feathers are packed and ruffled under the probe. Second, oblique (slanting) measurements not only induce error in the elevation angle, but they also introduce error from variation in the rotational angle of the probe in relation to the orientation of barbs and rachis (one angle of which may create strong gloss—see above). To summarize, the proposition here is to use CN measuring whenever possible.

For particularly shiny plumage (i.e., with strong spectrally unselective surface reflection) or other integument such as bill or feet, the gloss at CN measuring may strongly desaturate or conceal coloration (see above). Using the standard reflection probe, and given that the specularity is largely directional in relation to the plane of the vanes (and thus the plumage surface), one solution to this is to use CO (Figure 2.3) achieved by a probe pointer cut at, for example, 45°. If this does not help, one can in addition rotate the measuring axis around the plumage normal until roughly parallel with either barbs or rachis, depending on which of these structures seems to contribute most of the specular reflectance. If none of these options is sufficient, specular components can be almost entirely removed by using separate illumination and reading axes (Oblique Normal [ON]; Figure 2.3). Finally, one can use an integrating sphere to obtain diffuse illumination and/or reading over the entire 180° solid angle (Diffuse Normal [DN]; Figure 2.3), but this is impractical in many cases and, in addition, usually a misrepresentation of natural signaling conditions.

Another largely ignored problem is the modification of the reflectance properties of the measured integument that may be caused by the physical contact with the probe, the abnormal heat (IR) and tissue damage (UV) from the artificial illumination, or simply the handling of the animal. Bird plumage is dead and dry tissue (e.g., nails), which is why such effects usually are assumed to be negligible, but for long and/or frequent exposures, especially with intense and UV-rich light sources, some bleaching of pigments or disruption of fine structures might occur. The problems may be acute for live integument like skin and fleshy structures such as combs and wattles, which are not only glossier than most feathers, requiring oblique or diffuse geometry (CO, DN; Figure 2.3), but may also change color due to the handling and treatment. A quantitative investigation of such effects on different kinds of coloration would be very valuable.

Choice of Measuring Patch

In many avian coloration projects, one of the first things to consider are which of a number of alternative colored plumage regions to measure, and then which and how many patches within these regions. For complex patterns with a variety of colors, it is usually necessary to select a few (three to five) or even a single region to reduce handling time and subsequent analyses (Chapter 3).

If previous results or general predictions already point to a particular pattern element (location or color), this of course should be the main target for reflectance measurements, but quite commonly there are several alternative patches to consider. Obvious candidates are plumage regions that have empirically been confirmed to function in signaling or other behavioral adaptations, in the study species or in phylogenetical and/or ecological relatives, as well as those for which the proximate mechanism of color is of particular interest. Following (or in the absence of) such presumptions, there are also practical considerations that may improve the quality of measurements, including the size, uniformity and topographic location of the plumage patch. To maximize repeatability and comparability of measurements, the chosen patch should be small, ideally just larger than the probe, and without fine patterning. For the same reasons, regions with larger contour feathers and sleek plumage, such as wing coverts and mantle, are preferable to small, erectible feathers on the nape and chin, for example.

Sampling Plumage Patches

Related to the choice of which plumage patches to measure are the questions of whether and how to repeat measurements within the selected patches. One important question is whether to aim at a single, representative spot or to move the probe within the patch between measurements. For plumage regions with no visible patterning or within-patch color variation, we use the former approach (i.e., aiming for the same spot but removing the probe between scans—typically three or five). Thus we assume that the spot is representative of the entire patch, whereas the measurement error primarily estimates the repeatability of the measuring technique (e.g., how the probe is aligned and pressed against the plumage).

Another reason for not confounding technical measuring error with within-patch measuring error is that full independence between scans is virtually impossible to obtain anyway, for the simple reason that the first scan unavoidably influences the expectations of the following scans. Solutions to this could be to blindly stop scanning (for example after a standardized time), or having several persons independently take one scan each: both methods would be error-prone and too slow and impractical for most projects, but could be used on a smaller sample in order to quantify the various sources of measurement error. These issues are discussed in more detail in Chapter 3.

File Organization and Naming

Most researchers have experienced the frustration and waste of time (or even data) caused by poor management and inconsistent naming of computer files in various stages from raw data to the finally executed stats files. Reflectance data can become a nightmare in this respect because they pass through so many processing stages (Chapter 3). Our general advice is therefore to plan the entire procedure from raw data to final color variables before deciding how to name and organize collected spectra. File names need to be short and consistent for automated import to spreadsheets or databases, yet there must be information on *when, how,* and from *what* a scan was obtained. *When,* however, can be left out, because date and time are automatically included in the file heading. *How* can usually also be omitted from the file name, to the extent that methods are standardized within the project.

The crucial information to associate with a scan is thus *what* it was obtained from, and the safest solution is to include both individual ID (e.g., ring number) and the patch name or acronym (see below), followed by a scan number (e.g., 1–5). This method produces such file names as "1kj34604gpc3" for the third scan of the greater primary coverts of a Blue Tit. Not only will this naming convention allow retrieval of scans that have been saved in the wrong place, but it will also appear in the top row of scans assembled into a spreadsheet, not having to be entered again while cross-referencing against a protocol. However, even with such a name, there are too many keys to press with a furious bird in the hand and several more waiting in bags. One remedy is to reduce the ring number to the last three digits, but this is not to be recommended if there are different ring prefixes in use. Although file creation dates can be used to correct mix-ups, it takes time and, worse, the mix-up may go unnoticed.

The solution we have settled on is to create a new directory (folder) for each individual, labeled with the ring number, and to name files only with patch and scan number (e.g., gpc3). This procedure is somewhat risky if files escape from directories, but it speeds up measuring considerably, and (in our case) relies on an import procedure (to Microsoft® Access™) that extracts the directory name together with each underlying file name. Moreover, to associate scans with their corresponding dark and reference files, which we strongly recommend, individual directories must be created anyway.

Consistent naming of patches simplifies things, especially in collaborative projects that are pooling or comparing reflectance data from different sources. Figure 2.10 shows a general topography of plumage regions with acronyms

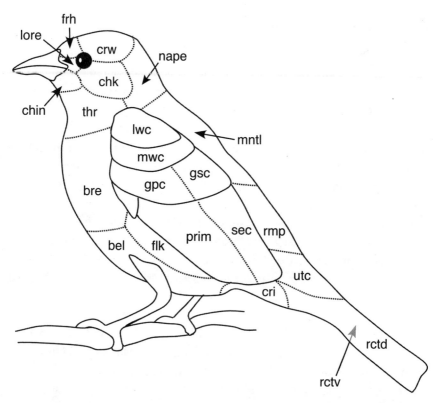

Figure 2.10. Topography of plumage regions and suggested acronyms for file naming. See Table 2.1 for a key to the acronyms.

explained in Table 2.1. Several of these are adopted by other projects, and we encourage others to use them as well.

Reflectance and Objective Colorimetrics

The acquisition of consistent and high-quality reflectance data may seem technically difficult, but the main confusion and most widely different opinions concern how reflectance variation, once obtained, should best be represented, analyzed, and interpreted. The next chapter (Chapter 3) reviews and compares many of the theoretical and statistical techniques for analyzing plumage reflectance, including the application of the generalized avian tetrachromacy model by Vorobyev and Osorio (1998). In the final part of this chapter, we continue to focus on reflectance variation as such (disregarding

Table 2.1. Key to Plumage Region Acronyms from Figure 2.10

Acronym	Plumage region
bel	Belly
bre	Breast
chin	Chin
chk	Cheek (auriculars)
cri	Crissum
crw	Crown
flk	Flank
frh	Forehead
gpc	Greater primary coverts
gsc	Greater secondary coverts
lore	Lore
lwc	Lesser wing coverts
mntl	Mantle
mwc	Median wing coverts
nape	Nape
prim	Primaries
rctd	Rectrices, dorsal side
rctv	Rectrices, ventral side
rmp	Rump
sec	Secondaries
thr	Throat
utc	Upper tail coverts

both light and receiver), and make some observations of how it should relate to the two main types of plumage coloration in birds: pigmentary and structural color mechanisms. This relation has implications for how properties of the sender (signal content) can be derived from reflectance shape variation, leading to the suggestion of a few simple objective colorimetrics presented in Table 2.2. Here color perception is also brought into the picture, but not through any particular photoreceptor model or physiologically derived discrimination thresholds. Until there are species-specific and behaviorally derived color spaces available, our suggestion is to bypass visual physiology and go straight for receiver psychology by assuming Hailman's (1977) three universal dimensions of vertebrate color cognition: spectral intensity, spectral location, and spectral purity, corresponding to the human psychometrics brightness, hue, and chroma.

Let us first repeat once more that plumage reflectance is an optical property of the sender: contrary to how it sometimes is treated, it is not the emitted "signal spectrum" or "signal display." The latter corresponds instead to the

Table 2.2 Objective Colorimetric Measures for Reflectance Spectrometry

Relation to signal perception		Objective colorimetrics		Relation to signal content (sign of correlation)	
General	Human	Measure	Definition[a]	Pigmentary	Structural
Spectral intensity	Brightness	R_{avg}	R averaged over λ interval	Concentration (−)	Amount (+)
Spectral location[b]	Hue	λ_{R50}	λ halfway between R_{max} and R_{min}	Concentration (+) and composition	
		λ_{Rmax}	λ of R_{max}		Dimension (+)
Spectral purity	Chroma	C_{max}	$(R_{max} - R_{min})/R_{avg}$		Regularity (+)
	"Carotenoid chroma"	C_{car}	$(R_{700} - R_{450})/R_{700}$	Carotenoid concentration (unsaturated, +; saturated, 0)	

a. R and λ denote reflectance and wavelength, respectively.
b. Spectral location is estimated as λ_{R50} for step-function spectra (pigments), and λ_{Rmax} for peaks (structural colors).

reflected radiance discussed earlier (AR; Figure 2.1), and this is not what was measured in the procedures described above (unless the software by mistake was in scope mode). In reflectance/transmission mode, the software divides the signal spectrum with the illuminant (i.e., the reference spectrum), and one would thus need to multiply with that illuminant to recover the particular signal spectrum. That is, if one wants to know what the reflected radiance of the halogen or xenon light source looks like! Of more interest to behavioral ecologists, one can multiply the reflectance spectrum with a natural illuminant to assess the signal spectrum relevant to the signaling situation. For example, you may have obtained a reflectance spectrum from a surprisingly drab brown Curlew Sandpiper specimen in a museum, not at all like you remembered it from the wild. But then you multiply it with John Ender's early/late ambient light spectrum (Endler 1993a), and "alakazam"—there is the signal spectrum of the warm red color you marveled at on the tidal flat the other morning. This is the beauty of reflectance. It belongs to the sender or the object. Just like the linear measurement of tail length discussed earlier, its existence does not depend on light conditions, as reflected radiance does, or on a specific beholder, as cone capture and color perception do.

Objective Colorimetrics

"Objective colorimetrics" may sound like a contradiction in terms: how can one hope to objectively measure something as deeply subjective as a color sensation? The philosophical debate on this will surely go on for decades (e.g., Thompson et al. 1992), but in evolutionary studies of animal coloration, the problem can largely be avoided by (1) sticking close to the physical variation that *can* be measured (i.e., light and light reception); (2) distinguishing carefully between sender and receiver aspects of coloration; and, when involving a receiver, (3) imposing few and universal assumptions about color cognition.

Note that these rules do not necessarily exclude the occasional use of human color psychometrics as estimates of signal spectra and reflectance. Trusting the human eye or a camera to quantify something real about animal integument is not the same as assuming that animal eyes and brains perceive it as humans do. It is the latter fallacy that was the main target of Bennett et al. (1994), and it was indeed important to stress just how stuck we humans are in our own color cognition. As an illustration, consider the not uncommon questions "What color is ultraviolet to birds?" or "What are the colors that birds can see in addition to the colors humans can see?" With a little extra thought, one soon realizes how (literally) senseless such questions are. Color names like "red" and "blue" are entirely rooted in human cognition. So are experience-based color names like "chestnut" or "lavender," although one might argue a kind of objectivity in assuming that birds also would agree on the similarity. In any case, asking for the color of UV radiation, to which we are blind, is just as meaningless as speculating on the hue of radio waves or the sound of ultrasounds, or any other signal or signal modality outside our sensory capabilities.

Colorimetrics and Signal Production

Regardless of whether any creature perceives it, reflectance variation holds a lot of information about the morphology and chemistry of the reflective material. For the biology of plumage coloration, such information is pertinent to numerous questions, both mechanistic and adaptive (e.g., signal functions), that need not make any assumptions about avian color vision. One need only assume that the color variation that humans see or measure has a materialistic basis. Consequently, this is also where human-subjective color quantification, although inferior to spectrometry in range and resolution (a bit like measuring pH by taste), is less of a philosophical problem. Proximate mechanisms of avian coloration are treated elsewhere (Chapters 5–9), but we emphasize a few important points relevant to the choice and interpretation of objective colorimetrics.

Carotenoids and Other Pigments are Subtractive Colorants

First we must stress that pigments, which in birds are primarily carotenoids and melanins, modify color (spectral shape) by absorptance of light, and thereby reduce the reflectance of the material in which they are deposited. This is the principle of subtractive color mixing (Billmeyer and Saltzman 1981), and the reason that the pretty red and green watercolors were such a disappointment to mix in preschool, removing most of the uniform reflectance from the white paper! To illustrate how increasing pigment concentration affects color, we use a simple simulation of increasing absorptance from a carotenoid (i.e., lutein; Figure 2.11b). Starting off with an unpigmented feather (uppermost curve), we assume that the reflectance of increasingly pigmented feathers (successively lower curves) is a power function of the absorptance caused by an arbitrary unit of pigment concentration (Figure 2.11a). The associated color change is shown in Figure 2.12, in which we use the colorimetric variables suggested

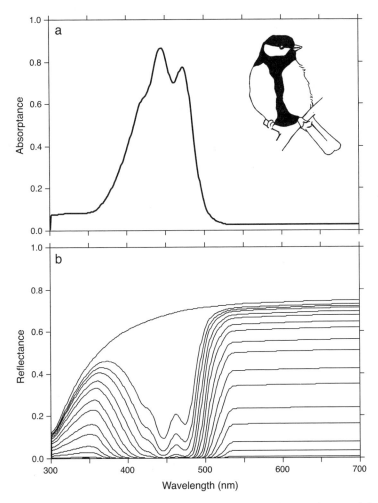

Figure 2.11. Carotenoid absorptance and simulated reflectance change. (a) Spectral absorptance of lutein, isolated from yellow breast feathers of a Great Tit (*Parus major*). (b) Simulation of reflectance change due to increased lutein concentration. The uppermost curve shows reflectance of an unpigmented feather; successively lower curves represent reflectance of increasingly pigmented feathers.

and summarized in Table 2.2. For avian pigmentary coloration and signaling, the simulation shows that, all else being equal, spectral intensity (brightness) should be negatively related to pigment concentration (Figure 2.12).

Furthermore, it is important to distinguish between unsaturated and saturated pigment colors. Pigments absorb light according to some spectral

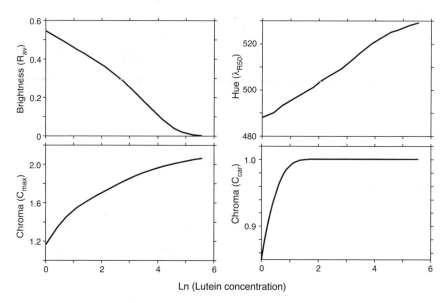

Figure 2.12. Simulated effects of increasing concentration (deposition) of pigment (lutein) on colorimetrics: brightness (R_{av}), hue (λ_{R50}), and two measures of chroma (C_{max} and C_{car}). See Table 2.2 for an explanation of these colorimetrics.

absorptance function that can be nearly uniform, like melanins, or have dramatic peaks, like carotenoids (Figure 2.11a), but in the latter, there is still some absorbance even in spectral regions where it is proportionally very weak. Therefore, in unsaturated colors, that is, those for which not all reflectance has been reduced in the spectral region of maximum absorptance (e.g., ~450 nm for lutein), it is primarily spectral purity (chroma and saturation) that varies with carotenoid concentration (Figure 2.12). Consequently, if pigment concentration is condition-dependent, spectral purity colorimetrics will contain the "honest" signal dimension (Andersson 1999a; McGraw and Gregory 2004). At saturation from a particular pigment or pigment combination, however, where reflectance in the peak absorptance region is maximally reduced, the increasing peak height (and thus positive correlation with chroma) reverses and starts to decrease instead. Thus, in saturated pigment colors, chroma will be unrelated to pigment concentration! In fact, because the chroma of the subsequent reflectance variation is positively rather than negatively related to brightness (i.e., not diluted by any reflectance before the cut-on), the perceived chroma should even decrease with pigment concentration in this interval. This situation, however, is not captured by the chroma estimates C_{max} and C_{car} (Table 2.2) used here, because both measures are controlled for brightness (by division) and thus level out at saturation (Figure 2.12). The direct estimate of carotenoid chroma, C_{car}, is still to be preferred, because it should capture the carotenoid concentration under a variety of circumstances (e.g., variation in melanin content).

The most important implication of the above is that measures of hue or spectral position (Table 2.2) increases monotonically with pigment concentration, also after saturation (Figure 2.12), and thus should be the primary "honest dimension" of saturated carotenoid displays (Andersson 1999a). This illustrates how highly saturated yellow pigmentation can produce orange-red hues.

Whereas our illustration concerns only a single pigment (lutein), most pigmentary plumage colors derive from a mixture of co-deposited carotenoids (e.g., Stradi 1998) and/or melanins (e.g., McGraw et al. 2004) with different absorptance properties. The principles of subtractive color mixing remain the same, with the combined absorptance of a specific pigment mixture determining the spectral shape of the reflectance and how it changes with increasing concentration. Depending on the degree of saturation of the respective pigments, however, the effects of some components may be partially or completely masked by the absorptance of others. For example, a moderate concentration of eumelanin (uniform absorptance across UV/VIS, creating the—to humans—"achromatic" grayscale) together with a "yellow" carotenoid, such as lutein, will produce a "shallower" (less chromatic) and less intense (duller) version of the unsaturated spectra in Figure 2.12, perceived by humans as a dull green color. For saturated eumelanin (black) or phaeomelanin (brown) pigmentation, there will simply be no shortwave light left for the carotenoid to absorb. In addition, red carotenoids and brown melanin may interact, also at saturation, to produce various reflectance shapes intermediate between the typical spectral signatures of each pigment type.

Structural Color Mechanisms Enhance Reflectance

Unlike the subtractive pigments, structural coloration derives from additive and spectrally selective modifications of feather reflectance, primarily in short wavelengths. Structural coloration is caused by nanometer-scale layers or "spongy structures" producing directional laminar interference, or iridescence, typically in barbules, or largely diffuse coherent scattering, typically in barbs

(Dyck 1976). Note that such basic reflective structures are a prerequisite also for pigmentary colors, as without them there would be no reflectance for pigments to subtract from (Shawkey and Hill 2005). In fact, because the whiteness of an unpigmented and structurally unmodified feather also depends on multiple internal (but incoherent) scattering, as in snow or paper, white is, in a sense, also a structural "color." In the following discussion, however, we limit our definition to include only those colors that result from structures causing spectrally selective modifications of incident light in their own right, albeit sometimes co-occurring with pigmentary absorption.

The potential ("honesty-maintaining") costliness of such structures has merely been suggested by a few studies showing correlations between color and individual condition or quality (reviewed in Chapter 12), and remains to be clearly demonstrated. Production costs that have been speculated to be incurrred include time and energy expenditure for accumulating structure thickness, and developmental stability for achieving regularity and precision of structure dimensions (Fitzpatrick 1998; Andersson 1999b).

Accordingly, the colorimetrics containing the honest signal dimension of structural colors should be the ones that best reflect variations in structure thickness and/or regularity. In contrast to pigment concentration, structure thickness is expected to be positively correlated to spectral intensity (brightness). Brightness alone is not likely to be a reliable signal in itself, however, as it could also be increased through loss of the absorbing layer pigments underneath. Color intensification due to increased structure thickness only will be accompanied by an increase in spectral purity (chroma), at least to a certain degree. Again, as discussed for carotenoid pigmentation above, the increase in chroma measures should theoretically cease and even reverse at saturation of a structural color (i.e., when perfect [100%] reflectance of the dominant wavelengths has been attained). Subsequent addition (i.e., increased thickness) to the scattering structure will add reflectance to other parts of the spectrum, thus continuing to increase brightness but lowering (desaturating) chroma.

The wavelengths of light that are primarily affected by scattering or constructive interference—in other words the spectral location (hue) of a structural color—are determined by the dimensions and relative alignment of structural components (Dyck 1976; Chapter 7). Simply put, reflection of shorter wavelengths requires more fine-scaled structures. If costs and/or constraints are more pronounced for smaller structures, the costs will consequently be inversely related to hue. Furthermore, increased regularity or precision of structures should cause reflectance to be more wavelength-specific and, in effect, increase spectral purity.

In summary, whereas condition-dependent variation in structural coloration may affect both spectral intensity (brightness) and position (hue), it is likely to be best captured by chroma colorimetrics, conveying differences in structure thickness and regularity (see Table 2.2).

Summary

Physical measures of optical signals were proposed and pioneered long ago, but it is only in the past decade, thanks to the interest in receiver psychology and the quest for UV signals, that spectrometry has become the principal method for avian color quantification. With the advent of portable fiber-optic CCD spectroradiometers, UV/VIS reflectance analysis will soon replace human-based colorimetrics and photometry.

However, the sensitive and psychophysically calibrated human color vision system and associated technologies can still be useful and are not necessarily more flawed than other signal measurements for which avian cognition is unknown. The main limitation is the neglect of UV-A (320–400 nm) variation, but UV reflectance seems sufficiently predictable from proximate mechanisms and human-visible color variation to allow some extrapolation from VIS measurements. Apart from the above considerations of human color technology, however, the remainder of the chapter has been devoted to the principles, instrumentation, and techniques of reflectance spectrometry.

Color communication is a product of ambient light (A), reflectance (R), transmittance (T), and receiver sensitivity (S). All of these are aspects of color and potential sources of variation and selection on plumage color. Which to quantify and which to generalize or ignore in a given study depends on the question asked and particularly on whether the focus is on the sender (i.e., proximate mechanisms and fitness consequences of color variation) or on the receiver, describing signal selection exerted by, for example, mate choice or ecological signal conditions.

Reflectance is or should be the central measure in any study of color variation and evolution. It is a physical property of senders or objects that describes the ratio of reflected to incident light (AR/A), as a function of wavelength, and is thus independent of receivers (objective) and of light conditions (context-independent). Light, or radiant power, used to derive this ratio can in principle be measured in any unit, provided that it is recorded per wavelength.

In addition to obvious qualities of a field spectrometer—such as range, resolution, portability, and power consumption—considerations for selecting an instrument also include the distribution and smoothness of detector sensitivity in relation to that of the light source for a maximally uniform signal across the measuring range. The most important feature to consider is the quality of the acquisition software, which ideally should be tested for speed and ease of use under the relevant field conditions. For illumination sources, the main choice is between (1) tungsten-halogen lamps, giving a strong, stable, and spectrally smooth signal in the human-visible range (these are suitable for pigmentary colors with minimal UV reflectance, which should be confirmed by some full-spectrum measurements); or (2) xenon strobes or bulky deuterium-halogen combination lamps, giving a weaker and noisier signal but over the complete range (300–700 nm) necessary for most structural colors. For measuring birds in the hand, the standard, bifurcated fiber-optic reflection probes are convenient and can be fitted with a home-made probe pointer for adjusting and standardizing measuring spot size and geometry. For acquisition of reliable white standard spectra and dark currents, we describe a probe socket fitted on the reference tile, and a simple cardboard dark box.

In practice, plumage reflectance (R) is obtained as reflected sample radiance (AR) divided by white standard radiance, both subtracted by a dark current that corrects for noise: $R = (AR - D)/(AR_r - D)$. To avoid desaturating gloss and to facilitate standardization of acquisition angle and distance, we propose the coincident normal (CN) measuring geometry, easily achieved and standardized with the described probe pointer. Plumage patches thus measured should ideally be small, uniform, and smooth.

A few objective colorimetrics (reflectance shape variables) are suggested that capture universal dimensions of color cognition (spectral intensity, location, and purity) as well as potentially costly (honest) features of structural and pigmentary colors. In short, spectral intensity (brightness) should be positively related to the amount (thickness) of structural mechanisms, but negatively related to pigment concentration. Estimates of spectral purity (chroma) should indicate the regularity (variance) of structural mechanisms but, importantly, be positively related to pigment concentration only in unsaturated pigmentary colors. Finally, spectral location (hue) should indicate the nanoscale dimensions of scattering structures and be positively related to pigment concentration in both unsaturated and saturated colors.

References

Andersson, S. 1999a. Efficacy and content in avian colour signals. In Y. Espmark, T. Amundsen, and G. Rosenqvist, ed., Adaptive Signals: Signalling and Signal Design in Animal Communication, 47–60. Trondheim: Royal Norwegian Society of Arts and Letters, Tapir Publishers.

Andersson, S. 1999b. Morphology of UV reflectance in a whistling-thrush: Implications for the study of structural colour signalling in birds. J Avian Biol 30: 193–204.

Andersson, S., and T. Amundsen. 1997. Ultraviolet colour vision and ornamentation in bluethroats. Proc R Soc Lond B 264: 1587–1591.

Auber, L. 1957a. The distribution of structural colours and unusual pigments in the class Aves. Ibis 99: 463–476.

Auber, L. 1957b. The structures producing "non-iridescent" blue colour in bird feathers. Proc Zool Soc Lond 129: 455–486.

Bennett, A. T. D., I. C. Cuthill, and K. J. Norris. 1994. Sexual selection and the mismeasure of color. Am Nat 144: 848–860.

Bennett, A. T. D., I. C. Cuthill, J. C. Partridge, and E. J. Maier. 1996. Ultraviolet vision and mate choice in zebra finches. Nature 380: 433–435.

Billmeyer, F. W., and M. Saltzman. 1981. Principles of Color Technology, second edition. New York: John Wiley and Sons.

Bowers, D. 1956. A study of methods of colour determination. Syst Zool 5: 147–160.

Bradbury, J. B., and S. L. Vehrencamp. 1998. Principles of Animal Communication. Sunderland: Sinauer.

Burkhardt, D. 1982. Birds, berries and UV. Naturwissenschaften 69: 153–157.

Burkhardt, D. 1989. UV vision: A bird's eye view of feathers. J Comp Phys A 164: 787–796.

Burtt, E. H. J. 1986. An analysis of physical, physiological, and optical aspects of avian coloration with emphasis on wood-warblers. Ornithol Monogr 38: 1–126.

CIE 1971. Colorimetry: Official recommendations of the International Commission on Illumination (CIE). Paris: Bureau central de la CIE.

CIE 1978. Recommendations on uniform color spaces, color-difference equations, and psychometric color terms. Paris: Bureau central de la CIE.

de Repentigny, Y., H. Ouellet, and R. McNeil. 1997. Quantifying conspicuousness and sexual dimorphism of the plumage in birds: A new approach. Can J Zool 75: 1972–1981.

Dyck, J. 1966. Determination of plumage colours, feather pigments and structures by means of reflection spectrophotometry. Dansk Orn For Tidskr 60: 49–76.

Dyck, J. 1976. Structural colours. Proc Int Orn Congr 16: 426–437.

Eaton, M. D., and S. M. Lanyon. 2003. The ubiquity of avian ultraviolet plumage reflectance. Proc R Soc Lond B 270: 1721–1726.

Endler, J. A. 1978. A predator's view of animal color patterns. Evol Biol 11: 319–364.

Endler, J. A. 1990. On the measurement and classification of colour in studies of animal colour patterns. Biol J Linn Soc 41: 315–352.

Endler, J. A. 1992. Signals, signal conditions, and the direction of evolution. Am Nat 139: S125–S153.

Endler, J. A. 1993a. The color of light in forests and its implications. Ecol Monogr 63: 1–27.

Endler, J. A. 1993b. Some general comments on the evolution and design of animal communication systems. Phil Trans R Soc Lond B 340: 215–225.

Endler, J. A., and M. Théry. 1996. Interacting effects of lek placement, display behavior, ambient light, and color patterns in three neotropical forest-dwelling birds. Am Nat 148: 421–452.

Enquist, M. A., A. 1998. Neural representation and the evolution of signal form. In R. Dukas, ed., Cognitive Ecology: The Evolutionary Ecology of Information Processing and Decision Making, 21–87. Chicago: University of Chicago Press.

Fitzpatrick, S. 1998. Colour schemes for birds: Structural coloration and signals of quality in feathers. Ann Zool Fenn 35: 67–77.

Greenewalt, C. H., W. Brandt, and D. D. Friel. 1960. Iridescent colours of hummingbird feathers. J Opt Soc Am A–Opt Image Sci Vis 50: 1005–1016.

Guilford, T., and M. S. Dawkins. 1991. Receiver psychology and the evolution of animal signals. Anim Behav 42: 1–14.

Hailman, J. P. 1977. Optical Signals. Bloomington: Indiana University Press.

Hardin, C. L. 1993. Color for Philosophers: Unweaving the Rainbow. Indianapolis, IN: Hackett.

Hill, G. E. 2002. A Red Bird in a Brown Bag: The Function and Evolution of Colorful Plumage in the House Finch. Oxford: Oxford University Press.

Hunt, S., R. M. Kilner, N. E. Langmore, and A. T. D. Bennett. 2003. Conspicuous, ultraviolet-rich mouth colours in begging chicks. Proc R Soc Lond B 270: S25–S28.

Isaacs, A., J. Daintith, and E. A. Martin. 1999. Oxford Dictionary of Science. Oxford: Oxford University Press.

Jacobs, G. H. 1981. Comparative Color Vision. New York: Academic Press.

Jansson, L., B. Forkman, and M. Enquist. 2002. Experimental evidence of receiver bias for symmetry. Anim Behav 63: 617–621.

Juhn, M. 1964. Spectrophotometric identification of feather pigments in the brown leghorn fowl. Nature 202: 507–508.

Lubnow, E., and G. Niethammer. 1964. Zur Methodik von Farbmessungen für taxonomische Untersuchungen. Verh D Dtsch Ges München: 646–663.

Lythgoe, J. N. 1979. The Ecology of Vision. Oxford: Clarendon Press.

Maier, E. J. 1994. To deal with the invisible—on the biological significance of ultraviolet sensitivity in birds. Naturwissenschaften 80: 476–478.

McGraw, K. J., and A. J. Gregory. 2004. Carotenoid pigments in male American Goldfinches: What is the optimal biochemical strategy for becoming colourful? Biol J Linn Soc 83: 273–280.

McGraw K. J., K. Wakamatsu, A. B. Clark, and K. Yasukawa. 2004. Red-winged blackbirds *Agelaius phoeniceus* use carotenoid and melanin pigments to color their epaulets. J Avian Biol 35:543–550.

Ocean Optics. 2003. USB2000 Fiber Optic Spectrometer Operating Instructions. Dunedin, FL: Ocean Optics.

Ornborg, J., S. Andersson, S. C. Griffith, and B. C. Sheldon. 2002. Seasonal changes in a ultraviolet structural colour signal in Blue Tits, *Parus caeruleus*. Biol J Linn Soc 76: 237–245.

Osorio, D., and A. D. Ham. 2002. Spectral reflectance and directional properties of structural coloration in bird plumage. J Exp Biol 205: 2017–2027.

Peiponen, V. A. 1963. Experimentelle Untersuchungen über das Farbensehen beim Blaukehlchen, *Luscinia svecica* (L.) und Rotkehlchen, *Erithacus rubecula* (L.). I. Ann Zool Soc "Vanamo" 24: 1–49.

Pryke, S. R., S. Andersson, M. J. Lawes, and S. E. Piper, 2002. Carotenoid status signaling in captive and wild Red-collared Widowbirds: Independent effects of badge size and color. Behav Ecol 13: 622–631.

Real, L. A. 1992. Information processing and the evolutionary ecology of cognitive architecture. Am Nat 140: S108–S145.

Shawkey, M. D., and G. E. Hill. 2005. Carotenoids need nanostructures to shine. Biol Lett 1:121–124.

Siitari, H., J. Honkavaara, E. Huhta, and J. Viitala. 2002. Ultraviolet reflection and female mate choice in the Pied Flycatcher, *Ficedula hypoleuca*. Anim Behav 63: 97–102.

Stradi, R. 1998. The Colour of Flight. Milan, Italy: Solei Gruppo Editoriale Informatico.

Swaddle, J. P., and D. A. Ruff. 2004. Starlings have difficulty in detecting dot symmetry: Implications for studying fluctuating asymmetry. Behaviour 141: 29–40.

Thompson, E., A. Palacios, and F. J. Varela. 1992. Ways of coloring: Comparative color vision as a case study for cognitive science. Behav Brain Sci 15: 1–25.

Villafuerte, R., and J. J. Negro. 1998. Digital imaging for colour measurement in ecological research. Ecol Lett 1: 151–154.

Vorobyev, M., and D. Osorio. 1998. Receptor noise as a determinant of colour thresholds. Proc R Soc Lond B 265: 351–358.

Vorobyev, M., D. Osorio, A. Bennett, N. Marshall, and I. Cuthill. 1998. Tetrachromacy, oil droplets and bird plumage colours. J Comp Phys A 183: 621–633.

Wyszecki, G., and W. S. Stiles. 1982. Color Science, 2. New York: John Wiley and Sons.

Zuk, M., and J. G. Decruyenaere. 1994. Measuring individual variation in colour—A comparison of two techniques. Biol J Linn Soc 53: 165–173.

3

Analyzing Colors

ROBERT MONTGOMERIE

Virtually all studies of bird coloration require some means of objectively quantifying the colors of individuals, populations, or species in a manner that allows those colors to be compared, their variation to be assessed, or their properties to be analyzed in relation to other variables. Although there has long been a scientific interest in bird coloration, the first serious attempts to compare the colors of bird plumages did not occur until the 1970s, using color-ranking methods (e.g., Baker and Parker 1979), and it was not until the 1990s that instrumentation was commonly used to measure reflectance spectra from feathers and bare parts (Chapter 2). To answer some questions about bird coloration, details of such things as ambient light spectra, plumage reflectance spectra, environmental conditions, and observer perception are needed, but other questions can be answered with less sophisticated measurements. In this chapter, I review all of the methods used to date to quantify and analyze bird colors. I begin with a brief synopsis of the various reasons for measuring colors, as these determine what measurements are actually needed. I then summarize five very different methods for analyzing colors using a variety of measurement techniques (color-ranking, color swatch matching, photography, digital color meters, and reflectance spectrometry). The chapter ends with a short discussion of three statistical issues that often arise in the analysis of color data. Reflectance spectrometry is currently the most widely used method for measuring colors, and Chapter 2 outlines the equipment and techniques used to gather spectral data.

I assume from the outset that the reader knows how to use computer applications for data management (e.g., Microsoft® Excel™, Microsoft® Access™, Filemaker™, Paradox™), statistical analysis (e.g., JMP™, SAS™, SPSS™, S-Plus™, Statistica™, SYSTAT™), and graphing (e.g., Kaleidagraph™, Deltagraph™, SigmaPlot™, Origin™) and has a good working knowledge of basic statistics (Sokal and Rohlf 1995; Zar 1999; Quinn and Keough 2002). My goal here is to summarize the steps needed to decide which analysis is best for the research question of interest, and to provide guidelines to ensure that data are handled carefully and that statistical results are interpreted and presented correctly. Endler (1990) covered similar issues 15 years ago and much of what he said then is still valid and valuable. In the intervening years, however, portable spectrometry has become the instrument of choice for measuring animal colors (as Endler 1990 predicted it would), and the methods of analyzing data from these instruments have proliferated. As a result, researchers in this field do not always agree on which technique is best, so I have tried to present both the strengths and weaknesses of each method. There are, however, different reasons for wanting to quantify a bird's colors, and these should be considered at the outset of any study, as they will have some influence on the analytical methods chosen.

Reasons for Measuring Colors

Birds do not see colors in the same way that we do, for a variety of reasons (Bowmaker et al. 1997; Chapter 1), and much recent attention has been given to quantifying a bird's-eye view of their color world (e.g., Maddocks et al. 2001). To understand the signal value of bird colors, we must know how those colors are perceived by the observer (Box 3.1), but we are still far from achieving that goal because of the dearth of information available on avian perception and cognition. Fortunately, more than one of the different levels at which bird colors can be quantified (Box 3.1) are useful for behavioral and evolutionary ecologists. Clearly, the methods used to quantify colors will depend to some extent on the reasons for measuring those colors in the first place.

To estimate what colors birds perceive (Box 3.1), the photon or quantum catch method of analysis and other procedures using some of the same basic principles (e.g., segment classification; Endler 1990) provide the closest approximation. At the other end of the scale of complexity and objectivity, color-ranking methods can be applied to compare the colors of species and individuals. Between these two extremes in color quantification are the currently

Box 3.1. A Hierarchy of Color Signals

Using reflectance spectrometry and a number of other measurements or estimates and assumptions, we can calculate the signal from a patch of color at a variety of different levels (Figure B3.1), with increasing complexity as shown in Figure B3.1.

The signal potential of a bird's colors is determined by the macro- and microstructure of the feathers and bare parts (Chapter 7), pigments deposited inside feather barbs during molt (Chapters 5, 6, 8), and the dirt, waxes, and abrasion applied to or acquired by the tissues over time (Chapter 9). This signal is what is measured with a spectrometer under standardized lighting conditions and is often treated as if it were the signal that other birds perceive, but this must rarely be the case (Endler 1990), as outlined below. Endler (1990) called this measured color the "inherent color" of a surface.

The actual color (signal display) that is available to other organisms (usually members of its own species) is determined by the effects of ambient light on the color of the signaler's plumage, bare parts, and any objects that it uses in display (Chapter 9). Thus the natural reflectance spectrum from any patch of color is the product of its signal potential and the spectral properties of the ambient light (Figure B3.1). Endler (1990) and Théry (Chapter 4) give detailed treatments of the effects of incident light (irradiance) and angle of reflection on the spectral properties of reflected light.

The color signal received at the observer's eye is modified from the signal display by both the properties of the air and the distance to the receiver. Both dust and water droplets (e.g., fog) in the air reflect light (veiling light; Lythgoe 1979) and thus influence the signal that arrives at the receiver. In addition, the temperature of the air can influence spectral shape, and the distance from the signal to the observer affects the perceived brightness of the signal (Endler 1990).

The signal filtered through the ocular media (cornea, aqueous humor, lens, and vitreous humor) finally reaches the rods and cones in the retina of the observer's eye, where the photons are absorbed by visual pigments and the ensuing signal transduction events produce an electrical current (Kandel et al. 2001). That current is picked up by the optic nerve and transmitted to the brain. Bird lenses do not permit much light transmission outside wavelengths from 320 to 700 nm (Hart et al. 1998a; Chapter 1). In addition, the single cones each contain one of four types of oil droplets that further filter the light that has passed through the ocular media into one of four narrow

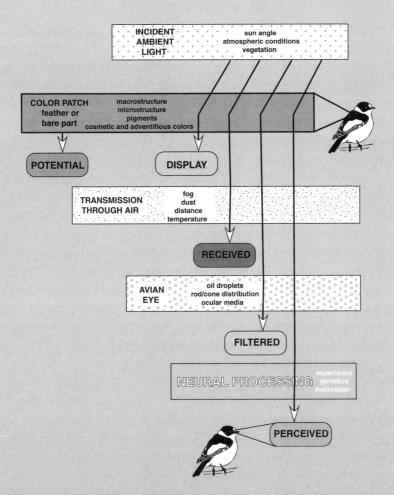

Figure B3.1. A hierarchy of signals that can be measured or calculated from the color of a bird's plumage or bare parts. Large rectangular boxes show the different influences on the signal as it passes from the signaler to the observer; small rounded boxes show the signal measured or calculated at each level. Arrows indicate the cumulative effect of all of the factors that influence the signal at each level.

---

wavelength segments (Chapter 1). The relative number of photons absorbed by the cones is sometimes called the "photon" or "quantum" catch (Vorobyev and Osorio 1998). Although it might be expected that this filtered signal is what the bird actually sees, the brain further interprets this signal, such that color perception is even more complicated.

Thus the color signal perceived by the observer can be influenced by the observer's experience, environment, and physiological state, as well as by any colors that might be adjacent to the focal color patch. These four factors all influence color perception in humans, and their effects have been described as a wide variety of phenomena (e.g., Stevens and Hunt effects, simultaneous contrast, crispening, surround, color constancy, color memory, discounting-the-illuminant, as well as light, dark, and chromatic adaptation; Boff et al. 1986). For example, one form of perceptual bias, color constancy, occurs when an object is perceived as having the same color under different ambient light conditions, when in fact the reflectance spectra are very different. Thus color patches with the same signal potential but very different signal displays (due to very different ambient lighting) may actually be perceived as being the same by the observer due to this color constancy (Chapter 1). Although there is limited information about such phenomena in birds, it is probably reasonable to assume that at least some of them are common to all vertebrate visual perception (e.g., Maddocks et al. 2001).

Endler (1990) referred to the perceived signal as the "apparent color," and it is clearly the best color to quantify if we want to fully understand the nature and function of avian color signals. The perceived signal has never been measured for birds, and so little is known about perceptual bias in birds that there is no immediate prospect of making such measurements (Chapter 1). In the meantime, it is necessary for the researcher to decide which of the other measures of the color signal (potential, display, received, or filtered; Figure B3.1) is most appropriate to address the question at hand, and to be careful about the interpretation of results from each of these analyses.

most commonly used techniques—in part, these methods are popular because they are computationally relatively straightforward and give an accurate measure of signal potential (Box 3.1), or radiance, of a color display. Even though tristimulus color variables are based on the human model of color perception (see below), they can provide a reasonable measure of signal potential. Remember, however, that measures of signal potential may yield little information about what colors the viewer might actually see (Box 3.1).

Despite this limitation, measures of signal potential can be quite useful for the following reasons. First, they are the only unbiased measures of feather or bare-part color that can be used to compare individuals and species independent of lighting environment and the observer's ocular properties and perceptual psychology. As a result, measures of signal potential can reveal a great deal about the quality of the bearer, regardless of whether that aspect of quality can be perceived by other members of the same species. In this way, an estimate of signal potential is analogous to such measures as bilateral asymmetry (e.g., fluctuating asymmetry; van Valen 1962), because it can tell researchers something useful about the individuals they are studying, even if the measured trait does not function as a signal. Thus, in the absence of good information about ambient light, cone sensitivities, and perceptual processes, a measure of signal potential is probably the only index of color display that can be reliably estimated for most bird species at present. Second, an estimate of signal potential provides a baseline for the calculation of actual signal displays and how they are received and perceived (Box 3.1). Finally, a "signal potential" index is easiest to calculate and can be determined for any patch of color on any bird.

Analyzing Color Ranks

During the 1970s and 1980s, the method most commonly used by behavioral and evolutionary ecologists for the analysis of bird colors was simply to rank the colors of either individuals or species on some arbitrary, predetermined scale (e.g., dull to bright or vivid; Baker and Parker 1979; Hamilton and Zuk 1982; Butcher and Rohwer 1988; Flood 1989; Promislow et al. 1992; Greene et al. 2000). Although this method has largely been replaced by more quantitative, objective, and repeatable methods using reflectance spectrometry, it still has some utility in certain situations (e.g., Greene et al. 2000) and will be widely encountered in the older literature. For those reasons, I provide a brief summary here both of the scoring methods and the statistical analyses used.

There are two main methods used to rank bird colors. First (and usually when comparing species or sexes), birds are ranked on a scale of colorfulness or vividness (sometimes confusingly called "brightness"; e.g., Flood 1989; Promislow et al. 1992). Usually several observers are asked to score a number of bird specimens, photos, or paintings on this scale, ranking each one on a scale from 1 (dullest) to 5 (most vivid), for example, based on total plumage color, and an average colorfulness score is calculated. Repeatability is actually reasonably high with this method, as long as the observers are coached in advance on the rating system (Promislow et al. 1992), even though birds of different sizes, colors, patterns, and body conformation are sometimes compared.

In my experience, repeatability between research groups using published criteria but with no coaching is much lower but usually still significant.

Second (and usually when comparing individuals within a species, sex, or age class), specific color patches on individual birds are ranked on an explicit scale, based on photographs, specimens, or color swatches. Typically a researcher might photograph different individuals, then create a rank series of photographs for observers to match with the birds being studied in the field. Greene et al. (2000) used this method to characterize the colors of Lazuli Bunting (*Passerina amoena*) males during the breeding season and documented an interesting example of male-male cooperation driven by sexual selection.

Once rank color scores have been obtained, a variety of largely nonparametric statistics are used for analysis. The use of nonparametric statistics is wise, as the underlying distribution of color ranks is generally unknown. Of course, the relations between color ranks and reflectance spectra, tristimulus color variables, and bird perception are also largely unknown. For example, three birds ranked 1, 2, and 3 on such a scale might not be ranked the same way with respect to hue, saturation, or brightness, and the quantitative difference in colors between ranks 1 and 2 might well be much larger or smaller than the difference between ranks 2 and 3. It would be very useful to compare the results from such rankings to a more quantitative scale based on reflectance spectrometry.

Analyzing Tristimulus Color Variables

Every one of the millions of colors that humans with normal vision are able to distinguish can be described by only three color variables that can be combined mathematically into a color model. The colors predicted by such a model form a tristimulus color space, such that all colors can be illustrated as a region in three-dimensional Euclidean space (Plate 4). Interestingly, the complete tristimulus color space of human perception contains only about half of the 16.8 million colors that we can set our 24-bit computer monitors to display (Kandel et al. 2001). Although the human tristimulus color space must also be smaller than the color space perceived by birds (Chapter 1), tristimulus color variables have been very popular for the description of bird colors (and the colors displayed on computer monitors), for reasons given below. I begin, however, with a brief history and description of human tristimulus color models to provide some background to the analytical details that follow.

In the 1800s, Thomas Young (1773–1829), John Maxwell (1831–1879), and Hermann von Helmholz (1821–1894) developed the idea that all human-visible colors could be created by combining three primary colors (red, green, and blue) in different combinations. Then, during the 1900s, several other tristimulus color systems were developed to suit a variety of purposes (Box 3.2) based on the principles estabished by Young, Maxwell, and von Helmholz. The modern standard for human-perceived colors was first defined in 1931, when the Commission Internationale de l'Éclairage (CIE) developed the CIE XYZ color space, using a tristimulus model based on a Standard Observer and a Standard Illuminant (Box 3.2; Plate 4). This model was updated in 1976 to create the Luv and Lab standards (Box 3.2), both of which are more perceptually uniform than the XYZ color model. Nonetheless, colors defined by all of the other tristimulus models can be calculated from the CIE XYZ color space (Plate 4) if both the light source and the illuminant are specified. It is therefore possible to transform variables from one tristimulus color model to another.

Hue, saturation, and brightness (HSB) are by far the most commonly reported tristimulus color variables used in the study of bird coloration. The HSB color space was derived from the hue, chroma, value (HCV) color space (Box 3.2) originally defined by Albert Munsell (1858–1918). Munsell was an artist who wanted to develop a system that could be used by artists for defining and organizing colors, based on the kinds of color differences and relations that were perceived by humans. The Munsell color system, which is still in widespread use today, separates the value (or brightness) dimension from both hue and chroma (or saturation) to provide an unambiguous notation for characterizing colors (Table 3.1).

The basis of the original Munsell color system is a color wheel (Box 3.2; Plate 4) that separates the chromatic (hue, chroma) from the achromatic (value) components of color. This system represents hue and chroma (saturation) on a color wheel and value (brightness) along an axis perpendicular to the plane of the wheel. Six basic hues are evenly spaced around the circumference of the color wheel, whereas chroma increases from the center of the wheel outward (Plate 4). Because the maximum chroma perceived by humans depends on the hue, the Munsell color space is asymmetrical (Box 3.2). A key feature of this color space is that it was designed to be perceptually uniform; that is, the distances between defined colors corresponds to the differences that we perceive. The HSB and HSL color models (Box 3.2) are very similar to the HCV model from which they were derived. The tristimulus color variables in the HSB model used in most bird research are defined as follows.

Box 3.2. Human Tristimulus Color Spaces

Several different tristimulus color spaces have been developed to describe the colors that humans can see. In this box, I summarize their characteristics.

Color space	History	Features	Usage	Graphical representation
RGB	Originally developed in the 1800s by Young, Maxwell, and von Helmholz	Red, green and blue additive primaries, starting with black; device dependent	Television and computer monitors	
CMY(K)	Developed in the early twentieth century	Subtractive primaries; cyan (C; redless), magenta (M; greenless), and yellow (Y; blueless) subtracted from white; black (K) added to get pure black as C, M, Y together cannot achieve this color	Process colors, printing	
HSB (HSV)	Developed by A. Munsell to describe relations among and differences between colors	Separates value (V; which is color independent) from hue and chroma (H, C; which are color-related); distances between colors correspond to perceived differences	Color matching, art	

HSV	Created in 1978 by A. R. Smith	Nonlinear transformation of RGB color space but more similar to the way humans perceive colors	Computer graphics applications
HSL	Developed in the 1980s from HSB	Nonlinear transformation of RGB color space; similar to HSV but with different definitions for saturation and brightness (lightness or luminance)	Computer graphics applications
CIE XYZ	Developed in 1931 by Commision Internationale de l'Éclairage	Specifies Standard Observer, Standard Illuminants; XYZ primary system related to RGB but better as computational standard; xyZ color space derived from XYZ but separates color-related (x, y) from luminance-related (Z) attributes	Paints, dyes, fabrics; to facilitate communication about the colors of manufactured products
CIELAB, CIELuv	Developed in 1976 from CIE XYZ	Perceptually uniform; distances between colors correspond to perceived differences; L (lightness) defined in 300 steps from black (0) to white (300)	Luv used in the design of TV and computer monitors, controlled light sources; Lab used in graphics arts, reflective products

Table 3.1. Color Swatches, Atlases, Charts, Bars, and Chips Used for Matching Bird Colors

System	Number of colors	Notation[a]	Number of levels			Reference
			Hue	Saturation[b]	Brightness[b]	
Methuen	1260	H-B-C [10-A-8]	30	6 (intensity)	8 (tone)	Kornerup and Wanscher (1983)
Munsell	1000s	H B/C [8.75R 4.5/16.5]	100	≥15	10 (value)	Parkkinen et al. (1989)
Ostwald	~1000	HWB[c] [8pa]	24	N/A	>100 (lightness)	Ostwald (1916)
Palmer	21	Names [scarlet]	N/A	N/A	N/A	Palmer (1962)
Pantone[d]	1000s	Pantone code [Pantone 485 C]	8	N/A	(value)	Eiseman and Herbert (1990)
Ridgway	1115	Names [scarlet]	36	5 (saturation)		Ridgway (1912)
Smithe	194	Numbers, names [color 14, scarlet]	N/A	N/A	N/A	Smithe (1974, 1975, 1981)
Villalobos	7279	H-B-C [SSO-10-12°]	38	12 (chromaticity)	19	Villalobos-Domínguez and Villalobos (1947)

a. H, hue; C, chroma; B, brightness; W, whiteness. An example, equivalent to Ridgway's "scarlet" is given in square brackets for each system.
b. Terminology used by each system in brackets.
c. B, blackness; hue is represented by a number from 1 to 24; both whiteness and blackness are each represented by a lower case letter from a to n in 13 equal steps.
d. Pantone colors are determined by the mixing of eight basic color inks in different proportions; notation is always preceded by the word "Pantone."

Hue is the technical term for what we call "color" in everyday speech; thus, pure "colors," such as red, blue, green, and yellow are all different hues. More technically, hue indicates which wavelengths contribute most to the total radiance, and thus, when estimating the perceived signal, it is determined by the degree of activation of each cone type in the retina (three cone types in humans, four in birds; Chapter 1). Hue is often referred to as "spectral location" because it is usually measured as a position (wavelength) along the visible spectrum or an angle on a color wheel. The hue determined from the signal potential, or radiance, of a surface can be very different from the hue of the perceived signal, which must pass through both environmental and ocular media (Box 3.1). In the HSB model, hue is represented by an angle on the color wheel from 0° (red) through yellow (60°), green (120°), cyan (180°), blue (240°), magenta (300°), and back to red at 360° (Plate 4). Note that the additive primary colors (red, green, blue [RGB]) are equidistant from one another, with the subtractive primary colors (cyan, magenta, and yellow [CMY]) located exactly half way between them. This distribution of hues in the Munsell color system is not perfectly correlated with the wavelength of different hues, but instead is set according to perceptual uniformity. Recall that Munsell designed this system so that the distances between colors would match our perception of how different they were, and would thus be useful in talking about color balance and complementarity. More recent work has resulted in modifications of the Munsell System to make the color model more perceptually uniform (e.g., McCamy 1993). Although some instruments provide an estimate of hue on this scale, various indices of hue that are correlated with this measure can be readily calculated from reflectance spectra, as described below.

Saturation is a measure of the degree to which a color appears to be pure, and thus composed of a single wavelength of light. More technically, saturation is the radiance in a specified part of the spectrum in relation to the radiance from the whole visible spectrum. When estimating the perceived signal, saturation is determined by the differences in stimulation between pairs or sets of retinal cone types. Saturation is often referred to as "spectral purity," because it measures how much of the reflectance (or perceived color) in the visible spectrum is coming from the region of interest. The result, for an observer, is that a highly saturated color stimulates one cone type more strongly than the others. Munsell's chroma is formally the same as saturation and the two terms are used interchangeably in the bird color literature. The term "saturation" has some intuitive appeal and for that reason might be preferred as a general term, reserving "chroma" to describe the relative saturation in specific regions of the spectrum (e.g., UV chroma). In the HSB color space, saturation is measured as the percentage distance from the center (0%) of the color wheel to its circumference (100%), where the circumference of the circle represents pure spectral colors (Plate 4). Some instruments provide a measure of saturation on this scale, but various other indices of saturation have been derived in studies of bird coloration, as described below.

Brightness is a measure of the total amount of light (radiance) coming from a unit area of a surface (either reflected or transmitted) at a particular angle. More technically, the brightness of the signal coming from a surface (i.e., signal potential; Box 3.1) is the integral of radiance over all visible wavelengths, which is the same as the area under the spectral radiance curve. Thus brightness is often referred to as "spectral intensity," because it measures the total amount (intensity) of radiance from a surface. When estimating the perceived signal, brightness is determined by the total stimulation of all cone types. Unfortunately, use of the term "brightness" causes some confusion, as we tend to speak of colors that are more saturated as being brighter or more vivid in everyday speech, even when there is no change in brightness, whereas colors that have higher brightness at the upper end of the brightness scale actually look washed out (Plate 4). To add to the confusion, "lightness" is actually the correct technical term for the area under the percentage reflectance curve that we actually measure with a spectrometer (Chapter 1). Thus, in practice, we do not actually calculate radiance from a surface, but instead calculate reflectance (see below) in relation to the radiance from a near perfect, spectrally flat reflector (Chapter 2). I prefer the term "lightness," because it is both technically correct and intuitively easy to visualize—in common parlance, a light-colored and a bright-colored bird look very different. The term "brightness," however, is so entrenched in the animal-coloration literature and is so easy to relate to the Munsell color system that it continues to be used. Whatever term researchers decide to use, it should be defined clearly to minimize confusion.

The HSB tristimulus model thus provides three variables that are intuitively easy to visualize (Plate 4), which probably accounts for their popularity among behavioral ecologists, even though the HSB color space is a human-oriented rather than a bird-oriented view of color (see Chapter 1). Thus the HSB model cannot describe all the colors in a bird's visible spectrum, because this model excludes the UV portion of the spectrum (320–400 nm), which birds can see but humans cannot. To get around that problem, behavioral ecologists have derived various indices of hue, saturation, and brightness that include UV wavelengths, as described below. Note, however, that the ways that these indices map onto the bird's ability to distinguish colors has only been studied in a few species (Peiponen 1992 and references therein), so they do not necessarily reveal anything useful about the signals that are actually perceived.

Because of the various inaccuracies and approximations in the use of tristimulus color models based on human vision, many researchers deplore their use with other animals, such as birds and fishes (J. Endler, D. Osorio, pers. comm.). They recommend instead that color variables be calculated directly from a combination of reflectance spectra, ambient conditions, and the bird's spectral sensitivities. It seems likely, however, that the intuitive appeal and ease of calculation of tristimulus color variables will ensure their usage for some time. To date, traditional tristimulus variables have provided many apparently useful insights (e.g., Hill 2002, many of the examples in this book), and there is as yet no clear evidence that the results from studies that use them are flawed. Nevertheless, a detailed, quantitative comparison of results obtained by using tristimulus color models compared to other, more rigorous methods (see below) would be very useful in this regard.

Quantifying Hue, Saturation, and Brightness

In this section, I provide details on how hue, saturation, and brightness are calculated in bird-color research, both as a guide to the published literature and as a primer for those who wish to use these methods in their own studies. Indices of the hue, saturation, and brightness of a given patch of plumage or bare-part color can be determined by matching the patch to a color swatch, by photographing the patch, by using instruments specifically designed for this purpose, or by the direct analysis of reflectance spectra. Once these variables have been determined, they can then be used as continuous variables in both univariate and multivariate statistical analyses. Some statistical issues that arise from the use of these variables are dealt with in a later section.

Using Reference Color Swatches

Prior to the advent of portable reflectance spectrometry (Chapter 2), color matching was the most widespread method of characterizing bird colors, usually accomplished by the simple visual matching of a reference color swatch (Table 3.1) held against the bird's color patch (e.g., Hill 1990) under standardized lighting conditions. One serious problem with this method is its subjectivity, depending critically on the perceptual biases of the user and the quality of the ambient light. These issues can be alleviated somewhat by using a standard reference illuminant (e.g., SoLux® light box) and a single observer, but the data obtained from field studies, where such controls are difficult to apply, can be highly variable due to differences among observers (Wyszecki and Stiles 1982). Moreover, biologically important variation may be hard to detect when the color sensitivities of the bird are different from those of the human observer (Bennett et al. 1997).

Despite the subjectivity of this method, Hill (1998) found that, for House Finches (*Carpodacus mexicanus;* Plate 31), hue and saturation determined by matching patches of plumage color with swatches in the Methuen Handbook of Color (Kornerup and Wanscher 1983) were reasonably well correlated with color variables estimated more objectively using a visible-light spectrophotometer. Thus hue estimated from reflectance spectra explained 92% of the variation in hue determined by color matching and 42% of the variation in saturation, but almost none (1%) of the variation in brightness (Hill 1998). Note, however, that the high correlation for hue in this study might have been the result of considerable intraspecific variation in hue in this species (Hill 2002), and may not be so high for the majority of avian species. In this research, Hill (2002) was also analyzing colors in the long-wavelength (yellow-red) region of the spectrum, where the ability to match the hue or saturation of a bird to reference color swatches is more repeatable (J. Endler, pers. comm.).

The Methuen Handbook of Color (Kornerup and Wanscher 1983) has so far been used with some success to quantify the red, carotenoid-based plumage colors of House Finches (Hill 2002) and Northern Cardinals (*Cardinalis cardinalis;* Wolfenbarger 1999a,b,c; Plate 25, Volume 2). Hill (2002) effectively used this color-scoring method to look at geographic variation in plumage color, and the relations between plumage color and nest initiation, mate choice, extrapair paternity, and food quality in House Finches. To reduce variation in scoring techniques and perception among observers, Hill did all of the plumage scoring himself (Hill 2002: 48).

In the Northern Cardinal, the redness of male plumage was related to female choice, reproductive success, and their dominance status in winter (Wolfenbarger 1999a,b,c). Although using color swatches to quantify Northern Cardinal plumage color has yielded some interesting correlations (e.g., redder males paired earlier and nested on higher-quality territories), it is difficult to know when the relatively weak relations detected in much of this work were simply due to the problems associated with this method of scoring color. For example, color swatch matching cannot detect variation in the UV range, and it is clear that there is both reflectance and variation in the UV in this species (Wolfenbarger 1999b).

Munsell color chips have also been used to score the red colors of Northern Cardinals (Linville and Breitwisch 1997; Linville et al. 1998; Jawor et al. 2003). In these studies, the tristimulus Munsell scores were converted to a rank order (e.g., Burley and Coopersmith 1987) from vivid to dull red, and then used as a continuous variable. The ranking gave most weight to value (brightness), then hue, and least to chroma, such that saturated red-orange colors were given the highest rank (Linville and Breitwisch 1997). Using this method of color scoring, male breast color was shown to vary with winter food conditions: birds were found to have redder plumage during years with better winter fruit crops (Linville and Breitwisch 1997). In addition, the relative effort invested in feeding nestlings was correlated with the redness of both male breast and female underwing color scores (Linville et al. 1998), and these variables were correlated within pairs, suggesting assortative mating (Jawor et al. 2003). The latter result was supported by measurement of reflectance spectra from a smaller number of pairs using the X-Rite® Digital Swatchbook™ (see below) (Jawor et al. 2003).

Using Photography

User-defined regions on digital (or digitized) photographs can be analyzed using graphic software applications like Adobe Photoshop™, Canvas™, SigmaScan Pro™, Corel Paint™, or ImageJ to give mean values of hue, saturation, and brightness for the color patch. To make these measurements, photographs are simply imported into the graphics application, the area of interest selected, and a tool or plug-in is used to reveal the tristimulus color scores of each pixel in the region selected. Usually the colors of pixels even within a small region (10–100 pixels) vary considerably, so it is wise to select a region of relatively uniform-looking color of 100 pixels or more, then calculate average scores for each tristimulus color variable. Using ImageJ (available at no cost from http://rsb.info.nih.gov/ij/), this is accomplished by selecting the region of interest with one of the "selections" tools, then choosing the Color Histogram plug-in to reveal the distribution, mean and standard deviation of the selected pixels in RGB color space. Similarly, using Adobe Photoshop CS2™, a small region of interest is selected, and the Histogram tool is used to reveal these statistics for the selected pixels.

This technique suffers from some of the same limitations, as does quantification with reference color swatches, in that it cannot quantify reflectance below 400 nm and may lack both precision and accuracy. Moreover, the color accuracy of photographs is highly dependent upon ambient lighting and the quality of the image (determined by the camera's lens and the method used to record colors—either film or, in digital cameras, the charge-coupled device [CCD]). Because of this dependence, it is essential that standardized color swatches be included in each photograph, so that inter-photograph variation in these confounding variables can be controlled statistically. It is also advisable to adjust the computer monitor, so that the colors visible on the reference swatches in the photos are similar to the real colors of those swatches. This adjustment can be done using a digital color meter (see below) to calibrate the monitor. Note, however, that adjusting the monitor has no effect on the measured colors using any of the software mentioned above, as these programs measure the colors in the original digital file and not what is actually displayed.

Very few studies have used photography to quantify bird colors, probably because of the potential inaccuracy of this technique compared to others. Nonetheless, Villafuerte and Negro (1998) have developed some useful computer software for correcting the changes in illumination that are difficult to control in the field (see also Yam and Papadakis 2004), and their method has been used to quantify human-visible colors in the plumage of Red-legged Partridges (*Alectoris rufa;* Villafuerte and Negro 1998), Northern Flickers (*Colaptes auratus;* Wiebe and Bortolotti 2001, 2002), and House Sparrows (*Passer domesticus;* Plate 23; McGraw et al. 2002), in the iris of American Kestrels (*Falco sparverius;* Bortolotti et al. 2003), and on the eggs of the South American Tern (*Stern hirundinacea;* Blanco and Bertellotti 2002). For photographs of color standards, this method had high repeatability and matched the human perception of differences among colors (Villafuerte and Negro 1998). Even with this method of analysis, however, the best results are obtained when the lighting can be standardized; for example, by using an electronic flash at a set distance from the color patch.

Even without using color swatches to standardize measurements, a few studies have used digital photography to quantify bird colors simply by ensuring that lighting and photography conditions were held constant (Kilner 1997; Dale 2000; Fitze and Richner 2002; Tschirren et al. 2003; Alonso-Alvarez 2004). Although this technique is less likely to measure real colors accurately, the rank order of colors measured this way may not be affected. Dale (2000) also showed that measures of Red-billed Quelea (*Quelea quelea;* Plate 6, Volume 2) plumage colors obtained from photographs were highly repeatable ($R > 0.97$) and strongly correlated ($r \geq 0.96$) with color variables calculated from reflectance spectra.

Although for several reasons reflectance spectrometry is a better method for measuring bird colors than is digital photography, there are three situations in which photography might still be more practical. First, digital photography provides a means for quantifying colors in very small or very large patches, and at some distance, when reflectance spectrometry cannot be readily used. For example, because of the size of the area that portable spectrometers measure (~3 mm^2), very small or very large patches of plumage or bare-part color cannot be readily or reasonably quantified, whereas digital photography of a feather through a dissecting microscope, or of a whole bird, might be a viable alternative. Likewise, the colors of birds that are difficult to capture might be quantifiable at a distance using a telephoto lens, although standardizing lighting conditions and photographing color standards might prove problematic.

Second, photography is a much easier method for quantifying the colors of wet, hard to reach, and irregular surfaces (gapes, Kilner 1997; irises, Bortolotti et al. 2003; bills, Alnso-Alvarez et al. 2004), although measurements cannot easily be taken in the UV range. Thus Kilner (1997) used photography to quantify the gape colors of Common Canary (*Serinus canaria*) nestlings, converting the photos into digital images and using Adobe PhotoShop™ to quantify the hue, saturation, and brightness of 10 randomly chosen spots on photos of their mouth linings. Even within the limits of this technique, Kilner (1997) found that mouth colors changed with time since feeding, thereby providing the parents with a potentially useful signal of nestling hunger.

Third, digital photography is relatively less expensive and more portable than spectrometry and may well fill a need when money and mobility are considerations. Except for the limitation that measurements can only be taken readily within the human-visible spectrum, this method has some promise and should be further investigated.

Using Digital Color Meters

Tristimulus color variables can also be determined directly, and more objectively, with hand-held digital color meters, such as the Colortron™ (originally available from Light Source Computer Images Inc., Larkspur, CA, but this company is no longer in business) or the Digital Swatchbook™ (X-Rite®, Grandville, MI), designed specifically to calculate tristimulus variables from measured reflectance spectra. Although these instruments measure and provide spectral data, I call them "digital color meters" here to distinguish them from portable reflectance spectrometers (Chapter 2) that measure spectra outside the human-visible spectrum and usually have a greater level of precision (wavelengths <1 nm). Digital color meters are self-contained, in that they provide their own illumination and are easily held in one hand, but most of them

Table 3.2. Tristimulus Color Variables Used in the Analysis of Bird Colors

Color variable	Names used	Formula	Reference
Brightness	Total brightness, total reflectance, spectral intensity[a,b]	$B_1 = B_T = \int_{\lambda_{320}}^{\lambda_{700}} R_i = \sum_{\lambda_{320}}^{\lambda_{700}} R_i$	1–4,6,8,10,13
	Mean brightness[a,b,c]	$B_2 = \sum_{\lambda_{320}}^{\lambda_{700}} R_i / n_w = B_1/n_w$	5,11
	Intensity[d]	$B_3 = R_{max}$	1,7,9
Saturation	Chroma, reflectance ratio, spectral purity[a,b,e]	$S_1 = \sum_{\lambda_a}^{\lambda_b} R_i / \sum_{\lambda_{320}}^{\lambda_{700}} R_i = \sum_{\lambda_a}^{\lambda_b} R_i / B_1$	3,4,6,11–13
	Spectral saturation[d]	$S_2 = R_{max}/R_{min}$	1
	Chroma[a,f]	$S_3 = \sum_{\lambda_{Rmax}-50}^{\lambda_{Rmax}+50} R_i / B_1$	13
	Spectral purity[g]	$S_4 = |bmax_{neg}|$	1
	Chroma[h]	$S_5 = \sqrt{(B_r - B_g)^2 + (B_y - B_b)^2}$	8
	Contrast, amplitude[d]	$S_6 = R_{max}/R_{min}$	7,9
	Spectral saturation[a,i]	$S_7 = \left(\sum_{\lambda_{320}}^{\lambda_{Rmid}} R_i - \sum_{\lambda_{Rmid}}^{\lambda_{700}} R_i\right) / B_1$	2,10
	Chroma[d]	$S_8 = (R_{max} - R_{min})/B_2$	2,6
	Carotenoid chroma[a]	$S_9 = (R_{\lambda 450} - R_{\lambda 700})/R_{\lambda 700}$	12
	Peaky chroma[d,g]	$S_{10} = [(R_{max} - R_{min})/B_2] \times |bmax_{neg}|$	3
Hue	Hue, peak wavelength, spectral location[f]	$H_1 = \lambda_{Rmax}$	1,3–7,11,13
	Hue[g]	$H_2 = \lambda_{bmaxneg}$	4,6
	Hue[i]	$H_3 = \lambda_{Rmid}$	2,6,10
	Hue[h]	$H_4 = \arctan\{[(B_y - B_b)/B_1]/[(B_r - B_g)/B_1]\}$	8
	Hue[g]	$H_5 = \lambda_{bmaxpos}$	9

Notes: Variable names from the original sources unless other wise noted; notation has been changed to be consistent with usage in this chapter. Numerical subscripts on B, S, and H merely distinguish the different formulas.

References: 1. Andersson (1999); 2. Andersson et al. (2002); 3. Örnborg et al. (2002); 4. Andersson et al. (1998); 5. Delhey et al. (2003); 6. Smiseth et al (2001); 7. Keyser and Hill (2000); 8. Saks et al. (2003); 9. Keyser and Hill (1999); 10. Pryke et al. (2001); 11. Siefferman and Hill (2005); 12. Peters et al. (2004); 13. Shawkey et al. (2003).

a. $R_{\lambda i}$ = Percentage (or proportional) reflectance at the ith wavelength (λ_i).
b. Upper and lower limits of bird-visible wavelengths indicated in this formula as 700 and 320 nm (Hart 2002).
c. n_w = Number of wavelength intervals used to calculate B_T.
d. R_{max}, R_{min} = Maximum and minimum percent reflectances, respectively.
e. Wavelengths from a to b are a segment of the bird-visible spectrum.
f. λ_{Rmax} = Wavelength at maximum reflectance.
g. $bmax_{neg}$ and $bmax_{pos}$ = Maximum negative and positive slopes of reflectance curve in a region of interest.
h. Subscripts refer to red (r = 625–700 nm), yellow (y = 550–625 nm), green (g = 475–550 nm), and blue (b = 400–475 nm) segments of the spectrum.
i. λ_{Rmid} = Wavelength at the reflectance midpoint between R_{max} and R_{min} (i.e., [$R_{max} + R_{min}$]/2).

need to be connected to a computer for calculating and viewing results. They typically provide spectral data in increments of 10–20 nm in the range of 400–700 nm and convert these data (with accompanying software) to a wide variety of tristimulus variables (Table 3.2).

The Colortron™, for example, has been used extensively by Hill and his coworkers to quantify the HSB tristimulus values from House Finch plumage (Hill 2002; see also McGraw and Hill 2000 and McGraw and Ardia 2003 for its use in scoring beak color in American Goldfinches [*Carduelis tristis;* Plate 30] and Zebra Finches [*Taeniopygia guttata;* Plate 27], respectively). This handy instrument supplies objective measures of tristimulus color variables in the human-visible spectrum (380–700 nm) under a variety of Standard Illuminants, although it is limited to measurements with both incident and reflected light perpendicular to the feather surface. The Colortron™ is no longer manufactured but is sometimes available in the used equipment marketplace (e.g., eBay).

X-Rite® (Grandville, MI; http://www.xrite.com) now makes a variety of hand-held spectrophotometers similar to the Colortron, in that they provide measurements over the human-visible spectrum (400–700 nm) in increments of 10–20 nm. X-Rite® instruments, however, come in a much wider variety of configurations than the Colortron™ to suit different purposes and measurement needs, including more angles of incident and reflected light, and more Standard Illuminants and Standard Observers. The X-Rite® Digital Swatchbook™ has been used by Jawor et al. (2003) to quantify the plumage colors of Northern Cardinals, using the spectral data obtained from this instrument to calculate HSB from the formulas in Andersson et al. (2002), as summarized in Table 3.2.

Remember that these digital color meters cannot measure in the UV portion of the spectrum that birds can see, and thus may not be very useful for quantifying the colors of many species. This limitation may not be a problem for such species as the House Finch that apparently have little plumage reflectance below 400 nm (Hill 2002: Figure 3.2). In the Northern Cardinal, however, there does appear to be a secondary peak of reflectance in the UV range (Wolfenbarger 1999b) that would have a large effect on the calculation of saturation, for example. This measurement error may have relatively little effect on the conclusions from studies using digital color meters (e.g., Jawor et al. 2003; Jawor and Breitwisch 2004), but to date this effect has not been quantified. One simple way to do this is to measure the same individuals with both a digital color meter and a reflectance spectrometer that measures

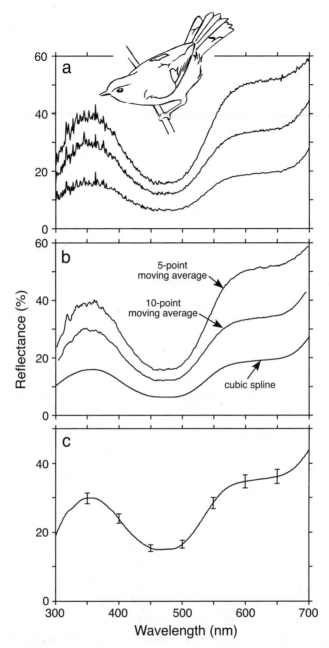

Figure 3.1. Reflectance spectra from the orange patch on the outer vane of rectrix 3 from male American Redstarts (*Setophaga ruticilla*). (a) Raw data from a single spectrometer reading from each of three males, showing both noise in the curves and the range of variation. (b) Same curves as in (a), but smoothed using 5- or 10-point moving averages, or a cubic spline function (lambda = 5000). (c) An average curve calculated from the mean curve recorded from five measurements from each of 21 males, smoothed using a cubic spline (lambda = 1000) and showing 95% CL at 50-nm intervals. From R. Montgomerie (unpubl. data).

into the bird-visible UV (Chapter 2) and compare HSB calculated using both methods.

Using Reflectance Spectrometers

By far the most common method of measuring bird colors in use today is to determine a complete reflectance spectrum in the bird-visible range (Figure 3.1) using a portable spectrometer (Chapter 2) and then to calculate some simple indices of hue, saturation, and brightness from those spectral data. For most researchers, three daunting challenges emerge from the use of reflectance spectrometry for measuring bird colors: (1) how to handle the vast amounts of data obtained, (2) what measures of color to compute, and (3) what statistical analyses to use to answer the questions of interest. Andersson (Chapter 2) provides useful information on managing the huge number of data files obtained when using reflectance spectrometry; the importance of this step should not be underestimated.

Both an advantage, and a potential disadvantage, of reflectance spectrometry is the large number of color variables that can be calculated (Table 3.2). Particularly for complex (e.g., multimodal) spectral curves, more than one index of saturation is sometimes calculated to capture some of this complexity (Örnborg et al. 2002). In practice, at least one index of each tristimulus variable (HSB) should be calculated from each spectrum as a minimal summary of the signal display from a patch of color. The difficulty is often in picking which ones to calculate, so here is a rough guide.

First, to calculate an index of total brightness (Figure 3.2), most researchers add up the percentage reflectance values measured by the spectrometer at each wavelength between 320 (or 300) and 700 nm (e.g., Andersson et al. 1998). Despite its popularity in the bird color literature, this sum is difficult to compare among species, because different spectral ranges and wavelength increments are often used to calculate this index of total brightness. Instead, one can calculate mean reflectance as the sum of all the percentage reflectance values in the spectral region of interest, divided by the number of values summed (Doucet et al. 2004). Mean reflectance is thus a standardized index of brightness that can readily be compared among individuals and species.

Second, an index of hue is most often calculated as the wavelength at which reflectance is highest (λ_{Rmax}; Table 3.2; H_1 on Figure 3.2), as this is an easy variable to calculate for most reflectance spectra with a clear peak and a unimodal distribution (e.g., Figure 3.2). The main problem with λ_{Rmax} is that it tends to be quite variable, even when hue is constant, due to random noise in reflectance curves from portable spectrometers (Figure 3.1a). This noise can be reduced by various curve-smoothing algorithms (Figure 3.1b), but

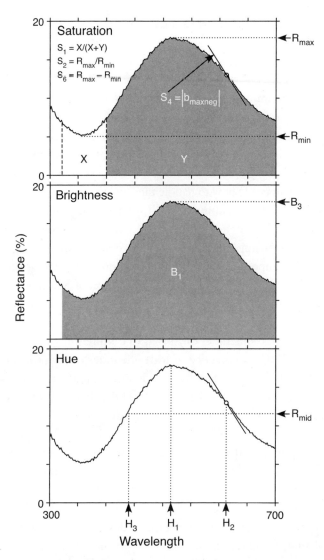

Figure 3.2. Representative reflectance curve from the blue-green back of a male Tree Swallow, showing various ways to calculate hue, saturation, and brightness (see Table 3.2 for formulas). From R. Montgomerie (unpubl. data).

$\lambda_{R\max}$ may still be a less reliable measure of hue than one of the alternatives (Table 3.2). Hue measured as $\lambda_{R\mathrm{mid}}$ (Table 3.2; H_3 on Figure 3.2) may turn out to be more useful because it depends on more than one reflectance estimate (both $R_{\max}$ and $R_{\min}$). Random fluctuations in reflectance at each wavelength measured are unlikely to be systematically biased toward shorter or longer wavelength, potentially making λ_{Rmid} a more reliable measure than λ_{Rmax}. Moreover, hue estimated as λ_{Rmid} explained 78% of the variation in human-perceived hue measured in the CIELAB color space (Pryke et al. 2001), suggesting that λ_{Rmid} is capturing a lot of the variation in hue that humans can see.

Unfortunately, none of these indices of hue (Table 3.2) are very useful when the reflectance curve has more than one peak (Figure 3.1), as is the case for certain iridescent (e.g., Rock Pigeons [*Columba livia*]; McGraw 2004) and UV-yellow colors (e.g., American Goldfinches; MacDougall and Montgomerie 2003). In such cases, the problem is that these indices are rarely influenced by the number, height, and position of secondary peaks in a curve, even though the hue clearly varies. Thus a different measure of hue, like that provided from segment classification (see below), may be required.

Finally, for saturation, the most commonly used index is calculated by dividing the total reflectance in a region of interest by the total reflectance across the bird-visible spectrum (S_1 in Table 3.2). For this index, it is best to use a numerator calculated from the color range that is likely to be relevant to the signal at hand. Thus, for example, for the red-to-yellow plumage of House Finches, an index of red chroma would seem to be most appropriate, whereas for the blue throat patch of male Bluethroats (*Luscinia svecica;* Plate 19, Volume 2), blue or possibly UV chroma should be quantified (Smiseth et al. 2001). Often two saturation (chroma) values are needed to capture the relevant variation, particularly when a reflectance spectrum is bimodal. The yellow plumage of male American Goldfinches, for example, has a bimodal reflectance curve, and MacDougall and Montgomerie (2003) have used an index of both UV and yellow/red chroma to characterize the color saturation of this plumage. Both indices were highly variable and correlated within pairs, suggesting assortative mating (MacDougall and Montgomerie 2003). One problem with this index of saturation is that it is sometimes highly correlated with brightness (Delhey et al. 2003), and with λ_{Rmax} (Sheldon et al. 1999; Shawkey et al. 2003; Delhey et al. 2003), an index of hue, and may thus not capture much variation in that it is independent of those variables. In such cases, principal components analysis (see below) can help to reveal the variation in each of the three tristimulus variables (HSB) that is independent of variation in the other two (Sheldon et al. 1999).

Alternatively, an index of saturation based on other aspects of the reflectance spectrum can be calculated (Table 3.2; Figure 3.2). Spectral saturation (S_7 in

Table 3.2), for example, has been shown to be highly correlated ($r = 0.94$) with variation in human-perceived chroma in the CIELAB color space (Pryke et al. 2001), suggesting that it captures some variation relevant at least to human perception. Similarly "peaky chroma" (S_{10} in Table 3.2) explained 75% of the variation in CIELAB chroma and was significantly correlated ($r = 0.57$) with UV chroma (Örnborg et al. 2002).

Univariate Statistical Analysis

Hypothetically, in a typical study, one might calculate indices of HSB from five haphazardly chosen locations within each of five plumage patches of the same color (e.g., crown, mantle, breast, belly, and rump on a sample of 25 individuals). The resulting dataset comprises 625 reflectance spectra with about 1100 reflectance measures per spectrum, yielding 15 tristimulus color variables (three per patch) calculated for each individual in the sample. If each of the variables is not significantly different among patches, then they might be used to calculate mean HSB values for each bird, averaging over all patches measured. In such cases, the analysis is usually quite straightforward. However, it is often useful to consider each tristimulus color variable from each plumage region separately, so that color patches that are the most informative signals are identified (e.g., Safran and McGraw 2004). In this case, however, statistical problems may arise because a large number of variables is measured and some correction for multiple tests may be needed (see below in the statistical recommendations section).

Another approach is to choose to analyze only the one or two measured variables, either those that show the highest variance among individuals or those that are known from previous research to be correlated with other quality indicators. If we assume that the most variable signals contain the most information for receivers (Chapter 2, Volume 2), then choosing tristimulus color variables with the highest among-individual variance makes sense. In some studies (e.g., McGraw and Hill 2000; Mennill et al. 2003), coefficients of variation (CV) have been calculated and compared, often as a means of assessing whether traits might be sexually versus naturally selected (Alatalo et al. 1988). As pointed out by Dale (Chapter 2, Volume 2), however, the use of CVs for tristimulus color variables is statistically invalid.

Alternatively, we sometimes know a priori that certain color variables tend to be correlated with pigment deposition or feather microstructures and thus might be useful indices of the underlying color-producing mechanism. For example, in Eastern Bluebirds (*Sialia sialis;* Plate 32), the regularity of nanoscale structuring of the keratin rods that produce structural colors is correlated with UV chroma (Shawkey et al. 2003). Also, in European Greenfinches (*Carduelis chloris*) and American Goldfinches, the concentration of carotenoids that are responsible for the yellow color of male plumage is positively correlated with yellow chroma but not with hue or brightness (Saks et al. 2003; McGraw and Gregory 2004). Although such correlations seem logical, the study of such relations is still in its infancy and is very much in need of more extensive empirical data, particularly for melanin-based colors.

Multivariate Statistical Analysis

Multivariate techniques, like multiple regression, logistic regression, multivariate analysis of variance (MANOVA), canonical variates analysis (CVA), discriminant function analysis (DFA), and principal components analysis (PCA), can also be used to analyze tristimulus color variables, both to reduce the number of variables for analysis and to control for confounding or correlated variables. For example, when birds can be classified a priori by sex, age, species, and the like, CVA will generate component scores that maximize the difference between categories. Mennill et al. (2003), for example, used DFA to analyze 18 patch/color variables in Black-capped Chickadees (*Poecile atricapilla*) with respect to sex and dominance rank, revealing clear differences for each. Logistic regression is also useful for such analyses and has the added advantage that multivariate normality and linearity among the independent variables is not required. All of these multivariate procedures are described in detail in Grimm and Yarnold (1995) and Tabachnick and Fidell (2001).

Multivariate techniques are particularly useful when the colors of several patches have been measured and the number of potentially interesting traits (patches × tristimulus color variables) to be analyzed is large. Particularly in the early stages of research on a given species, it is best not to pool data from different plumage and bare-part regions, even if they appear to the eye to be the same color. In such cases, each color patch measured will provide three or more tristimulus color variables for analysis, and the total number of traits (patch/color variables) to be analyzed can exceed 20. Multivariate analyses can handle all of these traits simultaneously, identifying the single traits that contribute most to color variation and allowing the researcher to avoid the complexities of correcting for multiple tests (see the statistical recommendations section below). Because PCA is the most commonly used multivariate method in bird color research, I outline some of the important aspects of its application here.

PCA is particularly useful for summarizing and identifying the sources of variation in tristimulus color variables. PCA results in a number of principal components (PCs) that are orthogonal (i.e., not correlated with one another), where PC1 explains the largest proportion of the variance in the whole sample, PC2 the next most, and so on. Factor loadings reveal which of the original variables are most highly correlated with each PC (from factor loadings) and thus provides a means to interpret each axis (i.e., component).

The general procedure for using this technique is reasonably straightforward, and the analysis is included in most statistics software. Normally for PCA, a good rule of thumb is to ensure that there are at least 100 subjects and that the subjects-to-variables (STV) ratio is at least 5 (Grimm and Yarnold 1995). Axis (component) loadings may not be stable if this rule of thumb is violated, but the stability of the solution can always be checked by jackknifing (repeating the analysis as many times as there are subjects, each time with a different subject left out). The confidence limits of factor loadings can then be calculated and used to determine whether the loadings are significantly greater than zero.

PCA assumes both multivariate normality and a linear relation between all pairs of variables. In practice, these assumptions are usually met well enough when analyzing spectral data, using simple transformations when necessary. Variance-covariance matrices should be analyzed so that the scale of variation is incorporated into the analysis, and the output should be limited to two or three factors. Even if only three tristimulus variables are analyzed, PCA can be useful for collapsing these into a single variable (PC1) that accounts for most of the variation in color, and the factor loadings will indicate how much each of the input (tristimulus) variables contributed to this variation. Also, factor rotation can be used to simplify the interpretation of axis loadings, if these look to be complicated in the unrotated factors (Tabachnick and Fidell 2001). For example, the VARIMAX rotation is often useful, as it compromises between allocating all of the variance to one axis (QUARTIMAX rotation) and distributing it equally among all of the axes (EQUIMAX rotation). The goal is to achieve factor loadings such that each component (e.g., PC1, PC2) has high loadings with some of the input (patch/color) variables, and that these loadings make biological sense.

PCA of tristimulus color variables has been used to generate a single color score (PC1) that explains most of the variation in plumage color in Blue Tits (*Parus caeruleus;* Sheldon et al. 1999; Plate 18, Volume 2), Northern Cardinals (Jawor et al. 2004), American Goldfinches (MacDougall and Montgomerie

2003), and Barn Swallows (*Hirundo rustica*; Safran and McGraw 2004; Plate 7), as well as variation in the beak color of Zebra Finches (McGraw et al. 2003). In each case, PC1 explained a large proportion of variation in the original color variables (≥85% in cardinals and goldfinches), and could be interpreted as a UV (in tits), redness (in cardinals), or carotenoid (in goldfinches) color score. One advantage of this technique is that it reduces the number of variables for analysis and thus makes the interpretation of relations to other (non-color) variables relatively straightforward.

Analyzing Spectral Shape with PCA

One useful and interesting alternative to the analysis of tristimulus color variables is to use PCA to generate orthogonal variables directly from reflectance spectra (Cuthill et al. 1999; Grill and Rush 2000). This method does not make a priori assumptions about the sources of variation in reflectance spectra, but instead accounts for and describes that variation statistically. PCA is useful in this regard because it allows the researcher to (1) summarize all of the information about the shape of even complex reflectance spectra into a few PCs that are independent of one another, (2) categorize the spectral characteristics of each PC in terms of hue, saturation, and brightness, (3) identify those sections of the spectrum (wavelength regions) that are contributing to the observed variation (Figure 3.3), and (4) generate PC scores that can be used as variables in other statistical analyses. Moreover, the analysis of spectral shape using PCA requires no particular assumptions about the color vision of the bird species being studied, except that the bird-visible spectrum spans 320–700 nm.

The basic principles and statistical assumptions of PCA are described in the previous section. To apply PCA to the analysis of reflectance spectra, divide the data from each spectrum into bins of equal size, and then calculate the mean or median reflectance for each bin. Binning is necessary to reduce the large number of data points obtained from each reflectance measurement (typically 300–1,000 data from each measurement in the 320–700 nm part of the spectrum) to a relatively small dataset that can then be used as model (independent) variables in a PCA. Medians are usually a better choice than means, as they are less sensitive to outliers, such as the spikes that sometimes occur in reflectance measurements due to electrical anomalies and artificial lighting (Figure 3.1a). Bin sizes of 2–20 nm have be used in published studies using this technique (Cuthill et al. 1999; Grill and Rush 2000), to reduce the number of variables and simplify computation.

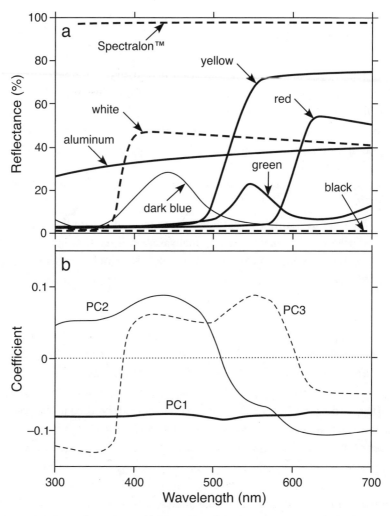

Figure 3.3. PCA analysis of spectral shape. (a) Reflectance spectra across the bird-visible range for six colors of plastic leg bands, plus aluminum bands, and a Spectralon™ standard. (b) Principal component coefficients in relation to wavelength for the first three principal components from a PCA on all the spectra in (a). PC1 accounts for 84% of the variation in the original spectra. The coefficients for PC1 are all negative and of similar magnitude, so PC1 accounts for achromatic variation in the original spectra. PC2 accounts for an additional 12% of the variation in the original spectra. The coefficients relating PC2 are all positive, whereas those above 500 nm are all negative, showing that PC2 represents variation in the relative ratio of short- to long-wavelength reflectance. PC3 accounts for 3% of the total variation, and represents variation in the relation of medium to both long and short wavelengths. Redrawn from Cuthill et al. (1999).

Next, perform PCA on the median or mean values from each bin, using bins as variables and each reflectance spectrum as a separate case. Use variance-covariance matrices (as above) for analysis, apply factor rotations as needed, limit the output to two or three factors, and determine the proportion of variance explained by each principal component (e.g., PC1, PC2). PC1 from a PCA on raw reflectance scores is often highly correlated with total reflectance or brightness (Cuthill et al. 1999), whereas the remaining PCs describe variation in spectral shape. PC2 and PC3 (and so on) can also be interpreted by looking at the coefficients between the PCs and the raw scores. Thus, if PC2 has negative coefficients associated with short (blue) wavelengths and positive coefficients associated with long (red) wavelengths, then variation among spectra is largely due to the relative amount of blue versus red light that is reflected; this pattern could be interpreted as variation in blue chroma. In general, PC2 is often correlated with saturation, whereas PC3 may frequently be interpreted as a hue axis (Grill and Rush 2000). PC2 and PC3 may not always be interpretable in this fashion, so it is recommended that each analysis be examined in relation to the PC coefficients to characterize the sorts of variation that the PCA is capturing.

Given that PC1 is often strongly correlated with variation in brightness (Cuthill et al. 1999), it is sometimes desirable to factor out this variation, so that all PCs represent spectral shape. To do this, simply analyze spectral data from which the overall mean reflectance is subtracted from the reflectance of each wavelength segment (e.g., see Cuthill et al. 1999). In practice, it is probably best to begin analysis without factoring out brightness in this fashion, to ensure that any interesting variation in mean reflectance is included in subsequent analyses.

When several spectra are measured from each individual, separate PCAs can be run on sets of spectra from patches of the same color to reduce the potential number of variables for later analysis. If this is done, however, plots of PC3 versus PC2 from each patch should be compared to ensure that analyses of data from different patches yield similar results. It is not usually wise to analyze spectra from different colors in the same PCA because the axes are likely to be uninterpretable and may capture color variation that is not relevant to signaling.

Bennett et al. (1997) were the first to use this method for analyzing bird coloration in their study of mate preferences in the European Starling (*Sturnus vulgaris;* Plate 23). In that study, they calculated PC1, 2, and 3 after standardizing for brightness, and found that UV components of the reflectance spectra predicted female preferences. Cuthill et al. (1999) provide many more details on the analytical techniques used by Bennett et al. (1997) and use the method to analyze reflectance spectra from both sexes of starlings, finding some evidence for dimorphism that could not be detected by human observers. Cuthill et al. (1999) should be consulted by anyone wishing to apply this technique, as it contains both useful methodological details and a thoughtful discussion.

PCA has also been used to generate PC scores from the reflectance spectra of (1) the plumage patches on Blue Tits (Hunt et al. 1998), Long-tailed Finches (*Poephila acuticauda;* Langmore and Bennett 1999), Yellow-breasted Chats (*Icteria virens;* Plate 5; Mays et al. 2004), and Wire-tailed Manakins (*Pipra filicauda;* Heindl and Winkler 2003); (2) the bill color of Mallards (*Anas platyrhynchos;* Peters et al. 2004; Plate 15, Volume 2); and (3) the egg colors of the Red-chested Cuckoo (*Cuculus solitarius*) and its hosts (Cherry and Bennett 2001). In each of these studies, spectra were divided into small bins (1–5 nm), resulting in as many as 100 or more variables for each PCA. Because sample sizes were relatively small in each study (<50), the use of such small bins seriously violates the STV rule for PCA. I suggest, therefore, selecting much larger bin sizes and using jackknifing analysis to double-check the stability of component loadings when sample sizes are small.

Mennill et al. (2003) also used PCA to analyze the reflectance spectra from six plumage color patches on male and female Black-capped Chickadees. To do this, they divided the reflectance spectra (300–700 nm) into 40 bins that were each 10 nm wide, and calculated the mean reflectance in each bin. The PCA was then performed on these 40 variables from 73 chickadees. Although this analysis clearly violates the STV rule of thumb for PCA, jackknifing showed that the results were robust. PC1 explained the majority of the variance in reflectance spectra from each patch, and the first three PCs explained 96–99.6% of the total variance, depending upon the patch. PC2 and PC3 were associated with the shapes of the reflectance curves. PC2 had high positive loadings from UV chroma and wavelengths in the UV range, whereas PC3 was correlated (had high loadings) with both very short and very long wavelengths. Mennill et al. (2003) then used these PC scores in a DFA to document color differences between the sexes, and among males of different dominance status. Doucet et al. (2005) analyzed the same plumage reflectance dataset calculating tristimulus variables instead. They found the same patterns as Mennill et al. (2003), but also found that the UV chroma of a male's black plumage and the brightness of his white plumage predicted his within-pair paternity and total reproductive success.

Unfortunately, one real drawback of this technique for analyzing color data is that PC scores are not comparable between species, studies, or even between patches analyzed separately with PCAs. Thus, each PCA run on a different set of tristimulus variables creates a different set of PCs with different loadings so that, unlike HSB indices, PC1 from one analysis is not directly comparable to PC1 from another. The only way around this problem is to include all measured variables in a single PCA, usually an impossibility when comparing among studies.

Analyzing Colors by Segment Classification

Endler (1990) introduced a general method of segment classification for the analysis of animal coloration, based on estimating the quanta of light energy reaching the observer's eye and using this estimate to calculate hue, saturation (chroma), and brightness. This method was developed to be independent of the properties of the observer's visual system while incorporating some environmental variables, and thus capturing some properties of the color signal that would be common to many animals. As a result, segment classification estimates what I have called the "received" signal in Box 3.1. Although the method was developed initially for species with vision more like humans (based on three cones and a visible spectrum from 400 to 700 nm), it can be readily adapted to the bird-visible spectrum, as outlined below. To date this method has rarely been applied to the analysis of bird colors (Endler and Théry 1996), possibly because researchers seldom have enough information to make all of the necessary calculations.

To apply this method, first calculate a measure of total brightness as the sum of the reflectances (B_1 in Table 3.2) at all relevant wavelengths (320–700 nm for birds). For consistency, this is best done by calculating the reflectance (R_i) in 1.0-nm increments across this range, where R_i is the proportion of incident light that is reflected. Second, Endler (1990) recommended dividing the (human) visible spectrum into four segments of 75 nm each, with b = 400–474 nm, g = 475–549 nm, y = 550–624 nm, and r = 625–699 nm, corresponding (roughly) to the blue, green, yellow, and red segments of the visible spectrum, respectively. Endler (1990) divided the visible spectrum into four segments because of the way that humans perceive differences in saturation. Because birds can see into the UV, this classification would have to be modified by adding a fifth segment (u = 325–399) to keep the size of segments approximately equal, or by redefining the four segments as u = 320–415 nm, b = 415–510 nm, g = 510–605 nm, r = 605–700 nm. The latter is probably preferable, because it corresponds rather closely to the four cone sensitivities in birds (Hart 2001) and allows the use of Endler's (1990) formulas for chroma (C) and hue (H), directly, substituting the appropriate subscripts:

$$C = [(Q_r - Q_b)^2 + (Q_g - Q_u)^2]^{1/2}, \text{ and}$$

$$H = \arctan([(Q_g - Q_u)/Q_T]/[(Q_r - Q_b)/Q_T]),$$

where Q_i is the intensity of light (or quantum flux, or photon flux) that reaches the eye in the r, g, b, and u segments of the bird-visible spectrum, and Q_T is the total photon flux.

Endler (1990) defined the total quantum (or photon) flux, Q_T, at a distance x between signaler and observer (bird in this case), as

$$Q_T(x) = \int_{320}^{700} A(\lambda) R(\lambda) T(\lambda) d\lambda,$$

where $A(\lambda)$ represents ambient light, $R(\lambda)$ represents the fraction of incident light that is reflected, and $T(\lambda)$ represents the fraction of reflected light that reaches the observer's eye after passing through the air, each measured as photon flux at all wavelengths λ in the bird-visible spectrum. Thus Q_T is the area under the curve obtained by multiplying $A(\lambda) \times R(\lambda) \times T(\lambda)$. Q_T can be approximated, and the calculation simplified, by calculating the Riemann integral —photon flux over small discrete wavelength intervals (e.g., 1 nm):

$$Q_r(x) = \sum_{605}^{700} A(\lambda) R(\lambda) T(\lambda),$$

which is quite accurate when wavelength intervals are small (J. Endler, pers. comm.). The photon flux of the ith segment is calculated by restricting the analysis to the wavelengths in that segment. Thus, for example,

$$Q_r(x) = \sum_{605}^{700} A(\lambda) R(\lambda) T(\lambda)$$

is the photon flux in the red segment of the bird-visible spectrum. Note that this model does not consider the transmission properties of the ocular media, $O(\lambda)$, or the sensitivities of the four color-filtering cones, $C(\lambda)$, in birds, both of which are dealt with in the analysis of photon catch (see the next section).

Grill and Rush (2000) measured reflectance spectra from more than 700 Munsell color chips so that they could compare the results of a segment classification analysis to that from a PCA of spectral shape (both as described above). For each analysis, they generated values for hue, saturation, and brightness and found that both methods performed well when estimating brightness or saturation (compared to the Munsell scores), but were less good at predicting hue in some instances. Thus the PCA method often gave better results when accurate estimates of hue were important, especially when samples were all of the same general color category (e.g., reds). As pointed out by Grill and Rush (2000), these differences are most likely a function of the specific shapes of color curves. When they applied both of these techniques to reflectance spectra from a poeciliid fish (*Limia perugia*), both revealed differences that were expected between sexes and populations (Grill and Rush 2000). From the findings of this study, I conclude that neither method of analysis is superior in all situations, and they might be used interchangeably, with the caveat that in some situations one will outperform the other. In a given study, it might be advisable to test the performance of both in revealing patterns that are expected a priori.

In their study of three lekking bird species in a tropical forest, Endler and Théry (1996) used both plumage reflectance and ambient light spectra to make some estimates of signal display (Box 3.1) using segment classification methods, and compared these to background colors at locations on and off the leks. To make these comparisons, they plotted the locations of these measurements in two-dimensional color spaces, with axes comprising either brightness versus saturation or segment axes that were calculated from the standardized distances between nonadjacent segments. Their use of the segment method allowed them to quantify the colors of both the birds and their lek sites in a comparable fashion. They showed that each species was selective in the location of its leks, maximizing the apparent visual contrast between the male's plumage color and the surrounding environment.

Analyzing Photon Catch

The only method of analyzing bird colors that comes close to assessing what birds actually perceive (Box 3.1) requires measurements of (1) the irradiance spectra of incident light, (2) the reflectance properties of feathers, (3) the transmission properties of air and the bird's ocular media, and (4) the spectral sensitivities of the bird's retinal cones. These data are then used to estimate the

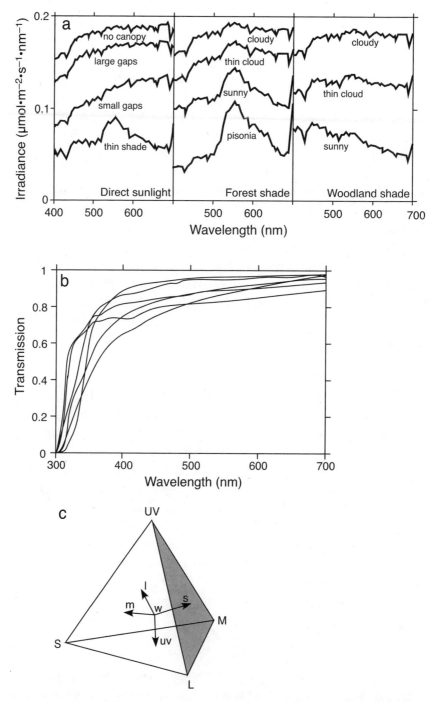

Figure 3.4. Analysis of photon catch. (a) Irradiance curves in the human-visible spectrum from different environments and lighting conditions. Adapted from Endler (1993). (b) Transmission spectra from the ocular media of various bird species. Redrawn from Hart et al. (1998a,b, 1999, 2000a,b); Hart (2004). (c) Tetrahedral color space. Redrawn from Goldsmith (1990).

quantum or photon catch of each of the four single cone receptors that appear to influence color perception in birds (Chapter 1). Although such a measure is highly desirable when we want to understand the signals that birds actually perceive, the quantity and quality of information needed to make such calculations has so far been prohibitive. Nonetheless, recent advances both in the study of cone sensitivities (Hart 2001; Ödeen and Håstad 2003; Chapter 1) and in portable spectrometry (Chapter 2) make this method feasible in some cases, especially when comparing colors of one individual, sex, age class or species to another, or of birds and other objects to their environmental background (see the next section). Many assumptions are made in such analyses, the details of which are outlined by Vorobyev et al. (1998).

To calculate the photon catch of a color patch by a specific bird observer, the spectrum of incident light in the bird-visible wavelengths (320–700 nm) is first measured using an integrating sphere (Chapter 2). The result is $I(\lambda)$, the irradiance spectrum of the ambient light (Figure 3.4a). Second, the reflectance spectrum, $R_p(\lambda)$, of the color patch (relative to a reflectance standard; Figures 3.1, 3.2) is measured at the angles of incidence and reflection of interest (Chapter 2; Figure 2.9). Third, the transmission properties of the receiver's ocular media, $M(\lambda)$, are estimated, as shown in Figure 3.4b. Fourth, the cone sensitivities of the receiver are determined as:

$$S_i(\lambda) = P_i(\lambda) D_i(\lambda) M(\lambda),$$

where $S_i(\lambda)$ is the spectral sensitivity of the ith cone, $P_i(\lambda)$ is the normalized absorptance of the pigment in the ith cone, and $D_i(\lambda)$ is the transmission of that cone's oil droplet. Recent work has shown that avian cone sensitivities probably fall into two general categories (Ödeen and Håstad 2003; Chapter 1), such that the sensitivity of cones at all wavelengths can be readily estimated for any bird if we know into which of these classes it falls. The sensitivity of each cone is influenced by reflectance from both the bird's color patch, $R_p(\lambda)$, and the background in the visual field, $R_b(\lambda)$. This effect is accounted for by normalizing the spectral sensitivity of each cone by:

$$P_i(\lambda) = [R_b(\lambda) I(\lambda) S_i(\lambda)]^{-1}.$$

Fifth, the photon catch for each cone is calculated as:

$$Q_i = \sum_{\lambda=320}^{700} R_p(\lambda) S_i(\lambda) I(\lambda).$$

Finally, using the relative photon catch from each of the four cones, the location of the color patch in a color tetrahedron (Figure 3.4c) is calculated. Each point in this tetrahedron can be specified by four coordinates perpendicular to the sides of the tetrahedron, as shown for the patch of color marked "W" in Figure 3.4c (Goldsmith 1990). Begin by normalizing the maximum excitation of each cone to 1, and calculating the relative excitation of each cone (E_i):

$$E_i = Q_i/(Q_i + 1).$$

Then, using the relative excitation of each cone, calculate the three-dimensional coordinates of each color in the color space tetrahedron:

$$x = \frac{2\sqrt{2}}{3} \cos 30°(E_{MW} - E_{LW}),$$

$$y = E_{UV} - \frac{1}{3}(E_{SW} + E_{MW} + E_{LW}), \text{ and}$$

$$z = \frac{2\sqrt{2}}{3} [\sin 30°(E_{MW} + E_{LW}) - E_{SW}],$$

where LW, MW, SW, and UV are the four different cone types (Chapter 1).

The photon catch method recognizes that the reflectance spectrum measured from bird color patches under standardized conditions might be very different from the relative flux of photons at different wavelengths that reach the retinal cones and are transduced by the visual pigments to the observer's nervous system. This model of photon catch depends critically on a large number of assumptions about ambient light, the transmission properties of air and ocular media, and the effects that the four single cones have on the perception of reflected light (Vorobyev et al. 1998). For some species, model assumptions based on a four-dimensional avian color space make fairly accurate predictions about measured spectral sensitivities (Figure 3.5) under controlled lighting conditions.

Using the photon-catch method, the coordinates (in the color space tetrahedron; Figure 3.4c) of any color patch can be calculated for use in comparing different colors (see the next section). Alternatively, coordinates can be converted into HSB indices with respect to the coordinates of a neutral gray color

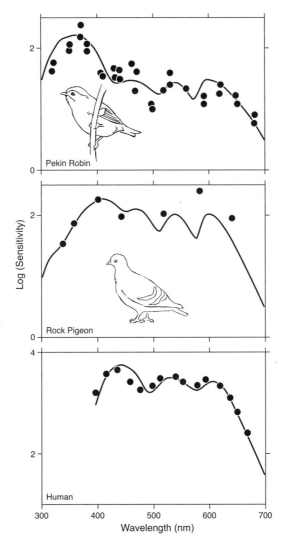

Figure 3.5. Examples of predicted spectral sensitivities (lines) of Pekin Robins (*Leiothrix lutea*), domestic Rock Pigeons, and humans, with empirical data marked as dots. Redrawn from Vorobyev at al. (1998).

that would be perceived by the observing bird. Avian color vision is thus assumed to be influenced by neural comparison of the output from those four single cones (Osorio et al. 1999).

The photon catch method for analyzing the colors of birds has so far been applied in field studies only in two unrelated analyses of manakin (family

Pipridae) coloration (Heindl and Winkler 2003; Uy and Endler 2004) and an analysis of European Starling reproductive performance in relation to age and ornamentation (Komdeur et al. 2005). All three of these papers should be consulted for further details, although a brief summary is presented below.

Heindl and Winkler (2003) used this technique to quantify the conspicuousness (both chromatic and achromatic contrast) of four manakin species in relation to the background colors at their rainforest lekking sites. They found that each species' lek sites were located in the vertical dimension in the forest to maximize the contrast of their colors against the background. Uy and Endler (2004) also used the photon catch method to compare the golden plumage patches of the Golden-collared Manakin (*Manacus vitellinus*) to the color of its lekking courts and the adjacent forest floor. These manakins clear a court in the forest litter, thereby increasing both the chromatic and achromatic contrast of their golden patches against the court background compared to the uncleared forest floor. Cleared courts also provided a less variable color background for the manakin's displays.

Komdeur et al. (2005) measured reflectance spectra from the iridescent throat feathers of both sexes sampled from 89 mated pairs of starlings. Then, using these spectra, as well as data on the visual pigments, oil droplets, and optical media transmission of starlings (Hart et al. 1998a), and assuming a Standard Illuminant, they calculated the photon catches of each of the four single cone types in the starling's retina. They used the color space model in Maddocks et al. (2001) to calculate the following three color variables: (1) "hue index" as the ratio of SW + LW cone catches to UV + MW cone catches, (2) "chromatic index" as the Euclidean distance from the achromatic center of the bird color space tetrahedron (Figure 3.4c) to the locus of the reflectance spectrum from each feather, and (3) "brightness index" as the sum of the four single cone catches. They found significant positive correlations between a feather's length (a noncolor ornament) and its hue and brightness indices, as well as significant sex and age differences in these color variables. Age and hue index were both significant predictors of mate ornamentation (suggesting assortative mating) and reproductive success, but it was impossible to disentangle the independent effects of age and color in this study.

Comparing Colors

One of the most useful applications of the photon catch and segment classification methods is in the comparison of colors, for example when comparing sexes, ages, and treatments, or when comparing both the colors of birds to their backgrounds (see the previous section) as well as adjacent but differently colored patches on the same individual. Colors can be compared simply by calculating the Euclidean distance between any two points (x and y) in color space. Using the segment classification method, this Euclidean distance is:

$$D_{xy} = \sqrt{(H_x - H_y)^2 + (C_x - C_y)^2 + (Q_{Tx} + Q_{Ty})^2},$$

where H and C are hue and chroma (saturation), respectively, and Q_T is the total photon catch (Endler 1990). For the photon catch method, the Euclidean distance between points (e.g., color patches) a and b is:

$$D_{ab} = \sqrt{(\Delta x)^2 + (\Delta y)^2 + (\Delta z)^2},$$

where x, y, and z are the standardized coordinates of each color patch in the color-space tetrahedron (Figure 3.4c; see the previous section).

Both the chromatic and achromatic (brightness) contrasts between two color patches, or between a color patch and the background, can also be calculated using similar techniques (Uy and Endler 2004). Chromatic contrast (C_c) is calculated as the Euclidean distance between the cone stimulus score from a color patch (subscript p) and that from the visual background (subscript b) for all four cone types:

$$C_c = \sqrt{(X_{UVp} - X_{UVb})^2 + (X_{SWp} - X_{SWb})^2 + (X_{MWp} - X_{MWb})^2 + (X_{LWp} - X_{LWb})^2},$$

where X_i is the relative stimulus of each of the four cone types (Uy and Endler 2004). Objects that are further apart in the tetrahedral color space have higher values of C_c. Achromatic contrast (C_a) can be calculated as:

$$C_a = (B_p - B_b)/(B_p + B_b),$$

where B_p is the perceived brightness of the color patch, and B_b is the perceived brightness of the background (Uy and Endler 2004). The quantity C_a varies from −1 to +1, with patches brighter than the background having $C_a > 0$, and those darker than the background having $C_a < 0$.

Maddocks et al. (2001) used the photon catch method to quantify the colors of red and white millet seeds presented to Zebra Finches on a brown hardboard background with and without some colored stone distractors, under different lighting conditions (using filters). They showed that foraging preferences were influenced by the relative Euclidean distances between seeds and the background in the color space measured, particularly when long wavelengths of light were filtered out.

Eaton (2005) used a similar method to examine the degree of sexual dichromatism in 139 passerine species that appear to be monomorphic to the human eye. This analysis was performed by calculating the Euclidean distance between males and females within species in a color space determined by the photon catch. The photon catch was estimated using only the reflectance spectrum from a color patch on each bird and the spectral sensitivity of each cone type for a representative passerine bird, the Blue Tit (Chapter 1). This distance was evaluated in relation to the presumed ability of birds to distinguish colors to estimate whether the sexes would appear to be different from one another within species. About 85% of the supposedly monomorphic species were shown to be sexually dichromatic by this method.

Statistical Recommendations

The analysis of color data does not require any special statistical treatment in that most parametric, nonparametric, multivariate, and univariate tests can be applied, as described in any modern statistics textbook. However, three statistical issues relevant to most color analyses are worth describing in some detail here, as they are either frequently ignored (outlier analysis), currently controversial (Bonferroni correction), or inconsistently dealt with (presentation of results) in the behavioral ecology literature.

Dealing with Outliers

Statistical outliers are an important reason that the parametric assumption of normality is often violated, and thus are one of the most common sources of statistical error. Even though the use of nonparametric statistics can often circumvent this problem, there are some instances where even these statistical tests will fail to provide correct answers if outliers are not properly handled. Here I outline some procedures for finding and dealing with the kinds of outliers most often encountered in the analysis of bird color data. There are a few formal statistical methods for detecting outliers (Tabachnick and Fidell 2001), but these are beyond the scope of this chapter and are generally not needed.

The most convenient methods for outlier detection are simply graphical techniques, such as Tukey box plots (also called box and whisker plots), that show potential outliers. In Tukey box plots (Figure 3.6b), potential outliers are identified as those data beyond the beyond the upper (75th percentile + 1.5·IQR) and lower (25th percentile – 1.5·IQR) ends of the whiskers, with points more than 3·IQR beyond the box considered to be extreme outliers (Tukey 1977), where IQR is the interquartile range extending from the 25th to the 75th percentile. For multivariate analyses, jackknifed Mahalabonis distances (Tabachnick and Fidell 2001) can be calculated and plotted for each case (i.e., bird or color patch), as shown in Figure 3.6c. The Mahalabonis distance is defined as the distance of each datum from the multivariate centroid, with points farther from the centroid being more likely to be outliers. Potential multivariate outliers can be identified statistically (Tabachnick and Fidell 2001) or simply by comparison to a reference line (Figure 3.6c), as is done by JMP™ statistics software.

Once outliers have been identified, the next and most difficult task is to decide what to do with them. Often, the source of the outliers will be clear, as they are caused by some error or anomaly in data collection. When that is the case, remove all data points that are consistently identified as being the result of such errors, then replot the data to see whether any additional outliers are revealed. Repeat this process until all erroneous data points are removed. It is also good practice to remove all data points that can be operationally associated with any outlier. Thus if more than one outlier is associated with a particular sampling condition (e.g., a rainy day that might have affected electrical connections on the spectrometer), it is sometimes wise to remove all data associated with that condition, even if some of those data are not clear outliers, in case those data are also biased. When the reasons for outliers cannot be clearly identified, they should not be removed from the dataset, though it is sometimes useful to analyze data with and without outliers and to report both analyses.

Outliers can have a serious effect on the results of both parametric and nonparametric tests and need to be dealt with carefully. They are a common feature of bird color measurement for at least three reasons. First, as the equipment used for measuring spectra becomes more complex, the chances of equipment failure and measurement anomalies can be quite high. In work on Tree Swallows (*Tachycineta bicolor;* Plate 27, Volume 2; R. Montgomerie, unpubl. data), for example, about 2% of spectral readings failed (i.e., resulted in anomalously low readings) as a result of voltage fluctuations from a car battery or loose

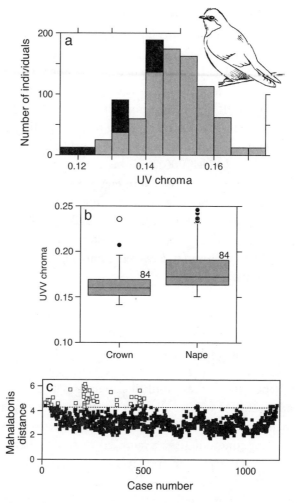

Figure 3.6. Outlier analysis. Plots of data from measurements taken from the blue-green plumage of male Tree Swallows (R. Montgomerie unpubl. data). (a) Frequency distribution of UV chroma (black bars indicate feathers identified as being brown by inspection of the reflectance curve). (b) Tukey box plots of mean UV-violet (UVV) chroma from the crown and nape. These box plots show a box bounded by the interquartile range (IQR; 25th to 75th percentile), with the median (50th percentile) inside the box; whiskers extending either to 1.5·IQR above and below the box or to the most extreme data points, whichever is less; outliers beyond the whiskers as either solid dots or open circles (extreme outliers that are >3·IQR above or below the box); and sample sizes are shown above each box. (c) Mahalabonis distances calculated from the correlation matrix of color variables (open symbols indicate cases that may be outliers). Color variables were calculated from a total of 1145 measurements taken from five blue-green plumage regions of different patches on 84 males.

connections between light source, probe, spectrometer, and computer. Second, when measuring reflectance from a color patch in the field, it is virtually impossible to ensure that the angles of incident and reflected light are held constant. Although such anomalies will most often result in random error, they are sometimes severe enough to produce a statistical outlier. Finally, colors within a patch are often far from uniform. Thus readings taken within a patch, even when they are not subject to equipment or operator error, can sometimes vary widely, due to microscale variation in feather arrangement, dirtiness, barbule damage, and the like (Figure 3.6a).

Correcting for Multiple Tests

When analyzing datasets comprising a large number of tristimulus color variable × patch combinations, it is sometimes necessary to correct P-values to account for the possibility that statistically significant results have arisen by chance alone. Although there has been some discussion and confusion about the need for such corrections in different situations (Moran 2003; Nakagawa 2004), they may be required in color analysis when a large number of either dependent (e.g., morphological and behavioral traits) or independent (patch × color variable combinations) variables are analyzed.

Sequential Bonferroni methods provide a way to make such corrections (Rice 1989). These methods are best applied by calculating a new alpha level against which to compare calculated P-values, so that the experiment-wise error rate remains at 0.05. By reporting the corrected alpha rather than the corrected P-values, as is sometimes recommended, the raw P-values are available to readers, who can then make their own conclusions about the significance and biological relevance of the results.

There is an ongoing controversy over the use of Bonferroni corrections (Cabin and Mitchell 2000; Moran 2003; Nakagawa 2004). Most behavioral and evolutionary ecologists today follow Rice's (1989) guidelines for this procedure, but it is increasingly clear that these guidelines are sometimes overly stringent and likely to result in Type II errors, especially when sample sizes are small (Nakagawa 2004). The result is that biologically meaningful patterns are sometimes overlooked. As Moran (2003) has pointed out, it might be wise to accept a higher Type I error rate than usual in studies of an exploratory nature (see also Stoehr 1999), lest interesting biological patterns are missed by the strict application of Bonferroni corrections. For example, when correcting for many correlational analyses performed on a single dataset, the sequential

Bonferroni is applied by dividing the initial alpha level by $(m - 1)$, where m is the number of correlations analyzed. Thus when examining correlations between a focal variable and color variables derived from several patches, the value of m can sometimes be quite large (e.g., >20), requiring very small P-values to achieve statistical significance (e.g., <0.0025 for $m = 20$).

Most important, when different analyses each address a specific hypothesis, the Bonferroni correction need not be applied. For example, if a dependent variable is tested separately against a single tristimulus color variable (e.g., hue) from separate carotenoid- and melanin-colored patches, then these could potentially be considered to be testing independent hypotheses and no correction is necessary. However, examining the correlations between a dependent variable and the hue, saturation, and brightness from a single patch or related patches would require a Bonferroni correction, as these would all be testing the same hypothesis.

In the early stages of a study, when many analyses of color variables are often required to explore a dataset for possible relations, even the correct application of Bonferroni procedures can result in prohibitively small alpha levels. In such cases, there are two options, although each comes with a cost. First, simply collect more data, replicating the study, using the results of the first analysis to indicate which correlations in the replicated study are likely to be meaningful. Perform only those analyses identified as meaningful on the new dataset. Second, before performing such an exploratory analysis in the first place, divide the sample in two, using the first exploratory analysis to guide the second toward the relations most likely to be important.

Presenting Results

Once an analysis has been completed, some final concerns about presentation need to be considered before publication. In this section, I deal with two issues —reporting effect sizes and confidence limits, and plotting color data—that I think could be better handled in the bird color literature, in particular, and in behavioral and evolutionary ecology, in general.

In behavioral and evolutionary ecology, it is standard practice to report test statistics, P-values, and sample sizes for each analysis, but there tends to be little information provided about effect sizes, relying instead upon P-values and significance tests to tell the story. Many psychology journals, however, require or at least strongly encourage researchers to report effect sizes and to provide information that will allow the reader to put these effect sizes into a biological context (Wilkinson and Task Force on Statistical Inference 1999). Because *P*-values depend on sample sizes and effect sizes (among other things), they alone are really not very useful indices of the biological relevance of a result. By reporting effect sizes, researchers provide useful information about the magnitude of the effect and provide data for subsequent metaanalyses. I prefer effect sizes that are not standardized (e.g., correlation coefficients and differences between means) in preference to standardized estimates, like Cohen's D, because I find the former easier to interpret. Wilkinson and Task Force on Statistical Inference (1999) lists a number of references that summarize the various measures of effect size used in the psychology literature.

It is particularly important to report effect sizes in bird color research, so that the biological significance of trends and differences can be assessed. Statistically significant differences in brightness, for example, might be so small that the signal receiver cannot detect them. We know relatively little about the ability of birds to detect differences in color variables, except that they do not appear to be as good as humans at discriminating differences in brightness and saturation (Peiponen 1992; Chapter 1), at least in the human-visible part of the spectrum.

For many statistics, like randomization tests, appropriate effect sizes are not available, so in such cases, the reporting and plotting of confidence limits (CL) is recommended (Wilkinson and Task Force on Statistical Inference 1999). In general, CLs are underused in behavioral ecology (Colgrave and Ruxton 2003), which is unfortunate, as they are much more informative than the usually-reported standard error or standard deviation. I strongly suggest that 95% CL be reported with all descriptive statistics, as it provides useful information in the same spirit as the effect size.

When plotting results, it is both common practice and useful to plot representative reflectance spectra for each color patch analyzed. Such curves should be plotted from 320–700 nm only, and can either be a representative curve from a "typical" individual or a median (or mean) curve calculated from all measurements in the sample. The statistical error associated with a mean curve tends to vary relatively little across the bird-visible spectrum, so it is not necessary to plot more than a few error bars or, preferably, 95% CLs, on such curves, for example at 50-nm intervals (Figure 3.1c). For plots of tristimulus color variables, I prefer box plots (Figures 3.6b, 12.5), as these illustrate the entire distribution of the variable, and can indicate potential outliers (Figure 3.6b). There are a few different ways to construct box plots, so it is necessary to describe the plot itself whenever these are used (see the captions of Figures 3.6b and 12.5).

Conclusions

Over the past two decades, there has been a clear trend toward increasing sophistication in the analysis of bird color data, in part because of the increasing availability of reflectance spectrometry (Chapter 2) and computer applications (Box 3.3) for handling and analyzing the data obtained with such instruments. Indeed, almost all of the analytical methods described in this chapter have been used effectively in studies of bird coloration, although the application of segment classification and photon catch methods has so far been limited. Less sophisticated techniques, such as color matching and the calculation of tristimulus variables, can still be quite useful, as long as the researcher is aware of the signal that is being estimated (Box 3.1) and the limitations of the techniques.

Although digital photography and color swatch matching are easy and inexpensive techniques for quantifying birds colors, reflectance spectrometry is now inexpensive and portable enough that it should be used whenever possible. The careful use of reflectance spectrometry (Chapter 2) ensures that measurements are not biased by the subjectivity of the human observer's judgment about colors, and that measurements are taken into the UV region where birds can see. There is also increasing evidence that even species that are not blue or purple might have significant signal variation in the UV range that is invisible to humans (Doucet and Montgomerie 2003; Eaton and Lanyon 2003; MacDougall and Montgomerie 2003; Chapter 7).

Analyses of photon catch are clearly needed and will be extremely useful in helping to understand the role of ambient light and the observer's ocular characteristics in signal perception (Box 3.1). Nonetheless, we are probably a long way from seeing this method regularly applied in the analysis of bird colors, due to the complexities of the light environment during most avian signaling. Indeed, most species display their colors in a wide variety of lighting environments due to variation in times of day, microhabitats, weather conditions, and temperature that all influence the way that colors appear to the observer (Box 3.1). Thus it is not surprising that this method of analyzing bird colors has so far been applied only to lekking species that display in a fixed location at certain times of day (Heindl and Winkler 2003; Uy and Endler 2004), in lab environments where the lighting could be precisely controlled and measured (Maddocks et al. 2001), or when comparing the sexes and assuming a Standard Illuminant (Eaton 2005; Komdeur et al. 2005). Thus a simpler application of this method of analyzing colors is to consider only plumage and/or bare-part reflectance and cone sensitivities in a model, to capture some of the interplay between the signal production and reception, but ignoring the complexities of the light environment (e.g., Komdeur et al. 2005).

One potential drawback of reflectance spectrometry is that it allows measurement and calculation of signal properties in a way that reveals much more variation than birds can actually perceive (Peiponen 1992; Chapter 1). This is not a problem when measuring signal potential (Box 3.1) as an index of phenotypic or even genetic quality. However, it is a big problem if we incorrectly assume that the signal variation that we measure can be detected and used by the receiver of the signal (Swaddle 1999). The relation between signal variation and signal detection is clearly an area that needs more work. Researchers should certainly exercise caution in the interpretation of statistically significant results with small effect sizes.

At this early stage in the exploration of patterns related to variation in plumage and bare-part colors, it would be counterproductive to limit discovery by overzealous application of statistical corrections (e.g., sequential Bonferroni) that could result in type II errors (Moran 2003; Nakagawa 2004). I suggest, therefore, that both researchers and reviewers try to avoid conservative application of statistics, but at the same time express caution in the interpretation of results. In this way, we are more likely to reveal some interesting general trends worthy of more in-depth analysis at the (slight) cost of finding some patterns that do not hold up to more scrutiny.

Summary

Bird colors can be quantified by (1) matching color patches against standard color swatches, (2) ranking colors with reference to a predetermined scale, (3) analyzing digital or digitized photographs, and (4) performing reflectance spectrometry using a digital color meter to generate tristimulus variables, or a portable spectrometer to calculate reflectance spectra directly. Although each of these methods has been used successfully for the analysis of bird colors during the past 30 years, reflectance spectrometry appears to be the most objective method and allows the quantitative analysis of colors in the UV range (320–400 nm) that birds, but not humans, can perceive. Each of these methods is outlined in this chapter, with some details of statistical analyses that can be applied to the data obtained.

Once the spectral properties of color patches on a bird's plumage or bare parts have been measured using reflectance spectrometry, the often voluminous

Box 3.3. Computer Resources for Analyzing Bird Colors

Several software applications and websites are available to assist in the analysis of color variables directly, or to provide guidance for the statistical procedures described in this chapter. A website associated with this book (http://biology.queensu.ca/~color) provides links to all of these and more, and will be updated as this information changes.

Software

ColoR (v1.7) is a series of Excel workbooks for summarizing and collating data generated by reflectance spectrometry, as well as for calculating a number of tristimulus color variables from these data. Reflectance spectra can also be plotted, anomalies detected, and reflectance curves smoothed with this package. Written in Visual Basic for Applications (VBA), ColoR works within Microsoft Excel on any platform.

SPEC is a program written in R (see below) for the analysis of data files generated by the OOIBase32 or Spectrawin software applications that are used with reflectance spectrometers. SPEC calculates photon catch based on cone sensitivity data, three different irradiance spectra (daylight, blue sky, and forest shade), and the transmission spectra of some ocular media (Blue Tit, chicken, and human) using built-in libraries, although other libraries can be readily added. SPEC also calculates UV chroma and total brightness (Table 3.2), as well as the discriminability of two stimuli based on the receptor-noise-limited color opponent model of Vorobyev et al. (1998).

OOIBase32 is Windows-based application that is supplied free with Ocean Optics® reflectance spectrometers. OOIBase32 determines percentage reflectance at <1 nm increments from 200 to 800 nm and generates data files that can be plotted and analyzed by other programs (e.g., ColoR, SPEC). A more versatile version, OOIBase32 Platinum (U.S. $999), is scriptable using VBA to add more control and flexibility, and can be modified to interface with spectrometers from other manufacturers. OOIColor (U.S. $299) is a similar Excel-based, user-programmable application that interfaces with Ocean Optics spectrometers and calculates variables in a variety of color spaces. OOIBase is currently being rewritten as a Java application that will be platform independent and should be available late in 2005.

SpectraWin2 is a Windows-based application that, like OOIBase32, generates percentage reflectance data from spectrometers and provides tristimulus (both CIELAB and RGB) color variables calculated from the spectra. SpectraWin2 is supplied with spectrometers from Photo Research® and can be purchased in versions that allow the user to add scripts for automating analysis.

R is an open-source language and statistical environment for the analysis and plotting of data and is particularly useful for randomization tests and jackknifing, as well as for customizing statistical analyses. R is available at http://www.r-project.org/ for MacOSX, Windows, Unix, and Linux operating systems.

WEBSITES

Statistics and Statistical Graphics Resources

The website http://www.math.yorku.ca/SCS/StatResource.html is an excellent portal to a wide variety of statistical information and programs available on the web.

GraphPad QuickCalcs (http://graphpad.com/quickcalcs/index.cfm) provides some simple calculators for statistical analysis, interpreting P-values, and generating random numbers.

Simple Interactive Statistical Analysis (http://home.clara.net/sisa/) provides a wide variety of user-friendly, web-based forms for statistical analysis.

Web Pages that Perform Statistical Calculations

The website http://members.aol.com/johnp71/javastat.html provides links to hundreds of such web pages.

data obtained can be analyzed by a variety of methods. Most studies to date have analyzed tristimulus color variables (HSB) based on the human visual system, in part because these variables are relatively easily understood. Tristimulus color variables can be provided by the measuring device or calculated in a variety of ways from raw reflectance data measured across the bird-visible spectrum (320–700 nm), using simple formulas, PCA, or segment classification methods. Tristimulus variables can then be analyzed using common parametric and nonparametric statistical methods. Reflectance spectra can also be readily analyzed using PCA to reduce data to factors related to HSB. PC scores can be used in further statistical analyses.

Recently, more sophisticated methods of color analysis have taken into account both the irradiance spectra of ambient light and the spectral sensitivities of the four single cones in birds, in addition to the reflectance spectrum measured from a color patch, to calculate the photon catch of the cone receptors. These data are then used to determine the position of each perceived color in the avian tetrahedral color space. Colors so determined can be compared between individuals, sexes, and species by calculating the Euclidean distance, which can then be used in other statistical analyses. This method is undoubtedly the way of the future, but too few data are available for most species in nature to make this technique generally useful in field studies, although it has already had some interesting applications.

Three statistical procedures (outlier analysis, correcting for multiple tests, and presenting results) are outlined in some detail here, as they are sometimes controversial in the analysis and presentation of color data. Particular caution is suggested when applying Bonferroni corrections to P-values when many tests have been performed, as this correction is sometimes erroneously applied and can often obscure interesting trends.

References

Alatalo, R. V., J. Höglund, and A. Lundberg. 1988. Patterns of variation in tail ornament size in birds. Biol J Linn Soc 34: 363–374.

Alonso-Alvarez, C., S. Bertrand, G. Devevey, M. Gaillard, J. Prost, B. Faivre, and G. Sorci. 2004. An experimental test of the dose-dependent effect of carotenoids and immune activation on sexual signals and antioxidant activity. Am Nat 164: 651–659.

Andersson, S. 1999. Morphology of UV reflectance in a whistling-thrush: Implications for the study of structural colour signalling in birds. J Avian Biol 30: 193–204.

Andersson, S., J. Örnborg, and M. Andersson. 1998. Ultraviolet sexual dimorphism and assortative mating in Blue Tits. Proc R Soc Lond B 265: 445–450.

Andersson, S., S. R. Pryke, J. Örnborg, M. J. Lawes, and M. Andersson. 2002. Multiple receivers, multiple ornaments, and a trade-off between agonistic and epigamic signaling in a Widowbird. Am Nat 160: 683–691.

Baker, R. R., and G. A. Parker. 1979. The evolution of bird coloration. Phil Trans R Soc Lond B 287: 63–130.

Bennett, A. T. D., I. C. Cuthill, J. C. Partridge, and K. Lunau. 1997. Ultraviolet plumage colors predict mate preferences in starlings. Proc Natl Acad Sci USA 94: 8618–8621.

Blanco, G., and M. Bertellotti. 2002. Differential predation by mammals and birds: Implications for egg-colour polymorphism in a nomadic breeding seabird. Biol J Linn Soc 75: 137–146.

Boff, K. R., L. Kaufman, and J. P. Thomas, ed. 1986. Handbook of Perception and Human Performance. Volume I: Sensory Processes and Perception. New York: John Wiley and Sons.

Bortolotti, G. R., J. E. Smits, and D. M. Bird. 2003. Iris colour of American Kestrels varies with age, sex, and exposure to PCBs. Physiol Biochem Zool 76: 99–104.

Bowmaker, J. K., L. A. Heath, S. E. Wilkie, and D. M. Hunt. 1997. Visual pigments and oil droplets from six classes of photoreceptor in the retinas of birds. Vision Res 37: 2183–2194.

Burley, N., and C. B. Coopersmith. 1987. Bill color preferences of Zebra Finches. Ethology 76: 133–151.

Butcher, G. S., and S. Rohwer. 1988. The evolution of conspicuous and distinctive coloration for communication in birds. Curr Ornithol 6: 51–108.

Cabin, R. J., and R. J. Mitchell. 2000. To Bonferroni or not to Bonferroni: When and how are the questions. ESA Bull 81: 246–248.

Cherry, M. I., and A. T. D. Bennett. 2001. Egg colour matching in an African Cuckoo, as revealed by ultraviolet-visible reflectance spectrophotometry. Proc R Soc Lond B 268: 565–571.

Colegrave, N., and G. D. Ruxton. 2003. Confidence intervals are a more useful complement to nonsignificant tests than are power calculations. Behav Ecol 14: 446–447.

Cuthill, I. C., A. T. D. Bennett, J. C. Partridge, and E. J. Maier. 1999. Plumage reflectance and the objective assessment of avian sexual dichromatism. Am Nat 160: 183–200.

Dale, J. 2000. Ornamental plumage does not signal male quality in Red-billed Queleas. Proc R Soc Lond B 267: 2143–2149.

Delhey, K., A. Johnsen, A. Peters, S. Andersson, and B. Kempenaers. 2003. Paternity analysis reveals opposing selection pressures on crown coloration in the Blue Tit (*Parus caeruleus*). Proc R Soc Lond B 270: 2057–2063.

Doucet, S. M., and R. Montgomerie. 2003. Multiple sexual ornaments in Satin Bowerbirds: Ultraviolet plumage and bowers signal different aspects of male quality. Behav Ecol 14: 503–509.

Doucet, S. M., M. Shawkey, M. Rathburn, H. Mays, and R. Montgomerie. 2004. Concordant evolution of plumage colour, feather microstructure, and a melanocortin receptor gene between mainland and island populations of a fairy-wren. Proc R Soc Lond B 271: 1663–1670.

Doucet, S. M., D. J. Mennill, R. Montgomerie, P. T. Boag, and L. M. Ratcliffe. 2005. Achromatic plumage reflectance predicts reproductive success in male Black-capped Chickadees. Behav Ecol 16: 218–222.

Eaton, M. D. 2005. Human vision fails to distinguish widespread sexual dichromatism among sexually "monochromatic" birds. Proc Natl Acad Sci USA 102: 10942–10946.

Eaton, M. D., and S. M. Lanyon. 2003. The ubiquity of avian ultraviolet plumage reflectance. Proc R Soc Lond B 270: 1721–1726.

Eiseman, L., and L. Herbert. 1990. The Pantone Book of Color: Color Basics & Guidelines for Design, Fashion, Furnishings . . . and More. New York: Harry N. Abrams.

Endler, J. A. 1990. On the measurement and classification of colour in studies of animal colour patterns. Biol J Linn Soc 41: 315–352.

Endler, J. A. 1993. The color of light in forests and its implications. Ecol Monogr 63: 1–27.

Endler, J. A., and M. Théry. 1996. Interacting effects of lek placement, display behavior, ambient light, and color patterns in three neotropical forest-dwelling birds. Am Nat 148: 421–452.

Fitze, P. S., and H. Richner. 2002. Differential effects of a parasite on ornamental structures based on melanins and carotenoids. Behav Ecol 13: 401–407.

Flood, N. J. 1989. Coloration in New World orioles. 1. Tests of predation-related hypotheses. Behav Ecol Sociobiol 25: 49–56.

Goldsmith, T. H. 1990. Optimization, constraint, and history in the evolution of eyes. Q Rev Biol 65: 281–322.

Greene, E., B. E. Lyon, V. R. Muehter, L. Ratcliffe, S. J. Oliver, and P. Boag. 2000. Disruptive sexual selection for plumage coloration in a passerine bird. Nature 407: 1000–1003.

Grill, C. P., and V. N. Rush. 2000. Analysing spectral data: Comparison and application of two techniques. Biol J Linn Soc 69: 121–138.

Grimm, L. G., and P. Y. Yarnold, ed. 1995. Reading and Understanding Multivariate Statistics. Volume 1. Washington, DC: American Psychological Association.

Hamilton, W. D., and M. Zuk. 1982. Heritable true fitness and bright birds: A role for parasites? Science 218: 384–387.

Hart, N. S. 2001. The visual ecology of avian photoreceptors. Progr Ret Eye Res 20: 675–703.

Hart, N. S. 2004. Microspectrophotometry of visual pigments and oil droplets in a marine bird, the Wedge-tailed Shearwater *Puffinus pacificus:* Topographic variations in photoreceptor spectral characteristics. J Exp Biol 207: 1229–1240.

Hart, N. S., J. C. Partridge, and I. C. Cuthill. 1998a. Visual pigments, oil droplets and cone photoreceptor distribution in the European Starling (*Sturnus vulgaris*). J Exp Biol 201: 1433–1446.

Hart, N. S., J. C. Partridge, and I. C. Cuthill. 1998b. Visual pigments, ocular media and predicted spectral sensitivity in the Domestic Turkey (*Meleagris gallopavo*). Vision Res 39: 3321–3328.

Hart, N. S., J. C. Partridge, I. C. Cuthill, and A. T. D. Bennett. 2000a. Visual pigments, oil droplets, ocular media and cone receptor distribution in two species of passerine bird: The Blue Tit (*Parus caeruleus* L.) and the Blackbird (*Turdus merula* L.). J Comp Physiol A 186: 375–387.

Hart, N. S., J. C. Partridge, A. T. D. Bennett, and I. C. Cuthill. 2000b. Visual pigments, cone oil droplets and ocular media in four species of estrildid finch. J Comp Physiol A 186: 681–694.

Heindl, M., and H. Winkler. 2003. Interacting effects of ambient light and plumage color patterns in displaying Wire-tailed Manakins (Aves, Pipridae). Behav Ecol Sociobiol 53: 153–162.

Hill, G. E. 1990. Female house finches prefer colorful males: Sexual selection for a condition-dependent trait. Anim Behav 40: 563–572.

Hill, G. E. 1998. An easy, inexpensive means to quantify plumage colouration. J Field Ornithol 69: 353–363.

Hill, G. E. 2002. A Red Bird in a Brown Bag: The Function and Evolution of Colorful Plumage in the House Finch. Oxford: Oxford University Press.

Hunt, S., A. T. D. Bennett, I. C. Cuthill, and R. Griffiths. 1998. Blue Tits are ultraviolet tits. Proc R Soc Lond B 265: 451–455.

Jawor, J. M., and R. Breitwisch. 2004. Multiple ornaments in male Northern Cardinals, *Cardinalis cardinalis,* as indicators of condition. Ethology 110: 113–126.

Jawor, J. M., S. U. Linville, S. M. Beall, and R. Breitwisch. 2003. Assortative mating by multiple ornaments in Northern Cardinals (*Cardinalis cardinalis*). Behav Ecol 14: 515–520.

Jawor, J. M., N. Gray, S. M. Beall, and R. Breitwisch. 2004. Multiple ornaments correlate with aspects of condition and behaviour in female Northern Cardinals, *Cardinalis cardinalis*. Anim Behav 67: 875–882.

Kandel, E. R., J. H. Schwartz, and T. M. Jessell. 2001. Principles of Neural Science, fourth edition. New York: McGraw-Hill.

Keyser, A. J., and G. E. Hill. 1999. Condition-dependent variation in the blue-ultraviolet coloration of a structurally based plumage ornament. Proc R Soc Lond B 266: 771–777.

Keyser, A. J., and G. E. Hill. 2000. Structurally based plumage coloration is an honest signal of quality in male blue grosbeaks. Behav Ecol 11: 202–209.

Kilner, R. 1997. Mouth colour is a reliable signal of need in begging canary nestlings. Proc R Soc Lond B 264: 963–968.

Komdeur, J., M. Oorebeek, T. van Overveld, and I. C. Cuthill. 2005. Mutual ornamentation, age, and reproductive performance in the European starling. Behav Ecol 16: 805–817.

Kornerup, A., and J. H. Wanscher. 1983. Methuen Handbook of Colour. London: Methuen.

Langmore, N. E., and A. T. D. Bennett. 1999. Strategic concealment of sexual identity in an estrildid finch. Proc R Soc Lond B 266: 543–550.

Linville, S. U., and R. Breitwisch. 1997. Carotenoid availability and plumage coloration in a wild population of Northern Cardinals. Auk 114: 796–800.

Linville, S. U., R. Breitwisch, and A. J. Schilling. 1998. Plumage brightness as an indicator of parental care in northern cardinals. Anim Behav 55: 119–127.

Lythgoe, J. N. 1979. The Ecology of Vision. Oxford: Oxford University Press.

MacDougall, A. K., and R. Montgomerie. 2003. Assortative mating by carotenoid-based plumage colour: A quality indicator in American Goldfinches, *Carduelis tristis*. Naturwissenschaften 90: 464–467.

Maddocks, S. A., A. T. D. Bennett, S. Hunt, and I. C. Cuthill. 2001. Context-dependent visual preferences in starlings and Blue Tits: Mate choice and light environment. Anim Behav 63: 69–75.

Mays, Jr., H. L., K. J. McGraw, G. Ritchison, S. Cooper, V. Rush, and R. S. Parker. 2004. Sexual dichromatism in the Yellow-breasted Chat *Icteria virens*: Spectrophotometric analysis and biochemical basis. J Avian Biol 35: 125–134.

McCamy, C. S. 1993. The primary hue circle. Color Res Application 18: 3–10.

McGraw, K. J. 2004. Multiple UV reflectance peaks in the iridescent neck feathers of pigeons. Naturwissenschaften 91: 125–129.

McGraw, K. J., and D. R. Ardia. 2003. Carotenoids, immunocompetence, and the information content of sexual colors: an experimental test. Am Nat 162: 704–712.

McGraw, K. J., and A. J. Gregory. 2004. Carotenoid pigments in male American Goldfinches: What is the optimal biochemical strategy for becoming colourful? Biol J Linn Soc 83: 273–280.

McGraw, K. J., and G. E. Hill. 2000. Differential effects of endoparasitism on the expression of carotenoid- and melanin-based ornamental coloration. Proc R Soc Lond B 267: 1525–1531.

McGraw, K. J., E. A. Mackillop, J. Dale, and M. E. Hauber. 2002. Different colors reveal different information: How nutritional stress affects the expression of melanin- and structurally based ornamental plumage. J Exp Biol 205: 3747–3755.

McGraw, K. J., A. J. Gregory, R. S. Parker, and E. Adkins-Regan. 2003. Diet, plasma carotenoids, and sexual coloration in the Zebra Finch (*Taeniopygia guttata*). Auk 120: 400–410.

Mennill, D. J., S. M. Doucet, R. Montgomerie, and L. M. Ratcliffe. 2003. Achromatic color variation in Black-capped Chickadees, *Poecile atricapilla:* Black and white signals of sex and rank. Behav Ecol Sociobiol 53: 350–357.

Moran, M. D. 2003. Arguments for rejecting the sequential Bonferroni in ecological studies. Oikos 100: 403–405.

Nakagawa, S. 2004. A farewell to Bonferroni: The problems of low statistical power and publication bias. Behav Ecol 15: 1044–1045.

Ödeen, A., and O. Håstad. 2003. Complex distribution of avian color vision systems revealed by sequencing the SWS1 Opsin from total DNA. Molec Biol Evol 20: 855–861.

Örnborg, J., S. Andersson, S. C. Griffith, and B. C. Sheldon. 2002. Seasonal changes in a ultraviolet structural colour signal in Blue Tits, *Parus caeruleus.* Biol J Linn Soc 76: 237–245.

Osorio, D., M. Vorobyev, and C. D. Jones. 1999. Colour vision of domestic chicks. J Exp Biol 202: 2951–2959.

Ostwald, W. 1916. Die Farbenfibel. Leipzig: Unesma.

Palmer, R. S. 1962. Handbook of North American Birds. Volume 1. New Haven, CT: Yale University Press.

Parkkinen, J. P. S., J. J. Hallikainen, and T. Jaaskelainen. 1989. Characteristic spectra of Munsell colors. J Opt Soc Am 6: 318–322.

Peiponen, V. A. 1992. Colour discrimination of two passerine bird species in the Munsell system. Ornis Scand 23: 143–151.

Peters, A., A. G. Denk, K. Delhey, and B. Kempenaers. 2004. Carotenoid-based bill colour as an indicator of immunocompetence and sperm performance in male mallards. J Evol Biol 17: 1111–1120.

Promislow, D. E. L., R. Montgomerie, and T. Martin. 1992. Mortality costs of sexual dimorphism in birds. Proc R Soc Lond B 250: 143–150.

Pryke, S. R., M. J. Lawes, and S. Andersson. 2001. Agonistic carotenoid signalling in male Red-collared Widowbirds: Aggression related to the colour signal of both the territory owner and model intruder. Anim Behav 62: 695–704

Quinn, G., and M. Keough. 2002. Experimental Design and Data Analysis for Biologists. Cambridge, UK: Cambridge University Press.

Rice, W. R. 1989. Analyzing tables of statistical tests. Evolution 43: 223–225.

Ridgway, R. A. 1912. Color Standards and Color Nomenclature. Washington, DC: R. A. Ridgway.

Safran, R. J., and K. J. McGraw. 2004. Plumage coloration, not length or symmetry of tail-streamers, is a sexually selected trait in North American Barn Swallows. Behav Ecol 15: 455–461.

Saks, L., K. McGraw, and P. Hörak. 2003. How feather colour reflects its carotenoid content. Funct Ecol 17: 555–561.

Shawkey, M. D., A. M. Estes, L. M. Siefferman, and G. E. Hill. 2003. Nanostructure predicts intraspecific variation in ultraviolet-blue plumage colour. Proc R Soc Lond B 270: 1455–1460.

Sheldon, B. C., S. Andersson, S. C. Griffith, J. Ornborg, and J. Sendecka. 1999. Ultraviolet colour variation influences Blue Tit sex ratios. Nature 402: 874–877.

Siefferman, L., and G. E. Hill. 2005. UV-blue structural coloration and competition for nestboxes in male Eastern Bluebirds. Anim Behav 69: 67–72.

Smiseth, P. T., J. Örnborg, S. Andersson, and T. Amundsen. 2001. Is male plumage reflectance correlated with paternal care in Bluethroats? Behav Ecol 12: 164–170.

Smithe, F. B. 1974. Naturalist's Color Guide Supplement. New York: American Museum of Natural History.

Smithe, F. B. 1975. Naturalist's Color Guide. New York: American Museum of Natural History.

Smithe, F. B. 1981. Naturalist's Color Guide. Part III. New York: American Museum of Natural History.

Sokal, R. R., and F. J. Rohlf. 1995. Biometry: The Principles and Practice of Statistics in Biological Research, third edition. New York: W. H. Freeman and Co.

Stoehr, A. M. 1999. Are significance thresholds appropriate for the study of animal behaviour? Anim Behav 57: F22–F25.

Swaddle, J. P. 1999. Limits to length asymmetry detection in starlings: Implications for biological signalling. Proc R Soc Lond B 266: 1299–1303.

Tabachnick, B. G., and L. S. Fidell. 2001. Using Multivariate Statistics, fourth edition. Boston: Allyn and Bacon.

Tschirren, B., P. S. Fitze, and H. Richner. 2003. Proximate mechanisms of variation in the carotenoid-based plumage coloration of nestling Great Tits (*Parus major* L.). J Evol Biol 16: 91–100.

Tukey, J. 1977. Exploratory Data Analysis. Reading, MA: Addison-Wesley.

Uy, J. A. C., and J. A. Endler. 2004. Modification of the visual background increases the conspicuousness of Golden-collared Manakin displays. Behav Ecol 15: 1003–1010.

van Valen, L. 1962. A study of fluctuating asymmetry. Evolution 16: 125–142.

Villafuerte, R., and J. J. Negro. 1998. Digital imaging for colour measurement in ecological research. Ecol Lett 1: 151–154.

Villalobos-Dominguez, C., and J. Villalobos. 1947. Atlas de los Colores. Buenos Aires: Libreria El Ateneo Editorial.

Vorobyev, M., and D. Osorio. 1998. Receptor noise as a determinant of colour thresholds. Proc R Soc Lond B 265: 351–358.

Vorobyev, M., D. Osorio, A. T. D. Bennett, N. J. Marshall, and I. C. Cuthill. 1998. Tetrachromacy, oil droplets and bird plumage colours. J Comp Physiol A 183: 621–633.

Wiebe, K. L., and G. R. Bortolotti. 2001. Variation in colour within a population of Northern Flickers: A new perspective on an old hybrid zone. Can J Zool 79: 1046–1052.

Wiebe, K. L., and G. R. Bortolotti. 2002. Variation in carotenoid-based color in Northern Flickers in a hybrid zone. Wilson Bull 114: 393–400.

Wilkinson, L., and Task Force on Statistical Inference. 1999. Statistical methods in psychology journals: Guidelines and explanations. Am Psychol 54: 594–604.

Wolfenbarger, L. L. 1999a. Red coloration of male Northern Cardinals correlates with mate quality and territory quality. Behav Ecol 10: 80–90.

Wolfenbarger, L. L. 1999b. Is red coloration of male Northern Cardinals beneficial during the nonbreeding season?: A test of status signaling. Condor 101: 655–663.

Wolfenbarger, L. L. 1999c. Female mate choice in Northern Cardinals: Is there a preference for redder males? Wilson Bull 111: 76–83.

Wyszecki, G., and W. S. Stiles, ed. 1982. Color Science: Concepts and Methods, Quantitative Data and Formulas. New York: John Wiley and Sons.

Yam, K. L., and S. E. Papadakis. 2004. A simple digital imaging method for measuring and analyzing color of food surfaces. J Food Eng 61: 137–142.

Zar, J. H. 1999. Biostatistical Analysis, fourth edition. Upper Saddle River, NJ: Prentice-Hall.

4

Effects of Light Environment on Color Communication

MARC THÉRY

A precise knowledge of light environments and visual systems is required to understand the evolution of color signals. Color displays largely depend on the type, intensity, and directionality of ambient light. As for actors on the stage of a theater, for birds in the wild, only particular light environments can fully reveal the elaborate character of their coloration. In this chapter, I focus on the nature and effects of light environments on birds. I show that the primary manner in which environmental light has driven the evolution of avian color displays is by constraining both the conspicuousness of colors used as intraspecific signals and the design of cryptic coloration meant to prevent detection by predators. Consequently, because environmental light is heterogeneously distributed in time and space, sensory drive could lead to species divergence through habitat selection.

Bird Communication and Light Environments

Why Does Ambient Light Affect Visual Communication?

In the late 1970s, Hailman (1977) and Lythgoe (1979) published the first syntheses of visual communication in animals. Among several other important principles of animal communication, Hailman (1977:247) wrote that "the major variables affecting conspicuousness are the light available for reflection by the sender, the optical nature of the background against which the sender is viewed, and the types of optical disruption the signal suffers during transmission." He presented spectral measurements of ambient light in tropical forests and considered the influence of visual systems and viewing angle on signal perception. Lythgoe (1979) thoroughly discussed the effects of light environments in both terrestrial and aquatic ecosystems.

In the early 1990s, Endler published a techniques paper on how to measure and analyze animal colors (Endler 1990; Box 4.1). He showed that the perception of an element of a color pattern depends on (1) the physics of light reflectance and transmission to the eye of the viewer; (2) the physics of light transmission, refraction, and photoreception within the eye; and (3) the neural processes in the retina and brain that interpret the perceived image. Ambient light affects the perception of a color pattern, independently of the color vision of the viewer, because the image of the color pattern reaching the eye is the product of the reflectance spectrum of patches in the pattern and the spectrum of ambient light (Figure 4.1). Many of the effects of ambient light on perception were also described by Endler (1978) and Lythgoe (1979).

Ambient light affects the perception of a particular color pattern against the visual background because the color patterns of both the object and background may change in different ways when exposed to different ambient light spectra (Endler 1991, 1993; Endler and Théry 1996). Therefore some color patterns may be highly conspicuous in one light environment and cryptic in another light environment. The most conspicuous colors are those that are rich in the colors of ambient light but poor in the colors reflected by the visual background (Endler 1990, 1992, 1993).

Five Forest Light Environments

Light intensity is strongly reduced by vegetation. Direct sunlight is rapidly reduced by the upper canopy layer and then progressively decreases further in the understory (reviewed in Théry 2001). Mean light intensities of 0.1–4.2% of direct sunlight have been measured at ground level in tropical forests, where sun flecks provide the most intense light environment (Chazdon and Fetcher 1984; Bongers et al. 2001). Forest light is also greatly altered by the angle of the sun and by cloud cover. The result is five distinct types of forest light environments (Endler 1993):

1. *Early/Late.* At dawn and dusk, sun angles of less than 10° above the horizon induce longer trajectories of light through the atmospheric ozone layer,

Box 4.1. Measuring Light Environments

Endler (1990) presented the most comprehensive methods for how to measure ambient light spectra. Because the appearance of any object depends upon the spectra of ambient light, transmission of light along the light path, and veiling of light during its transmission, light environments should ideally be measured where and when the animal views the received signal. Different components of ambient light should be measured with a spectrometer sensitive to UV and visible wavelengths:

Irradiance is measured over a 180° solid angle (2π steradians) with a special sensor called a "cosine receptor," which takes into account the decrease in the intensity of light reaching a plane with the cosine of the incident angle. Ambient light measurements should always represent the number of photons received by the eye of the viewer, because animal (and plant) photoreceptors respond to the number of photons rather than to their energy, and energy is related to wavelength. Consequently, spectra in watts (energy) overestimate the visual effects of light at short wavelengths and underestimate it at long wavelengths. Photographic light meters are even worse. To convert a watt spectrum $W(\lambda)$ to quanta, $Q(\lambda) = 0.0083519 \cdot \lambda \cdot W(\lambda)$ gives the correct irradiance spectrum in $\mu mol/m^2/s/nm$.

Radiance, reflectance, and transmission of light are measured with a radiance receptor that usually has a narrow angle of acceptance (usually <1°) and thus collects light in the form of a narrow beam of either ambient light reflected by an object or background (radiance measured in $\mu mol/m^2/s/nm/str$), reference light reflected by the measured surface (reflectance; no unit), or ambient or reference light transmitted through a surface (transmission; no unit, but distance specific).

Reflectance is measured as the ratio:

$$\frac{\text{Radiance from object}}{\text{Irradiance on object}} = \frac{\text{Radiance from object}}{\text{Radiance from standard}}.$$

Transmission is measured as the ratio:

$$\frac{\text{Radiance output at } x}{\text{Radiance input at source}}.$$

> Absorption is measured as:
>
> $$1 - \frac{\text{Radiance output}}{\text{Irradiance input}} \cong 1 - \text{reflectance}.$$

which absorbs middle wavelengths, particularly around 604 nm. Therefore, early and late in the day, ambient light is deficient in middle wavelengths, giving an apparent purple hue. This purple light may diminish when yellowish and reddish light is reflected by clouds (Endler 1993; Figure 4.2).

2. *Forest Shade*. Ambient light under a forest canopy consists of direct sunlight filtered and reflected by vegetation. It shows a broad peak in the green around 550 nm, and a steady increase above 680 nm, appearing yellow-green to human eyes. Typical forests have a mostly continuous canopy with few canopy gaps, whereas in woodlands, the crowns of trees do not touch, creating more canopy gaps and leading to a higher proportion of sky visible from the floor. *Forest Shade* is thus more common in forest with closed or dense canopies.

3. *Woodland Shade*. Because more blue sky is visible in woodlands, the *Woodland Shade* light environment is bluish when the open sky is blue. When blue sky is obscured by clouds, fog, haze, smoke, or air pollution, the light environment may be bluish-gray to gray. *Woodland Shade* is more common in forest with more open canopies, such as pine, eucalyptus, or dry forests. As more blue skylight is visible with increasing height in forest, a progressive transition from greenish *Forest Shade* to bluish *Woodland Shade* occurs from understory to canopy (Endler 1993).

4. *Small Gaps*. This light environment results from small sun flecks, which are rich in longer wavelengths and appear yellow-orange. *Small Gaps* are most common in forests with closed canopy and few holes caused by branch falls or near regenerated treefalls.

5. *Open/Cloudy*. Ambient light in open areas and large gaps (light passing though holes in the canopy that subtend a solid angle of less than 2° from the vertical [e.g., a 1-m-diameter patch for a 30-m-high canopy; Endler 1993]) is relatively evenly distributed along all wavelengths, except wavelengths below about 470 nm, and appears whitish. The spectrum of large gaps is very similar to that of open areas, except for a slight attenuation of shorter wavelengths.

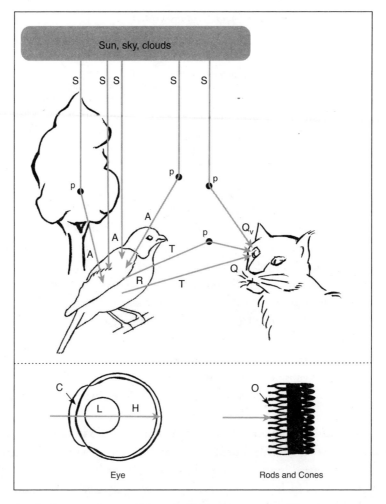

Figure 4.1. In the upper diagram, light paths (thick gray arrows) reaching the viewer's eye. S = sunlight, direct or diffuse through clouds. A = ambient light striking a color patch on the bird, integrated over a 180° solid angle. Ambient light may come from light reflected from other objects (tree), through objects (leaves), directly from the sky, and reflected from water or dust particles (p, black dots). R = fraction of photon flux reflected from the patch. T = fraction of photons transmitted to the predator's eyes, some of which may be absorbed or scattered by particles (p) in the air. Q = light beam striking the predator's eyes. Q_v = light striking the predator's eyes after being scattered by water or dust particles (p). In the lower diagram, light striking predator's eye is filtered by cornea (C), aqueous filling the anterior chamber, lens (L), and humors (H), before passing through the layers of cells comprising the retina, being filtered by oil droplets (O) and finally absorbed by the rods and cones, the receptor cells that initiate vision. The effects of veiling light, which can be absorbed or scattered in the air by water or dust particles, can often be neglected under normal conditions. However, those effects are important in terrestrial environments for vultures and other raptors viewing objects from great distances, and other species that regularly experience fog (e.g., shorebirds) or dust (desert birds). Adapted from Hailman (1977), Lythgoe (1979) and Endler (1990).

Effects of Light Environment on Color Communication

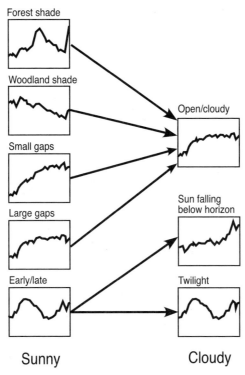

Figure 4.2. Forest light environments in both sunny and cloudy weather. Graphs show irradiance (*y*-axes) as a function of wavelength (*x*-axes) in the human-visible range (400–700 nm). Redrawn from Endler (1993).

When the weather is cloudy, colors of *Forest Shade*, *Woodland Shade*, and *Small Gaps* light environments converge on the whitish spectrum of large gaps and open areas.

As a consequence of heterogeneous light distribution, ambient light and visual backgrounds show considerable variation among forest habitats. Forest canopy has large holes that lead to white *Large Gaps* and blue *Woodland Shade*, whereas the understory has small holes that lead to yellow-red *Small Gaps* and green *Forest Shade* (Endler 1993, 1997; Gomez and Théry 2004). This heterogeneous distribution of light habitats in the canopy and understory is indeed used by foraging birds (Walther 2002). In forest canopies, the background is green and poor in ultraviolet (UV) wavelengths, and ambient light is rich in blue and UV. By contrast, the understory light environment is set against brown or gray backgrounds poor in UV, and ambient light is green with little UV

Box 4.2. Tests of Predictions about Bird Coloration on the Basis of Light Environments, and Their Outcomes

Tested hypothesis	Outcome
Cryptic color pattern is a random sample of the background seen by predators	Confirmed by Endler and Théry (1996) in females and juveniles; opposite for adult males (Gomez and Théry 2004)
Quality of background matching is better at low than at high predation intensities	Confirmed by Endler and Théry (1996) at different times and places in the same species
Conspicuous patterns with adjacent patches that have complementary radiance spectra stimulate combinations of cones in opposite ways	Confirmed by Endler and Théry (1996) for cryptic patterns of females and juveniles, and for conspicuous patterns of males during display, and by Heindl and Winkler (2003a,b) for conspicuous males
Mean brightness or color contrast of cryptic plumage pattern is similar to brightness or contrast of background	Confirmed by Endler and Théry (1996), Gomez and Théry (2004), Heindl and Winkler (2003a,b)
Contrast among patches changes with ambient light, affecting contrast within and between birds and their backgrounds	Confirmed by Endler and Théry (1996) and Heindl and Winkler (2003a,b)
A greater number of color patches and brighter coloration enhance visual display in dark and closed habitats	Confirmed by Marchetti (1993), but not by McNaught and Owens (2002) nor by Gomez and Théry (2004)
Brighter plumage in open habitats maximizes contrast over long distances.	Confirmed by McNaught and Owens (2002)
Darker plumage in open habitat increases contrast against bright backgrounds	Confirmed by Gomez and Théry (2004) in males but not females
UV signals are more prevalent in the canopy for both crypsis and conspicuousness	Confirmed by Gomez and Théry (2004)
Red and orange is most common in *Forest Shade*	Confirmed by Endler and Théry (1996), McNaught and Owens (2002), Gomez and Théry (2004), Heindl and Winkler (2003a)
Blue is most common in *Woodland Shade*	Confirmed by Gomez and Théry (2004) in canopy.
Yellow, orange or red are most common in *Small Gaps*	Confirmed by Endler and Théry (1996), Heindl and Winkler (2003b)
There is a greater diversity of colors in canopy than forest floor	Confirmed by Gomez and Théry (2004)

and with a few small yellow-red patches. The intensity of *Forest Shade* has been shown to vary among forest types in Neotropical rainforest, whereas the color (hue angle) of *Forest Shade* is comparable among study sites for undisturbed terra firme reference plots (Bongers et al. 2001).

Because avian vision of polarized light remains controversial in the research community (see review by Hunt et al. 2001), I will not detail variations of polarized light in terrestrial ecosystems (but see, e.g., Shashar and Cronin 1998 for its natural variation in tropical forest).

Light Environments, Visual Communication, and Optimal Signals

The illumination of both color pattern and visual background will greatly affect the degree of bird conspicuousness. Using our knowledge of background colors and ambient lighting in forests, specific predictions can be made for optimal color signals to be used for camouflage or display in different forest habitats (Endler 1978, 1990, 1992, 1993; Marchetti 1993; Endler and Théry 1996; Zahavi and Zahavi 1997; Théry 2001; Gomez and Théry 2004; Box 4.2).

For example, signaling for conspicuousness at short distances should cause birds that dwell in *Forest Shade* under continuous canopy to use red or orange signals against the green background. On the contrary, blue and blue-green signals would be most conspicuous in *Woodland Shade*. Species that inhabit *Small Gaps* should use red, orange, or yellow hues for increased conspicuousness, and to be conspicuous in the *Early/Late* light environment species should use blue, red, or purple signals. Conspicuousness should also select for color patterns strongly reflecting UV, both in the canopy and in the understory. In addition, because there is more UV contrast between light and vegetation in the canopy, the use of UV signals may be more prevalent for both crypsis and conspicuousness in canopy species (Gomez and Théry 2004). Reverse strategies are expected for crypsis.

Another strategy to increase conspicuousness is to create strong color contrast between elements of the color pattern (Hailman 1977; Endler 1993; Endler and Théry 1996; Gomez and Théry 2004). For instance, to be most conspicuous, red signals in *Forest Shade* should also have a small amount of blue or blue-green in their pattern. Conversely, in *Woodland Shade* blue or blue-green signals will be more conspicuous if they include small red or orange patches. Purple or blue should be used to increase color contrast within patterns in *Small Gaps*, and yellow, yellow-green, or green should be used under *Early/Late* conditions. There is no prediction for crypsis or conspicuousness in *Open/Cloudy*, because whitish ambient light will not affect the radiance spectra of color signals.

Because forest canopy is a mosaic of *Woodland Shade* and *Large Gaps*, whereas the understory is a mosaic of *Forest Shade* and *Small Gaps*, more blue (for conspicuousness) or green (for crypsis) is expected in the plumages of birds that live in the canopy, and more orange or brown is expected in the understory. Because of the broader range of colors in the canopy compared with the understory, more diverse colors are expected in the canopy.

Opposite predictions have been made with respect to achromatic brightness of light environments. Marchetti (1993) hypothesized that brighter coloration would enhance visual displays in dark and closed habitats. On the contrary, both Zahavi and Zahavi (1997) and Gomez and Théry (2004) expected brighter plumages in open habitats, to maximize contrast over long distances and to enhance crypsis against the bright background.

Some examples of variation of the radiance of color signals exposed to typical forest light environments have been given by Endler (1997). An orange-red color patch reflecting 29% under white light will reflect 36% more under yellow-red light of *Small Gaps,* but 19% less under bluish light of *Woodland Shade* (Figure 4.3). On the contrary, a yellow-green patch also reflecting 29% of white light would reflect 6% less in *Small Gaps,* and 24% less in *Woodland Shade.*

Using virtual color signals, Endler (1997) showed that colorimetric variables were differently affected by ambient light environments. For this study, he used three patches with the same blue-violet hue but with different chromas (saturation of color; Figure 4.4). He found that patches with high chroma were relatively insensitive to varying light environments, whereas patches with low chroma were affected more by changing light environments. The hue of color patches can change dramatically between light environments, especially patches with low chroma. Therefore, to be conspicuous, a species must either evolve low-chroma colors for specific light environments or high-chroma colors that retain their contrast and hue independent of light environment.

Light environments affect the crypsis and conspicuousness of a bird and consequently communication with predators, prey, conspecifics, and other animals. They also influence the efficiency of communication between birds and plants, which are usually mediated by visual signals of flowers or fruit.

From Description to Spectral Analysis of Light Environments

The study of light environments progressed together with spectrometric techniques. Initially, biologists could only describe striking bird displays, noting that environmental light apparently influenced the location and timing of displays. With the development of portable light meters, pioneering studies con-

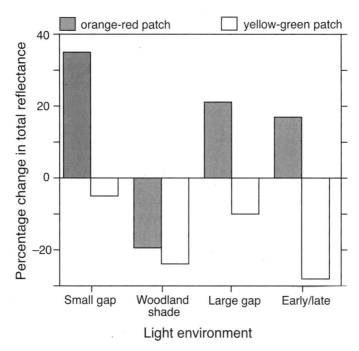

Figure 4.3. Variation in the total reflectance of ambient light by two color patches illuminated by four different light environments. Both orange-red (600–700 nm) and yellow-green (520–620 nm) patches reflect 29% of white light. The bars show the percentage change in total reflectance (350–700 nm) when the illuminated spectrum shifts from *Forest Shade* to the light environments shown here. Adapted from Endler (1997).

sidered the influence of light intensity on the location and quality of display sites and on the role of sensory drive in species divergence. As portable spectrometers became available, biologists were able to study more precisely the effects of environmental light spectra on plumage colors, timing of display, territory location, and display behavior. Beginning with tropical lekking birds, these studies have more recently been conducted at large interspecific scales to test the effects of sensory drive on the evolution of bird colors under forest cover, in the canopy, or in open habitats.

Early Descriptions of Natural Light Environments in Tropical Birds

Before biologists really began to study light environments, several ornithologists noted that tropical birds displaying at fixed locations were actively clearing their arenas of leaves or debris or were selecting display sites for optimal light

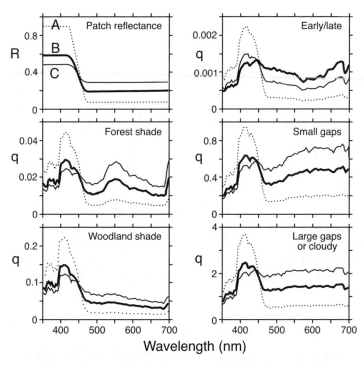

Figure 4.4. Reflectance spectra of three color patches (A, B, and C in patch reflectance graph) and their resulting radiance spectra when illuminated by each of the five light environments present in forests. Note how the shape (color) of the radiance spectrum of patch A changes little with ambient light, whereas that of C changes almost as much as the ambient light spectra. The flat reflectance spectrum of a gray patch would radiate almost the same color as that of ambient light. Redrawn from Endler (1997).

exposure. In 1940, Rand described the behavior of the Magnificent Bird of Paradise (*Diphyllodes magnificus*) clearing ground debris beneath display stages and ripping off leaves on and above them, opening a kind of skylight window into the forest where the male displays his mirrorlike ventral plumage by reflecting skylight in the direction of the female perched above him. About 20 years later, Gilliard (1959:8) noted that male Blue-backed Manakins (*Chiroxiphia pareola*) were clearing their dance perches by constant stripping and pecking, and he speculated that blue and red ornamental plumages "are probably enhanced by increased light resulting from the reduction of leaves surrounding the bower." He also suggested that the clearing of leaves or scratching of "ground arenas as in certain other manakins, birds of paradise and bowerbirds may be convergent examples of the development of ingenious defense mechanisms" (Gilliard 1959:9), preventing the close approach of unobserved predators. These initial hypotheses remained untested for years, but similar observations were made for several other species of manakins, cotingas, and cocks-of-the-rock, showing that males clear or select well-exposed display sites (see reviews in Théry 1990a; Théry and Vehrencamp 1995). A few biologists also mentioned that lek displays tend to occur during the tropical dry season, when leaves are shed, which allows more light to reach bird display sites (Gilliard 1959, 1962; Robbins 1983). Most interestingly, Gilliard (1962:44) observed on leks of the Guianan Cock-of-the-Rock (*Rupicola rupicola*) that leaf fall "permitted an unusual amount of light from the sky to penetrate to the forest floor". He was also the first to describe males positioning their spectacularly colored bodies in shafts of sunlight appearing on the lek (*Small Gap* light environment, Plate 3), strikingly changing their appearance compared to *Forest Shade* illumination. His report foreshadowed modern studies of light environment, when he wrote that "the cocks were relatively inactive in the duller parts of the day, and they tended to become excited and active when a shaft of sunlight penetrated to their leks, which suggest that light intensity on the floor of the forest may be important in the regulation of the reproductive cycles" (Gilliard 1962:61).

Early Studies Considering Light Environment

The first studies to consider light environments for their effects on the location and timing of bird color communication measured the brightness of plumage and/or light. Probably the first study to relate light intensity levels to display activity was that of Benalcazar and Silva de Benalcazar (1984), who showed in the Andean Cock-of-the-Rock (*Rupicola peruviana*) that morning display activity ends when the light intensity is similar to that at the beginning of the afternoon display session, suggesting a positive relationship between the bird's activity and light intensity.

In another study measuring light intensity on leks, Théry (1987, 1990a) looked at the influence of spatial and temporal patterns of illumination on the habitat selection and courtship performance of three species of manakins. Male displays were shown to coincide temporally with light intensity levels on their respective arenas. Males apparently use their courtship arenas only when these are properly illuminated and, in this way, enhance the brightness of their own color markings. Locations of courts, microstructure of ornamental feathers, and color patterns were also analyzed in relation to light environments. However, the implications in terms of color conspicuousness, female visitation rates, or predation risk were not considered.

In a study of *Phylloscopus* warblers (Plate 29, Volume 2), Marchetti (1993) found a negative correlation between an indirect measurement of light intensity (derived from camera shutter speed) and brightness of feather tips (measured by reflectance spectrometry), as well as a negative correlation between light intensity and the number of color patches. This correlation was interpreted as a response of color patterns to different light environments, leading to species divergence and speciation through sensory drive (but see Irwin et al. 2001). Elaborating on earlier studies, Théry and Vehrencamp (1995) examined the possible role of site differences, male attributes, and male dominance as cues for mate choice in the lekking White-throated Manakin (*Corapipo gutturalis*). Their results suggested that light properties of the display site and attendance of the male determine the attractiveness of male displays to females, whereas male interactions subsequently mediate copulatory success. Females clearly preferred to visit males displaying at sites with bright illumination that contrasted with a dark background and that enhanced black and white flashes from the male's white wing and throat patches (Théry 1990b). Indeed the shiny blue structural color of plumage on the males' upper parts is conspicuous only in direct sunlight, but not in *Forest Shade* (Endler and Théry 1996). Clearly an understanding of this signaling system required measurement of the reflectance spectra of color signals and visual background, and quantification of the irradiance spectra of ambient light using spectrometry.

Effect of Light Environments on Tropical Lekking Birds

The first studies of the interactions among habitat selection, display behavior, conspicuousness of displays, and light environments were conducted in the lekking Guianan Cock-of-the-Rock, White-throated Manakin, and White-fronted Manakin (*Lepidothrix serena;* Endler and Théry 1996; Théry and Endler 2001). Like all studies to date on light environments and color displays in birds, these studies focused on lekking species dwelling in tropical forests. In general, lek display sites are fixed for generations, and the relative stability of display conditions may have favored the evolution of elaborate strategies for crypsis and conspicuousness though the use of light environments on leks. It is also obvious that the fixed locations of bird display sites may have facilitated the study of light environments by evolutionary ecologists.

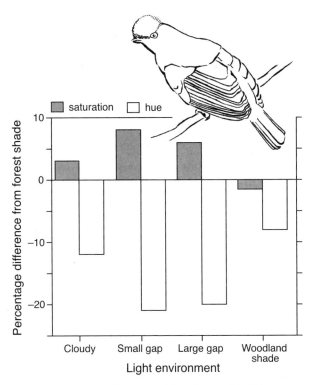

Figure 4.5. Variation in saturation (chroma) and hue (hue angle) of the orange plumage on the back and shoulder of adult male Guianan Cocks-of-the-Rock, using *Forest Shade* as a reference, between the light environments used during the display or on the lek but not displaying. Adapted from Endler and Théry (1996).

The Guianan Cock-of-the-Rock locates its leks in patches of forest that show different vegetation composition and spatial structure than the surrounding high forest (Théry and Larpin 1993). The lek vegetation is mostly composed of trees whose fruits are eaten by cocks, and whose seeds are selectively dispersed on lekking sites. This particular vegetation has a low canopy and thin leaves that produce more chromatic *Forest Shade* than in the surrounding forest, where birds forage. Males are mostly orange-red (including iris, beak and legs). Males stay in *Forest Shade* when not displaying, and displays are directly stimulated by the appearance of *Small Gaps* on or just above display perches. Males are often observed moving into sun flecks as they appear on their defended perches, which strikingly increases plumage reflectance (Figure 4.5; Plate 3).

When male cocks-of-the-rock are in *Forest Shade,* there is very little radiance from their orange plumage, because *Forest Shade* has very few photons with wavelengths above 650 nm, and feathers show little reflectance below that wavelength. In addition, plumage brightness is not significantly different from that of most background elements exposed to *Forest Shade.* Even though hue and chroma of plumage are different than the visual background, the bird is highly cryptic with respect to brightness contrast, which is the aspect of color that is used for long-distance detection. During displays, males expose parts of their bodies in *Small Gaps* (most often the yellow-orange strings, the rump, and the back), keeping the rest of the body in *Forest Shade.* Light from *Small Gaps* is very rich in the color of orange and yellow-orange ornamental feathers, creating high color contrast with the background.

The extraordinarily conspicuous display of males is caused by the juxtaposition of bright saturated orange on wing, rump, and back plumage in *Small Gaps,* with the rest of the plumage appearing relatively dark in *Forest Shade* (Plate 3). In addition, the plumages of males are far less visible in the *Forest Shade* on the lek sites than in *Forest Shade* of the surrounding higher forest, where they show significantly higher chroma. Therefore males can be highly cryptic in *Forest Shade,* particularly on the lek, and highly conspicuous when partly moving into sun flecks. Females, however, are highly cryptic because (1) their color spectra represent a random sample of the visual background colors, (2) the spectral properties of the color pattern are similar to those of background in all light environments, and (3) colors within their pattern contrast less than the colors in males patterns, making females more cryptic than males at short distances (Endler and Théry 1996).

I will not detail the different results obtained for the two manakin species by Endler and Théry (1996). In all three species studied for the effects of light environment on visual contrast, color, and/or brightness, contrast of plumage against the background is stronger at the time and places of courtship displays; it is reduced when birds are not displaying or when they move away from leks. Indeed breeding appears adapted to coincide with optimum light environment for display (dry season for the Guianan Cock-of-the-Rock and White-throated Manakin, wet season for the White-fronted Manakin). Each species maximizes its conspicuousness to conspecifics and minimizes its conspicuousness to predators by different combinations of display postures, plumage colors, lek placement, and light habitats. This adaptation affects the conspicuousness of presumed carotenoid colors of the Guianan Cock-of-the-Rock, melanin colors of the White-throated manakin, as well as structural colors of both White-throated and White-fronted Manakins (Endler and Théry 1996 and references therein).

A comparable study was conducted on the Wire-tailed Manakin (*Pipra filicauda*) by Heindl and Winkler (2003a). These vividly colored males have bright red crowns, yellow undersides, black upper parts, and a white wing band. Contrary to Guianan Cocks-of-the-Rock and White-throated Manakins that display in sun flecks and White-fronted manakins that display in *Forest Shade* when the sun is blocked by clouds (*Open/Cloudy*), Wire-tailed manakins mostly display in *Forest Shade* when the weather is sunny. This behavior allows them to simultaneously reduce their conspicuousness to potential predators at longer viewing distances, while at shorter distances, they maximize visual contrast within the plumage color pattern and increase the background contrast. As in the study by Endler and Théry (1996), in Heindl and Winkler (2003a), contrast within the plumage pattern is enhanced by the effects of different light environments on the visibility of each bird color. The Wire-tailed Manakin differs from species studied by Endler and Théry (1996) in that it manages to optimize simultaneously crypsis (at long distance) and conspicuousness (at short distance) while staying in the same light patch.

Heindl and Winkler (2003b) also studied vertical lek placement in relation to variations of ambient light in Wire-tailed, Golden-headed (*Pipra erythrocephala*), Blue-crowned (*Lepidothrix coronata*), and White-crowned (*Dixiphia pipra*) Manakins. Light environments were studied away from leks from a mobile canopy crane, which allowed three-dimensional measurements. They showed that each species placed its lek at the height at which light environments maximize chromatic contrast (used at short distance) and/or achromatic contrast (used at long distance) of the plumage signals against the background. For example, male Golden-headed Manakins displayed in the mid-story, where the contrast was increased between elements of the plumage pattern and where the chromatic contrast of the yellow crown was higher than in the understory (Figure 4.6; Plate 3). However, male Golden-headed Manakins that stay in *Forest Shade* are rather cryptic (lowest achromatic contrast of crown and within-plumage contrast).

Interestingly, Heindl and Winkler (2003b) found that, if differences in ambient light and background characteristics had played a dominant role in the evolution of color signals in those four species, then each species should look like the Wire-tailed Manakin, given that its red crown and yellow under

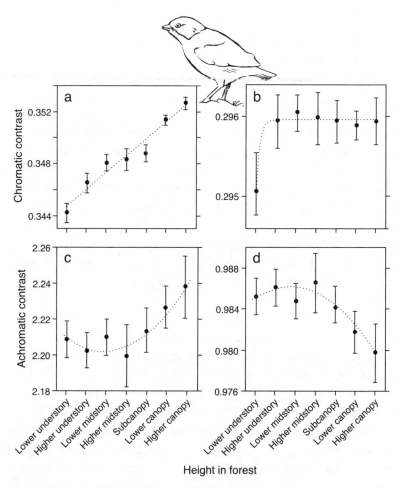

Figure 4.6. Chromatic (a,b) and achromatic (c,d) contrasts (mean ± standard error) of the yellow crown (a,c) and within the red and white thighs and black underparts (b,d) of Golden-headed Manakins at different heights in the forest. Forest environments are shown left to right from lowest to highest. Redrawn from Heindl and Winkler (2003b).

parts show highest chromatic and achromatic contrasts of all plumage colors irrespective of height. Clearly, other constraints are acting on the evolution of colors, like differences in predation rates, intensity of sexual selection, phylogenetic relationships, and lek organization. However, studies by Endler and Théry (1996) and Heindl and Winkler (2003a,b) consistently demonstrate that species evolve, through sensory drive, a particular strategy for conspicuousness to mates and crypsis to predators using different combinations of light environments, plumage colors, display behaviors, and lek placement in the forest. Similar effects of light environments have also been found in fish (Endler 1991; Boughman 2001, 2002; Fuller 2002; Fuller and Travis 2004).

Interspecific Variation in Plumage Color with Ambient Light

Some studies have examined the showiness of birds in relation to foraging stratum and therefore indirectly in relation to sensory drive and variations of light environments with height in forests. For example, Garvin and Remsen (1997) demonstrated that canopy birds are significantly showier than understory birds, a relation found by several other biologists (see the review in Walther et al. 1999). This relationship, however, was generally interpreted as resulting from the influence of an unmeasured third variable such as greater abundances of predators on the ground or more prevalent parasites in the canopy. McNaught and Owens (2002) explicitly tested the hypothesis that light environment drives the evolution of color pattern by examining associations between plumage coloration and habitat use. They used spectrometry to measure pairs of closely related Australian bird species, one of which lived in open habitat and the other in closed habitat. They expected to find red and orange colors (Endler 1993) and brighter birds (Marchetti 1993) in closed habitats. They confirmed that red colors were more likely to occur in closed habitats, but they found that species from closed habitats were darker than those from relatively open habitats, contrary to the results of Marchetti.

The relationship between feather coloration and light environment was again studied by Gomez and Théry (2004), who used spectrometry and comparative analyses to examine colors of 40 bird species living at ground level or in the canopy of a Neotropical rainforest. In agreement with the predictions of natural selection for crypsis, they found that canopy species had brighter coloration than ground-dwelling species (again contrary to results of Marchetti 1993, who compared species from open and closed habitats). They also found that understory birds had more dark brown or reddish-brown colors and that canopy species more frequently had green coloration (Figure 4.7). Similarly, they found that canopy birds have more varied color patterns than do ground birds, which may result in increased crypsis, because light is spectrally more diverse in the canopy. To increase conspicuousness, species in the canopy use a greater range of hues and more contrasting color patterns than do species in the understory, which can also be related to the broader range of wavelengths found in canopy compared to understory. In addition, canopy

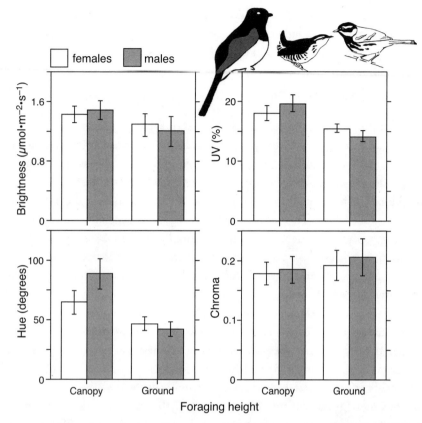

Figure 4.7. Mean (± standard error) brightness, percentage UV, hue, and chroma in relation to foraging height for males and females of 40 mature primary rainforest bird species. Brightness is expressed relative to an incident light of 10 μmol/m²/s. UV is measured from 300 to 400 nm. Hue angle is expressed in degrees with the Munsell colors 5 Red 4/14, 5 Yellow 8/12, 5 Green 5/8, 5 Blue 4/8 and 5 Purple 4/12 respectively at 12°, 51°, 170°, 220°, and 331°. Chroma (saturation) varies from 0 to 1. Redrawn from Gomez and Théry (2004).

---

birds display more UV coloration, which makes sense, given the greater availability of UV in forest canopy. Although females of canopy species are cryptic, males are much more conspicuous. They are brighter, with more colors rich in short wavelengths than females in the canopy. Males in the understory, in contrast, are darker and have more orange and red than do females. These results suggest that general patterns of plumage coloration in tropical birds are governed mainly by natural selection for crypsis. In terms of both mean coloration and color diversity, light environments appear to constrain hue and UV content of plumage more than brightness or chroma.

Light Environments and Mouth Colors in Begging Chicks

Most studies of light environments have considered light environments at male display sites, but recently, Hunt et al. (2003) used the same approach to study the effect of light environment on mouth coloration of begging chicks (Plate 13) in eight species of European passerines. They found that the mouth and flange of every species was highly contrasting against the nest background in the long wavelengths and also particularly in the UV. The contrasting UV reflectance increases the conspicuousness of begging chicks against nest backgrounds, although in this study, nest light environments were relatively poor in the UV. Nestlings with UV-reflecting skin are likely detected more easily by parents and thus gain relatively more mass than nonreflecting nestlings (Jourdie et al. 2004). An alternative explanation is that parents preferentially feed UV-reflecting nestlings because they have stronger cell-mediated immune responses (Jourdie et al. 2004). The light environments of nests and UV patches displayed by nestlings during begging deserve further study, particularly in species positioning their nests in UV-rich light environments.

Experimental Manipulation of UV and Visible Wavelengths

With the surge of interest in UV color signals and vision in birds, which followed the publication of reviews by Maier (1993) and Bennett, Cuthill, and colleagues (Bennett and Cuthill 1994; Bennett et al. 1994), several biologists began to manipulate light environments using UV-blocking filters or different lights to determine if UV is critical for mate choice and foraging in birds (Chapter 4, Volume 2). Even though the use of UV-blocking filters not only alters the appearance of UV signals, but also the appearance of the entire body and that of the surrounding environment (Viitala et al. 1995; Bennett et al. 1996, 1997; Hunt et al. 1997, 1999; Church et al. 1998; Chapter 4, Volume 2), such experiments clearly show that UV is an important component of bird light environments. UV light is crucial to the perception of food and reflectance of UV patches, but also to induce UV fluorescence of parrot colors (Pearn et al. 2001, 2003; Arnold et al. 2002; Plate 16). But is UV reflectance more important than other colors in the "visible" spectrum?

As the importance of UV for bird light environments was repeatedly confirmed, the question arose whether UV might serve as a "private" communication channel for birds (e.g., Guilford and Harvey 1998) or whether it was comparable to other regions of the bird-visible spectrum. The first explicit test of this hypothesis was conducted on Zebra Finches (*Taeniopygia guttata*) by Hunt et al. (2001), using filters that could selectively remove UV wavelengths or wavelengths corresponding to peak sensitivity of short-, medium-, and long-wavelength cones. Results showed that removing visible wavelengths could have similar or even stronger effects than removing UV. In relation to the importance of the red beak for sexual selection in Zebra Finches (Burley and Coopersmith 1987), removal of the red wavelengths from the light environment has the strongest effect on female mate choice, demonstrating at least for Zebra Finches that UV is not a private communication channel. Analogous experiments were conducted on Zebra Finches to investigate the effect of different regions of the light environment spectrum on seed choice (Maddocks et al. 2001). The removal of UV had no significant effect on foraging behavior, whereas the removal of long wavelengths had the strongest effect on food choice, increasing the number of white versus red seeds eaten.

Evidence is accumulating to show that UV and fluorescent signals are not part of a private communication channel in birds. Although UV and fluorescent signals are frequently associated with sexual signaling in birds (Hausmann et al. 2003), it is still unknown if this channel is accessible to predators (for evidence in fish, see Cummings et al. 2003). Further studies in birds will need to precisely identify avian predators and the sensitivity of their visual systems.

Unanswered Questions for Future Research

Light environments are just beginning to garner attention, and several questions are not answered, if even asked. For example, color patterns are rarely taken into account. It would be interesting to study the mechanisms and functions of stripes, small spots, patches, and surrounding colors. This problem also raises questions about the function of multiple color (and acoustic) signals. Similarly, the existence of countershading to improve camouflage has not been studied at large, interspecific scales in birds. Probably because of the traditional locations of lekking sites, most field studies have been conducted in tropical forests. Future studies will have to determine whether sensory drive of color signaling is also active in birds with different mating systems and habitats; for example, temperate songbirds or colonial seabirds. All these studies will benefit from better knowledge of visual systems and neural processing.

Summary

A consideration of light environment is crucial to an understanding of the perception and evolution of optical signals. Light environments were first studied for their stimulating effect on courtship displays and for their potential role in species divergence through sensory drive. With the development of portable spectrometers, biologists began to characterize light environments and to study the interactions between habitat selection, display behaviors, and light environments in lekking birds. All studies to date have shown that bird species evolved a particular strategy for conspicuousness to mates and crypsis to predators using different combinations of light environments, plumage color patterns, and placement of display sites. Interspecific studies also analyzed variation in bird coloration in relation to sensory drive and light environments, either by contrasting open with closed habitats or by comparing colors of species that live in the understory and canopy. As expected from the spectra of light environments, red or orange color displays have most often been found in *Forest Shade,* blue and blue-green in *Woodland Shade,* red, orange or yellow in *Small Gaps* light environments. In relation to the greater availability of UV in the canopy, and to the broader range of light colors at that height, more diverse colors and greater UV reflectance have been found in the plumages of birds that dwell in the canopy. The conspicuousness of plumage patterns is also expected to be enhanced by surrounding patches already contrasting with the background with complementary colors, like blue or blue-green in *Forest Shade,* red or orange in *Woodland Shade,* purple or blue in *Small Gaps,* yellow or green in *Early/Late.* Another confirmed prediction relates to the use of blue to increase conspicuousness or green to increase crypsis in forest canopy, whereas orange or brown, respectively, are used for the same purpose in the understory. Different predictions have been made concerning plumage brightness. The most frequently confirmed prediction is that plumages are brighter in open habitats. The primary manner in which environmental light has driven the evolution of avian color displays is by constraining both the conspicuousness of color displays used as intraspecific signals and the design of cryptic coloration meant to prevent detection by predators. Although light environments have mostly been studied at foraging or display sites of birds, two recent studies examined the light environments of nests, showing that mouths of begging chicks strongly contrasted in UV and red against the visual background of the nest. Experimental manipulation of ambient lights in the context of food or mate discrimination have shown that, although UV is an important component of bird light environments, it is not more important than wavelengths of the "visible" spectrum. Future research should not only study the most spectacular color patches, but also patterns, stripes, countershading, and multiple signals. Studies of sensory drive should also be conducted in a variety of ecosystems and mating systems to determine whether conclusions based on observations of tropical lekking bird studies can be widened to other groups.

References

Arnold, K. E., I. P. F. Owens, and N. J. Marshall. 2002. Fluorescent signaling in parrots. Science 295: 92.

Benalcazar, C. E., and F. Silva de Benalcazar. 1984. Historia natural del Gallo de roca andino (*Rupicola peruviana sanguinolenta*). Cespedesia 13: 59–92.

Bennett, A. T. D., and I. C. Cuthill. 1994. Ultraviolet vision in birds: What is its function? Vision Res 34: 1471–1478.

Bennett, A. T. D., I. C. Cuthill, and K. J. Norris. 1994. Sexual selection and the mismeasure of color. Am Nat 144: 848–860.

Bennett, A. T. D., I. C. Cuthill, J. C. Partridge, and E. J. Maier. 1996. Ultraviolet vision and mate choice in Zebra Finches. Nature 380: 433–435.

Bennett, A. T. D., I. C. Cuthill, J. C. Partridge, and K. Lunau. 1997. Ultraviolet plumage colors predict mate preferences in starlings. Proc Natl Acad Sci USA 94: 8618–8621.

Bongers, F., P. J. van der Meer, and M. Théry. 2001. Scales of ambient light variation. In F. Bongers, P. Charles Dominique, P.-M. Forget, M. Théry, ed., Nouragues: Dynamics and Plant-Animal Interactions in a Neotropical Rainforest. Dordrecht: Kluwer Academic.

Boughman, J. W. 2001. Divergent sexual selection enhances reproductive isolation in sticklebacks. Nature 411: 944–948.

Boughman, J. W. 2002. How sensory drive can promote speciation. Trends Ecol Evol 17: 571–577.

Burley, N., and C. B. Coopersmith. 1987. Bill color preferences of Zebra Finches. Ethology 76: 133–151.

Chazdon, R. L., and N. Fetcher. 1984. Light environments of tropical forests. In E. Medina, H. Moony, and C. Vasquez-Yanes, ed., Physiological Ecology of Plants in the Wet Tropics, Tasks for Vegetation Science 12. The Hague: Junk.

Church, S. C., A. T. D. Bennett, I. C. Cuthill, and J. C. Partridge. 1998. Ultraviolet cues affect the foraging behaviour of Blue Tits. Proc R Soc London B 265: 1509–1514.

Cummings, M. E., G. G. Rosenthal, and M. J. Ryan. 2003. A private ultraviolet channel in visual communication. Proc R Soc London B 270: 897–904.

Endler, J. A. 1978. A predator's view of animal colour patterns. Evol Biol 11: 319–364.

Endler, J. A. 1990. On the measurement and classification of colour in studies of animal colour patterns. Biol J Linn Soc 41: 315–352.

Endler, J. A. 1991. Variation in the appearance of Guppy color patterns to Guppies and their predators under different visual conditions. Vision Res 31: 587–608.

Endler, J. A. 1992. Signals, signal conditions and the direction of evolution. Am Nat 139: 5125–5153.

Endler, J. A. 1993. The color of light in forests and its implications. Ecol Monogr 63: 1–27.

Endler, J. A. 1997. Light, behavior, and conservation of forest-dwelling organisms. In J. R. Clemmons, and R. Buchholz, ed., Behavioral Approaches to Conservation in the Wild. Cambridge: Cambridge University Press.

Endler, J. A., and M. Théry. 1996. Interacting effects of lek placement, display behavior, ambient light, and color patterns in three neotropical forest-dwelling birds. Am Nat 148: 421–452.

Fuller, R. C. 2002. Lighting environment predicts the relative abundance of male colour morphs in Bluefin Killifish (*Lucania goodei*) populations. Proc R Soc London B 269: 1457–1465.

Fuller, R. C., and J. Travis. 2004. Genetics, lighting environment, and heritable responses to lighting environment affect male color morph expression in Bluefin Killifish, *Lucania goodei*. Evolution 58: 1086–1098.

Garvin, M. C., and J. V. Remsen, Jr. 1997. An alternative hypothesis for heavier parasite loads of brightly colored birds: Exposure at the nest. Auk 114: 179–191.

Gilliard, E. T. 1959. Notes on the courtship behavior of the Blue-backed Manakin (*Chiroxiphia pareola*). Amer Mus Novitates 1942: 1–19.

Gilliard, E. T. 1962. On the breeding behavior of the Cock-of-the-Rock (Aves, *Rupicola rupicola*). Bull Am Mus Nat Hist 124: 31–68.

Gomez, D., and M. Théry. 2004. Influence of ambient light on the evolution of colour signals: Comparative analysis of a Neotropical rainforest bird community. Ecol Lett 7: 279–284.

Guilford, T., and P. H. Harvey. 1998. The purple patch. Nature 392: 867–879.

Hailman, J. P. 1977. Optical Signals, Animal Communication and Light. Bloomington: Indiana University Press.

Hausmann, F., K. E. Arnold, N. J. Marshall, and I. P. F. Owens. 2003. Ultraviolet signals in birds are special. Proc R Soc London B 270: 61–67.

Heindl, M., and H. Winkler. 2003a. Interacting effects of ambient light and plumage color patterns in displaying Wire-tailed Manakins (Aves, Pipridae). Behav Ecol Sociobiol 53: 153–162.

Heindl, M., and H. Winkler. 2003b. Vertical lek placement of forest-dwelling manakin species (Aves, Pipridae) is associated with vertical gradients of ambient light. Biol J Linn Soc 80: 647–658.

Hunt, S., I. C. Cuthill, J. P. Swaddle, and A. T. D. Bennett. 1997. Ultraviolet vision and band-colour preferences in female Zebra Finches, *Taeniopygia guttata*. Anim Behav 54: 1383–1392.

Hunt, S., I. C. Cuthill, A. T. D. Bennett, and R. Griffiths. 1999. Preferences for ultraviolet partners in the Blue Tit. Anim Behav 58: 809–815.

Hunt, S., I. C. Cuthill, A. T. D. Bennett, S. C. Church, and J. C. Partridge. 2001. Is the ultraviolet waveband a special communication channel in avian mate choice? J Exp Biol 204: 2499–2507.

Hunt, S., R. M. Kilner, N. E. Langmore, and A. T. D. Bennett. 2003. Conspicuous, ultraviolet-rich mouth colours in begging chicks. Proc R Soc London B 270 (Suppl. 1): S25–S28.

Irwin, D. E., S. Bensch, and T. D. Price. 2001. Speciation in a ring. Nature 409: 333–337.

Jourdie, V., B. Moureau, A. T. D. Bennett, and P. Heeb. 2004. Ultraviolet reflectance by the skin of nestlings. Nature 431: 262.

Lythgoe, J. N. 1979. The Ecology of Vision. Oxford: Oxford University Press.

Maddocks, S. A., S. C. Church, and I. C. Cuthill. 2001. The effects of the light environment on prey choice by Zebra Finches. J Exp Biol 204: 2509–2515.

Marchetti, K. 1993. Dark habitats and bright birds illustrate the role of the environment in species divergence. Nature 362: 149–152.

McNaught, M. K., and I. P. F. Owens. 2002. Interspecific variation in plumage colour among birds: Species recognition or light environment? J Evol Biol 15: 505–514.

Pearn, S. M., A. T. D. Bennett, and I. C. Cuthill. 2001. Ultraviolet vision, fluorescence and mate choice in a parrot, the Budgerigar *Melopsittacus undulatus*. Proc R Soc London B 268: 2273–2279.

Pearn, S. M., A. T. D. Bennett, and I. C. Cuthill. 2003. The role of ultraviolet-A reflectance and ultraviolet-A-induced fluorescence in budgerigar mate choice. Ethology 109: 961–970.

Rand, A. L. 1940. Results of the Archbold expeditions. No. 26. Breeding habits of the birds of paradise: *Macregoria* and *Diphyllodes*. Amer Mus Novitates 1073: 1–14.

Robbins, M. B. 1983. The display repertoire of the Band-tailed Manakin (*Pipra fasciicauda*). Wilson Bull 95: 321–342.

Shashar, N., and T. W. Cronin. 1998. The polarization of light in a tropical rain forest. Biotropica 30: 275–285.

Théry, M. 1987. Influence des caractéristiques lumineuses sur la localisation des sites traditionnels, parade et baignade des manakins (Passériformes: Pipridae). CR Acad Sci Paris ser. III 304: 19–24.

Théry, M. 1990a. Influence de la lumière sur le choix de l'habitat et le comportement sexuel des Pipridae (Aves: Passériformes) en Guyane française. Rev Ecol (Terre Vie) 45: 215–236.

Théry, M. 1990b. Display repertoire and social organization of the White-fronted and White-throated Manakins. Wilson Bull 102: 123–130.

Théry, M. 2001. Forest light and its influence on habitat selection. Plant Ecol 153: 251–261.

Théry, M., and J. A. Endler. 2001. Habitat selection, ambient light and colour patterns in some lek-displaying birds. In F. Bongers, P. Charles Dominique, P.-M. Forget, and M. Théry, ed., Nouragues: Dynamics and Plant-Animal Interactions in a Neotropical Rainforest. Dordrecht: Kluwer Academic.

Théry, M., and D. Larpin. 1993. Seed dispersal and vegetation dynamics at a Cock-of-the-Rock's lek in the tropical forest of French Guiana. J Trop Ecol 9: 109–116.

Théry, M., and S. L. Vehrencamp. 1995. Light patterns as cues for mate choice in the lekking White-throated Manakin (*Corapipo gutturalis*). Auk 112: 133–145.

Viitala, J., E. Korpimaki, P. Palokangas, and M. Koivula. 1995. Attraction of kestrels to vole scent marks visible in ultraviolet light. Nature 373: 425–427.

Walther, B. A. 2002. Vertical stratification and use of vegetation and light habitats by Neotropical forest birds. J Ornithol 143: 64–81.

Walther, B. A., D. H. Clayton, and R. D. Gregory. 1999. Showiness of Neotropical birds in relation to ectoparasite abundance and foraging stratum. Oikos 87: 157–165.

Zahavi, A., and A. Zahavi. 1997. The Handicap Principle. New York: Oxford University Press.

II

Mechanisms of Production

5

Mechanics of Carotenoid-Based Coloration

KEVIN J. MCGRAW

Among the various ways in which birds and other animals become colorful, the carotenoid-based colors have attracted the most attention of late, from both an evolutionary and a mechanistic standpoint. Carotenoids are the second most prevalent pigment in the avian integument, melanins (Chapter 6) being the first. Initial support for the signaling role and sexually selected benefit of being colorful came from studies of the carotenoid colors of fishes (Endler 1983) and birds (Hill 1990, 1991). Because they play diverse and important roles in photosynthesis, free radical protection, and animal coloration (Vershinin 1999), the chemical properties of and methods for analyzing carotenoid pigments are comparatively well described (Goodwin 1980, 1984; Krinsky et al. 1990; Britton et al. 1995). Their links to diet and to health also make them attractive models for understanding the honesty-reinforcing mechanisms that underlie costly yet beneficial ornamental traits (Lozano 1994; Hill 2002).

Several reviews of the mechanisms of carotenoid colors in birds have been published in the past decade (e.g., Olson and Owens 1998; Stradi 1998; Hill 1999a, 2002; Møller et al. 2000). My aims here are to (1) update the information that has accumulated over the past few years in this rapidly growing subdiscipline, (2) add a historical perspective to the development of certain aspects of this field, and (3) summarize some of the deeper biochemical aspects of these pigments that typically are left out of reports in the evolutionary and behavioral literature and yet factor critically into—and may catalyze future lines of—research on the function of carotenoid-based coloration in animals.

Identity, Origin, and Distribution of Carotenoids

Occurrence as Animal Colorants

Carotenoids confer color to members of most invertebrate phyla (e.g., sponges, mollusks, crustaceans; Matsuno 2001) and to all the vertebrate classes—a characteristic shared only by the melanins, among all the major animal pigments (Needham 1974). They are also found in nearly every form of integumentary tissue in vertebrates, including the skin, scales, and eggs of fishes, amphibians, and reptiles; the feathers, beaks, facial wattles/combs, eyes, and tarsal skin of birds; and the skin of mammals (Fox and Vevers 1960). Among vertebrate integumentary tissues, only mammalian hair has never been shown to be colored by carotenoid pigments.

Carotenoids typically give a red, orange, or yellow hue to animals (discussed in more detail below). The orange color of Goldfish (*Carassius auratus*) and the yellow color of Common Canaries (*Serinus canaria*) are classic examples. However, carotenoids are not the only class of natural pigment to bestow these particular colors on animals. At least six other groups of pigments can also generate red, orange, or yellow colors in birds (see Chapters 8 and 9). Examples include (1) the pterin pigments in the eyes and wings of many insects and butterflies, the skin of many poikilothermic vertebrates, and the eyes of certain birds (Plate 14); (2) hemoglobin in the blood-filled tissues of avian skin; and (3) the psittacofulvin pigments that color the red, orange, and yellow plumage of parrots (Plate 15).

Structures, Classifications, and Light-Absorbing Properties

Carotenoids are 40-carbon tetraterpenoid molecules, consisting of a series of eight 5-carbon isoprene residues whose sequence is reversed at the center (Figure 5.1). Their linear hydrocarbon skeleton of conjugated double bonds can exist alone or can be cyclized at one or both ends, and these end-rings are often substituted by different functional groups. Those unsubstituted carotenoids containing only carbon and hydrogen are referred to as the "carotenes" (e.g., β-carotene). The absence of functional groups leaves carotenes quite nonpolar and lipid soluble. In contrast, those carotenoids that contain oxygen are much more polar and are collectively termed the "xanthophylls" (e.g., zeaxanthin, canthaxanthin). Xanthophylls can be further subdivided into two classes, depending on whether the oxygen-containing substituent is a hydroxl (hydroxycarotenoids; e.g., lutein, canary xanthophyll) or a ketone (ketocarotenoids or oxocarotenoids; e.g., astaxanthin). Most natural carotenoids exist in the all-trans (E) stereoisomeric form (Goodwin 1980, 1984; Figure 5.1).

Carotenoids are colored because of their conjugated double-bond system, known as the "chromophore," which absorbs particular wavelengths of light and gives color to the molecules based on the degree of conjugation in the hydrocarbon chain and end-rings. Carotenoids that impart bright colors on bird feathers and bare parts range from yellow through orange to red, and molecules with more conjugated double bonds absorb more short wavelengths of light and thus are redder in color (Figure 5.1). The color of carotenoid-enriched tissues is also determined by the concentration of pigment displayed, as well as the biological material in which the carotenoid is presented (discussed in more detail below). Not all carotenoids in nature are colored, including such non- or poorly conjugated forms as the acyclic carotenes (e.g., phytoene, phytofluene; Rodriguez-Amaya 1999).

Sources of Carotenoids

We cannot begin to understand the chemical and biological roles of carotenoids in birds without identifying the origin of these pigments. Carotenoids evolved first in archaebacteria as lipophilic biochemicals that reinforced cell membranes (and may still do so in certain fungi and animals), but later became valued as accessory light-harvesting pigments to chlorophyll in such photosynthetic organisms as plants, algae, fungi, and bacteria (the pigments also function as photoprotectants and hormone precursors; Vershinin 1999). Animals lack the enzyme (e.g., phytoene synthase) to manufacture carotenoids from the latter's common precursor (phytoene), so their only source of carotenoids is through diet (or very rarely, through a symbiotic association with photosynthetic organisms, as in the Sea Slug [*Elysia chlorotica*]; Rumpho et al. 2000). Birds obtain carotenoids for display in feathers and bare parts either by directly consuming algae, fungi, and plant parts or by ingesting prey (e.g., insects, crustaceans, vertebrates) that harbor carotenoids they acquired from their diet.

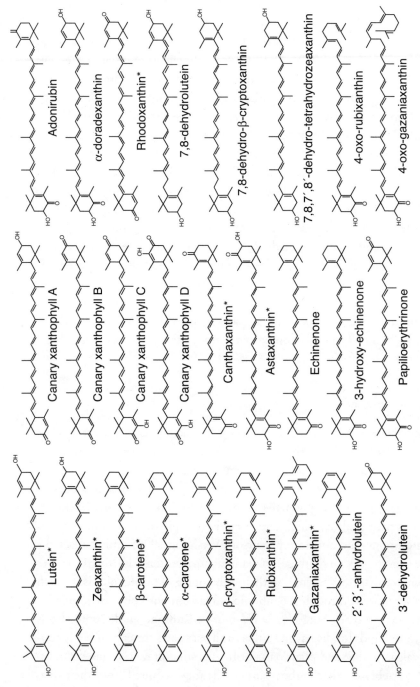

Figure 5.1. Chemical structures of the carotenes and xanthophylls common to the food (indicated with an asterisk) and tissues (all pigments shown) of birds.

Brockmann and Völker (1934) first demonstrated that birds cannot synthesize but must ingest carotenoids at the very time that carotenoids were initially confirmed as pigments of bird plumage. They formulated and fed a carotenoid-free diet to canaries during the molt period and found that the birds grew white feathers lacking carotenoids. These researchers went on to characterize canary xanthophyll as the yellow carotenoid in canary feathers. Another classic example of the dietary acquisition of color is, of course, the Greater Flamingo (*Phoenicopterus ruber*), which gets the pigments for its rich pink feathers from the carotenoid-containing aquatic organisms (e.g., brine shimp) on which they filter-feed (Fox 1962a).

Although birds and other vertebrates cannot synthesize carotenoids de novo, it is important to recognize from the outset that they can metabolize dietary carotenoids into different forms (Brush 1990). For example, ingested hydroxycarotenoids (e.g., zeaxanthin) are converted by some birds into more oxidized ketocarotenoids (e.g., astaxanthin; McGraw et al. 2001; discussed in more detail below). Thus many of the carotenoids found as plumage, leg, and beak colorants are not those present in the diet (Plate 5). These metabolic processes are thought to factor prominently into the information content, signal function, and evolutionary history of carotenoid-based ornamental coloration (see Volume 2). To establish the mechanistic framework for carotenoid coloration, I review the carotenoids present in the food, feathers, and bare parts of birds, the ways in which birds locate and assimilate carotenoids from food to feathers, and the relevant biological and chemical properties of carotenoid-based pigmentation in birds.

Carotenoid Color—A Pigment's Perspective and a Bird's-Eye View

More than 600 carotenoids have been characterized from living things (Vershinin 1999). Less than 30 of these have been described from the diet and tissues of birds (Figure 5.1). This small number likely reflects the limited set of dietary carotenoids available to birds, the physiological affinities that birds have for carotenoids of a specific polarity, the extent to which analytical techniques have detected or characterized trace quantities of carotenoids (e.g., see the list of carotenoids present in human breast milk and serum, many of which are present in minute amounts; Khachik et al. 1997), as well as the small number of avian species for which we have described dietary and tissue carotenoids.

Carotenoids were first described as colorants of avian egg yolk (in chickens; Willstatter and Escher 1912) and later of feathers (Brockmann and Völker 1934), bare parts (e.g., legs, beak; Lonnberg 1938), and eyes (Hollander and Owen 1939). Some authors also attribute nestling mouth colors in birds to carotenoids (Saino et al. 2000, 2003a), but we still await biochemical tests to determine the extent to which carotenoids versus hemoglobin are responsible for nestling mouth flushes and other red, orange, or yellow mouth colors (see Chapter 8; Plate 13).

To date, carotenoids have been biochemically characterized from the integument of roughly 150 avian species (Table 5.1) spanning seven orders. In several other species and orders, it has been presumed or hypothesized that carotenoids give color to red, orange, and yellow feathers and bare parts (e.g., Bortolotti et al. 1996, 2003; Massaro et al. 2003), and in several studies, there is good reason to suspect this (e.g., based on diet manipulations), but as of yet, the types and amounts of carotenoids have not been identified. We should resist speculating on which groups of birds display carotenoids in the integument based on color appearance alone, because several other classes of pigments can also impart the colors commonly attributed to carotenoids (see Chapters 8 and 9). Carotenoids are believed to be present in the egg yolks of all birds, but because riboflavin can also bestow yellow color on yolk (Chapter 8), biochemical tests should be conducted before any conclusions are drawn.

To unravel the process by which birds use carotenoids to become colorful, it is important to trace these molecules from food to feathers and consider the relevant steps along the way. Below I review the dietary, physiological, and morphological characteristics of carotenoid coloration, along with the factors that contribute to inter- and intraspecific variation in the avian colors that we see.

Foods and Pigments that Birds Eat

Animals cannot color their integument with carotenoids without adequate pigment supplies from the diet. Less than 10 colorful carotenoids have been described from the avian diet (Figure 5.1). For comparison, humans—the animal for which we best know carotenoid composition of the diet—ingest and accumulate more than 30 carotenoids from foods (Rodriguez-Amaya 1999). The low number of carotenoids detected in birds comes about because (1) for very few birds have the types and amounts of food carotenoids been described and (2) for many species, we are not even sure of the actual foods from which carotenoids are derived.

The bird for which the most dietary carotenoid studies have been conducted is the Domestic Chicken (*Gallus gallus;* Plate 26). Lutein and zeaxanthin are the dietary carotenoids commonly administered to chickens, which they use as yellow colorants of the yolk, retina, legs, and skin (Marusich and Bauernfeind 1981; Tyczkowski and Hamilton 1986a). These carotenoids are often accompanied by canthaxanthin (an orange carotenoid), which provides a more intense yellow-orange hue to flesh and egg yolks (Latscha 1990). Lutein and zeaxanthin are ubiquitous in plant tissues (Goodwin 1980) and are likely to be the historical dietary carotenoids that Red Junglefowl (*Gallus gallus*)—the wild ancestors of domesticated chickens—encountered. Canthaxanthin, however, has never been documented in plants, and, among potential bird-food sources, is found only in aquatic/marine invertebrates and vertebrates (Matsuno 2001). Thus we must cautiously interpret the evolutionary information gleaned from poultry studies that use canthaxanthin. The same is true for studies that use artificial colorants in other birds and animals, although the ability of chickens and songbirds (e.g., American Goldfinches [*Carduelis tristis*]; McGraw et al. 2001; Plate 30) to accumulate canthaxanthin does speak to the physiological affinities for carotenoids of various polarities (discussed in more detail below).

For decades, aviculturists and pet fanciers also have kept close tabs on avian nutritional requirements and have contributed much to what we know about dietary carotenoids in birds. Because it is such a valuable vitamin-A precursor, β-carotene is added as a supplement to many pet foods. It rarely if ever appears as an integumentary colorant in birds (Table 5.1), however, perhaps because of rapid conversion to vitamin A in the intestine and liver or due to low assimilation efficiency (see below). Some birds (e.g., coots, moorhens, gulls) do sequester high β-carotene levels in tissues (Surai et al. 2001), but it is not known whether their aquatic diets are comparatively enriched with carotenes or they more effectively extract it from food than do such birds as chickens and songbirds (Marusich and Bauernfeind 1981; McGraw et al. 2002a).

With such a paucity of information on avian diets, researchers typically infer the types of carotenoids acquired from the diet by measuring carotenoids in serum (Slifka et al. 1999). Lutein and zeaxanthin are again recovered as major plasma carotenoids in several wild-caught estrildid finches (McGraw et al. 2002a; McGraw and Schuetz 2004), American Goldfinches (McGraw and Gregory 2004), and zoo animals (Slifka et al. 1999), with β-carotene a minor

Table 5.1. Bird Species for Which Carotenoids Have Been Identified from Colorful Feathers and Bare Parts

Order/Family	Species	Trait	Major carotenoids	Reference
Anseriformes	Mallard (*Anas platyrhynchos*)	Orange tarsi	1, 2, 4, 12, 15[a]	Czeczuga (1979)
	Greylag Goose (*Anser anser*)	Orange tarsi	3, 4, 12[b]	Czeczuga (1979)
Falconiformes	Egyptian Vulture (*Neophron percnopterus*)	Yellow facial skin	1	Negro et al. (2002)
Galliformes	Domestic Chicken (*Gallus gallus*)	Red comb	1, 4, 12[b]	Czeczuga (1979)
		Yellow tarsi	1, 2, 3, 4, 12, 15[b]	Czeczuga (1979)
	Wild Turkey (*Meleagris gallopavo*)	Red wattle	3, 4, 12[b]	Czeczuga (1979)
		Orange-brown tarsi	1, 2, 3, 4[b]	Czeczuga (1979)
	Grey Partridge (*Perdix perdix*)	Orange-brown tarsi	1, 3, 12, 15	Czeczuga (1973)
	Ring-necked Pheasant (*Phasianus colchicus*)	Orange-brown tarsi	1, 3, 12, 15	Czeczuga (1973)
		Red facial comb	12	Brockmann and Völker (1934)
	Capercaillie (*Tetrao urogallus*)	Red supraocular comb	1, 2, 12, 14	Egeland et al. (1993)
Ciconiiformes	Greater Flamingo (*Phoenicopterus ruber*)	Pink plumage	12, 14, 15	Fox et al. (1967); Stradi (1999)
	Chilean Flamingo (*Phoenicopterus chilensis*)	Pink plumage	12, 14, 15	Fox et al. (1967); Stradi (1999)
	Lesser Flamingo (*Phoenicopterus minor*)	Pink plumage	12, 14, 15	Fox et al. (1967); Stradi (1999)
	James's Flamingo (*Phoenicopterus jamesi*)	Pink plumage	12, 14, 15	Stradi (1999)
	Scarlet Ibis (*Eudocimus ruber*)	Scarlet plumage	15[c]	Fox (1962b); Stradi (1999)
	Roseate Spoonbill (*Ajaja ajaja*)	Scarlet body plumage	12, 14, 15	Stradi (1999)
	White Stork (*Ciconia ciconia*)	Red bill and tarsi	12	Negro and Garrido-Fernandez (2000)
Charadriiformes	Elegant Tern (*Sterna elegans*)	Pink plumage flush	2, 12, 15	Hudon and Brush (1990)
	Ring-billed Gull (*Larus delawarensis*)	Pink plumage flush	12	McGraw and Hardy (in press)
	Franklin's Gull (*Larus pipixcan*)	Pink plumage flush	12	McGraw and Hardy (in press)
Piciformes	White Woodpecker (*Melanerpes candidus*)	Yellow plumage	22, 23	Stradi et al. (1998); Stradi (1999)
	Lewis's Woodpecker (*Melanerpes lewis*)	Red plumage	12, 13, 14, 15	Stradi (1999)
	Yellow-bellied Sapsucker (*Sphyrapicus varius*)	Yellow plumage	1, 22	Stradi et al. (1998); Stradi (1999)
	Great Spotted Woodpecker (*Dendrocopos major*)	Red plumage	12, 13, 14, 15	Stradi et al. (1998); Stradi (1999)
		Red plumage	12, 13, 14	Stradi et al. (1998); Stradi (1999)
	Hairy Woodpecker (*Picoides villosus*)	Yellow plumage	1, 2, 22, 23	Stradi et al. (1998); Stradi (1999)
		Red plumage	12, 13, 14, 15	Stradi et al. (1998); Stradi (1999)
	Three-toed Woodpecker (*Picoides tridactylus*)	Yellow plumage	1, 2	Stradi et al. (1998); Stradi (1999)

	Green-barred Woodpecker (*Colaptes melanochlorus*)	Yellow plumage	22	Stradi et al. (1998); Stradi (1999)
	Northern Flicker (*Colaptes auratus*)	Red plumage	12, 13, 14	Stradi et al. (1998); Stradi (1999)
		Yellow plumage	1, 2, 3, 22	Stradi et al. (1998); Stradi (1999)
		Red plumage	12, 13, 14	Stradi et al. (1998); Stradi (1999)
	Campo Flicker (*Colaptes campestris*)	Yellow plumage	1, 2, 3	Stradi et al. (1998); Stradi (1999)
	Pileated Woodpecker (*Dryocopus pileatus*)	Yellow plumage	1, 22, 23	Stradi et al. (1998); Stradi (1999)
		Red plumage	12, 13, 14	Stradi et al. (1998); Stradi (1999)
	Cream-backed Woodpecker (*Campephilus leucopogon*)	Red plumage	12, 13, 14	Stradi et al. (1998); Stradi (1999)
	Scaly-bellied Woodpecker (*Picus squamatus*)	Yellow plumage	1, 22	Stradi et al. (1998); Stradi (1999)
	Green Woodpecker (*Picus viridis*)	Yellow plumage	1, 2, 22, 24	Stradi et al. (1998); Stradi (1999)
		Red plumage	12, 13, 14, 15	Stradi et al. (1998); Stradi (1999)
	Toco Toucan (*Ramphastos toco*)	Red vent plumage	1, 13	Stradi (1999)
Passeriformes				
Corvidae	Golden Oriole (*Oriolus oriolus*)	Yellow plumage	1, 2	Stradi (1998, 1999)
	Black-hooded Oriole (*Oriolus xanthornus*)	Yellow plumage	1, 2	Stradi (1998, 1999)
Campephagidae	Scarlet Minivet (*Pericrocotus flammeus*)	Red plumage (m)	13, 15	Stradi (1998, 1999)
		Yellow plumage (f)	1, 2	Stradi (1998)
Malaconotidae	Sulfur-breasted Bushshrike (*Telophorus sulfureopectus*)	Yellow plumage	6, 7	Stradi (1999)
		Red plumage	12, 13	Stradi (1999)
Bombycillidae	Cedar Waxwing (*Bombycilla cedrorum*)	Yellow/orange rectrix tips	6, 7, 18[d]	Hudon and Brush (1989); Stradi (1998)
		Red waxwing	12	Brush (1963)
	Bohemian Waxwing (*Bombycilla garrulus*)	Yellow tail band	6, 7	Stradi (1998, 1999)
	Japanese Waxwing (*Bombycilla japonica*)	Yellow plumage	6, 7	Stradi (1999)
		Red plumage	12	Stradi (1998, 1999)
Muscicapidae	Korean Flycatcher (*Ficedula zanthopygia*)	Yellow plumage	1, 2	Stradi (1999)
	Robin (*Erithacus rubecula*)	Orange plumage	1, 10	Stradi (1999)
	Siberian Rubythroat (*Luscinia calliope*)	Red plumage	12, 14, 15	Stradi (1998, 1999)
	Golden Bush-robin (*Tarsiger chrysaeus*)	Yellow plumage	1, 2, 10	Stradi (1999)
Sittidae	Wallcreeper (*Tichodroma muraria*)	Red plumage	12	Stradi (1998, 1999)
Paridae	Great Tit (*Parus major*)	Yellow plumage	1, 2	Partali et al. (1987); Stradi (1998, 1999)
	Blue Tit (*Parus caeruleus*)	Yellow plumage	1, 2	Stradi (1998, 1999)
	Yellow-cheeked Tit (*Parus spilonotus*)	Yellow plumage	1, 2	Stradi (1999)
Aegithalidae	Long-tailed Tit (*Aegithalos caudatus*)	Pink plumage	16	Stradi (1998, 1999)

(*continued*)

Table 5.1. (continued)

Order/Family	Species	Trait	Major carotenoids	Reference
Regulidae	Goldcrest (*Regulus regulus*)	Yellow plumage	1, 2	Stradi (1998, 1999)
Silviidae	Pekin Robin (*Leiothrix lutea*)	Red plumage	12, 13, 14	Stradi (1998, 1999)
		Yellow plumage	1, 10	Stradi (1998, 1999)
	Silver-eared Mesia (*Leiothrix argentauris*)	Red plumage	12, 13	Stradi (1998, 1999)
		Yellow plumage	1, 10	Stradi (1999)
Coerebidae	Bananaquit (*Coereba flaveola*)	Red plumage	12, 13	Stradi (1999)
Passeridae	Yellow Wagtail (*Motacilla flava*)	Yellow plumage	1, 6, 7	Hudon et al. (1996)
	Cape Weaver (*Ploceus capensis*)	Yellow plumage	1, 2	Stradi (1998, 1999)
	African Masked-weaver (*Ploceus velatus*)	Yellow plumage	1, 2	Stradi (1999)
	Village Weaver (*Ploceus cucullatus*)	Yellow plumage	1, 2	Stradi (1999)
		Yellow plumage	1, 2	Brockmann and Völker (1934); Stradi (1999)
	Nelicourvi Weaver (*Ploceus nelicourvi*)	Yellow plumage	1, 2	Stradi (1999)
	Sakalava Weaver (*Ploceus sakalava*)	Yellow plumage	1, 2	Stradi (1999)
	Baya Weaver (*Ploceus philippinus*)	Yellow plumage	1, 2	Stradi (1999)
	Forest Weaver (*Ploceus bicolor*)	Yellow plumage	1, 2	Stradi (1999)
	Cardinal Quelea (*Quelea cardinalis*)	Red plumage[f]	12, 13, 14, 15	Stradi (1999)
	Red-headed Quelea (*Quelea erythrops*)	Yellow plumage	1, 2	Stradi (1999)
		Red plumage	12, 13, 14, 15	Stradi (1999)
	Red-billed Quelea (*Quelea quelea*)	Red plumage[f]	12, 13, 14, 15	Stradi (1999)
	Red Fody (*Foudia madagascariensis*)	Red plumage	12, 13, 14, 15	Stradi (1999)
	Yellow-crowned Bishop (*Euplectes afer*)	Yellow plumage	1, 2	Kritzler (1943); Stradi (1999)
	Red Bishop (*Euplectes orix*)	Yellow plumage	1, 2	Stradi (1999)
		Red plumage	12, 13, 14, 15	Stradi (1999)
	Yellow Bishop (*Euplectes capensis*)	Yellow plumage	1, 2	Stradi (1999)
Estrildidae	Red-headed Parrot Finch (*Erythrura psittacea*)	Red plumage	12, 13	Stradi (1999)
	Gouldian Finch (*Chloebia gouldiae*)	Yellow plumage	1, 2, 10	Stradi (1999)
		Red plumage	12, 13, 21	Stradi (1999)
	Zebra Finch (*Taeniopygia guttata*)	Red-orange beak	12, 13, 14, 15	McGraw et al. (2002a)
	Star Finch (*Neochmia ruficauda*)	Yellow plumage	1, 2	McGraw and Schuetz (2004)
	Zebra Waxbill (*Amandava subflava*)	Yellow plumage	1, 2, 10, 11	McGraw and Schuetz (2004)
	Red Avadavat (*Amandava amandava*)	Yellow plumage	1, 2, 10, 11	McGraw and Schuetz (2004)

Fringillidae	Chaffinch (*Fringilla coelebs*)	Yellow plumage	1, 2	Stradi (1998, 1999)
	Brambling (*Fringilla montifringilla*)	Yellow plumage	1	G. E. Hill and K. J. McGraw (unpubl. data)
	Collared Grosbeak (*Mycerobas affinis*)	Yellow plumage	1, 2	Stradi (1999)
	Black-and-yellow Grosbeak (*Mycerobas icteroides*)	Yellow plumage	1	G. E. Hill and K. J. McGraw (unpubl. data)
	Spot-winged Grosbeak (*Mycerobas melanozanthos*)	Yellow plumage	1	G. E. Hill and K. J. McGraw (unpubl. data)
	White-winged Grosbeak (*Coccothraustes carnipes*)	Yellow plumage	1	G. E. Hill and K. J. McGraw (unpubl. data)
	Hooded Grosbeak (*Hesperiphona abeillei*)	Yellow plumage	1	G. E. Hill and K. J. McGraw (unpubl. data)
	Evening Grosbeak (*Hesperiphona vespertinus*)	Yellow plumage	1	McGraw et al. (2003b)
	Gold-naped Finch (*Pyrrhoplectes epauletta*)	Yellow plumage	1, 10	G. E. Hill and K. J. McGraw (unpubl. data)
	Golden-winged Grosbeak (*Rhynchostruthus socotranus*)	Yellow plumage	6, 7	G. E. Hill and K. J. McGraw (unpubl. data)
	Orange Bullfinch (*Pyrrhula aurantiaca*)	Orange plumage	6, 7	G. E. Hill and K. J. McGraw (unpubl. data)
	Beavan's Bullfinch (*Pyrrhula erythraca*)	Orange plumage	6, 7	G. E. Hill and K. J. McGraw (unpubl. data); Stradi (1999)
	Red-headed Bullfinch (*Pyrrhula erythrocephala*)	Red plumage	15, 16	Stradi (1999)
		Orange-brown plumage	6, 7	G. E. Hill and K. J. McGraw (unpubl. data)
	Eurasian Bullfinch (*Pyrrhula pyrrhula*)	Red plumage	12, 13, 14, 15, 21	Stradi et al. (1995a, 2001); Stradi (1998, 1999)
	White-winged Crossbill (*Loxia leucoptera*)	Yellow plumage (f)	6, 7	Stradi et al. (1996); Stradi (1998, 1999)
		Red plumage (m)	16, 19, 20	Hudon (1991); Stradi et al. (1996); Stradi (1998, 1999)
	Red Crossbill (*Loxia curvirostra*)	Yellow plumage (f)	6, 7	Stradi et al. (1995a,b, 1996, 2001); Stradi (1998, 1999)
		Red plumage (m)	16, 19, 20	Stradi et al. (1995a, 1996, 2001); Stradi (1998, 1999)

(*continued*)

Table 5.1. (continued)

Order/Family	Species	Trait	Major carotenoids	Reference
	Pine Grosbeak (*Pinicola enucleator*)	Yellow plumage (f)	1, 10	Stradi et al. (1995a, 1996, 2001); Stradi (1998, 1999)
		Red plumage (m)	13, 14, 15, 16, 19	Stradi et al. (1995a, 1996, 2001); Stradi (1998, 1999)
	Scarlet Finch (*Haematospiza sipahi*)	Red plumage	12, 13, 14, 15, 16, 19	Stradi (1999)
	Yellow-fronted Canary (*Serinus mozambicus*)	Yellow plumage	6, 7	Stradi (1998, 1999)
	Citril Finch (*Serinus citrinella*)	Yellow plumage	6, 7	Stradi et al. (1995b); Stradi (1998, 1999)
	Common Canary (*Serinus canaria*)	Yellow plumage	6, 7	Brockmann and Völker (1934); Hudon (1991); Stradi 1999)
	European Serin (*Serinus serinus*)	Yellow plumage	6, 7	Stradi et al. (1995a,b); Stradi (1998, 1999)
	Red-fronted Serin (*Serinus pusillus*)	Yellow and red plumage[g]	6, 7	Stradi et al. (1995b); Stradi (1998, 1999)
	American Goldfinch (*Carduelis tristis*)	Yellow plumage	6, 7	McGraw et al. (2001, 2002b)
	European Greenfinch (*Carduelis chloris*)	Yellow plumage	1, 6, 7	Stradi et al. (1995a,b); Stradi (1998, 1999); Saks et al. (2003a)
	Yellow-breasted Greenfinch (*Carduelis spinoides*)	Yellow plumage	1, 6, 7	Stradi et al. (1995b)
	Oriental Greenfinch (*Carduelis sinica*)	Yellow plumage	1, 6, 7	Stradi et al. (1995b)
	European Goldfinch (*Carduelis carduelis*)	Yellow and red plumage[g]	6, 7, 8, 9	Hudon (1991); Stradi et al. (1995b); Stradi (1998, 1999)
	Eurasian Siskin (*Carduelis spinus*)	Yellow plumage	6, 7	Stradi et al. (1995a,b); Stradi (1998, 1999)
	Red Siskin (*Carduelis cucullata*)	Red plumage	12, 13, 14, 15	Stradi (1998, 1999)
	Black Siskin (*Carduelis atrata*)	Yellow plumage	6, 7	Stradi (1998, 1999)
	Common Redpoll (*Carduelis flammea*)	Red plumage	12, 14, 15, 16	Stradi et al. (1997, 2001); Stradi (1998, 1999)
	Linnet (*Carduelis cannabina*)	Red plumage	12, 14, 15, 16	Stradi et al. (1997, 2001); Stradi (1998, 1999)
	Hoary Redpoll (*Carduelis hornemanni*)	Red plumage	12, 14, 15, 16, 19	Stradi et al. (1997)
	Long-tailed Rosefinch (*Uragus sibiricus*)	Red plumage	12, 13, 14, 19	Stradi et al. (1997, 2001); Stradi (1998, 1999)

	Species	Plumage	Numbers	References
	Desert Finch (*Rhodopechys obsoleta*)	Pink plumage	12, 13, 14, 15	Stradi (1998, 1999)
	Trumpeter Finch (*Rhodopechys githaginea*)	Pink plumage	12, 13, 14, 15	Stradi (1999)
	Streaked Rosefinch (*Carpodacus rubicilloides*)	Red plumage	16, 19	Stradi et al. (1997, 2001); Stradi (1999)
	White-browed Rosefinch (*Carpodacus thura*)	Red plumage	13, 14, 15, 16, 19	Stradi (1999)
	Three-banded Rosefinch (*Carpodacus trifasciatus*)	Red plumage	12, 14, 16, 19	Stradi et al. (1997); Stradi (1999)
	Pallas's Rosefinch (*Carpodacus roseus*)	Red plumage	12, 13, 14, 16, 19, 21	Stradi et al. (1995b, 1997, 2001); Stradi (1998, 1999)
	Beautiful Rosefinch (*Carpodacus pulcherrimus*)	Red plumage	12, 13, 14, 16, 19, 21	Stradi (1999)
	House Finch (*Carpodacus mexicanus*)	Yellow plumage	1, 6, 7, 10	Stradi (1999); Inouye et al. (2001)
		Red plumage	12–17, 19, 20	Stradi (1999); Inouye et al. (2001)
	Dark-breasted Rosefinch (*Carpodacus nipalensis*)	Yellow plumage (f)	1, 6, 7	Stradi (1999)
		Red plumage (m)	12, 16	Stradi (1999)
Emberizidae	Yellowhammer (*Emberiza citrinella*)	Yellow plumage	1, 2	Stradi (1998, 1999)
	Black-headed Bunting (*Emberiza melanocephala*)	Yellow plumage	1, 2	Stradi (1998, 1999)
	Northern Cardinal (*Cardinalis cardinalis*)	Red plumage	6, 7, 12, 13, 14, 15	Hudon (1991); McGraw et al. (2001)
	Rose-breasted Grosbeak (*Pheucticus ludovicianus*)	Red plumage	12, 15	Hudon (1991)
	Saffron Finch (*Sicalis flaveola*)	Yellow plumage	1	G. E. Hill and K. J. McGraw (unpubl. data)
Parulidae	Tristan Bunting (*Nesospiza acunhae*)	Yellow plumage	1, 2, 6, 7	Ryan et al. (1994)
	Common Yellowthroat (*Geothlypis trichas*)	Yellow plumage	1	McGraw et al. (2003b)
	Yellow Warbler (*Dendroica petechia*)	Yellow plumage	1	McGraw et al. (2003b)
	Yellow-rumped Warbler (*Dendroica coronata*)	Yellow plumage	1	G. E. Hill and K. J. McGraw (unpubl. data)
	Palm Warbler (*Dendroica palmarum*)	Yellow plumage	1	G. E. Hill and K. J. McGraw (unpubl. data)
	Yellow-breasted Chat (*Icteria virens*)	Yellow plumage	1	Mays et al. (2004)
	Nashville Warbler (*Vermivora ruficapilla*)	Yellow plumage	1	Brush and Johnson (1976)
	Virginia's Warbler (*Vermivora virginiae*)	Yellow plumage	1	Brush and Johnson (1976)
	American Redstart (*Setophaga ruticilla*)	Orange plumage	6, 7	K. J. McGraw (unpubl. data)
			6, 7, 15	K. J. McGraw (unpubl. data)
Icteridae	Red-winged Blackbird (*Agelaius phoeniceus*)[e]	Red plumage	1, 2, 6, 12, 15	McGraw et al. (2004d)
		Yellow plumage (juv.)	1	K. J. McGraw (unpubl. data)
	Eurasian Blackbird (*Turdus merula*)	Orange-yellow bill	1, 2, 3, 4, 5	Faivre et al. (2003)
	Northern Oriole (*Icterus galbula*)	Orange plumage	1, 6, 7, 12, 13, 15, 17	Hudon (1991)

(*continued*)

Table 5.1. (continued)

Order/Family	Species	Trait	Major carotenoids	Reference
Zosteropidae	Japanese White-eye (*Zosterops japonica*)	Yellow plumage	1	G. E. Hill and K. J. McGraw (unpubl. data)
Pipridae	Golden-headed Manakin (*Pipra erythrocephala*)	Yellow plumage	1, 2	Hudon et al. (1989)
	Red-headed Manakin (*Pipra rubrocapilla*)	Red plumage	12, 13, 14, 15	Hudon et al. (1989)
	Round-tailed Manakin (*Pipra chloromeros*)	Red plumage	12, 13, 14, 15	Hudon et al. (1989)
		Red plumage	12, 13, 14, 15, 18	Hudon et al. (1989)
Thraupidae	Sooty-capped Bush Tanager (*Chlorospingus pileatus*)	Yellow plumage	1	Johnson and Brush (1972)
	Scarlet Tanager (*Piranga olivacea*)	Red plumage (m)	12, 13, 14, 15	Brush (1967); Hudon (1991)
		Yellow plumage (f)	1, 6, 7	Hudon (1991)
	Western Tanager (*Piranga ludoviciana*)	Yellow plumage	1, 6, 7	Hudon (1991)
		Red plumage	18	Hudon (1991)
	Summer Tanager (*Piranga rubra*)	Red plumage	15	Hudon (1991)
	Hepatic Tanager (*Piranga flava*)	Red plumage	6, 12, 14, 15	Hudon (1991)
	Crimson-backed Tanager (*Ramphocelus dimidiatus*)	Red plumage	12, 14, 15	Hudon (1991)

Notes: f, female; juv, juvenile; m, male. Key to carotenoids: 1 = lutein; 2 = zeaxanthin; 3 = β-cryptoxanthin; 4 = β-carotene; 5 = α-carotene; 6 = canary xanthophyll A; 7 = canary xanthophyll B; 8 = canary xanthophyll C; 9 = canary xanthophyll D; 10 = 3′-dehydrolutein; 11 = 2′,3′-anhydrolutein; 12 = astaxanthin; 13 = α-doradexanthin; 14 = adonirubin (formerly known as phoenicoxanthin); 15 = canthaxanthin; 16 = 3-hydroxy-echinenone; 17 = echinenone; 18 = rhodoxanthin ;19 = 4-oxo-rubixanthin; 20 = 4-oxo-gazaniaxanthin 21 = papilioerythrinone; 22 = 7,8-dehydrolutein; 23 = 7,8,7′,8′-tetrahydro-zeaxanthin; 24 = 7,8-dehydro-β-cryptoxanthin.

a. Astaxanthin and beta-carotene was found in captive birds only.
b. Only captive birds were studied.
c. Accompanied by an unidentified carotenoid.
d. Canary xanthophylls are the naturally occurring yellow feather pigments, but by adding the berries of an introduced honeysuckle to their diet, waxwings obtain rhodoxanthin that gives feathers their orange coloration.
e. Carotenoids are only found in red feathers. Yellow border is colored by melanins. Melanins are also found in red feathers.
f. Yellow plumage is not created by the presence of carotenoid or melanin pigments, but by an as-of-yet unidentified type of pigment.
g. Red plumage gets its color from the means by which the canary xanthophylls are bound to feather keratin.

or absent component. However, because some birds, such as estrildid finches (McGraw et al. 2002a) and parrots (McGraw and Nogare 2004), metabolize carotenoids that appear in circulation (e.g., anhydrolutein from lutein), measuring serum levels should not be considered a reliable technique for identifying carotenoids in the diet.

Carotenoids from Plant Diets

Plants produce hundreds of carotenoids that theoretically could be taken up and used physiologically and morphologically by animals (Goodwin 1980), but again, very few are used by birds. With the exception of the roots, carotenoids are present in modest to high concentrations in nearly all plant parts (Goodwin 1980). Dandelions (Su et al. 2002), marigolds (Bosma et al. 2003), and corn (Quackenbush et al. 1961) get their rich yellow colors from lutein and zeaxanthin. Fruits like grapes and berries like raspberries contain lutein and zeaxanthin as primary pigments as well (see the thorough review of human fruit and vegetable foods in Mangels et al. 1993). In these instances, however, the red pigments of fruit and berry skins are not due to carotenoids but to anthocyanins (as in wine; Carreño et al. 1997; Heinonen et al. 1998; Liu et al. 2002).

Studies on natural plant-derived carotenoids in birds have focused on songbirds. Inouye (1999) extracted gut contents from granivorous House Finches (*Carpodacus mexicanus;* Plate 31) during the molt period and isolated the xanthophylls lutein and zeaxanthin as major components (>60% of total carotenoids); β-carotene and β-cryptoxanthin constituted the remainder. McGraw et al. (2001) characterized these same four carotenoids, again with lutein and zeaxanthin dominant, from seeds (sunflower, white millet, red millet) commonly fed to captive birds (see also Su et al. 2002 for carotenoids in thistle). Rose hips are an interesting source of carotenoids for songbirds (Stradi 1998), in that the hips contain the pigments gazaniaxanthin and rubixanthin (Hornero-Mendez and Minguez-Mosquera 2000) that are thought to be precursors of certain rare rose-red pigments (4-oxo-rubixanthin and 4-oxo-gazaniaxanthin) and colors in feathers (Stradi 1999). McGraw (2004) also identified gazaniaxanthin and rubixanthin in the plasma of molting wild House Finches.

There may be sexual, age, seasonal, or other ecologically relevant variations in carotenoid intake that has important consequences for ornamental coloration (see also Chapter 12). New food sources introduced into the diet, such as honeysuckle berries to Cedar Waxwings (*Bombycilla cedrorum*), can contribute new pigments (e.g., rhodoxanthin) and generate new color displays (e.g., orange tail bands as opposed to yellow; Hudon and Brush 1989; Witmer 1996; Plate 30). Such environmental events as exceptionally cold winters that reduce fruit availability have been reported to decrease plumage redness in Northern Cardinals (*Cardinalis cardinalis;* Plate 25, Volume 2; Linville and Breitwisch 1997). At the individual level, dietary carotenoid concentration positively predicted the color of plumage acquired by male House Finches, for example (Hill et al. 2002). Beal (1907) reported an increase in fruit intake during molt in House Finches, which has been used to explain this shift in carotenoid status (Hill 2002). However, there were no corresponding age or sex differences in food carotenoid levels (Inouye et al. 2001), despite dramatic differences in coloration among these groups. Interspecifically, carotenoid-based plumage seems to have evolved in a phylogenetic context in concert with a shift to a fruit diet in pigeons and doves (Columbiformes; Mahler et al. 2003). Frugivory was also associated with higher plasma carotenoid levels in a larger comparative study of 80 bird species from 25 families and eight orders (Tella et al. 2004; Figure 5.2).

Carotenoids from Animal Diets

The diets of insectivores are among the most poorly characterized of avian diets with regard to carotenoid content, but one of the earliest and classic studies that traced carotenoids through the food chain was done on insectivorous nestling tits (*Parus* spp.), which are fed mostly on caterpillars (Partali et al. 1987). The carotenoid content of larvae from deciduous and coniferous forests was analyzed for comparison with the intensity of yellow, carotenoid-derived plumage displayed in nestling tits from the two habitats. Lutein was present in higher concentrations in both the larvae and the nestling plumage from the deciduous forest (Partali et al. 1987). More recently, Ninni (2003) analyzed the carotenoid content of food and nonfood invertebrate prey in Barn Swallows (*Hirundo rustica;* Plate 7). Flies (Diptera); beetles (Coleoptera); bees, wasps, and ants (Hymenoptera); and some butterflies and moths (Lepidoptera) were identified in food boluses. Lutein and zeaxanthin again made up more than half of all dietary carotenoids, with carotene and cryptoxanthin also present. Although this is one of the best studies on food carotenoids in birds, we learn little about their role in carotenoid-based integumentary coloration because, despite early reports to this effect (Stradi 1998), these swallows do not deposit carotenoids in feathers (McGraw et al. 2004a,b).

Mechanics of Carotenoid-Based Coloration

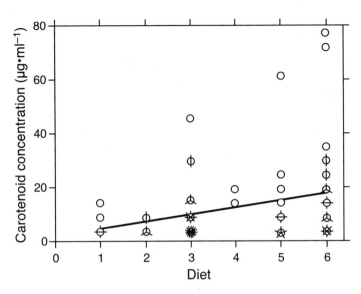

Figure 5.2. Relation between serum carotenoid concentration and diet for 80 species of birds found in Spain and Mexico. Diet type was scored on an integer scale from 1 to 6 according to likely carotenoid concentrations (1 = low-carotenoid carnivorous diet; 6 = high-carotenoid herbivorous diet). Carotenoid concentration in the serum was determined using high-performance liquid chromatography. The number of spokes on the symbols indicates the number of overlapping observations. Redrawn from Tella et al. (2004).

In contrast to terrestrial systems, quite a bit is known about the carotenoid content in the aquatic and marine food (e.g., fish, crustaceans) that birds may eat. Unlike in plant food or terrestrial herbivorous prey, the 4-oxo-carotenoids (e.g., astaxanthin, canthaxanthin) are the dominant carotenoids in aquatic animals (Matsuno 2001). These carotenoids serve as precursors or direct plumage colorants in their predators (e.g., flamingos; Fox 1962b; Fox et al. 1965, 1967, 1970). Variations in the levels of carotenoids across prey populations are known to contribute to population-level variations in carotenoid status and integumentary color. White Storks (*Ciconia ciconia*; Plate 5), for example, that inhabit areas with large populations of an introduced, astaxanthin-rich crayfish (*Procambarus clarkii*) deposit more astaxanthin in the feathers (which are normally white), beaks, and legs (Negro and Garrido-Fernandez 2000; Negro et al. 2000). Gulls (e.g., Ring-billed Gulls [*Larus delawarensis*]) and terns in certain regions show a hint of pink, astaxanthin-derived coloration in their usually white plumage as well (Hudon and Brush 1990), which may be attributed to uncharacteristically high concentrations of dietary carotenoids. In the case of Ring-billed Gulls, farmed salmon that are now being raised on carotenoid-supplemented diets may be the source of the pigments appearing in feathers (Hardy 2003).

Carotenoid colors have also been studied in carnivorous species, such as American Kestrels (*Falco sparverius*), that consume land vertebrates. These birds are thought to predominantly circulate lutein through blood (Bortolotti et al. 1996, 2003). In a study of a breeding kestrel population in Saskatchewan, Canada, it was found that the density of their main prey, Red-backed Voles (*Clethrionomys gapperi*), on the territories of mated pairs was correlated with blood carotenoid levels of the resident adults (Bortolotti et al. 2000); no link was made to sexual coloration. Mammal prey are generally considered to harbor different types of carotenoids (carotenes over xanthophylls) and fewer carotenoids overall in tissues than do birds (Hill 1999b), so it would be interesting to compare how mammal composition in the diet explains inter- and intraspecific variations in carotenoid status and coloration in birds.

Unusual Diets

Perhaps the most unique of all food sources that offer carotenoids for integumentary coloration in birds is the ungulate feces that Egyptian Vultures (*Neophron percnopterus;* Plate 30) consume (Negro et al. 2002). Although it is otherwise generally nutritionally unrewarding (with the possible exception of salts), ungulate excrement contains measurable amounts of carotenoids, and the rotting carcasses that vultures typically consume are thought to be carotenoid-deficient. Negro et al. (2002) showed in wild-caught vultures from different populations that blood carotenoid levels are correlated with indices of local vegetative quality, and in captive feeding trials that only after switching vultures from a cow-meat to a cow-dung diet do carotenoid levels increase in the blood. It has yet to be determined whether yellow facial coloration is a visual signal of quality in these birds because coprophagy offers few nutrients and/or increases parasite exposure (sensu Negro et al. 2002).

Feeding and Foraging Strategies

If there are such strong selective advantages to being colorful (see Volume 2), in theory it would behoove animals to forage on those foods that serve as the best sources of carotenoids to become maximally colorful. This was the original hypothesis offered by Endler (1983) and Hill (1992) in fishes and birds,

Mechanics of Carotenoid-Based Coloration

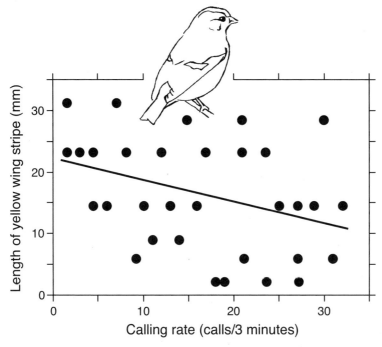

Figure 5.3. Correlation between the size of the carotenoid-pigmented color patch on the wing of male Eurasian Siskins and the frequency of long-distance contact calls that isolated individuals give. Males with smaller wing patches gave more calls, which was interpreted as a sign of reduced ability to locate food on their own. Redrawn from Senar and Escobar (2002).

respectively, for why carotenoid colors might evolve as honest advertisements of individual quality (i.e., foraging ability). The evidence for this "carotenoid-limitation" hypothesis is mounting (see Chapter 12). Grether et al. (1999) and Hill et al. (2002) provided the best support by showing significant positive correlations between the concentration of carotenoids in the guts and ornamental coloration in wild Guppies (*Poecilia reticulata*) and House Finches, respectively. Senar and Escobar (2002) went on to provide an interesting link between foraging and coloration by showing that Eurasian Siskins (*Carduelis spinus;* Plate 12, Volume 2) with more extensive carotenoid plumage coloration gave fewer contact calls when isolated from feeding flocks and were less attracted to conspecific decoys (Figure 5.3), both of which imply foraging independence and prowess.

These correlations show that more colorful fish and birds consume more carotenoids, but to what extent do they actively seek out carotenoid-rich foods? In controlled captive experiments, Guppies (Rodd et al. 2002), Three-spined (*Gasterosteus aculeatus*) and Nine-spined Sticklebacks (*Pungitius pungitius;* Smith et al. 2004), and House Finches (Stockton-Shields 1997), but not Red-winged Blackbirds (*Agelaius phoeniceus;* Plate 11, Volume 2; Avery et al. 1999), show strong preferences for artificially colored orange and red foods (or foodlike objects) over less colorful options. No comparable studies are available in wild animals, which is important, because it is still not clear (1) how much the natural foods of these animals vary in coloration, and (2) whether food coloration is a reliable proxy for the types or amounts of carotenoids that are valuable for integumentary coloration. Recall that reddish fruit and berry skins are colored not by carotenoids but by anthocyanin pigments. Birds can also use ultraviolet cues from carotenoid-containing fruits and berries (Burkhardt 1982; Altshuler 2001), flowers (Eisner et al. 1969), fish (Losey et al. 1999), and invertebrates (Silberglied 1979; Church et al. 1998a) to guide their foraging decisions (Church et al. 1998b, 2001; Siitari et al. 1999; Maddocks et al. 2001), but again, to what extent such cues predict carotenoid content is uncertain. To date, the only study that has tested for a carotenoid-specific food preference in birds was in Barn Swallows (a species lacking ornamental carotenoid coloration), and it was found that captured insect prey were signficantly less enriched in carotenoids than were random sweep-netted samples from the field (Ninni 2003).

Uptake from the Diet

Even after carotenoid-containing foods are procured from the environment, carotenoids must be absorbed from foods, and it is not guaranteed that all pigments present in the diet will be taken up by the body. In humans, for example, assimilation efficiencies for dietary carotenoids range from 1% to 99% across various studies (Parker et al. 1999). There are a number of factors that may govern the extent to which carotenoids are extracted from foods by animals. Most of these mechanisms have been described for humans; we await parallel studies in birds.

Vertebrates are generally thought to extract carotenoids from food, along with the general pool of lipids that diffuse passively through the intestine (Mitchell 1962; Parker 1996). This notion has been challenged in chickens, however. Littlefield et al. (1972) removed sections of chicken intestine and found removal of the ileum to most notably impair carotenoid uptake. Coccidian endoparasites that attack different portions of the intestine were also shown to differentially affect xanthophyll assimilation (Ruff and Fuller 1975; see more on carotenoids and coccidia in Chapter 12). Administering such toxins as ochratoxin and aflatoxin that disrupt carotenoid uptake also does not equally influence lipid assimilation (Osborne et al. 1982). Finally, Tyczkowski and Hamilton (1986a) demonstrated that carotenes and xanthophylls are absorbed at different intestinal sites in chickens (the ileal versus duodenal/jejunal regions, respectively). Thus there appear to be lipid-independent, region-specific absorption mechanisms for the main hydroxycarotenoids that chickens accumulate. No comparable studies have been done on carotenoid absorption in any other bird.

The physical matrix of food in which carotenoids are presented to digestive enzymes in the gut (i.e., their bioavailability) may affect the likelihood of carotenoid retrieval from food (Parker 1996). If carotenoids are not digested easily and released fully into the larger lipid droplets that form in the lumen, then they may not be incorporated into micelles that enter the musocal cells and will instead be either degraded (as in trout; Chavez et al. 1998) or voided with waste. High-fiber foods in particular can present a challenge for digestion and carotenoid absorption (Rock and Swendseid 1992). In contrast, high lipid content in the diet can aid in carotenoid uptake (Williams et al. 1998), although this is not required, given that humans can absorb carotenoids from fat-free diets (Prince and Frisoli 1993). We are just now gaining an understanding of the nutritional value of carotenoid-rich foods in birds. Fruits of varying color can differ in their nutritional rewards; for example, more saturated orange and yellow fruits indicate high protein levels, whereas more saturated blue fruits indicate a high carbohydrate load (Schaefer and Schmidt 2004). It would be interesting to couple this finding with fruit-pigment analyses to determine how colorful fruits may offer nutrients (e.g., fats) to optimize carotenoid accumulation by their avian consumers.

Although there may be strong selection pressures to accumulate as many carotenoids as possible for coloration purposes, there may be a physiological ceiling to carotenoid absorption, perhaps due to a diminished capacity to incorporate carotenoids into micelles (Parker 1996). In humans, β-carotene saturation occurs at approximately 100 mg (Parker 1996). American Goldfinches fed high doses of canthaxanthin voided a notable portion, as evidenced by their orange-colored droppings (McGraw et al. 2001). Whether wild birds ever reach or would benefit from reaching this ceiling on their natural food sources should be further tested. Zebra Finches (*Taeniopygia guttata;* Plate 27) fed a base diet of millet seed did not void any carotenoids in their feces (McGraw et al. 2002a).

In mammals, carotenoids are also known to interact with one another during uptake (Furr and Clark 1997) and potentially compete for space in lipid droplets, micelles, and chylomicrons that carry carotenoids to plasma lipoproteins for distribution throughout the body (e.g., Bierer et al. 1995; Clark et al. 1998; During et al. 2002). Carotenes and xanthophylls, which differ markedly in polarity and thus would be most likely to show different affinities for and solubilities in different lipid fractions, most commonly interact and negatively impact the accumulation of the less-abundant group of pigments in the diet (reviewed in van den Berg 1999). Two studies in colorful birds hint at such carotenoid interactions. Blount et al. (2002) suggested that the proportional increase in canthaxanthin, compared to β-carotene, in the eggs of carotenoid-supplemented Lesser Black-backed Gulls (*Larus fuscus*) might be because canthaxanthin outcompeted β-carotene in the gut. Also, despite an equal ratio in the diet, Zebra Finches and American Goldfinches accumulated more zeaxanthin than lutein in plasma, but only at quite high dietary concentrations (McGraw et al. 2004c; Figure 5.4). To what extent these interactions occur during uptake, as opposed to when carotenoids are locating binding sites on lipoproteins (see below), should be questioned. Several songbirds that naturally accumulate more of one type of colorful carotenoid accumulate more of all other types (e.g., McGraw et al. 2003a; Saks et al. 2003a; McGraw and Gregory 2004; but see Inouye et al. 2001), suggesting that competitive interactions among carotenoids during uptake may not always be strong.

How all of these parameters translate into differences in coloration among birds fed their natural diets remains to be seen. In far too many studies, next to nothing is known about carotenoid intake, or it is assumed that plasma carotenoids reflect assimilation from the diet, when in fact there are several body sources from which and to which plasma carotenoids come and go. Even in those studies in which dietary intake and plasma levels of carotenoids are measured together in relation to integumentary coloration (McGraw et al. 2003a), it is still unclear how assimilation of those ingested carotenoids predicts carotenoid status. Studies employing such stable isotope tracers as ^{13}C (Parker et al. 1993; Yao et al. 2000) will be invaluable in attempts to clarify the patterns and importance of carotenoid absorption in colorful birds.

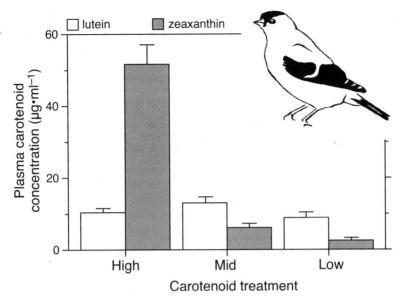

Figure 5.4. Effect of dietary carotenoid concentration on the accumulation of the two main xanthophylls in blood (lutein and zeaxanthin) in captive molting male American Goldfinches (mean + standard error). Three carotenoid treatments were used: high = 10 g/ml of a 70:30 mix of lutein:zeaxanthin, mid = 1 g/ml of a 70:30 mix of lutein:zeaxanthin, low = 0.1 g/ml of a 70:30 mix of lutein:zeaxanthin. Despite an equal ratio of the two pigments in all treatments (and one that mimics the ratio at which they naturally occur in goldfinch foods), birds fed the high-carotenoid dose accumulated disproportionately more zeaxanthin than lutein. Redrawn from McGraw et al. (2004c).

Circulation and Delivery

Once they are retrieved from food and fractioned with lipids (as micelles) in the intestinal mucosa, carotenoids are packaged into chylomicron fractions and enter the lymphatic system, where they are incorporated into lipoproteins that circulate through the bloodstream (Williams et al. 1998). Lipoproteins are the exclusive transporters of carotenoids in the body and, more generally, function to deliver a variety of lipids to internal cells and tissues through the blood. They are composed of a complex assembly of proteins (apoproteins) and lipids, including cholesterol, triglycerides, and phospholipids (Davis 1997). Three main classes of lipoproteins exist, varying in their molecular weight (or relative density, as separated by ultracentrifugation): high-density lipoprotein

(HDL), low-density lipoprotein (LDL), and very-low–density lipoprotein (VLDL; Parker 1996).

Carotenoids do not bind equally well to all lipoprotein types. In humans, the less-polar carotenes and hydrocarbons (e.g., lycopene from tomatos) are harbored mostly in LDL (60–70%), whereas the xanthophylls partition predominantly with HDL (Clevidence and Bieri 1993). This same analysis has only been done in two avian species. Trams (1969) found in the Scarlet Ibis (*Eudocimus ruber*) that greater than 90% of all carotenoids in the blood circulation were associated with HDL; these also appeared to be the presumed oxidized metabolites (e.g., canthaxanthin), as opposed to the nonpolar dietary carotenes that were bound to LDL. HDL also harbors the majority of plasma xanthophylls in chickens (Allen 1987; Attie et al. 2002). There are no solid data yet on whether certain of the xanthophylls have superior affinities for lipoproteins over others, but again, when correlations are examined (McGraw et al. 2003a; McGraw and Gregory 2004) or even when lipoprotein status is manipulated (K. J. McGraw and R. S. Parker, unpubl. data; discussed in more detail below), such songbirds as American Goldfinches and Zebra Finches tend to circulate more of all types of hydroxycarotenoids.

Carotenoids accumulate at highly variable concentrations in the serum of wild birds. Tella et al. (1998, 2004) sampled blood from 80 species spanning eight orders and detected a range of values from 0.5 to 75 µg/ml. In an ecological and evolutionary comparative analysis, the best predictors of plasma carotenoid concentration were diet, extent of carotenoid-based plumage, body mass, and phylogenetic relationships, the latter accounting for 65% of all variation (Tella et al. 2004). Species with the highest carotenoid values that fed on predominately plant-based diets exhibited more plumage area that was pigmented with carotenoids and tended to be smaller. Although this study did not consider which carotenoid types the various species circulated through the blood, it is by far the largest single dataset for any analysis of carotenoids in wild birds. Hill (1995) also showed in a smaller sample of songbird species that those taxa with carotenoid-based plumage exhibit redder (and thus likely more carotenoid-rich) blood plasma than do those birds with no carotenoids in their feathers.

A basic prediction of the link between carotenoid access and use is that dietary availability of carotenoids should correlate with levels in circulation. The recent interspecific study by Tella et al. (2004) supports this conclusion. However, intraspecific patterns are not as apparent. Birds must ingest pigments to circulate some through the body, and clearly, experimental changes in dietary carotenoid access influence plasma carotenoid status in chickens (Surai 2002) and Zebra Finches (Blount et al. 2003a; McGraw and Ardia 2003). However, in standardized food-choice tests, McGraw et al. (2003a) found no link between food intake and either plasma carotenoid circulation or beak pigmentation in captive male and female Zebra Finches. More studies are needed on carotenoid intake and circulation to better characterize the degree to which these two factors generally contribute to ornamental coloration.

Sex differences in carotenoid circulation are common in birds, both in species with carotenoid coloration (McGraw et al. 2003a; McGraw and Ardia 2005) and species without such color (e.g., shrikes, Bortolotti et al. 1996; parrots, McGraw and Nogare 2004). These differences are also not a function of diet in Zebra Finches (McGraw et al. 2003a), but instead likely come under physiological control, through the action of lipoproteins and such sex-steroid hormones as testosterone (K. J. McGraw and E. Adkins-Regan, unpubl. data), as well as the need for females to shunt carotenoid pigments to egg yolk (Blount et al. 2000). Clearly there may be added benefits, beyond coloration, to males for accumulating higher carotenoid levels than females (e.g., for their antioxidant properties, to combat testosterone-driven immunosuppression; McGraw and Ardia 2005; see below). Such additional benefits should be considered in future examinations of sex differences in avian carotenoid levels.

Levels of carotenoids in plasma predict integumentary coloration in at least six bird species (American Goldfinch, McGraw and Gregory 2004; American Kestrel, Negro et al. 1998; European Greenfinch [*Carduelis chloris*], Horak et al. 2004; Red-legged Partridge [*Alectoris rufa*], Negro et al. 2001a; White Stork, Negro et al. 2000; Zebra Finch, McGraw et al. 2003a). In two others (Cirl Bunting [*Emberiza cirlus*], Figuerola and Gutierrez 1998; House Finch, Hill et al. 1994), plasma coloration was positively correlated with plumage color. A recent study in Zebra Finches went on to consider how lipoprotein components co-vary with carotenoids and coloration. K. J. McGraw and R. S. Parker (unpubl. data) quantified and manipulated cholesterol levels—a reliable predictor of plasma carotenoid content in humans (e.g., Mayne et al. 1999)—to determine their relationship with and effects on plasma carotenoids and beak coloration in male Zebra Finches. Plasma cholesterol titers positively predicted both carotenoid status and beak color, and experimental in vivo additions and reductions of cholesterol induced predictable increases and decreases, respectively, in both carotenoids and coloration (Figure 5.5). This observation raises the possibility that a critical limiting factor in becoming colorful and sexually attractive is the ability to copiously synthesize these

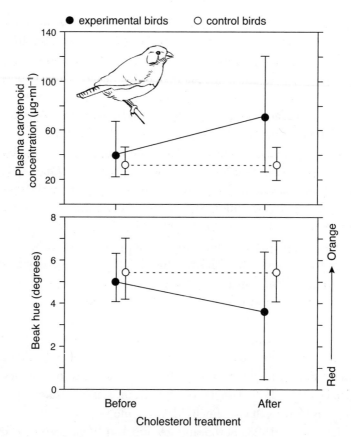

Figure 5.5. Effect (mean ± standard deviation) of cholesterol manipulation on plasma-carotenoid accumulation and beak color (hue) in male Zebra Finches. The addition of cholesterol to the diet for 4 weeks in experimental birds significantly increased both carotenoid circulation (t_{10} = 2.7, P = 0.02) and beak redness (t-test, t_{10} = 3.6, P = 0.006) compared to controls (t_{10} = 0.27 and 0.14, P = 0.80 and 0.87, respectively). Reducing systemic cholesterol using a cholesterol-lowering drug (Lipitor®) also depressed carotenoid levels in the blood and beak (data not shown). Redrawn from K. J. McGraw and R. S. Parker (unpubl. data).

cholesterol-rich molecules. Allen and Wong (1993) also showed that dietary administration of cholesterol increased plasma carotenoids in the HDL fraction of chickens. Such endoparasites as coccidia can also alter HDL profiles to disrupt carotenoid transport (Allen 1987; see also Chapter 12).

In addition to the ability to produce lipoproteins, one may also question the "quality" of lipoproteins—each molecule's ability to bind available carotenoids. There are no data yet on this subject in birds, but in vitro studies with Rainbow Trout (*Oncorhynchus mykiss*) indicated that serum lipoproteins did not saturate even at pharmacological canthaxanthin levels of 150 µg/ml (Chavez et al. 1998), which is nearly double the highest values recorded from the plasma of any wild bird (Tella et al. 2004). Thus the binding capacity of lipoproteins may not be a physiological limitation to carotenoid circulation in the body.

An intriguing and—until recently—unresolved mechanism for carotenoid accumulation in blood was elucidated in young Zebra Finches (Blount et al. 2003b) and chickens (Koutsos et al. 2003a). In these two species, chicks raised on low-quality diets, or by mothers that were fed no carotenoids and thus deposited none in yolk, were never able to circulate as high a concentration of plasma carotenoids later in life as did controls. Such early-life impairment has profound implications for the role of organizational effects on sexual signals, such as carotenoid coloration, much like the evidence for vocal traits in songbirds that involve complex neural machinery that birds develop early in life (Schlinger 1998). The physiological processes underlying these nutritional events in neonates remain mysterious; perhaps lipoprotein levels or the means of assimilating carotenoids from the diet were permanently depressed.

Among the more unusual and specialized mechanisms of carotenoid delivery may be that to egg yolks. As birds begin forming eggs for laying, VLDL levels increase in female chickens by nearly a factor of 200 (Yu et al. 1976), serving as the primary mechanism for lipid delivery to yolks (Walzem 1996), and extraordinarily high carotenoid concentrations appear in the serum (e.g., in American Kestrels and Zebra Finches; pers. obs.). Thus one might expect a tight link between plasma lipoproteins, plasma carotenoids, and egg yolk carotenoid composition in breeding female birds. In fact, McGraw et al. (2005a) and Bortolotti et al. (2003) found that female Zebra Finches and Red-legged Partridges, respectively, with higher plasma carotenoid titers deliver more carotenoids to egg yolks (lipoproteins were not measured in these studies). Interestingly, in Zebra Finches, these females also exhibited significantly redder beaks (McGraw et al. 2005a), suggesting that they may use their colors to signal to prospective mates their ability to defend embryos from oxidative damage (e.g., Blount et al. 2000).

Internal Tissue Sources and Storage

Lipoprotein-bound carotenoids are delivered by blood circulation to a myriad of internal tissues, in addition to the integument in carotenoid-colored birds. In the species that have been analyzed thoroughly (e.g., chicken, coot, gull, moorhen, quail), adipose tissue, liver, egg yolk, retina, and ovary are the tissues with the highest concentrations of carotenoids in the body (Surai 2002; Toyoda et al. 2002). Ovaries are said to contain about half of the body's carotenoid reserves in females (Nys 2000). Testes, semen, heart, kidney, breast/leg muscle, bile, and lung also house measurable amounts of carotenoids in these avian species (Surai and Sparks 2001; Surai et al. 2001). Only analyses of brain (in chicken, turkey, duck, and goose) have turned up tissue devoid of carotenoids (Surai et al. 1996, 1998); some carotenoids can cross the blood-brain barrier in fish (Czeczuga 1977) and humans (Craft et al. 1998), so it is possible that other birds deposit them in neural tissue.

To no surprise perhaps, the types of carotenoids found in the internal tissues of birds, from Zebra Finches (McGraw et al. 2002a), Japanese Quail (*Coturnix japonica;* Khachik et al. 2002), and White Storks (Negro and Garrido-Fernandez 2000) to coots, moorhens, gulls, and chickens (Surai et al. 2001), match those circulating through the blood. There is no evidence yet that a certain type or types of carotenoid(s) are allocated to one or a few tissues. However, there are subtle variations in concentrations and ratios that may reflect physiological discrimination and antioxidant/immune demands (discussed below).

Early work showed that birds were capable of using endogenous tissue carotenoids as reservoirs for coloring their feathers. Kritzler (1943) fed African weavers (*Ploceus* and *Euplectes* spp.) a carotenoid-free diet for 3 months prior to feather growth, but these birds still grew red and yellow feathers containing carotenoids, and autopsies revealed notable pigment stores in liver and fat. However, the ability to draw on carotenoid stores for integumentary pigmentation has not been universally noted. Canaries fed a carotenoid-free diet during plumage molt grew white feathers (Brockmann and Völker 1934). House Finches grew yellow plumage on a low-carotenoid diet, regardless of whether they were recently captured from the wild or housed in captivity on this same low-carotenoid diet for more than a year (Hill 2002). Thommen (1971) and Test (1969) also found no evidence that Ring-necked Pheasants (*Phasianus colchicus*) and Northern Flickers (*Colaptes auratus*), respectively, mobilize carotenoid reserves for coloration purposes. The potential reasons for these discrepancies across studies are many, including that (1) different dietary concentrations of carotenoids can lead to differences in storage levels and hence may affect availability for pigmentation, (2) species may store tissue carotenoids at different concentrations, (3) species may deplete tissue carotenoids at different rates, and (4) different carotenoids can be retained or expended at different rates. Researchers are just now beginning to investigate species- and pigment-specific differences in carotenoid storage, but there clearly remains a gap in studying adult birds (see below for data in chicks) and for each of the various internal tissues.

Adult tissue carotenoid levels can be sensitive to dietary input. White Stork populations that consume a crayfish- and carotenoid-rich diet have much higher adipose tissue carotenoid concentrations than do other populations (Negro and Garrido-Fernandez 2000; Negro et al. 2000). Moreover, carotenoid levels in the livers of House Finches fell to near-zero amounts only 7 days after the birds were placed on a carotenoid-deficient diet in captivity (Inouye 1999). Tissue levels can also be tied to total tissue mass. Greylag Geese (*Anser anser*) with the highest concentrations of carotenoids in adipose tissue have the smallest fat stores (Negro et al. 2001b). Greylag Geese and House Finches also show sex differences in internal-tissue carotenoids; male House Finches have more liver carotenoids than do females (Inouye 1999), and male geese have higher fat-carotenoid concentrations than found in females (Negro et al. 2001b). Interestingly, in both species, males have brighter carotenoid-based coloration than do females. There may be other important reasons why individuals and species vary in their ability to endogenously store and retrieve carotenoids from tissues. Age classes, sexes, and taxa may differ in their antioxidant needs (discussed in more detail below), for example. Moreover, some avian species are capital-breeders (e.g., Arctic waterfowl, waders) that rely solely on tissue reserves for nourishment of themselves and their offspring (in the form of egg yolk) during prebreeding and breeding. These species may be especially effective at accumulating and mobilizing tissue carotenoid repositories for self and offspring defense from oxidative damage. Mothers who raise precocial young may also prioritize carotenoid storage so they can channel carotenoids effectively to the offspring they do not provision after hatching.

Most of the recent attention to the mechanisms and functions of internal-tissue carotenoids in birds has been placed on developing young, particularly in such precocial species as chickens and waterfowl. In their earliest stages of development, embryos are exposed to and assimilate high concentrations of carotenoids in yolk that represent nearly 50% of their mothers' liver stores (Surai and Speake 1998; Surai et al. 1999). Neonates transfer maternally derived carotenoids from yolk to their tissues during development and retain high levels at and after hatching; the liver is the most carotenoid-enriched tissue of newly hatched chicken, coot, moorhen, and gull chicks (Surai et al. 1996,

2001; Surai and Speake 1998). Liver-carotenoid levels are known to rapidly decline within 2 weeks post-hatch (Surai et al. 1998, 2001), but some researchers have estimated that nearly one-quarter of all liver carotenoids in a 4-week-old chick are still from the yolk (Koutsos et al. 2003a). In several experiments on different tissues in developing chicks, carotenoid reserves have been shown to protect cells from oxidative damage (reviewed in Surai 2002; Surai et al. 2003). Thus mothers can adaptively boost offspring protection from free radicals by allocating carotenoids to yolk (Blount et al. 2000). In chickens, exposure to yolk carotenoids and intake of sufficient carotenoids after hatch also ensures that developing young adequately accumulate carotenoids in tissues later in life (Koutsos et al. 2003a; as occurs in plasma).

In sum, tissue carotenoid sources may serve antioxidant or immunoregulatory functions (see more below) and/or as storage depots for future use. Blount et al. (2001) offered the idea that allocating carotenoids to semen may serve this same cellular-protection function, but this hypothesis still awaits confirmation in birds. A recent and exciting advance in avian tissue carotenoids is that immune tissues, such as the thymus, bursa, and spleen, are now recognized as important carotenoid reservoirs in young birds (Koutsos et al. 2003a,b). In young chicks, levels in these tissues are differentially sensitive to dietary and/or maternal carotenoid sources (Koutsos et al. 2003a) and to the immune demands of individuals (Koutsos et al. 2003b). For example, thymic carotenoid levels responded to dietary intake in a dose-dependent fashion, whereas bursal carotenoids did not. Instead, bursal carotenoids were proportionally the highest when dietary carotenoids were low, suggesting that the bursa receives priority in carotenoid accumulation when dietary provisions are low (Koutsos et al. 2003a). Moreover, when immature chickens were induced to mount an acute-phase immune response, plasma and tissue carotenoid levels were generally depleted, but this condition depended on the tissue types and on dietary carotenoid intake (Koutsos et al. 2003b). Bursal and thymic carotenoid levels actually increased, perhaps to protect immune cells that form and proliferate in those tissues in young birds (Figure 5.6); liver levels declined only when dietary carotenoids were abundant, suggesting that there may be a mobilizable, expendable pool of carotenoids for use during infectious challenges and another that is more stable and required for local cell and tissue maintenance (Figure 5.6). These insightful perspectives on carotenoid accumulation and depletion should trigger future research aimed at elucidating the carotenoid requirements, priorities, and trade-offs of internal carotenoid pools as they relate to ornamental coloration (Lozano 1994).

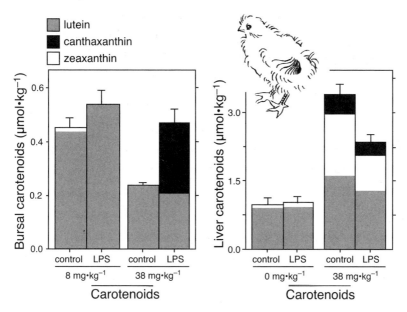

Figure 5.6. Carotenoid allocation patterns of immunologically challenged domestic chicks under two different dietary-carotenoid treatments. Chicks were raised on their respective diets for 4 weeks post-hatch, at which time "LPS" birds were injected with lipopolysaccharide to simulate an infectious challenge. Control and treatment (LPS) chicks were euthanized 24 h later and their tissue carotenoids analyzed via high-performance liquid chromatography. Bursal carotenoids increased significantly, but liver carotenoids decreased significantly in LPS-inoculated birds that were fed a high-carotenoid dose (38 mg/kg), indicating that tissue-carotenoid pools can be labile during immune challenges and that certain tissues seem to demand more carotenoids than others at these times. Data shown are mean + standard error. Redrawn from Koutsos et al. (2003b).

Incorporation and Display in Integumentary Tissues

Carotenoids can be incorporated into several different tissue types to give birds external color. Feathers, skin, scales, beaks, combs, wattles, and eyes round out the avian integumentary tissues known to contain carotenoids (Table 5.1). Carotenoid-endowed tissues most often come in red and yellow hues, which tend to be generated by different carotenoids. The 4-oxo-carotenoids, including astaxanthin, canthaxanthin, adonirubin, and α-doradexanthin (Figure 5.1), are the common red colorants of plumage and bare parts, whereas such xanthophylls as lutein, zeaxanthin, and the canary xanthophylls confer yellow colors (Table 5.1). Less commonly, carotenoid-based plumage appears orange or pink, and these colors can be a product of (1) a lower concentration of red ketocarotenoids, as in the orange beak of female Zebra Finches (Plate 27; unpubl. data) and the pink wing patches in the Desert Finch (*Rhodospiza obsoleta;* Stradi 1998); or (2) the presence of a blend of red and yellow carotenoids, as in the orange plumage of male House Finches (Plate 31; Inouye et al. 2001). Mixtures of red ketocarotenoids and yellow hydroxycarotenoids do not always result in orange appearances, however; the rich red epaulets of Red-winged Blackbirds contain both groups of carotenoids, but feathers appear red because of the higher concentration of red ketocarotenoids (McGraw et al. 2004d).

More than a single carotenoid type typically appears in any given integumentary tissue in birds (Table 5.1). Red feathers and bare parts (e.g., in Northern Cardinals; McGraw et al. 2001) often house a collection of 4-oxo-carotenoids, and the canary xanthophylls co-occur in yellow tissues. There are exceptions to this rule, of course. Lutein is the lone colorant of yellow feathers in songbirds like Yellow-breasted Chats (*Icteria virens;* Plate 5) and Evening Grosbeaks (*Hesperiphona vespertinus;* Plate 25; McGraw et al. 2003b). Under less natural conditions, Cedar Waxwings (Witmer 1996) and White Storks (Negro and Garrido-Fernandez 2000) accumulate rhodoxanthin and astaxanthin, respectively, from introduced food sources, and these appear in isolation in plumage. From a mechanistic perspective, birds likely deposit multiple carotenoids in the integument, because several are acquired from food and accumulated in the body. At a functional level, it is conceivable that multiple carotenoids play complementary chromatic roles, increasing overall concentration while also absorbing across a broader range of wavelengths to generate a more spectrally pure color.

Carotenoid colors can not only be affected by the types, amounts, and ratios of carotenoid pigments that occur in the integument, but also by their co-occurrence with other types of pigments or structural mechanisms and by the nature of their physical interactions with tissues. Greenfinches co-deposit yellow xanthophylls with brown melanins to generate an olive-colored plumage (Lucas and Stettenheim 1972). The purple feathers of Pompadour Cotingas (*Xipholena punicea*) are so colored because they contain red carotenoids within blue structurally colored feathers (Brush 1969). Depositing carotenoids in a matrix of structurally colored tissue may not always shift the hue away from what is typical for carotenoid colors, however; the yellow facial skin of the Toco Toucan (*Ramphastos toco;* Plate 11), for example, gets its color from a combination of presumed carotenoid pigments and a long-wavelength yellow microstructural mechanism (Prum and Torres 2003). Finally, carotenoids can form hydrogen bonds with feather keratin, and the nature of this interaction is thought to be responsible for the difference in coloration between the red facial and yellow wing feathers of European Goldfinches (*Carduelis carduelis*), both of which contain yellow canary xanthophylls (Stradi et al. 1995a,b).

Regarding the structural/pigmentary components of carotenoid coloration, several researchers have recently noted that carotenoid-based feathers reflect a substantial amount of ultraviolet light (e.g., in finches; MacDougall and Montgomerie 2003; Hill and McGraw 2004). It has been debated whether carotenoids alone are responsible for this feature (they typically absorb little UV light from 300 to 400 nm; Needham 1974) or whether the UV coloration is generated by structural components of feathers (Mays et al. 2004). A recent study by Shawkey and Hill (2005) showed that a microstructural mechanism is responsible for the UV reflectance of yellow plumage in American Goldfinches.

Birds nearly exclusively use xanthophylls, as opposed to carotenes, to color the integument. As carotenes accumulate in the body with difficulty, they almost never occur in the avian integument (although their derivatives, such as canthaxanthin, do). From only five species have integumentary carotenes been identified (Table 5.1), and it is interesting that they are always reported from bare parts (e.g., beak, comb, legs) and never from feathers. Where carotenes do occur at high titers in internal tissues, it will be interesting to test whether these species (e.g., gulls, moorhens) deposit them in feathers and bare parts.

The process of carotenoid incorporation into the integument is thought to begin by passive lipid diffusion into maturing cells, such as the skin keratinocytes and feather follicles (Lucas and Stettenheim 1972). These lipids accumulate in the cell in lipoidal droplets that ultimately provide nourishment as well as lipophilic pigments to the growing feather germ (Menon et al. 1986). It has even been suggested that, as precursors of retinoids, high carotenoid concentrations in primordial cells may play a direct role in feather development (Menon and Menon 2000). To date, however, there is little evidence for pro-vitamin–A carotenoid accumulation (e.g., carotenes, cryptoxanthins) in follicles. Only xanthophylls have been noted in the lipoidal droplets from the feather germs of American Goldfinches (McGraw 2004), Common Yellowthroats (*Geothlypis trichas*), and Yellow Warblers (*Dendroica petechia;* McGraw et al. 2003b).

Despite the passive nature of lipid uptake into follicles, however, birds have a remarkably fine-tuned ability to locally sequester specific carotenoids into feathers and bare parts. All birds naturally circulate carotenoids in the body and, in theory, could color all of their feathers with carotenoids. Yet only some species deposit carotenoids in plumage, and most often only in certain feather tracts or body regions. House Finches, for example, deposit carotenoids only in crown, breast, and rump feathers; the rest of their body plumage is white and brown (melanized). Moreover, only a small portion—typically the distal end—of individual feathers contains carotenoids. The basal portion is commonly colored gray or brown, also by melanins. How are these feats accomplished mechanically? These patterns are most likely the responsibility of binding proteins that shuttle carotenoids intracellularly to be incorporated into specific tissues at specific times (Brush 1990; McGraw et al. 2003b). Carotenoid-binding proteins have previously been described from the carapace of crustaceans (Zagalsky 1995), the mid-gut and silk gland of a silkworm (Jouni and Wells 1996; Tabunoki et al. 2002), and from the human retina (Yemelyanov et al. 2001). Further developments in the field of carotenoid deposition in feathers await characterization of such avian carotenoproteins. Only then will we be able to assess the importance of binding-protein distribution and expression for carotenoid pigmentation and better understand the hormonal, molecular, and genetic architecture that regulates their actions.

Beyond the spatial and temporal specificity of carotenoid accumulation in avian integumentary tissues, these tissues can also preferentially target certain carotenoids. McGraw et al. (2003b) showed for three North American songbirds with yellow coloration that, although the two xanthophylls lutein and zeaxanthin were assimilated from the diet and accumulated in blood and the lipid droplets of feathers, only lutein was present in feathers. Similar carotenoid-specific activity may occur in species for which neighboring feather tracts differ in carotenoid composition (e.g., the colorful yellow and red crown feathers of the Goldcrest [*Regulus regulus*]; Stradi 1998), although it is possible that these follicles may differ in enzyme activity and carotenoid synthesis as well (see below).

Although carotenoids are usually deposited in the free, all-trans form in feathers, pigments found in bare parts, such as the beak, legs, and wattle/comb, are often esterified (Czeczuga 1979). Esterification involves binding the hydroxyl or keto substituents on carotenoid end-rings with fatty acids (Tyczkowski and Hamilton 1986b), rendering the pigments more lipid soluble. The function and consequences of esterification for coloration in animals have not been considered. Fruits often house esterified carotenoids, and it is thought that esterification improves pigment stability and hinders pigment photodegradation, due to the long-chained nature of fatty acids (Minguez-Mosquera and Hornero-Mendez 1994). Perhaps carotenoids are protected in a similar manner in animal skin. Why they would not similarly be esterified, and thus afforded photoprotection, in feathers is unclear, however.

On this note, it has long been thought that, as dead tissue, carotenoid-based and other forms of feather coloration were fixed modes of sexual display, retained throughout a year after the annual molt or replaced twice a year in species with summer and winter plumages. Recent evidence, however, shows that plumages, including carotenoid-derived ornaments, are subject to color change over time. The breeding plumage of male House Finches was found to fade in color, becoming less red and more yellow, during the course of a breeding season (McGraw and Hill 2004; Figure 9.6; also see an example of a seasonal color change in Lawrence's Goldfinch [*Carduelis lawrencei*]; Willougby et al. 2002). The mechanism for this color change has yet to be determined and may involve feather soiling or wear, pigment degradation, or feather degradation by bacteria (Shawkey and Hill 2004; Chapter 9; Plate 25). Bare-part colors, in contrast, have long been known to fade in color, often on a much shorter time scale than feathers (e.g., in sick or recently deceased animals). The factors responsible for such fading are similarly still in question, because carotenoids in keratinized beak tissue should be equally irretrievable as are those in feathers. One likely possibility is that delivery of carotenoids to or metabolism of carotenoids at (McGraw 2004; see below) the actively growing tissue (beneath the keratinized portions) is compromised by infections (and then obviously shut down altogether at death). Long-term degradation of integument color (as in museum specimens) could still occur, as for feathers, from pigment bleaching or from tissue soiling/degradation.

Physiological Discrimination

Not all carotenoids are created equal. They differ in length of bond conjugation and end-ring substitution, and as a result, vary in color as well as antioxidant potential (discussed in more detail below). In theory, there may be very good reasons for birds to discriminate among the various ingested carotenoids and accumulate only those with the best chemoprotective or pigmentary potential (McGraw and Gregory 2004; McGraw et al. 2004c).

It has long been known that birds, like most fishes, amphibians, and reptiles (Schiedt 1989, 1998), but unlike mammals (Schweigert 1998; Slifka et al. 1999), selectively retain xanthophylls over carotenes in the body (Weis and Bisbey 1947). Carotenes are duodenally or hepatically converted to vitamin A in vertebrates (Wyss 2004), and thus may not accumulate at high levels or at all in internal avian tissues. Birds also circulate a higher percentage of HDL than do mammals, and the polar carotenoids, such as the xanthophylls, tend to associate more with HDL than with LDL in blood (Fox et al. 1965; Trams 1969; Allen 1987; Parker 1996). Zebra Finches (McGraw et al. 2002a) and chickens (Marusich and Bauernfeind 1981) fed a diet containing β-carotene circulated none of that carotene in the blood and deposited none in liver, adipose tissue, or egg yolk, for example. Songbirds like African Orioles (*Oriolus auratus*; Thommen 1971) and Canaries (Völker 1938) are also thought to poorly absorb nonpolar carotenoids, such as α- and β-carotene, lycopene, and violaxanthin. Serum from 14 zoo-housed species, spanning five orders and including penguins, pelicans, ibises, flamingos, and ducks, lacked β-carotene (Slifka et al. 1999). However, some birds do house large stores of carotenes in the body. Notable concentrations of β-carotene (up to 33 μg/g and up to 60% of total carotenoids) have been reported from liver in Greater Flamingos (Fox et al. 1970) and from liver, egg yolk, kidney, lung, heart, breast/leg muscle, skin, and bile in Lesser Black-backed Gulls, Common Moorhen (*Gallinula chloropus*), and American Coot (*Fulicula americana*; Surai et al. 2001; Blount et al. 2002). Mechanistically, this build-up may be a function of (1) the high β-carotene concentration in the aquatic diets of these birds, (2) different synthesis sites of or needs for vitamin A, or (3) a different capacity for assimilating β-carotene from the diet. Carotenes and xanthophylls are known to be processed by different portions of the intestinal tract in chickens (Tyczkowski and Hamilton 1986a), so one could envision digestive differences between groups or species that drive patterns of carotene use. From a chromatic standpoint, however, it is unclear why colorful birds would not pool carotene stores for pigmentation; the extinction coefficient of β-carotene is slightly higher than that for many xanthophylls (Rodriguez-Amaya 1999) and, at a given concentration, yields a deeper orange rather than yellow appearance. Moreover, it is believed to serve as a valuable precursor for the production of red plumage pigments in many songbirds (Stradi 1998; McGraw et al. 2001). Its efficacy as an antioxidant should also be considered (Mortensen and Skibsted 1997; discussed in more detail below).

Historically, there has been scant evidence that birds differentially take up or accumulate any of the hydroxycarotenoids. In chickens, for example, the lutein:zeaxanthin ratio in egg yolks reflects the relative abundance of these two pigments in the diet (Surai and Sparks 2001). Lutein and zeaxanthin are so similar in their molecular characteristics that they are largely thought to follow the same absorption and circulatory pathways (Parker 1996). A recent set of diet experiments on two finch species (American Goldfinches and Zebra Finches), however, sheds new light on the means by which these two carotenoids may behave as they are absorbed in the body. When fed at low dietary concentrations, these two pigments were accumulated at levels comparable to the ratio in which they appeared in the diet; at higher dietary concentrations, however, proportionally more zeaxanthin was present in blood, suggesting that zeaxanthin was outcompeting lutein for binding sites in the gut or in lipoproteins (McGraw et al. 2004c; Figure 5.4). Zeaxanthin is a more conjugated molecule and thus potentially a stronger chromophore and antioxidant (see below), so there is good reason to suspect preferential uptake of this molecule. In support of this preference, Zebra Finches and American Goldfinches supplemented with only zeaxanthin acquired redder and yellower coloration, respectively, than did males fed only lutein during molt (McGraw et al. 2004c). The generality of this form of differential xanthophyll accumulation remains to be determined.

Different internal tissues often harbor different concentrations and ratios of carotenoids. Surai et al. (2001) found liver and yolk to be the most carotenoid-rich tissues of newly hatched gulls, moorhens, and coots. Liver and egg-yolk are also enriched in chickens compared to immune tissues like the thymus and bursa (Koutsos et al. 2003a). Gull, moorhen, and coot yolk and liver are also especially well-endowed with β-carotene compared to other tissues (Surai et al. 2000, 2001). Again this concentration may be due to differences in rates of HDL- or LDL-facilitated carotenoid delivery to these tissues, restocking of the carotenoids into circulation (Surai et al. 2001), or use/degradation as antioxidants. In contrast, Zebra Finch tissues varied subtly in carotenoid content, with the most notable difference being in the proportions of anhydrolutein (a metabolically derived carotenoid) and its precursor (lutein; McGraw et al. 2002a; discussed in more detail below). Follow-up studies of the causes and consequences for such variability, both among species and among different tissues within species, will illuminate these possible means of physiological fractionation.

Certain carotenoids can be selectively excluded from feathers as well. Völker (1938) reported that Common Canaries fed lycopene, violaxanthin, and β-carotene during molt grew colorless feathers. McGraw et al. (2001) reported the same phenomenon in captive American Goldfinches fed β-carotene during the spring molt. These instances are perhaps not surprising, because such birds are not expected to accumulate the straight-chain hydrocarbons as they do the xanthophylls. However, more subtle and refined forms of xanthophyll discrimination have recently been detected in the yellow plumage of several songbirds. Yellow Warblers, Common Yellowthroats, Yellow-breasted Chats, and Evening Grosbeaks all deposit only a single carotenoid—lutein—into their yellow feathers (Table 5.1). Blood samples acquired during molt and lipid fractions from growing feather follicles, however, reveal the presence of both lutein and zeaxanthin (McGraw et al. 2003b). Their presence demonstrates unprecedented selective hydroxycarotenoid assimilation into bird plumage, which presumably involves a xanthophyll-binding protein like those mentioned above. Again, we await an adaptive explanation for this pattern of carotenoid assimilation.

Carotenoid Metabolism

The ability to modify dietary carotenoids into alternate forms is a widespread feature of invertebrates and vertebrates, from apple snails (Tsushima et al. 1997) to humans (Khachik et al. 1997, 2002). The most basic and conserved means of metabolizing carotenoids is the cleavage of carotenes into vitamin A (Bauernfeind 1981). This reaction does not produce end-products that color the integument, however. Regarding colorful carotenoid metabolites, not all animals or birds can produce these, and even among those that can, the extent to which certain modifications occur, certain enzymes are present, and certain tissues perform these conversions varies considerably from taxon to taxon.

The studies that first documented the presence of carotenoids in bird feathers established that birds could metabolize ingested carotenoids for use in coloration. Brockmann and Völker (1934) fed lutein and zeaxanthin to canaries and described the "canary xanthophylls" as the pigments that gave yellow color to their feathers. Since that time, two general classes of metabolic transformations of carotenoids have been described in birds: (1) dehydrogenation and (2) 4-oxidation (Brush 1990; McGraw et al. 2003c). Dehydrogenation involves converting the hydroxyl group on one or both end-rings of such xanthophylls as lutein and zeaxanthin into a carbonyl group (Figure 5.7). Yellow carotenoids, such as the canary xanthophylls, anhydrolutein, and dehydrolutein, are made in this way (Stradi 1998). These reactions are quite common in birds, found more often in species with yellow integumentary colors but also occasionally present along with red carotenoids in certain red-colored plumages (e.g., in Red-winged Blackbirds and Northern Cardinals; Table 5.1). In contrast,

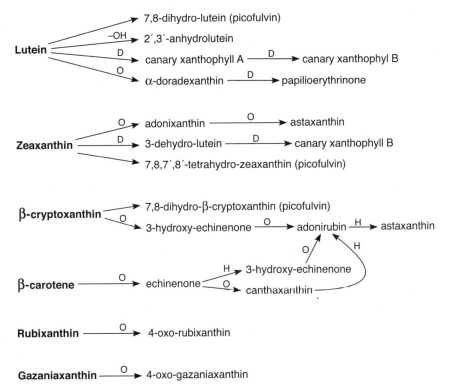

Figure 5.7. Hypothesized precursor-product relationships among avian carotenoids. Note that none of these pathways have been confirmed using appropriate pigment-labeling and pigment-tracing studies. They are simply based on the likely set of transformations in species for which dietary and plumage carotenoids have been identified. O, oxidation; H, hydroxylate; D, dehydrogenation. Compiled from Stradi (1998) and McGraw et al. (2001, 2002a).

4-oxidation occurs through the introduction of a keto (or oxo) group to the 4- or 4′ positions on the β-ionone rings (Figure 5.7). Unlike dehydrogenation, 4-oxidation reactions can occur with any dietary carotene or xanthophyll. Because this added functional group extends the conjugated bond system, more colorful red carotenoids, such as astaxanthin and canthaxanthin, are made by birds via the 4-oxidation process (from zeaxanthin and β-carotene, respectively; Stradi 1998).

Lutein, zeaxanthin, the carotenes (α and β), and the cryptoxanthins (α and β) are considered the primary dietary carotenoids in birds and thus the initial substrates for carotenoid metabolism. It was long thought that vertebrates could not make any of these four dietary carotenoids from other substrates, but recently some researchers have speculated that the cryptoxanthins and zeaxanthin may be derived from lutein (McGraw et al. 2002a; Khachik 2003). Interestingly, in Zebra Finches, lutein and β-cryptoxanthin are highly significantly negatively correlated in serum, suggestive of a possible precursor-product relationship (McGraw et al. 2002a, 2003b). This process would prove to be a valuable conversion in species like the House Finch, for which β-cryptoxanthin is the precursor for the valuable and sexually attractive red plumage carotenoids (discussed in more detail below). Nevertheless, work with radioactive- or stable-isotope-labeled material is needed to confirm these hypothetical relationships between dietary and integumentary carotenoids. Such studies have only been done for endogenously synthesized carotenoids in chickens (Schiedt 1998). These studies will also help to elucidate intermediates and pathways involved in the formation of various carotenoids. For example, Stradi (1998) proposed that lutein, unlike the direct precursor zeaxanthin, can serve as a substrate for canary xanthophyll B synthesis via production of canary xanthophyll A (Figure 5.2). Moreover, he hypothesized that β-carotene is the dietary substrate for astaxanthin production, through the formation of echinenone, canthaxanthin, and adonirubin as intermediates (all of which can appear in feathers; Table 5.1).

The anatomical sites of these metabolic reactions that generate colorful integumentary carotenoids in birds have remained elusive for decades. Many researchers have suspected that the liver is the main site of carotenoid metabolism (Brush 1990), largely because the liver harbors many enzymes similar to the cytochrome P450s that metabolize lipids, such as steroid hormones, for example (Danielson 2002). In fact, the liver, along with the small intestine, kidney, and lung, is a documented site of pro-vitamin-A carotenoid metabolism in birds and other animals (Olson and Hayaishi 1965; Wyss 2004). However, there are few data to support the idea that birds make their colorful display carotenoids in the liver. Studies on bishops (Kritzler 1943), House Finches (Inouye 1999), and several other species of estrildid and cardueline finch (McGraw 2004) have demonstrated that metabolically derived carotenoids present in feathers and bare parts are not found in the liver or serum of these birds at times when they should be forming pigments for deposition in the integument. Kritzler (1943) inferred that carotenoids must be produced at distant colorful tissues themselves, either in maturing feather follicles or beak/leg keratinocytes. Recently, I sampled lipid fractions from the growing yellow feathers of male American Goldfinches and detected a mix of dietary (lutein/

zeaxanthin) and metabolically derived (canary xanthophylls) carotenoids (McGraw 2004), providing the first direct support for the hypothesis that these carotenoid metabolites are formed in the integument.

Not all carotenoids that appear in the integument must be produced there, however. Some estrildid finches (e.g., avadavats, *Amandava* spp.) synthesize two lutein derivatives—anhydrolutein and dehydrolutein—that they circulate through the body and deposit into yellow feathers (McGraw and Schuetz 2004). Dehydrolutein, along with another presumed metabolite (canary xanthophyll A), also have been reported from the plasma of Japanese Quail (Khachik et al. 2002). In Zebra Finches, McGraw et al. (2002a) found that anhydrolutein was present in the duodenum and was enriched in the liver relative to other tissues. At present, however, there is no evidence that birds form red ketocarotenoids at tissues other than the integument. Clearly, more work is needed to determine whether only certain classes of metabolic reactions can occur in certain tissues and how common liver versus integumentary carotenoid metabolism is in the avian class.

The enzymes responsible for these dehydrogenation or 4-oxidation reactions have yet to be characterized. Brush (1967), Hudon (1994), and Stradi et al. (1996) suggested that a single generalist enzyme (referred to as a "mixed-function oxidase" or "4-oxygenase") may carry out the many of the 4-oxidation reactions seen in birds (Figure 5.2). Different enzymes (dehydrogenases) would be expected to catalyze the formation of dehydrogenated carotenoid products (McGraw et al. 2003c). Species for which individual displays of carotenoid pigmentation in feathers range from none to intense (e.g., the rumps of female House Finches; Hill 2002; the breast of male Red-billed Queleas [*Quelea quelea*]; Dale 2000; Plate 6, Volume 2) should be ideal candidates for feather-follicle enzyme screening.

Several birds do not metabolize carotenoids for coloration purposes and instead directly deposit dietary and serum carotenoids into feathers and bare parts. In fact, of the 112 passerines from 21 families listed in Table 5.1 for which the carotenoid composition of colorful integumentary structures is known, over a third of the species (37%) from over half of the families (62%) deposit only dietary carotenoids into feathers. In a comparative test using two clades—carduelline finch species (family Fringillidae, subfamily Carduelinae) and the passerine families (Passeriformes)—we found in both groups that the ancestral condition is the use of dietary carotenoids in feathers, and only later did birds evolve the capacity to metabolize carotenoids (Hill and McGraw, unpubl. data). Because of the complex enzymatic machinery needed to metabolize carotenoids, this finding offers support for the notion that more exaggerated color ornaments evolved from less exaggerated states (Hill and McGraw, unpubl. data). In these groups and in estrildid finches (family Estrildidae; McGraw and Schuetz 2004), however, there can be reversals in this character state, suggesting that this is a labile trait that can respond to selection based on differences in signal value and demands across species.

In other cases, only one sex (typically the male) is capable of metabolizing carotenoids. Female Scarlet Minivets (*Pericrocotus flammeus*) and Scarlet Tanagers (*Piranga olivacea*) incorporate dietary xanthophylls into feathers, whereas males synthesize oxocarotenoids for their red coloration (Brush 1967; Stradi 1998; Table 5.1). The sexes can also differ in the types of metabolites they produce. Male Red Crossbills (*Loxia curvirostra*), for example, deposit red ketocarotenoids in plumage, but females use dehydrogenated products (canary xanthophylls) to color their feathers yellow (Table 5.1). In still others, there may be age-dependent carotenoid metabolism. Unlike juvenile and female American Restarts (*Setophaga ruticilla*) that pigment their yellow plumage with canary xanthophylls, adult males additionally manufacture canthaxanthin to give their feathers an orange hue (Table 5.1). Juvenile Red-winged Blackbirds use only dietary lutein to acquire yellow plumage, but adult males and females with red coloration synthesize astaxanthin and canthaxanthin (Table 5.1). These species, sexes, and age-classes presumably lack the enzymes or enzyme expression necessary to metabolize dietary carotenoids in this fashion.

Strategies for Becoming Colorful

In theory, animals have a variety of ways of selectively accumulating different types and amounts of carotenoids in their various tissues. If the brightest colors are favored by sexual selection, what then is the biochemical strategy for becoming most colorful? In theory, birds could target very specific types of pigments that best impart the brightest color, or they may simply aim to accumulate the highest concentrations of all carotenoids (McGraw and Gregory 2004). Both sets of strategies seem to exist in the songbirds that have been studied to date.

American Goldfinches and European Greenfinches are congeneric cardueline finches that use canary xanthophylls A and B as major carotenoids to color their feathers yellow (Stradi 1998; McGraw et al. 2001). In both species, females prefer to mate with males having deeper yellow plumage (Eley 1991; Johnson et al. 1993). High-performance liquid chromatography and spectral reflectance data of feathers plucked from wild-caught birds showed that the most colorful finches were those that deposited the highest concentration of both canary xanthophylls (Saks et al. 2003a; McGraw and Gregory 2004). There was no particular ratio of pigments that generated the richest plumage. Thus, it is beneficial for males to accumulate as many xanthophylls as possible, which have very similar molecular structures and likely exhibit very similar light-reflectance properties.

So what is the optimal physiological route to becoming intensely pigmented with both canary xanthophylls? Canary xanthophylls are synthesized at the feather follicle from the common dietary xanthophylls (lutein and zeaxanthin) that are acquired from the diet and circulated throughout the body (McGraw 2004). It may be that these carotenoids are energetically costly to synthesize (sensu Hill 1996) or that the ability to procure precursors from the diet or accumulate them in the body limits plumage coloration. The most colorful male goldfinches in the wild circulate the highest level of dietary precursor carotenoids in blood during the period of feather growth (McGraw and Gregory 2004). Zebra Finches, a distantly related estrildid finch, show a similar pattern; those males and females that accumulate the highest concentrations of all carotenoids in blood develop the brightest red beaks (McGraw et al. 2003b). Food-deprivation experiments in goldfinches, designed to control for differences in carotenoid intake, indicate that it is most challenging to accumulate carotenoids in blood, and not necessarily to complete the metabolic steps to produce canary xanthophylls (McGraw et al. 2005b). It may be in these birds that the means by which they deliver carotenoids to peripheral tissues, bound to lipoproteins, controls ornamental coloration. In fact, studies aimed at quantifying and manipulating levels of lipoproteins in blood show that these transporter molecules can regulate the degree of carotenoid transport and color expression in Zebra Finches (McGraw and Parker unpubl. data).

A carotenoid-pigmented carduline finch from a different genus adopts a wholly different biochemical blueprint for maximal plumage coloration. Male House Finches display tremendous variation in carotenoid-based plumage color, from dull yellow to bright red (Hill 2002). Inouye et al. (2001) described a suite of 13 different types of carotenoids in these feathers, some of which typically confer a red hue (e.g., 3-hydroxy-echinenone), and others that are yellow in color (e.g., lutein). Male House Finches become most red (and thus sexually attractive) not by accumulating large amounts of all carotenoids in feathers, but by selectively targeting the richest red pigments, such as 3-hydroxy-echinenone (3HE; Inouye et al. 2001). Yellow males lack carotenoids in feathers and instead deposit yellow pigments. Ultimately, the physiological pathway to becoming the most colorful male House Finch involves both a high intake of dietary carotenoids as well as efficient metabolism of precursors. Hill et al. (2002) found a positive correlation between the concentration of carotenoids in the gut and plumage redness. Yet it also appears from food-deprivation studies that it is energetically challenging to manufacture these valued red metabolites. When food access was restricted during molt, males were less able to use the presumed precursor of 3HE (β-cryptoxanthin) to grow red plumage than were those fed food ad libitum (Hill 2000).

It is not yet clear why these finch relatives have evolved different optimal pigmentation strategies. House Finches may simply have a more diverse diet or superior capabilities to utilize different carotenoids for coloration. Pressures such as female mate choice (Hill and McGraw 2004) and antioxidant/immunomodulatory usage (see below) may also restrict the extent to which males of certain species exaggerate coloration and allocate pigments to plumage. More studies of non-Fringillid and non-songbird species are needed to clarify general strategies of carotenoid pigmentation in birds.

Important Biological and Chemical Actions of Carotenoids

Carotenoids perform a remarkable diversity of physiological roles in animals. When the functions of carotenoids are considered across all living things, their role as colorants in animals is clearly among their most derived functions. Carotenoids evolved initially in archaebacteria to provide support in lipid membranes and later in photosynthetic organisms as accessory light-harvesting pigments for photosynthesis (Vershinin 1999). That they can serve physiological functions other than integumentary pigments in animals is worthy of careful consideration, as such functions may bear critically on the costs as well as the benefits of being colorful.

Carotenoids as Antioxidants and Immunomodulators

As highly conjugated molecules, carotenoids can readily accept unpaired electrons from singlet oxygens, hydroxyl/peroxyl radicals, and other potentially damaging products formed during cellular metabolism and protect cells and tissues in the body from oxidative damage (Burton and Ingold 1984). For the past two decades, the nutrition and health literature on humans and other mammals has been stockpiled with reports of the antioxidant benefits of carotenoids

(Krinsky 2001). Most of the attention to the antioxidant role of carotenoids has centered on their protection of the immune system, due to the generation of radicals during immune-cell activation and proliferation (Chew 1993; Hughes 2001). This phenomenon prompted Lozano (1994) to speculate on the role that the immunomodulatory properties of carotenoids played in the evolution of carotenoid-based ornamental colors in animals. Animals may face a trade-off when allocating carotenoids acquired from the diet to physiological and coloration purposes, and only the highest-quality individuals (those who acquire the most carotenoids or are in the best health state) can devote more to the integument for advertisement (Lozano 1994).

A few tenets of this hypothesis have recently been tested. Many correlational and experimental studies positively link carotenoid coloration and the health or immune capacity of birds (e.g., Brawner et al. 2000; Lindstrom and Lundstrom 2000; McGraw and Hill 2000; Saks et al. 2003b; Chapter 12). More colorful birds tend to be those in the best health, and birds experimentally subjected to adverse health conditions (e.g., parasites) show a decrease in color expression. Only recently, however, was it shown that carotenoids can be the actual agents that stimulate the immune system in colorful birds. In their study of Common Moorhen chicks, Fenoglio et al. (2002) was the first to demonstrate an immunomodulatory role for carotenoids, although they used an unnaturally occurring dietary carotenoid (canthaxanthin). Moreover, it is not yet known whether integumentary coloration in these animals is carotenoid-based. Since that time, natural xanthophyll (lutein + zeaxanthin) supplements were found to elevate cell-mediated and humoral immunoresponsiveness in male Zebra Finches (Blount et al. 2003a; McGraw and Ardia 2003; see also McGraw and Ardia 2005 for evidence in females; Figure 5.8). However, carotenoid-facilitated immune enhancement may not be universal in colorful birds. A range of dietary xanthophyll concentrations administered to male American Goldfinches during molt had no effect on several aspects of immunity and disease resistance (Navara and Hill 2003). Clearly more studies of wild and captive birds are needed to ascertain how generally carotenoids exert this immunoregulatory effect. Moreover, no study has yet to actually demonstrated a true trade-off, where pigment investment in coloration sacrifices carotenoids for use as antioxidants (or vice versa). Studies to date that have shown a decline in carotenoid coloration as the result of an immune challenge (e.g., in Eurasian Blackbirds [*Turdus merula*]; Faivre et al. 2003) cannot conclusively demonstrate whether immune challenges decreased food intake, affected physiological utilization, or in fact demanded increased

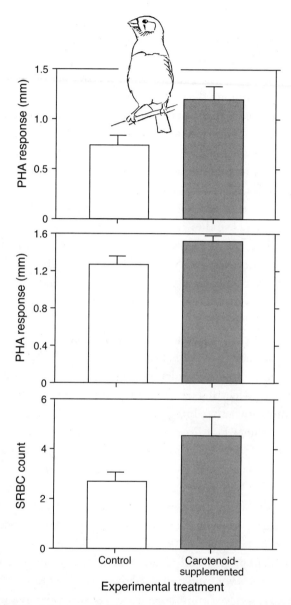

Figure 5.8. Evidence for immunomodulatory effects of carotenoids in Zebra Finches. In the top graph, males supplemented with dietary xanthophylls mounted a stronger response to a novel plant lectin (phytohemagglutinin [PHA]; redrawn from Blount et al. 2003a), whereas in the bottom two graphs, males supplemented with these pigments mounted significantly stronger responses both to PHA and to a humoral immune challenge (immunization with sheep red blood cells [SRBC]). Data shown are mean + standard error. Redrawn from McGraw and Ardia (2003).

use of carotenoids as antioxidants. A trade-off might be best illustrated experimentally by delivering a range of dietary carotenoid concentrations to immune-challenged birds and tracking the fate of carotenoids as integumentary colorants versus radical quenchers.

Their effects on adults aside, carotenoids are also thought to have potent protective effects on embryonic development and health (Blount et al. 2000). Rapid cell division and the rate of metabolic events during ontogeny subject embryos to unusually high levels of oxidative stress (Surai et al. 1996). Mothers, in turn, may defend their offspring from oxidative damage by enriching eggs with antioxidants, such as carotenoids. In studies on chickens, Surai and colleagues have shown that when dietary carotenoid concentrations are increased, mothers deposit more carotenoids in egg yolks, and the tissues (e.g., liver, yolk, yolk-sac membrane) from the chicks hatched from these eggs are particularly protected from lipid peroxidation (Surai et al. 1996, 2003; Surai and Speake 1998). Egg yolk produced by carotenoid-enriched mothers is also less susceptible to tissue peroxidation in Lesser Black-backed Gulls (Blount et al. 2002) and Zebra Finches (McGraw et al. 2005a). In these and other studies that aim to manipulate offspring nutrition and antioxidant status via the maternal diet, however, we cannot be certain whether carotenoids pigments transferred to the yolk by the hen or alternate physiological effects of the carotenoid-supplemented diet on the mother (e.g., an effect on vitamin E status) contribute most to antioxidant protection in young. To disentangle these effects, carotenoids should be injected directly into eggs. Saino et al. (2003b) has recently done this in Barn Swallows and found that experimentally elevated carotenoid levels in eggs boosted the T-cell-mediated immune performance of nestlings. In future studies, egg injections should be coupled with manipulations of maternal and offspring diet to understand the relative effects of these different carotenoid exposures on neonate development and health.

This idea of the carotenoid-based chemoprotection that mothers afford their young begs the questions: How far do the fitness effects extend? Do these molecules affect maternal production of young and the recruitment and fitness of offspring? High egg-carotenoid concentrations do appear to translate into improved survival and reproduction in two recent correlative studies. Lesser Black-backed Gull mothers that were fed supplemental carotenoids in the diet showed an increased propensity to produce a second clutch of eggs (Blount et al. 2004). Zebra Finch mothers fed a carotenoid-supplemented diet laid eggs that were more likely to hatch and fledge young (McGraw et al. 2005a). These offspring were also more often male than female and developed more brightly colored, sexually attractive beaks as adults. These results suggest that carotenoids may be a mechanism by which female birds bias offspring sex as well as modify offspring production, survival, and reproductive worth via improved health. Again, however, we await experiments injecting eggs with carotenoids to confirm whether the carotenoids in yolk per se are exerting these effects.

Other Actions of Carotenoids

Despite the current emphasis by behavioral ecologists on carotenoids as antioxidants and immunoenhancers, there are several other chemical and physiological actions of carotenoids that may factor into the information content and signaling function of bright colors. It was suggested above that birds might esterify bare-part carotenoids to improve photostability, but carotenoids themselves are thought of as photoprotectants in tissues like the eye. The macular region of the human retina, for example, is enriched with xanthophylls (Khachik et al. 2002) that arguably play a role in absorbing harmful short-wave light rays (Stahl and Sies 2002), protecting photoreceptors from oxidative damage and ultimately preventing age-related macular degeneration (Mozaffarieh et al. 2003). The only experimental evidence for a photoprotective role in the eye was provided recently in Japanese Quail (Thomson et al. 2002). Young quail were fed a carotenoid-deficient diet for 6 months, and then experimental groups either remained on this diet or were fed supplemental carotenoids (zeaxanthin) for 1, 3, or 7 days, at which point they were exposed to intermittant light and then euthanized. Retinas were removed and photoreceptor apoptosis was found to be significantly related to retinal zeaxanthin concentration (Figure 5.9). This observation raises the question of whether the retina is an important sink for carotenoids, away from other regions with feather and bare coloration, that may similarly leave coloration as an honest signal of antioxidant capacity (Lozano 1994). In birds with carotenoid-colored eyes (e.g., chickens, ducks; Hollander and Owen 1939; Oliphant 1988), birds may be signaling the direct photoprotection they are affording themselves with their retinal pigments. Finally, do species of bird with more colorful carotenoid-based integuments deposit more in the eyes? And what consequences does this concentration have for the visual system? By definition, based on the light absorbance properties of the pigments, this process would leave the visual systems of carotenoid-colored species optimally tuned to accept and perceive the light wavelengths that their integument reflects.

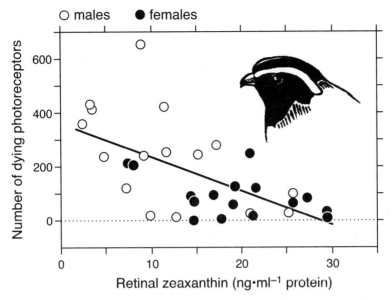

Figure 5.9. Photoprotective action of carotenoids in the retina of Japanese Quail. The retinas of birds with more zeaxanthin had fewer apoptotic photoreceptors after being exposed to 10 1-h intervals of white light at 6 months of age ($r = -0.62$, $P < 0.0001$). Redrawn from Thomson et al. (2002).

Throughout this chapter, I have highlighted the benefits of accumulating high carotenoid levels in birds, much like the recent literature on the topic has done. However, as with many nutrients acquired by the body, too much is not necessarily a good thing. The pro-oxidant potential of carotenoids cannot be ignored (Krinsky 2001). At very high concentrations or at high oxygen tensions, carotenoids are subject to autoxidation and may increase the production of or form reactive oxygen species that can damage cells (Palozza 1998; Young and Lowe 2001). Zahavi and Zahavi (1998) in fact argued that carotenoid colors in animals might be costly, honest signals of quality (or handicaps), because the pigments are toxic, and individuals advertise their quality—their ability to endure such toxicity, as well as their ability to detoxify themselves—by depositing these pigments in dead tissues, such as feathers. At present, no studies have demonstrated toxic effects of carotenoids accumulated by colorful birds. All data, collected either in vivo at physiological levels or in vitro under normal biological conditions (e.g., under the low oxygen tension of cells), support only beneficial physiological effects. A more thorough investigation of carotenoid action across a range of concentrations could, however, yield important insights into the spectrum and balance of costs and benefits that individuals face when accumulating high carotenoid levels for coloration.

One last and recently discovered biological property of carotenoids is their role in gap-junctional communication (Tapiero et al. 2004). One method of cell-to-cell signaling (e.g., the exchange of ions, electric currents, or nutrients) is accomplished by the transfer of information via gap junctions—pores that connect the cytosol of two neighboring cells. Proteins known as "connexins" are the governing body of gap-junctional communication, and recent studies have shown that carotenoids can bind to nuclear receptors in cells and upregulate gene expression and the production of connexins. This activity has been placed mostly in the context of cancer prevention, as carotenoids help overcome the loss of gap-junctional communication that often occurs during malignant cell transformation (Livny et al. 2002). However, it may also be placed in the content of sexual selection and individual quality, if carotenoids accumulated for coloration also serve these vital intercellular signaling roles. This avenue for future research highlights the interface between molecular biology and behavioral ecology that is receiving increased prominence by the day.

Summary

Among the pigments and structures that create ornamental colors in birds, behavioral ecologists have devoted the most attention to the carotenoids. Carotenoids are lipid-soluble hydrocarbons that generate a range of red, orange, and yellow hues in avian feathers and bare parts. Their link to nutrition has been paramount to the explosion of interest in these molecules and their colors. Vertebrates must accumulate carotenoids from plant or herbivorous-prey food sources, so to produce bright carotenoid coloration, animals must procure sufficient amounts from their diet. There are several physiological parameters, however, that may also influence carotenoid accumulation in the body and thus color intensity. These include absorption of carotenoids in the gut, transport of carotenoids bound to lipoproteins through the bloodstream, uptake of circulating carotenoids by maturing feathers, and metabolism of carotenoids at feather follicles. At present, there is evidence in birds that both carotenoid acquisition and certain aspects of carotenoid utilization (e.g., lipoprotein production) can regulate color expression among individuals within a given species, and more studies are needed to discern the strength and ubiquity of nutritional and physiological effects on carotenoid colors across bird taxa.

Birds generally capitalize on the use of xanthophylls rather than carotenes for coloration. This predilection may largely be a function of the carotenoids present in the diet; songbirds, for example, consume plant parts and insects that are more rich in xanthophylls than in carotenes; aquatic species, in contrast, consume carotene-laden crustacean and fish prey, and it is in these species where we see high carotene accumulation in the body. Diet does not explain all variation in the types of carotenoids birds use for color, however. Although some species simply deposit the carotenoids they acquire from food (e.g., lutein, β-carotene) directly into feathers and bare parts, many others are capable of modifying ingested carotenoids into oxidized or hydrogenated forms (e.g., canthaxanthin, canary xanthophyll) that are used as colorants. In the avian taxa studied to date, these metabolic transformations of carotenoids occur specifically within the follicles that give rise to colorful feathers.

Biochemical strategies for developing sexually selected carotenoid coloration can follow one of two alternate routes: the accumulation of as many types and amounts of carotenoids as possible, or the preferential accumulation of one or a few particularly chromatic form(s). Both tactics exist in songbirds. American Goldfinches and European Greenfinches, for example, use a relatively limited and similar set of carotenoids (canary xanthophylls) to become colorful, and males with the yellowest feathers in these species deposit high concentrations of all such xanthophylls. Male House Finches, in contrast, exhibit marked variation in plumage color and in the types and colors of carotenoids available, and bright red, sexually attractive males are those who produce the greatest amounts of a particular red carotenoid, as opposed to using yellow or orange forms.

Carotenoid colors have also had intuitive appeal to evolutionary biologists interested in sexual selection, because of the added physiological roles of the pigments themselves. Most notably, carotenoids can serve as antioxidants and immunomodulators, and thus more colorful birds might directly signal their somatic health and potential quality as a mate with their pigment-based colors. Recent correlational and experimental studies provide some support for the notion that carotenoid colors honestly signal immune capacity, but this notion is by no means universally accepted, and more work is needed on free-ranging birds to better understand carotenoid allocation to body maintenance versus sexual ornamentation. Moreover, depending on the type or concentration of carotenoids, these molecules can also have positive gene-regulatory effects on immune function (e.g., via gap-junctional communication) or detrimental pro-oxidant effects on cells and tissues in the body. Given that these effects are unstudied to date in wild birds, future mechanistic studies of carotenoid coloration are presented with the challenge of weighing a larger suite of costs and benefits of carotenoids than previously envisioned.

References

Allen, P. C. 1987. Effect of *Eimeria acervulina* infection on chick (*Gallus domesticus*) high density lipoprotein composition. Comp Biochem Physiol B 87: 313–319.

Allen, P. C., and H. Y. C. Wong. 1993. Effect of atherogenic diet on chicken plasma-lipids and lipoproteins. Poult Sci 72: 1673–1678.

Altshuler, D. L. 2001. Ultraviolet reflectance in fruits, ambient light composition and fruit removal in a tropical forest. Evol Ecol Res 3: 767–778.

Attie, A. D., Y. Hamon, A. R. Brooks-Wilson, M. P. Gray-Keller, M. L. E. MacDonald, et al. 2002. Identification and functional analysis of a naturally occurring E89K mutation in the ABCA1 gene of the WHAM chicken. J Lipid Res 43: 1610–1617.

Avery, M. L., J. S. Humphrey, D. G. Decker, and A. P. McGrane. 1999. Seed color avoidance by captive Red-winged Blackbirds and Boat-tailed Grackles. J Wildl Manage 63: 1003–1008.

Bauernfeind, J. C. 1981. Carotenoids as Colorants and Vitamin A Precursors: Technological and Nutritional Applications. New York: Academic Press.

Beal, F. E. L. 1907. Birds of California in relation to fruit industry. U.S. Dept Agric Biol Surv Bull 30: 13–23.

Bierer, T. L., N. R. Merchen, and J. W. Erdman, Jr. 1995. Comparative absorption and transport of five common carotenoids in preruminant calves. J Nutr 125: 1569–1577.

Blount, J. D., D. C. Houston, and A. P. Møller. 2000. Why egg yolk is yellow. Trends Ecol Evol 15: 47–49.

Blount, J. D., D. C. Houston, and A. P. Møller. 2001. Antioxidants, showy males and sperm quality. Ecol Lett 4: 393–396.

Blount, J. D., P. F. Surai, D. C. Houston, and A. P. Møller. 2002. Patterns of yolk enrichment with dietary carotenoids in gulls: the roles of pigment acquisition and utilization. Funct Ecol 16: 445–453.

Blount, J. D., N. B. Metcalfe, T. R. Birkhead, and P. F. Surai. 2003a. Carotenoid modulation of immune function and sexual attractiveness in zebra finches. Science 300: 125–127.

Blount, J. D., N. B. Metcalfe, K. E. Arnold, P. F. Surai, G. L. Devevey, and P. Monaghan. 2003b. Neonatal nutrition, adult antioxidant defences and sexual attractiveness in the Zebra Finch. Proc R Soc Lond B 270: 1691–1696.

Blount, J. D., D. C. Houston, P. F. Surai, and A. P. Møller. 2004. Egg-laying capacity is limited by carotenoid pigment availability in wild gulls *Larus fuscus*. Proc R Soc Lond B 271 (Suppl.): 79–81.

Bortolotti, G. R., J. J. Negro, J. L. Tella, T. A. Marchant, and D. M. Bird. 1996. Sexual dichromatism in birds independent of diet, parasites and androgens. Proc R Soc Lond B 263: 1171–1176.

Bortolotti, G. R., J. L. Tella, M. G. Forero, R. D. Dawson, and J. J. Negro. 2000. Genetics, local environment and health as factors influencing plasma carotenoids in wild American Kestrels (*Falco sparverius*). Proc R Soc Lond B 267: 1433–1438.

Bortolotti, G. R., J. J. Negro, P. F. Surai, and P. Prieto. 2003. Carotenoids in eggs and plasma of Red-legged Partridges: Effects of diet and reproductive output. Physiol Biochem Zool 76: 367–374.

Bosma, T. L., J. M. Dole, and N. O. Maness. 2003. Optimizing Marigold (*Tagetes erecta* L.) petal and pigment yield. Crop Sci 43: 2118–2124.

Brawner, III, W. R., G. E. Hill, and C. A. Sundermann. 2000. Effects of coccidial and mycoplasmal infections on carotenoid-based plumage pigmentation in male house finches. Auk 117: 952–963.

Britton, G., S. Liaaen-Jensen, and H. Pfander. 1995. Carotenoids. Basel, Switzerland: Birkhauser.

Brockmann, H., and O. Völker. 1934. Der gelbe Federfarbstoff des Kanarienvogels (*Serinus canaria canaria*) und das Vorkommen von Carotinoiden bei Vögeln. Hoppe-Seyl Z 224: 193–215.

Brush, A. H. 1963. Astaxanthin in the Cedar Waxwing. Science 142: 47–48.

Brush, A. H. 1967. Pigmentation in the Scarlet Tanager, *Piranga olivacea*. Condor 69: 549–559.

Brush, A. H. 1969. On the nature of "cotingin." Condor 71: 431–433.

Brush, A. H. 1990. Metabolism of carotenoids in birds. FASEB J 4: 2969–2977.

Brush, A. H., and N. K. Johnson. 1976. The evolution of color differences between Nashville and Virginia's warbler. Condor 78: 412–414.

Burkhardt, D. 1982. Birds, berries and UV. Naturwissenschaften 69: 153–157.

Burton, G. W., and K. U. Ingold. 1984. Beta-carotene: An unusual type of lipid antioxidant. Science 224: 569–573.

Carreño J., L. Almela, A. Martínez, and J. A. Fernández-López. 1997. Chemotaxonomical classification of red table grapes based on anthocyanin profile and external color. Lebensm Wiss Technol 30: 259–265.

Chavez, P. R. G., D. Rengel, R. Gomez, G. Choubert, and J. C. G. Milicua. 1998. Canthaxanthin saturation of serum lipoproteins from immature Rainbow Trout (*Oncorhynchus mykiss*). Comp Biochem Physiol B 121: 129–134.

Chew, B. 1993. Role of carotenoids in the immune response. J Dairy Sci 76: 2804–2811.

Church, S. C., A. T. D. Bennett, I. C. Cuthill, S. Hunt, N. S. Hart, and J. C. Partridge. 1998a. Does lepidopteran larval crypsis extend into the ultraviolet? Naturwissenschaften 85: 189–192.

Church S. C., A. T. D. Bennett, I. C. Cuthill, and J. C. Partridge. 1998b. Ultraviolet cues affect the foraging behaviour of Blue Tits. Proc R Soc Lond B 265: 1509–1514.

Church, S. C., A. S. L. Merrison, and T. M. M. Chamberlain. 2001. Avian ultraviolet vision and frequency-dependent seed preferences. J Exp Biol 204: 2491–2498.

Clark, R. M., L. Yao, L. She, and H. C. Furr. 1998. A comparison of lycopene and canthaxanthin absorption: Using the rat to study the absorption of non-provitamin A carotenoids. Lipids 33: 159–163.

Clevidence, B. A., and J. G. Bieri. 1993. Association of carotenoids with human plasma lipoproteins. Methods Enzymol 214: 33–46.

Craft, N., K. Garnett, E. T. Hedley-Whyte, K. Fitch, T. Haitema, and C. K. Dorey. 1998. Carotenoids, tocopherols, and vitamin A in human brain. FASEB J 12: 5601.

Czeczuga, B. 1977. Carotenoids in fish. 18. Carotenoids in brain of some fishes. Folia Histochem Cytochem 15: 343.

Czeczuga, B. 1979. Carotenoids in the skin of certain species of birds. Comp Biochem Physiol B 62: 107–109.

Dale, J. 2000. Ornamental plumage does not signal male quality in Red-billed Queleas. Proc R Soc Lond B 267: 2143–2149.

Danielson, P. B. 2002. The cytochrome P450 superfamily: Biochemistry, evolution and drug metabolism in humans. Current Drug Metab 3: 561–597.

Davis, R. A. 1997. Evolution of processes and regulators of lipoprotein synthesis: From birds to mammals. J Nutr 127: 795S–800S.

During, A., M. M. Hussain, D. W. Morel, and E. H. Harrison. 2002. Carotenoid uptake and secretion by CaCo2 cells: Beta-carotene isomer selectivity and carotenoid interactions. J Lipid Res 43: 1086–1095.

Egeland, E. S., H. Parker, and S. Liaaen-Jensen. 1993. Carotenoids in combs of Capercaillie (*Tetrao urogallus*) fed defined diets. Poult Sci 72: 747–751.

Eisner, T., R. E. Siblerglied, D. Aneshansley, J. E. Carrel, and H. C. Howland. 1969. Ultraviolet video-viewing: The television camera as an insect eye. Science 166: 1172–1174.

Eley, C. 1991. Status Signalling in the Western Greenfinch (*Carduelis chloris*). Ph.D. diss., University of Sussex, Brighton, U.K.

Endler, J. A. 1983. Natural and sexual selection on colour patterns in poeciliid fishes. Environ Biol Fishes 9: 173–190.

Faivre, B., A. Gregoire, M. Preault, F. Cezilly, and G. Sorci. 2003. Immune activation rapidly mirrored in a secondary sexual trait. Science 300: 103.

Fenoglio, E., M. Cucco, and G. Malacarne. 2002. The effect of a carotenoid-rich diet on immunocompetence and behavioural performances in moorhen chicks. Ethol Ecol Evol 14: 149–156.

Figuerola, J., and R. Gutierrez. 1998. Sexual differences in levels of blood carotenoids in Cirl Buntings *Emberiza cirlus*. Ardea 86: 245–248.

Fox, D. L. 1962a. Metabolic fractionation, storage and display of carotenoid pigments by flamingoes. Comp Biochem Physiol 6: 1–40.

Fox, D. L. 1962b. Carotenoids of the Scarlet Ibis. Comp Biochem Physiol 5: 31–43.

Fox, D. L., T. S. Hopkins, and D. B. Zilversmit. 1965. Blood carotenoids of the Roseate Spoonbill. Comp Biochem Physiol 14: 641–649.

Fox, D. L., V. E. Smith, and A. A. Wolfson. 1967. Carotenoid selectivity in blood and feathers of Lesser (African), Chilean and Greater (European) Flamingos. Comp Biochem Physiol 23: 225–232.

Fox, D. L., J. W. McBeth, and G. Mackinney. 1970. Some dietary carotenoids and blood-carotenoid levels in flamingos—II. γ-carotene and α-carotene consumed by the American Flamingo. Comp Biochem Physiol 36: 253–262.

Fox, H. M., and G. Vevers. 1960. The Nature of Animal Colours. London: Sidgwick and Jackson.

Furr, H. C., and R. M. Clark. 1997. Intestinal absorption and tissue distribution of carotenoids. Nutr Biochem 8: 364–377.

Goodwin, T. W. 1980. The Biochemistry of the Carotenoids. Volume I: Plants. London: Chapman and Hall.

Goodwin, T. W. 1984. The Biochemistry of the Carotenoids. Volume II: Animals. London: Chapman and Hall.

Grether, G. F., J. Hudon, and D. F. Millie. 1999. Carotenoid limitation of sexual coloration along an environmental gradient in Guppies. Proc R Soc Lond B 266: 1317–1322.

Hardy, L. 2003. The peculiar puzzle of the pink Ring-billed Gulls. Birding 35: 498–504.

Heinonen, I. M., P. J. Lehtonen, and A. I. Hopia. 1998. Antioxidant activity of berry and fruit wines and liquors. J Agric Food Chem 46: 25–31.

Hill, G. E. 1990. Female House Finches prefer colourful males: Sexual selection for a condition-dependent trait. Anim Behav 40: 563–572.

Hill, G. E. 1991. Plumage coloration is a sexually selected indicator of male quality. Nature 350: 337–339.

Hill, G. E. 1992. Proximate basis of variation in carotenoid pigmentation in male House Finches. Auk 109: 1–12.

Hill, G. E. 1995. Interspecific variation in plasma hue in relation to carotenoid plumage pigmentation. Auk 112: 1054–1057.

Hill, G. E. 1996. Redness as a measure of the production cost of ornamental coloration. Ethol Ecol Evol 8: 157–175.

Hill, G. E. 1999a. Mate choice, male quality, and carotenoid-based plumage coloration. Proc Int Ornithol Congr 22: 1654–1668.

Hill, G. E. 1999b. Is there an immunological cost to carotenoid-based ornamental coloration? Am Nat 154: 589–595.

Hill, G. E. 2000. Energetic constraints on expression of carotenoid-based plumage coloration. J Avian Biol 31: 559–566.

Hill, G. E. 2002. A Red Bird in a Brown Bag: The Function and Evolution of Colorful Plumage in the House Finch. New York: Oxford University Press.

Hill, G. E., and K. J. McGraw. 2004. Correlated changes in male plumage coloration and female mate choice in cardueline finches. Anim Behav 67: 27–35.

Hill, G. E., R. Montgomerie, C. Y. Inouye, and J. Dale. 1994. Influence of dietary carotenoids on plasma and plumage colour in the House Finch: Intra- and intersexual variation. Funct Ecol 8: 343–350.

Hill, G. E., C. Y. Inouye, and R. Montgomerie. 2002. Dietary carotenoids predict plumage coloration in wild House Finches. Proc R Soc Lond B 269: 1119–1124.

Hollander, W. F., and R. D. Owen. 1939. The carotenoid nature of yellow pigment in the chicken iris. Poult Sci 18: 385–387.

Horak, P., L. Saks, U. Karu, I. Ots, P. F. Surai, and K. J. McGraw. 2004. How coccidian parasites affect health and appearance of Greenfinches. J Anim Ecol 73: 935–947.

Hornero-Mendez, D., and M. I. Minguez-Mosquera. 2000. Carotenoid pigments in *Rosa mosqueta* hips, an alternative carotenoid source for foods. J Agric Food Chem 48: 825–828.

Hudon, J. 1991. Unusual carotenoid use by the Western Tanager (*Piranga ludoviciana*) and its evolutionary implications. Can J Zool 69: 2311–2320.

Hudon, J. 1994. Biotechnological applications of research on animal pigmentation. Biotech Adv 12: 49–69.

Hudon, J., and A. H. Brush. 1989. Probable dietary basis of a color variant of the Cedar Waxwing. J Field Ornithol 60: 361–368.

Hudon, J., and A. H. Brush. 1990. Carotenoids produce flush in the Elegant Tern plumage. Condor 92: 798–801.

Hudon, J., A. P. Capparella, and A. H. Brush. 1989. Plumage pigment differences in manakins of the *Pipra erythrocephala* superspecies. Auk 106: 34–41.

Hudon, J., H. Ouellet, E. Benito-Espinal, and A. H. Brush. 1996. Characterization of an orange variant of the Bananaquit (*Coereba flaveola*) on La Desirade, Guadeloupe, French West Indies. Auk 113: 715–718.

Hughes, D. A. 2001. Dietary carotenoids and human immune function. Nutrition 17: 823–827.

Inouye, C. Y. 1999. The physiological bases for carotenoid color variation in the House Finch, *Carpodacus mexicanus*. Ph.D. diss., University of California, Los Angeles, Los Angeles.

Inouye, C. Y., G. E. Hill, R. Stradi, and R. Montgomerie. 2001. Carotenoid pigments in male House Finch plumage in relation to age, subspecies, and ornamental coloration. Auk 118: 900–915.

Johnson, K., R. Dalton, and N. Burley. 1993. Preferences of female American Goldfinches (*Carduelis tristis*) for natural and artificial male traits. Behav Ecol 4: 138–143.

Johnson, N. K., and A. H. Brush. 1972. Analysis of polymorphism in the Sooty-capped Bush Tanager. Syst Zool 21: 245–262.

Jouni, Z. E., and M. A. Wells. 1996. Purification and partial characterization of lutein-binding protein from the midgut of the Silkworm *Bombyx mori*. J Biol Chem 271: 14722–14726.

Khachik, F. 2003. An efficient conversion of (3R,3′R,6′R)-lutein to (3R,3′S,6′R)-lutein (3′-epilutein) and (3R,3′R)-zeaxanthin. J Nat Prod 66: 67–72.

Khachik, F., C. J. Spangler, J. C. Smith, Jr., L. M. Canfield, A. Steck, and H. Pfander. 1997. Identification, quantification, and relative concentration of carotenoids and their metabolites in human milk and serum. Analyt Chem 69: 1873–1881.

Khachik, F., F. F. de Moura, D. Y. Zhao, C. P. Aebischer, and P. S. Bernstein. 2002. Transformations of selected carotenoids in plasma, liver, and ocular tissues of humans and in nonprimate animal models. Invest Ophthalmol Vis Sci 43: 3383–3392.

Koutsos, E. A., A. J. Clifford, C. C. Calvert, and K. C. Klasing. 2003a. Maternal carotenoid status modifies incorporation of dietary carotenoids into immune tissues of growing chickens (*Gallus gallus domesticus*). J Nutr 133: 1132–1138.

Koutsos, E. A., C. C. Calvert, and K. C. Klasing. 2003b. The effect of an acute phase response on tissue carotenoid levels of growing chickens (*Gallus gallus domesticus*). Comp Biochem Physiol A 135: 635–646.

Krinsky, N. I. 2001. Carotenoids as antioxidants. Nutrition 17: 815–817.

Krinsky, N. I., M. M. Mathews-Roth, and R. F. Taylor. 1990. Carotenoids: Chemistry and Biology. New York: Plenum Press.

Kritzler, H. 1943. Carotenoids in the display and eclipse plumages of Bishop birds. Physiol Zool 16: 241–255.

Latscha, T. 1990. Carotenoids—Their Nature and Significance in Animal Feeds. Basel, Switzerland: F. Hoffmann-La Roche.

Lindstrom, K., and J. Lundstrom. 2000. Male Greenfinches (*Carduelis chloris*) with brighter ornaments have higher virus infection clearance rate. Behav Ecol Sociobiol 48: 44–51.

Linville, S. U., and R. Breitwisch. 1997. Carotenoid availability and plumage coloration in a wild population of Northern Cardinals. Auk 114: 796–800.

Littlefield, L. H., J. K. Bletner, H. V. Shirley, and O. E. Goff. 1972. Locating the site of absorption of xanthophylls in the chicken by a surgical technique. Poult Sci 51: 1721–1725.

Liu, M., X. Q. Li, C. Weber, C. Y. Lee, J. Brown, and R. H. Liu. 2002. Antioxidant and antiproliferative activities of raspberries. J Agric Food Chem 50: 2926–2930.

Livny, O., I. Kaplan, R. Reifen, S. Polak-Charcon, Z. Madar, and B. Schwartz. 2002.

Lycopene inhibits proliferation and enhances gap-junction communication of KB-1 human oral tumor cells. J Nutr 132: 3754–3759.

Lonnberg, E. 1938. The occurrence and importance of carotenoid substances in birds. Proc Int Ornithol Congr 8: 410–424.

Losey, G. S., T. W. Cronin, T. H. Goldsmith, D. Hyde, N. J. Marshall, and W. N. McFarland. 1999. The UV visual world of fishes: A review. J Fish Biol 54: 921–943.

Lozano, G. A. 1994. Carotenoids, parasites, and sexual selection. Oikos 70: 309–311.

Lucas, A. M., and P. R. Stettenheim. 1972. Avian Anatomy: Integument. Agriculture Handbook 362. Washington, DC: U.S. Government Printing Office.

MacDougall, A. K., and R. Montgomerie. 2003. Assortative mating by carotenoid-based plumage colour: A quality indicator in American Goldfinches, *Carduelis tristis*. Naturwissenschaften 90: 464–467.

Maddocks, S. A., S. C. Church, and I. C. Cuthill. 2001. The effects of the light environment on prey choice by Zebra Finches. J Exp Biol 204: 2509–2515.

Mahler, B., L. S. Araujo, and P. L. Tubaro. 2003. Dietary and sexual correlates of carotenoid pigment expression in dove plumage. Condor 105: 260–269.

Mangels, A. R., J. M. Holden, G. R. Beecher, M. R. Forman, and E. Lanza. 1993. Carotenoid content of fruits and vegetables: An evaluation of analytic data. J Am Diet Assoc 93: 284–296.

Marusich, W. L., and J. C. Bauernfeind. 1981. Oxycarotenoids in poultry feeds. In J. C. Bauernfeind, ed., Carotenoids as Colorants and Vitamin A Precursors, 320–462. New York: Academic Press.

Massaro, M., L. S. Davis, and J. T. Darby. 2003. Carotenoid-derived ornaments reflect parental quality in male and female Yellow-eyed Penguins (*Megadyptes antipodes*). Behav Ecol Sociobiol 55: 169–175.

Matsuno, T. 2001. Aquatic animal carotenoids. Fish Sci 67: 771–783.

Mayne, S. T., B. Cartmel, F. Silva, C. S. Kim, B. G. Fallon, et al. 1999. Plasma lycopene concentrations in humans are determined by lycopene intake, plasma cholesterol concentrations and selected demographic factors. J Nutr 129: 849–854.

Mays, Jr., H. L., K. J. McGraw, G. Ritchison, S. Cooper, V. Rush, and R. S. Parker. 2004. Sexual dichromatism in the Yellow-breasted Chat (*Icteria virens*): Spectrophotometric analysis and biochemical basis. J Avian Biol 35: 125–134.

McGraw, K. J. 2004. Colorful songbirds metabolize carotenoids at the integument. J Avian Biol 35: 471–476.

McGraw, K. J., and D. R. Ardia. 2003. Carotenoids, immunocompetence, and the information content of sexual colors: An experimental test. Am Nat 162: 704–712.

McGraw, K. J., and D. R. Ardia. 2005. Sex differences in carotenoid status and immune performance in Zebra Finches. Evol Ecol Res 7: 251–262.

McGraw, K. J., and A. J. Gregory. 2004. Carotenoid pigments in male American Goldfinches: What is the optimal biochemical strategy for becoming colourful? Biol J Linn Soc 83: 273–280.

McGraw, K. J., and L. Hardy. Astaxanthin is responsible for the pink plumage flush in Franklin's and Ring-billed Gulls. J Field Ornithol (in press).

McGraw, K. J., and G. E. Hill. 2000. Differential effects of endoparasitism on the expression of carotenoid- and melanin-based ornamental coloration. Proc R Soc Lond B 267: 1525–1531.

McGraw, K. J., and G. E. Hill. 2004. Plumage color as a dynamic trait: Carotenoid pigmentation of male House Finches (*Carpodacus mexicanus*) fades during the breeding season. Can J Zool 82: 734–738.

McGraw, K. J., and M. C. Nogare. 2004. Carotenoid pigments and the selectivity of psittacofulvin-based coloration systems in parrots. Comp Biochem Physiol B 138: 229–233.

McGraw, K. J., and J. G. Schuetz. 2004. The evolution of carotenoid coloration in estrildid finches: A biochemical perspective. Comp Biochem Physiol B 139: 45–51.

McGraw, K. J., G. E. Hill, R. Stradi, and R. S. Parker. 2001. The influence of carotenoid acquisition and utilization on the maintenance of species-typical plumage pigmentation in male American Goldfinches (*Carduelis tristis*) and Northern Cardinals (*Cardinalis cardinalis*). Physiol Biochem Zool 74: 843–852.

McGraw, K. J., E. Adkins-Regan, and R. S. Parker. 2002a. Anhydrolutein in the Zebra Finch: A new, metabolically derived carotenoid in birds. Comp Biochem Physiol B 132: 811–818.

McGraw, K. J., G. E. Hill, R. Stradi, and R. S. Parker. 2002b. The effect of dietary carotenoid access on sexual dichromatism and plumage pigment composition in the American Goldfinch. Comp Biochem Physiol B 131: 261–269.

McGraw, K. J., A. J. Gregory, R. S. Parker, and E. Adkins-Regan. 2003a. Diet, plasma carotenoids, and sexual coloration in the Zebra Finch (*Taeniopygia guttata*). Auk 120: 400–410.

McGraw, K. J., M. D. Beebee, G. E. Hill, and R. S. Parker. 2003b. Lutein-based plumage coloration in songbirds is a consequence of selective pigment incorporation into feathers. Comp Biochem Physiol B 135: 689–696.

McGraw, K. J., G. E. Hill, and R. S. Parker. 2003c. Carotenoid pigments in a mutant cardinal: Implications for the genetic and enzymatic control mechanisms of carotenoid metabolism in birds. Condor 105: 587–592.

McGraw, K. J., K. Wakamatsu, S. Ito, P. M. Nolan, P. Jouventin, et al. 2004a. You can't judge a pigment by its color: Carotenoid and melanin content of yellow and brown feathers in swallows, bluebirds, penguins, and Domestic Chickens. Condor 106: 390–395.

McGraw, K. J., R. J. Safran, M. R. Evans, and K. Wakamatsu. 2004b. European Barn Swallows use melanin pigments to color their feathers brown. Behav Ecol 15: 889–891.

McGraw, K. J., G. E. Hill, K. J. Navara, and R. S. Parker. 2004c. Differential accumulation and pigmenting ability of dietary carotenoids in colorful finches. Physiol Biochem Zool 77: 484–491.

McGraw, K. J., K. Wakamatsu, A. B. Clark, and K. Yasukawa. 2004d. Red-winged Blackbirds use carotenoid and melanin pigments to color their epaulets. J Avian Biol 35: 543–550.

McGraw, K. J., E. Adkins-Regan, and R. S. Parker. 2005a. Maternally derived carotenoid pigments affect offspring survival, sex ratio, and sexual attractiveness in a colorful songbird. Naturwissenschaften (in press).

McGraw, K. J., G. E. Hill, and R. S. Parker. 2005b. The physiological costs of being colourful: Nutritional control of carotenoid utilization in the American Goldfinch, *Carduelis tristis*. Anim Behav 69: 653–660.

Menon, G. K., and J. Menon. 2000. Avian epidermal lipids: Functional considerations and relationship to feathering. Am Zool 40: 540–552.

Menon, G. K., B. E. Brown, and P. M. Elias. 1986. Avian epidermal differentiation: Role of lipids in permeability barrier formation. Tissue Cell 18: 71–82.

Minguez-Mosquera, M. I., and D. Hornero-Mendez. 1994. Changes in carotenoid esterification during the fruit ripening of *Capsicum annuum*. J Agric Food Chem 42: 640–644.

Mitchell, H. H. 1962. Comparative Nutrition of Man and Domestic Animals. New York: Academic Press.

Møller, A. P., C. Biard, J. D. Blount, D. C. Houston, P. Ninni, N. Saino, and P. F. Surai. 2000. Carotenoid-dependent signals: Indicators of foraging efficiency, immunocompetence, or detoxification ability? Avian Poult Biol Rev 11: 137–159.

Mortensen, A., and L. H. Skibsted. 1997. Importance of carotenoid structure in radical-scavenging reactions. J Agric Food Chem 45: 2970–2977.

Mozaffarieh, M., S. Sacu, and A. Wedrich. 2003. The role of the carotenoids, lutein and zeaxanthin, in protecting against age-related macular degeneration: A review based on controversial evidence. Nutr J 2: 20–27.

Navara, K. J., and G. E. Hill. 2003. Dietary carotenoid pigments and immune function in a songbird with extensive carotenoid-based plumage coloration. Behav Ecol 14: 909–916.

Needham, A. E. 1974. The Significance of Zoochromes. New York: Springer-Verlag.

Negro, J. J., and J. Garrido-Fernandez. 2000. Astaxanthin is the major carotenoid in tissues of White Storks (*Ciconia ciconia*) feeding on introduced Crayfish (*Procambarus clarkii*). Comp Biochem Physiol B 126: 347–352.

Negro, J. J., G. R. Bortolotti, J. L. Tella, K. J. Fernie, and D. M. Bird. 1998. Regulation of integumentary colour and plasma carotenoids in American Kestrels consistent with sexual selection theory. Funct Ecol 12: 307–312.

Negro, J. J., J. L. Tella, G. Blanco, M. G. Forero, and J. Garrido-Fernandez. 2000. Diet explains interpopulation variation of plasma carotenoids and skin pigmentation in nestling White Storks. Physiol Biochem Zool 73: 97–101.

Negro, J. J., J. L. Tella, F. Hiraldo, G. R. Bortolotti, and P. Prieto. 2001a. Sex- and age-related variation in plasma carotenoids despite a constant diet in the Red-legged Partridge (*Alectoris rufa*). Ardea 89: 275–280.

Negro, J. J., J. Figuerola, J. Garrido, and A. J. Green. 2001b. Fat stores in birds: An overlooked sink for carotenoid pigments? Funct Ecol 15: 297–303.

Negro, J. J., J. M. Grande, J. L. Tella, J. Garrido, D. Hornero, et al. 2002. An unusual source of essential carotenoids. Nature 416: 807–808.

Ninni, P. 2003. Carotenoid Signals in Barn Swallows *Hirundo rustica*. Ph.D. thesis, Laboratoire de Parasitologie Evolutive, Université Pierre et Marie Curie, Paris.

Nys, Y. 2000. Dietary carotenoids and egg yolk coloration: A review. Arch Gefluegelkd 64: 45–54.

Oliphant, L. W. 1988. Cytology and pigments of non-melanophore chromatophores in the avian iris. In J. T. Bagnara, ed., Advances in Pigment Cell Research, 65–82. New York: Alan R. Liss.

Olson, J. A., and O. Hayaishi. 1965. The enzymatic cleavage of beta-carotene into vitamin A by soluble enzymes of rat liver and intestine. Proc Natl Acad Sci USA 54: 1364–1370.

Olson, V. A., and I. P. F. Owens. 1998. Costly sexual signals: Are carotenoids rare, risky or required? Trends Ecol Evol 13: 510–514.

Osborne, D. J., W. E. Huff, P. B. Hamilton, and H. R. Burmeister. 1982. Comparison of ochratoxin, aflatoxin, and T-2 toxin for their effects on selected parameters related to digestion and evidence for specific metabolism of carotenoids in chickens. Poult Sci 61: 1646–1652.

Palozza, P. 1998. Prooxidant actions of carotenoids in biologic systems. Nutr Rev 56: 257–265.

Parker, R. S. 1996. Absorption, metabolism, and transport of carotenoids. FASEB J 10: 542–551.

Parker, R. S., J. E. Swanson, B. Marmor, K. J. Goodman, A. B. Spielman, et al. 1993. Study of β-carotene metabolism in humans using 13C-β-carotene and high precision isotope ratio mass spectrometry. Ann NY Acad Sci 691: 86–95.

Parker, R. S., J. E. Swanson, C. S. You, A. J. Edwards, and T. Huang. 1999. Bioavailability of carotenoids in human subjects. Proc Nutr Soc 58: 155–162.

Partali, V., S. Liaaen-Jensen, T. Slagsvold, and J. T. Lifjeld. 1987. Carotenoids in food chain studies—II. The food chain of *Parus* spp. monitored by carotenoid analysis. Comp Biochem Physiol 87: 885–888.

Prince, M. R., and J. K. Frisoli. 1993. Beta-carotene accumulation in serum and skin. Am J Clin Nutr 57: 175–181.

Prum, R. O., and R. Torres. 2003. Structural colouration of avian skin: Convergent evolution of coherently scattering dermal collagen arrays. J Exp Biol 206: 2409–2429.

Quackenbush, F. W., J. G. Firch, W. J. Rabourn, M. McQuistan, E. W. Petzold, and T. E. Kargl. 1961. Analysis of carotenoids in corn grain. J Agric Food Chem 9: 132–135.

Rock, C. L., and M. E. Swendseid. 1992. Plasma β-carotene response in humans after meals supplemented with dietary pectin. Am J Clin Nutr 55: 96–99.

Rodd, F. H., K. A. Hughes, G. F. Grether, and C. T. Baril. 2002. A possible non-sexual origin of mate preference: Are male Guppies mimicking fruit? Proc R Soc Lond B 269: 475–481.

Rodriguez-Amaya, D. B. 1999. A Guide to Carotenoid Analysis in Foods. Washington, DC: OMNI Research.

Ruff, M. D., and H. L. Fuller. 1975. Some mechanisms of reduction of carotenoid levels in chickens infected with *Eimeria acervulina* or *E. tenella*. J Nutr 105: 1447–1456.

Rumpho, M. E., E. J. Summer, and J. R. Manhart. 2000. Solar-powered sea slugs. Mollusc/algal chloroplast symbiosis. Plant Physiol 123: 29–38.

Ryan, P. G., C. L. Moloney, and J. Hudon. 1994. Color variation and hybridization among *Neospiza* buntings on Inaccessible Island, Tristan da Cunha. Auk 111: 314–327.

Saino, N., P. Ninni, S. Calza, R. Martinelli, F. De Bernardi, and A. P. Møller. 2000. Better red than dead: Carotenoid-based mouth coloration reveals infection in Barn Swallow nestlings. Proc R Soc Lond B 267: 57–61.

Saino, N., R. Ambrosini, R. Martinelli, P. Ninni, and A. P. Møller. 2003a. Gape coloration reliably reflects immunocompetence of Barn Swallow (*Hirundo rustica*) nestlings. Behav Ecol 14: 16–22.

Saino, N., R. Ferrari, M. Romano, R. Martinelli, and A. P. Møller. 2003b. Experimental manipulation of egg carotenoids affects immunity of Barn Swallow nestlings. Proc R Soc Lond B 270: 2485–2489.

Saks, L., K. J. McGraw, and P. Horak. 2003a. How feather colour reflects its carotenoid content. Funct Ecol 17: 555–561.

Saks, L., I. Ots, and P. Horak. 2003b. Carotenoid-based plumage coloration of male Greenfinches reflects health and immunocompetence. Oecologia 134: 301–307.

Schaefer, H. M., and V. Schmidt. 2004. Detectability and content as opposing signal characteristics in fruits. Proc R Soc Lond B 271 (Suppl.): 370–373.

Schiedt, K. 1989. New aspects of carotenoid metabolism in animals. In N. I. Krinsky, M. M. Mathews-Roth, and R. F. Taylor, ed., Carotenoids: Chemistry and Biology, 247–268. New York: Plenum Press.

Schiedt, K. 1998. Absorption and metabolism of carotenoids in birds, fish and crustaceans. In G. Britton, S. Liaaen-Jensen, and H. Pfander, ed., Carotenoids: Biosynthesis, Volume 3, 285–355. Basel, Switzerland: Birkhauser Verlag.

Schlinger, B. A. 1998. Sexual differentiation of avian brain and behavior: Current views on gonadal hormone-dependent and independent mechanisms. Annu Rev Physiol 60: 407–429.

Schweigert, F. J. 1998. Metabolism of carotenoids in mammals. In G. Britton, S. Liaaen-Jensen, and H. Pfander, ed., Carotenoids: Biosynthesis, Volume 3, 249–284. Basel. Switzerland: Birkhauser Verlag.

Senar, J. C., and D. Escobar. 2002. Carotenoid-derived plumage coloration in the siskin *Carduelis spinus* is related to foraging ability. Avian Sci 2: 19–24.

Shawkey, M. D., and G. E. Hill. 2004. Feathers at a fine scale. Auk 121: 652–655.

Shawkey, M. D., and G. E. Hill. 2005. Carotenoids need nanostructures to shine. Biol Lett 1: 121–124.

Siitari, H., J. Honkavaara, and J. Viitala. 1999. Ultraviolet reflectance of berries attracts foraging birds. A laboratory study with Redwings (*Turdus iliacus*) and Bilberries (*Vaccinium myritillus*). Proc R Soc Lond B 266: 2125–2129.

Silberglied, R. 1979. Communication in the ultraviolet. Annu Rev Ecol Syst 10: 373–398.

Slifka, K. A., P. E. Bowen, M. Stacewicz-Sapuntzakis, and S. D. Crissey. 1999. A survey of serum and dietary carotenoids in captive wild animals. J Nutr 129: 380–390.

Smith, C., I. Barber, R. J. Wootton, and L. Chittka. 2004. A receiver bias in the origin of Three-spined Stickleback mate choice. Proc R Soc Lond B 271: 949–955.

Stahl, W., and H. Sies. 2002. Carotenoids and protection against solar UV radiation. Skin Pharmacol Appl Skin Physiol 15: 291–296.

Stockton-Shields, C. 1997. Sexual selection and the dietary color preferences of House Finches. M.Sc. thesis, Auburn University, Auburn, AL.

Stradi, R. 1998. The Colour of Flight: Carotenoids in Bird Plumage. Milan: Solei Gruppo Editoriale Informatico.

Stradi, R. 1999. Pigmenti e sistematica degli uccelli. In L. Brambilla, G. Canali, E. Mannucci, et al., ed., Colori in volo: il piumaggio degli uccelli, 117–146. Milan: Università degli Studi di Milano.

Stradi, R., G. Celentano, and D. Nava. 1995a. Separation and identification of carotenoids in bird's plumage by high-performance liquid chromatography—diode-array detection. J Chromatogr B 670: 337–348.

Stradi, R., G. Celentano, E. Rossi, G. Rovati, and M. Pastore. 1995b. Carotenoids in bird plumage—I. The carotenoid pattern in a series of Palearctic Carduelinae. Comp Biochem Physiol B 110: 131–143.

Stradi, R., E. Rossi, G. Celentano, and B. Bellardi. 1996. Carotenoids in bird plumage: The pattern in three *Loxia* species and in *Pinicola enucleator*. Comp Biochem Physiol B 113: 427–432.

Stradi, R., G. Celentano, M. Boles, and F. Mercato. 1997. Carotenoids in bird plumage: The pattern in a series of red-pigmented Carduelinae. Comp Biochem Physiol B 117: 85–91.

Stradi, R., J. Hudon, G. Celentano, and E. Pini. 1998. Carotenoids in bird plumage: The complement of yellow and red pigments in true woodpeckers (Picinae). Comp Biochem Physiol B 120: 223–230.

Stradi, R., E. Pini, and G. Celentano. 2001. Carotenoids in bird plumage: The complement of red pigments in the plumage of wild and captive Bullfinch (*Pyrrhula pyrrhula*). Comp Biochem Physiol B 128: 529–535.

Su, Q., K. G. Rowley, C. Itsiopoulos, and K. O'Dea. 2002. Identification and quantitation of major carotenoids in selected components of the Mediterranean diet: Green leafy vegetables, figs and olive oil. Euro J Clin Nutr 56: 1149–1154.

Surai, P. F. 2002. Natural antioxidants in avian nutrition and reproduction. Nottingham, UK: Nottingham University Press.

Surai, P. F., and N. H. C. Sparks. 2001. Comparative evaluation of the effect of two maternal diets on fatty acids, vitamin E and carotenoids in the chick embryo. Brit Poult Sci 42: 252–259.

Surai, P. F., and B. K. Speake. 1998. Distribution of carotenoids from the yolk to the tissues of the chick embryo. J Nutr Biochem 9: 645–651.

Surai, P. F., R. C. Noble, and B. K. Speake. 1996. Tissue-specific differences in antioxidant distribution and susceptibility to lipid peroxidation during development of the chick embryo. Biochim Biophys Acta 1304: 1–10.

Surai, P. F., I. A. Ionov, E. F. Kuchmistova, R. C. Noble, and B. K. Speake. 1998. The relationship between the levels of α-tocopherol and carotenoids in the maternal feed, yolk and neonatal tissues: Comparison between the chicken, turkey, duck and goose. J Sci Food Agric 76: 593–598.

Surai, P. F., R. M. McDevitt, B. K. Speake, and N. H. C. Sparks. 1999. Carotenoid distribution in tissues of the laying hen depending on their dietary supplementation. Proc Nutr Soc 58: 30A.

Surai, P. F., N. K. Royle, and N. H. C. Sparks. 2000. Fatty acid, carotenoid and vitamin A composition of tissues of free living gulls. Comp Biochem Physiol A 126: 387–396.

Surai, P. F., B. K. Speake, N. A. R. Wood, J. D. Blount, G. R. Bortolotti, and N. H. C. Sparks. 2001. Carotenoid discrimination by the avian embryo: A lesson from wild birds. Comp Biochem Physiol B 128: 743–750.

Surai, A. P., P. F. Surai, W. Steinberg, W. G. Wakeman, B. K. Speake, and N. H. C. Sparks. 2003. Effect of canthaxanthin content of the maternal diet on the antioxidant system of the developing chick. Brit Poult Sci 44: 612–619.

Tabunoki, H., H. Sugiyama, Y. Tanaka, H. Fujii, and Y. Banno. 2002. Isolation, characterization, and cDNA sequence of a carotenoid binding protein from the silk gland of *Bombyx mori* larvae. J Biol Chem 277: 32133–32140.

Tapiero, H., D. M. Townsend, and K. D. Tew. 2004. The role of carotenoids in the prevention of human pathologies. Biomed Pharmacother 58: 100–110.

Tella, J. L., J. J. Negro, R. Rodriguez-Estrella, G. Blanco, M. G. Forero, et al. 1998. A comparison of spectrophotometry and color charts for evaluating total plasma carotenoids in wild birds. Physiol Zool 71: 708–711.

Tella, J. L., J. Figuerola, J. J. Negro, G. Blanco, R. Rodriguez-Estrella, et al. 2004. Ecological, morphological, and phylogenetic correlates of interspecific variation in plasma carotenoid concentration in birds. J Evol Biol 17: 156–164.

Test, F. H. 1969. Relation of wing and tail color of the woodpeckers *Colaptes auratus* and *C. cafer* to their food. Condor 71: 206–211.

Thommen, H. 1971. Metabolism. Basel, Switzerland: Birkhauser Verlag.

Thomson, L. R., Y. Toyoda, A. Langner, F. C. Delori, K. M. Garnett, et al. 2002. Elevated retinal zeaxanthin and prevention of light-induced photoreceptor cell death in quail. Invest Ophthalmol Vis Sci 43: 3538–3549.

Toyoda, Y., L. R. Thomson, A. Langner, N. E. Craft, K. M. Garnett, et al. 2002. Effect of dietary zeaxanthin on tissue distribution of zeaxanthin and lutein in quail. Invest Ophthalmol Vis Sci 43: 1210–1221.

Trams, E. G. 1969. Carotenoid transport in the plasma of the Scarlet Ibis (*Eudocimus ruber*). Comp Biochem Physiol 28: 1177–1184.

Tsushima, M., M. Katsuyama, and T. Matsuno. 1997. Metabolism of carotenoids in the Apple Snail, *Pomacea canaliculata*. Comp Biochem Physiol B 118: 431–436.

Tyczkowski, J. K., and P. B. Hamilton. 1986a. Evidence for differential absorption of zeacarotene, cryptoxanthin, and lutein in young broiler chickens. Poult Sci 65: 1137–1140.

Tyczkowski, J. K., and P. B. Hamilton. 1986b. Lutein as a model dihydroxycarotenoid for the study of pigmentation in chickens. Poult Sci 65: 1141–1145.

van den Berg, H. 1999. Carotenoid interactions. Nutr Rev 57: 1–10.

Vershinin, A. 1999. Biological functions of carotenoids—Diversity and evolution. BioFactors 10: 99–104.

Völker, O. 1938. The dependence of lipochrome-formation in birds on plant carotenoids. Proc Int Ornithol Congr 8: 425–426.

Walzem, R. L. 1996. Lipoproteins and the laying hen: Form follows function. Poult Avian Biol Rev 7: 31–64.

Weis, A. E., and B. Bisbey. 1947. The Relation of the Carotenoid Pigments of the Diet to Growth of Young Chicks and to Storage in Their Tissues. Resource Bulletin 405. Columbia: University of Missouri.

Williams, A. W., T. W. M. Boileau, and J. W. Erdman, Jr. 1998. Factors influencing the uptake and absorption of carotenoids. Proc Soc Exp Biol Med 218: 106–108.

Willoughby, E. J., M. Murphy, and H. L. Gorton. 2002. Molt, plumage abrasion, and color change in Lawrence's Goldfinch. Wilson Bull 114: 380–392.

Willstatter, R., and H. H. Escher. 1912. Lutein obtained from egg yolks. Zeit Physiol Chem 76: 214–225.

Witmer, M. 1996. Consequences of an alien shrub on the plumage coloration and ecology of Cedar Waxwings. Auk 113: 735–743.

Wyss, A. 2004. Carotene oxygenases: A new family of double bond cleavage enzymes. J Nutr 134: 246S–250S.

Yao, L., Y. Liang, W. S. Trahanovski, R. E. Serfass, and W. S. White. 2000. Use of a ^{13}C tracer to quantify the plasma appearance of physiological dose of lutein in humans. Lipids 35: 339–348.

Yemelyanov, A. Y., N. B. Katz, and P. S. Bernstein. 2001. Ligand-binding characterization of xanthophyll carotenoids to solubilized membrane proteins derived from human retina. Exp Eye Res 72: 381–392.

Young, A. J., and G. M. Lowe. 2001. Antioxidant and prooxidant properties of carotenoids. Arch Biochem Biophys 385: 20–27.

Yu, J. Y. L., L. D. Campbell, and R. R. Marquardt. 1976. Immunological and compositional patterns of lipoproteins in chicken (*Gallus domesticus*) plasma. Poult Sci 55: 1626–1631.

Zagalsky, P. F. 1995. Carotenoproteins. In G. Britton, S. Liaaen-Jensen, and H. Pfander, ed., Carotenoids: Isolation and Analysis. Volume 1a, 287–294. Basel, Switzerland: Birkhauser Verlag.

Zahavi, A., and A. Zahavi. 1998. The handicap principle: A missing piece of Darwin's puzzle. New York: Oxford University Press.

6

Mechanics of Melanin-Based Coloration

KEVIN J. MCGRAW

Melanin is the most abundant and widespread pigment in birds and other animals. With the rare exception of albinos, all birds deposit some melanin in the integument. One might imagine, therefore, that studies of the mechanisms and functions of avian melanism are the most prevalent and advanced in the field of bird coloration. On the contrary, biological and chemical investigations of melanin pigmentation in birds have generally lagged behind those on carotenoid coloration (Chapter 5), for example, both in number and depth. This lag likely stems from the difficulties that chemists have traditionally had in analyzing melanin, because of the localized, endogenous, and difficult-to-track processes by which melanin is formed, and perhaps due to the limited extent to which melanization varies seasonally, intraspecifically, and sexually relative to other forms of coloration. Nonetheless, there are many species of birds, such as House Sparrows (*Passer domesticus*) and Eastern Bluebirds (*Sialia sialis*), in which melanin expression is quite variable and is known to serve important behavioral or ecological functions (see Part I, Volume 2). Couple the emerging interest in melanin colors as signals with a series of recent advances in melanin biochemistry and physiology in other fields, and our sights are set on a productive future of research into how and why such a diversity of melanin colors and patterns have evolved in birds.

As was the case for the previous chapter on carotenoid coloration, there have been a few other recent reviews published on melanin coloration in birds

(Jawor and Breitwisch 2003; McGraw 2003). My goal here is to provide an updated, broader, more historical, and more biochemically centered review of the processes that generate integumentary melanization in birds. It is also through the incorporation of literature on melanin production in other organisms that I aim to highlight new avenues for future studies of the costs and information content of melanic signals in birds.

Types and Distribution of Melanins

Occurrence as Animal Colorants

Melanins yield many of the black, brown, gray, rufous, and buff shades and patterns we see in plants, fungi, and animals. The damaged skin of fruits (and humans!) turns brown as the result of melanization. *Daphnia* encase their eggs with a black protective shell that is laced with melanin (Gerrish and Caceres 2003). Among vertebrates, we find melanin in all of the main types of integumentary structures, from the black skin of Spotted Salamanders (*Ambystoma maculatum;* Lesser et al. 2001), the black–striped scale pattern in Zebrafish (*Danio* sp.; Parichy 2003), and the buff feathers of Japanese Quail (*Coturnix japonica;* Shiojiri et al. 1999), to the black, brown, red, and blonde hair in humans (Ito and Fujita 1985).

Although birds are perhaps best known for their unparalleled diversity of flashy integumentary colors (e.g., reds, yellows, blues), the varied means by which they use melanins as body colorants cannot be overlooked. Melanin colors in birds range from the fully black plumage in the Common Raven (*Corvus corax*) to the fully brown feathers of the Golden Eagle (*Aquila chrysaetos*) and the charcoal-gray feathers of Gray Catbirds (*Dumetella carolinensis*), with a tremendous array of intermediate patterns and hues within and among body regions and tissues. In fact, the selective and highly patterned occurrence of melanin pigments in feathers is unrivaled among the avian integumentary colorants. For no other form of coloration do we see the remarkable bars, stripes, and spots interspersed with regions of unpigmented feather (Plate 6). Mottled plumage appears especially common in the hawks, owls, grouse, rails, sandpipers, seed snipes, sandgrouse, goatsuckers, puffbirds, woodpeckers, and songbirds. Waterbirds (e.g., grebes, loons, ducks, gulls, terns, alcids) typically show a countershaded color pattern, developing dark, melanic upper parts but light lower parts. Other commonly melanized parts of the body include the head and neck plumage (of several ducks and finches), as well as feathers at the throat (typical of many songbird species) and breast (as in the brown colors of American Robins [*Turdus migratorius*] and Mallards [*Anas platyrhynchos*; Plate 15, Volume 2] or the horizontal black stripes of plovers; Plate 6), all of which are ideally situated for face-to-face displays and interactions that might be predicted for traits suspected to serve communicative functions. Individual feathers may also exhibit marked melanin bands or spots (e.g., the tail, flank, and breast feathers of the male Zebra Finch [*Taeniopygia guttata*]; Plate 27).

These complex color patterns largely speak to the localized control that maturing pigment-producing cells (melanocytes) have in turning melanin synthesis on and off during feather growth and keratinization (discussed in more detail below). Carotenoid pigments, by contrast, are taken up along with other circulating lipids by feather follicles and incorporated as a pool into single, contiguous patches of pigment (Chapter 5), never exhibiting any banding in individual feathers. Ultimately, it is via this production mechanism that melanins are selectively used in forms of disruptive and cryptic coloration that require precise arrangement of pigmented and unpigmented body regions to match the background characteristics of their microhabitat (Chapter 1, Volume 2). Interestingly, bare parts are rarely patterned, but instead are uniformly pigmented with melanin, like the black beak and legs of Tundra Swans (*Cygnus columbianus*; Plate 26, Volume 2).

Not all black, brown, gray, chestnut, or buff color in birds can be ascribed solely to melanin pigments, however (McGraw et al. 2004a). Porphyrin pigments give the brown or rufous hues to eggshells of all birds and to the feathers of bustards, owls, and goatsuckers (Chapter 8). Carotenoids (Chapter 5) and pterins (Chapter 8) confer yellowish and orangish colors that may be confused with melanins. In addition, although melanins are clearly present as the base pigment in feathers, microstructural arrangements of tissues can also modify the intensity of black plumage coloration and the absorbance of ultraviolet (UV) wavelengths (Finger et al. 1992). Melanins can also mask the presence of other brightly colored pigments (e.g., turacin in turacos; Moreau 1958) that may subtly contribute to perceived coloration. Thus, as is recommended for any type of pigment in animals, it is wise to withhold classification of a color as partially or wholly melanin-based before the appropriate biochemical tests are conducted (Box 6.1). If there is any safe assumption about avian coloration though, it is that black colors contain melanin, because no other black pigments have been identified in birds.

Box 6.1. Biochemical Methods for Melanin Analysis

Because of their large size and complex structure, melanins have historically been difficult to analyze biochemically (Ito et al. 2000). Some of the first sophisticated methods in melanin analysis, such as electron spin resonance spectroscopy, relied on chemical isolation of melanin, which was time-consuming and not always feasible for the alkali- and acid-insoluble eumelanin, and could not distinguish between eumelanin and phaeomelanin (Sealy et al. 1982). Melanins are, however, quite reactive, and their metabolites have been found to be much more manageable chemically. Initially devised by Ito and Fujita in 1985, a series of chemical tests, combined with chromatographic techniques, is now available for quantification of phaeomelanin and eumelanin from animal tissues. These tests capitalize on the formation of unique degradative products from eumelanin and phaeomelanin when they are oxidized and hydrolyzed, respectively (Figure B6.1). Eumelanin, when reacted with permanganate, degrades into pyrrole-2,3,5-tricarboxylic acid (PTCA). Phaeomelanin, in contrast, with its benzothiazine units, can be hydrolyzed with acid into aminohydroxyphenylalanine (AHP) isomers. PTCA and AHP can then be visualized using high-performance liquid chromatography (HPLC) and their concentrations determined. Conversion factors (e.g., multiplying PTCA by 50 and AHP by 9), which account for the incomplete yield of these degradation products from melanin, can then be used to estimate total eumelanin and phaeomelanin concentrations, respectively. This method has been revised over the years (Ito and Wakamatsu 1994; Wakamatsu and Ito 2002; Wakamatsu et al. 2002) to include chemical reactions (e.g., modification of permanganate oxidation) that improve yield and repeatability. It has been employed to analyze feather melanins in 13 avian species to date (Table 6.1).

The best method to store tissue for melanin analysis has not yet been systematically determined. At this stage, until definitive tests are conducted on the stability of melanins in animal tissues under a variety of storage conditions, I can only recommend conservative approaches to storing melanized tissues prior to analysis. Long-term freezing at ultracold temperatures is the preferred method, with storage at room temperature in the dark being the next most desired option if no freezer is immediately available. Serum has been analyzed for melanin-degradative products in humans and found to contain levels in proportion to intensity of melanoma in clinical patients (Ito and

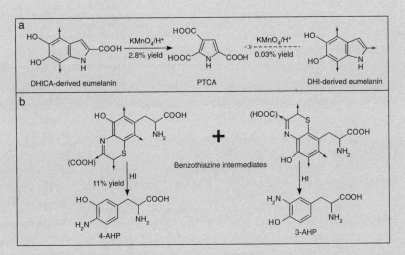

Figure B6.1. Chemical degradation pathways followed in HPLC-based analyses of melanin metabolites. (a) The two main forms of melanin are oxidized with $KMnO_4$ to yield PTCA. (b) The benzothiazine units of phaeomelanin are hydrolyzed using hydriodic acid to give two isomers of AHP. See Figure 6.2 for the structures of DHI and DHICA. Redrawn from Ito and Wakamatsu (2003).

Wakamatsu 2003), and it would be interesting to consider serum melanins in relation to the health and appearance of birds.

As with carotenoids (McGraw et al. 2005; Chapter 5), I point out that, for those biologists interested in simply characterizing a red, orange, yellow, or brown animal color as containing melanins, there is a simple and inexpensive set of tests that can be used instead of HPLC to identify phaeomelanins from animal tissues. First, ultraviolet-visible reflectance spectrophotometry can reveal a diagnostic melanin-containing spectral shape—a line that steadily increases in reflectance from 300 to 700 nm. Some feathers contain multiple types of pigments, however, so follow-up chemical tests are important. First, phaeomelanins can be extracted from tissue using a mild base (e.g., 0.1 N NaOH). The solution may need to be gently heated, or the tissue ground, to release the pigments into solution. Other pigments of similar color (e.g., yellow to brownish-red pterins) are soluble in these solutions, however. Thus a follow-up step is needed. The distinguishing feature of melanins and pterins is that only pterins fluoresce under long-wave UV light (Needham 1974), so the solution can then be held under a UV lamp to de-

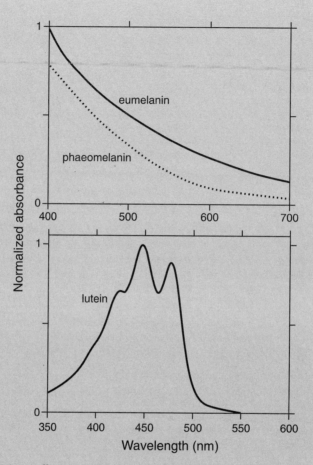

Figure B6.2. Differences in the absorbance spectra for melanins (eumelanin and phaeomelanin) and carotenoids (lutein) in solution. Note the generally linear spectrum for melanins. Melanin figure adapted from Krishnaswamy and Baranoski (2004). Lutein figure adapted from Zang et al. (1997).

termine its chemical nature. As further confirmation of the identity of the extracted study pigment, absorption spectrophotometry can be used to identify the spectral shape of your presumed melanin-containing solution; opposite of the reflectance spectrum, it should be linearly decreasing from low to high bird-visible wavelengths (Figure B6.2).

Mechanics of Melanin-Based Coloration 249

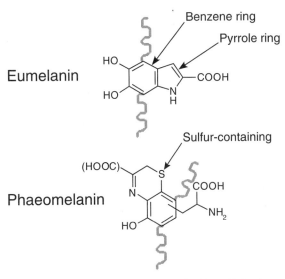

Figure 6.1. Chemical structure of the indole moiety ("backbone") for the two main forms of melanin: eumelanin and phaeomelanin. Diagnostic and bioactive features of melanin in general are highlighted in the eumelanin diagram. Note the unique presence of sulfur in phaeomelanin. The full chemical structures of melanins have never been identified, but are thought to consist of a series of connected indole moieties bound to various biochemicals (e.g., keratin, metals).

Chemical, Structural, and Chromatic Characteristics of Melanins

Melanins are biochemically classified as indole biochromes—a family of pigments in which nitrogen is incorporated into a core indole moiety that consists of a benzene and pyrrole ring (Fox and Vevers 1960; Figure 6.1). Melanin pigments contain several covalently linked indoles and are regarded as unusually large polymers compared to most natural pigments (Riley 1997). In fact, it is because of their large size and strong cross-linking properties that their precise structure(s) and characteristics remain incompletely described to this day. Depending on what melanins are bound to (e.g., proteins, metal ions), they can take on very different forms and exhibit very different properties (discussed in more detail below). They generally absorb more light than other pigments, with extinction coefficients two or three times greater than those of carotenoids at λ_{max} (Sarna and Swartz 1998).

There are two general categories of melanin pigments in animals—eumelanins and phaeomelanins. Eumelanin is the more prevalent form that confers dark black or brown hues in invertebrates and vertebrates. The high degree of conjugation and the large number of indole quinone groups in the irregular eumelanin biopolymer are responsible for the near-uniform absorption across the UV and visible spectrum, thus giving these pigments their dark black and brown appearance (Riley 1997). Eumelanin molecules are thought to be the larger of the two forms of melanin, occurring as large rodlike granules and leaving them insoluble in nearly all solvents. Phaeomelanins, in contrast, are the reddish-brown pigments that predominate in red human red hair, the red and yellow fur of mammals, and the chestnut and rufous feathers of birds. They are lower molecular-weight, smaller, globular granules that are soluble in alkaline solutions and have quite different light-absorbance and structural characteristics than eumelanins. Their characteristic absorbance spectra, which steadily decreases from low to high wavelengths, is due to the presence of fewer functional (carbonyl) groups than found in eumelanins (Nickerson 1941), which absorb more red light and leave them with a paler brown or buff appearance than eumelanins display. The structural properties of phaeomelanins are perhaps best described from the reddish feathers (containing gallophaeomelanin) of Domestic Chickens (*Gallus gallus*), in which sulfur is a major component, in the form of benzothiazoles (Minale et al. 1967, 1969; Prota and Nicolaus 1967; Fattorusso et al. 1968, 1969, 1970). Interestingly, invertebrates and poikilothermic vertebrates do not appear to be capable of manufacturing phaeomelanins (Prota 1992).

The Melanin Continuum: How Pigment Types Confer Color

Melanin was first described from feathers over a century ago (reviewed in Verne 1926), but it has only been in the past few decades that we have begun to fully appreciate the relative and absolute contributions that phaeomelanins and eumelanins make to the color displays of birds (see Ito and Wakamatsu 2003 for similar evidence in mammals). It took until the middle part of the twentieth century for biologists to first identify reddish-brown feather colors in such birds as chickens as melanin-based (Trinkaus 1948) and phaeomelanin-containing (Somes and Smyth 1965). However, throughout much of the literature on melanin coloration, biologists have been quick to characterize melanin color types as either eumelanin-based or phaeomelanin-based, with little consideration to the possibility that both types are present in tissues and simply vary in their densities or proportions to give the spectrum of shades and hues that typify melanin pigmentation. With the help of recently devised biochemical techniques that center on analyzing the degradation products of the two types of melanins (Box 6.1), we have now characterized the melanin profiles for colorful feathers in 13 avian species spanning four avian orders (Table 6.1). Even with such a small sample, several patterns emerge. First, it is clear that all melanin-containing feathers analyzed to date contain both eumelanin and phaeomelanin granules, even if one of them makes up a very small fraction of the total melanins present (e.g., 0.8% eumelanin in the rufous cheek patches of male Zebra Finches; Plate 27). Second, total melanin concentration is less meaningful in shaping color variability than is the relative proportion of the two pigment types. For example, black feathers in Zebra Finches (McGraw and Wakamatsu 2004) and yellow feathers in Red-winged Blackbirds (*Agelaius phoeniceus;* McGraw et al. 2004b) have roughly the same total melanin concentration (~15 mg/g) but exhibit markedly different colors, due to the comparative abundances of eumelanin (92% of total melanins in black feathers) and phaeomelanin (83% of total melanins in yellow feathers), respectively. Third, melanins can occur in feathers where other colorful pigments (e.g., carotenoids) are also present, although they usually are found in dilute quantities. In the end, results from this first set of analyses on melanin composition in feathers underscore the need for more analyses of this sort on bird feathers and bare parts of all hues and shades, as well as to cautiously classify colors as either "phaeomelanin-dominated" or "eumelanin-dominated" until full pigment analyses are completed.

Sites and Pathways of Melanogenesis

Unlike the other major source of pigments in the avian integument—the carotenoids—melanins are not derived directly from the diet and instead are manufactured by animals endogenously (see Chapter 8 for other endogenous sources of pigment in birds). Given the prevalence of melanin colors in animals and the interest in melanin synthesis in far-ranging fields from insect sclerotization (Sugumaran 2002) to human health (e.g., melanomas, stem-cell research, solar UV protection; Prota et al. 1994; Stinchcombe et al. 2004; also see below), there is an expansive literature on the cellular and molecular events that underlie melanogenesis and the development of the pigment-producing cells (melanocytes). The vast majority of the studies in vertebrates has been conducted in mammals (on mice and humans), but there is a recent and noteworthy body of in vitro work on pigment cells and production, particularly during ontogeny, in Japanese Quail (Dupin and LeDouarin 2002) and older

Table 6.1. Thirteen Avian Species for Which the Melanin Content of Plumage Has Been Characterized

Species	Plumage color and region	P	E	Total	Reference
Rock Pigeon (*Columba livia*)	Black morph ("spread")	0.70	25.35	26.05	Haase et al. (1992)
	Red morph ("red")	2.75	39.25	42.0	Haase et al. (1992)
Mallard (*Anas platyrhynchos*)	Green head	1.15	0.49	1.64	Haase et al. (1995)
	Brown breast	1.80	1.94	3.74	Haase et al. (1995)
	Gray flank	1.33	0.72	2.05	Haase et al. (1995)
Japanese Quail (*Coturnix japonica*)	Brown dorsal feathers	27.5	1.58	29.08	Shiojiri et al. (1999)
	Mutant black feathers	5.5	0.9	6.4	Shiojiri et al. (1999)
	Mutant yellow feathers	0.25	3.65	3.9	Shiojiri et al. (1999)
Macaroni Penguin (*Eudyptes chrysolophus*)	Yellow forehead plume[a]	0.18	0.3	0.48	McGraw et al. (2004a)
King Penguin (*Aptenodytes patagonicus*)	Orange auricular/chest plumage[a]	0.04	0.4	0.44	McGraw et al. (2004a)
Eastern Bluebird (*Sialia sialis*)	Rufous breast plumage	1.9	3.9	5.8	McGraw et al. (2004a)
Domestic Chicken (*Gallus gallus*)	Yellow downy feathers[a]		None detected		McGraw et al. (2004a)
Zebra Finch (*Taeniopygia guttata*)	Black breast feathers	1.1	13.6	14.7	McGraw and Wakamatsu (2004)
	Orange cheek patches	46.8	0.03	46.83	McGraw and Wakamatsu (2004)
	Brown flank feathers	55.3	3.8	59.1	McGraw and Wakamatsu (2004)
Barn Swallow (*Hirundo rustica rustica*)	Chestnut throat feathers	10.0	3.0	13.0	McGraw et al. (2004c)
(*H. r. erythrogaster*)	Chestnut throat feathers	6.6	3.2	9.8	McGraw et al. (2004a)
Red-winged Blackbird (*Agelaius phoeniceus*)	Red epaulet[b]	0.29	1.48	1.77	McGraw et al. (2004b)
	Yellow epaulet	12.4	2.6	15.0	McGraw et al. (2004b)
	Black plumage	0.03	21.0	21.03	McGraw et al. (2004b)
American Goldfinch (*Carduelis tristis*)	Yellow breast feather[b]	0.03	0.10	0.13	McGraw et al. (2004b)
Northern Cardinal (*Cardinalis cardinalis*)	Red breast feather[b]	0.06	0.14	0.20	McGraw et al. (2004b)
	Red breast feather[b]	0.33	0.37	0.70	McGraw et al. (2004b)
Red-billed Quelea (*Quelea quelea*)	Yellow breast feather[c]	0.44	0.28	0.72	J. Dale and K. McGraw (unpubl. data)

Notes: Both phaeomelanins (P) and eumelanins (E) were measured, in mg of pigment per g of pigmented feather. Modern chemodegradation and HPLC-based techniques were used in all cases to quantify melanin concentration (sensu Ito and Fujita 1985). It is worth pointing out that these degradation methods assume that all eumelanins and phaeomelanins across species are structurally similar and degrade at equivalent yields into their chemically analyzed by-products. This assumption has yet to be tested.
a. These feathers derive their rich yellow color from a different source of pigment, presumably related to pterins (Chapter 5).
b. These feathers derive their red colors from carotenoid pigments (Chapter 5).
c. This yellow color is derived from an as-of-yet undescribed pigment (Chapter 8).

work on Domestic Chickens (Pollock 1967). Caution must be exercised in extrapolating results from studies of melanin in other vertebrates to birds, given that there are a handful of unique differences in the cellular and hormonal triggers between birds and other taxa (discussed in more detail below). However, with the remarkable amount of information on melanogenesis in other animals (Schallreueter 1999; Fujii 2000), this may be the form of bird coloration that can best capitalize on research from other fields to elevate studies to a new level of understanding. Here, I review the cellular and molecular processes behind melanocyte and melanin formation in birds and draw comparisons to other systems where they provide interesting similarities, differences, or opportunities for future work.

Vertebrates synthesize melanin at peripheral tissues, such as the skin, for integumentary coloration. Specifically, it is the epidermis in birds and mammals that houses the melanocytes that manufacture melanin (Mason and Frost-Mason 2000). This configuration is unlike the cellular machinery behind melanin coloration in poikilothermic vertebrates, which use a layered pigment-cell system, made up of various types of chromatophores (e.g., xanthophores, erythrophores, iridophores, the melanin-containing melanophores), in the dermis to achieve a striking array of colors (Bagnara and Hadley 1973). These melanophores in fact are capable of actively sequestering or releasing melanin, via organelle dispersion or aggregation (discussed in more detail below, in the section on melanosomes) in response to hormonal cues, to rapidly change color (e.g., the darkening of reptiles, amphibians, and fish under aggressive or stressful conditions). This activity is in stark contrast with homeotherms, which use melanocytes to pigment dead, keratinized tissues (e.g., hair, fur, feathers, beaks, scales) with melanin as a form of fixed, morphological coloration. Some researchers have suggested that melanocytes are the only type of pigment cells in birds and mammals (Hach et al. 1993), but we have yet to characterize the cells that produce pterins for coloring the avian eye (Chapter 8).

Melanocytes are cell populations with complex ontogenetic trajectories (Hadley and Quevedo 1966). They originate as unpigmented melanoblasts that, like all pigment cells in vertebrates (DuShane 1934; Bagnara et al. 1979), are derived from embryonic neural crest cells. In the first few days of the embryonic period, melanoblasts migrate to the epidermis and, specifically, to newly forming feather germs (Rawles 1944). There the melanocyte precursors settle and, by the end of the first week of development, begin synthesizing melanin in unique, lysozyme-like organelles known as "melanosomes" (Schraermeyer 1996; Marks and Seabra 2001). At this time, they also begin differentiating, sending cytoplasmic processes out as branches that will eventually deliver melanosomes to the keratinocytes that give rise to the feather filaments (Yu et al. 2004). At this point they become known as "melanocytes" (Lucas and Stettenheim 1972).

Nearly all of the in vitro work on melanin formation in chicken and quail tissue has revolved around the intrinsic factors that drive melanoblast migration and proliferation (Dupin and LeDouarin 2003). Growth factors (e.g., insulin-like growth factor), fibroblast growth factors (e.g., see Schofer et al. 2001), stem cell factor (or Steel factor; Lahav et al. 1994; Nataf and LeDouarin 2000), and retinoic acid (Dupin and LeDouarin 1995) have been shown to direct the fate of melanoblasts and facilitate differentiation, migration toward epidermal sites for pigmentation, and even melanin production in chicken and quail embryos. Signal-transduction pathways like cyclic adenosine monophosphate–activated protein kinase A (Evrard et al. 2004) can also augment melanocyte populations and melanogenesis in quail neural crest cells. If the production of these factors is in any way tied to the health or quality of individuals during development, and thus prohibits or enhances melanoblast/melanocyte proliferation early in life, one could easily envision a previously overlooked organizational effect of pigment-cell formation on melanin coloration. There even appear to be genetic predispositions in quail (e.g., the Me1EM antigen; Niwa et al. 2002) to committing melanoblasts to melanocyte differentiation and thus to producing melanin in appropriate epidermal locations.

As the true ultracellular sites of synthesis, melanosomes contain all of the ingredients for melanogenesis (Hearing 2000; Sulaimon and Kitchell 2003). Melanins are formed from the aromatic amino acid tyrosine (Lerner and Fitzpatrick 1950; the effect of substrate availability on melanin accumulation is discussed in more detail below). Tyrosine is a semi-essential amino acid, in that it can be obtained from the diet, as well as synthesized from available phenylalanine (an essential amino acid) by a single hydroxylation step (Figure 6.2). Genetic disorders in humans (e.g., phenylketonuria) that knock out enzymatic conversion of phenylalanine from tyrosine (by phenylalanine hydroxylase) result in reduced melanin content in hair and skin (Cowie and Penrose 1951), so this indirect source of tyrosine is important for melanization in at least some systems. The sulfur-containing amino acid cysteine is also used as a substrate (Figure 6.2), but only in the formation of phaeomelanin. Cysteine is also semi-essential, as it can be synthesized from methionine (an essential amino acid). In mammals, specific proteins deliver these amino acids to melanosomes (Potterf et al. 1996, 1999).

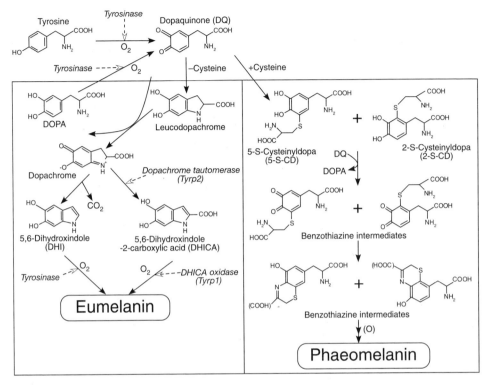

Figure 6.2. Pathways of melanin synthesis in animals. Note the action of tyrosinase at several points during eumelanogenesis, but its lack of participation in phaeomelanogenesis after its initial conversion of tyrosine to dopaquinone. Also note the input of cysteine early in phaeomelanogenesis. Redrawn from Wakamatsu and Ito (2002).

Central to melanogenesis in melanosomes is the presence of the enzyme tyrosinase—the main catalytic agent for melanin production in vertebrates (Sanchez-Ferrer et al. 1995). Tyrosinase completes the initial oxidation step that forms dopaquinone—an intermediate that can either generate phaeomelanin or eumelanin (Figure 6.2). This enzyme is aided by copper as a cofactor (melanins and minerals are discussed in more detail below) and is remarkably conserved both functionally and genetically across the major animal lineages (Sato et al. 2001). There have been several experiments showing that tyrosinase activity is the rate-limiting factor for melanogenesis in animals (reviewed in Prota 1992; Ito et al. 2000); these studies even trace the mechanism in quail embryos as deep as transcription factors (e.g., bone morphogenetic protein-2) that target the expression of tyrosinase-encoding genes (Bilodeau et al.

2001). There has been only one study linking tyrosinase to melanic plumage color in birds, however. Hall (1966) measured tyrosinase activity in the feather tracts of male and female Paradise Whydahs (*Steganura paradisaea*) at different times of the year and found that only the regions of nuptial plumage that contained melanin showed measurable tyrosine hydroxylation. With all of the recent interest in the honesty-reinforcing mechanisms of melanin ornamentation, it is quite striking that questions about tyrosinase activity and the factors that control it have not been revisited by the research community. For example, are the energy requirements of tyrosinase activity substantial enough to make this activity a viable cost to being colorful?

More recent work on melanin precursor-product relationships has shed light on the role of cysteine and tyrosinase in altering the paths to phaeomelanogenesis and eumelanogenesis. Ex vivo and in vitro experiments demonstrate that the presence of sufficient cysteine in melanosomes drives the reaction toward producing cysteinyldopa and other benzothiazine precursors of phaeomelanin (Smit et al. 1997; Land and Riley 2000). Phaeomelanin production is also generally thought to predominate when tyrosinase concentration or activity rate is low (Ozeki et al. 1997; Ito et al. 2000). In mice, a protein-mediated and gene-encoded pathway (agouti) blocks tyrosinase-directed formation of eumelanin in favor of phaeomelanin (Wolff 2003); this agouti signaling protein has also been documented for chickens (Takeuchi et al. 2000). With all of this research as a foundation, our work is now cut out for us to identify the substrate and enzymatic contributions to eumelanin and phaeomelanin production in birds.

Not only can the melanogenesis process proceed down two distinct paths, but there appears to be two types of melanosomes (eumelanosomes and phaeomelanosomes) that differ in appearance in avian melanized tissues (rod-shaped and spherical, respectively; Trinkaus 1948) and likely also differ in their chemical components. Phaeomelanosomes in mammals are thought to lack tyrosinase-protein (TRP)-1 and TRP-2 (Kobayashi et al. 1995)—enzymes that work in the later stages of eumelanin production to convert dopachrome to eumelanin via 5,6-dyhydroxindole-2-carboxylic acid, but play no role in phaeomelanogenesis (Jimbow et al. 2000). After melanins are formed, melanosomes bind to microtubules and migrate down the dendritic processes that lead to the periphery of the melanocyte, where they are taken up by neighboring keratinocytes in a phagocytosis-like fashion. Microtubule-dependent eumelanosome migration is powered by molecular motors, such as kinesin and dynein (Barral and Seabra 2004); nothing is known of phaeomelanosome transport.

Particularly in instances for which both yellow and rufous bird colors are attributed to a predominance of phaeomelanins (Table 6.1), it would be useful to better understand the chemical and chromatic properties of phaeomelanosomes.

At the level of the feather, all these cellular and molecular events must coincide with the maturation of the follicle during molt and the deposition of pigments into keratin at precise developmental timepoints (Yu et al. 2004). What little is known of this process in embryonic chickens (Lucas and Stettenheim 1972) is how melanocytes align themselves in longitudinal rows in the epidermis and innervate the barbule plates of the follicle with their melanosome-rich dendritic processes. Once all melanin is unloaded, the processes are retracted and the cells migrate and are replaced by new melanocytes, derived from a pool of melanoblast stem cells nearby or sent to the follicle. What remains to be demonstrated are the chemical signals that (1) send melanoblasts/melanocytes to particular follicles for melanin incorporation; then (2) direct melanocytes to traffic melanin to particular regions of the follicle, and at particular times (3) generate the complex patterns seen in melanic plumage (Price and Pavelka 1996; Prum and Williamson 2002). Melanin-based color signals vary both in the intensity of pigmentation and in the area of plumage that the melanin pigment covers, so it is important to uncover why pigment cells distribute themselves in differential fashion from individual to individual and then deliver phaeomelanins and eumelanins at varied concentrations.

Factors Controlling the Production and Accumulation of Melanin

Melanin-based color ornaments in birds come in a variety of seasonal, sexual, and integumentary forms, from the black throat in male Black-chinned Hummingbirds (*Archilochus alexandri*) to the ornate forehead plumes of male *Callipepla* quail. Among North American avian taxa, sexual dichromatism in melanin coloration is especially common in plumage as opposed to bare parts and in such groups as the ducks, quail, and several songbird families (e.g., orioles, warblers, sparrows). A challenge facing evolutionary biologists interested in the signal content of melanin-based colors in animals is to understand their costs of display. Unlike carotenoid-based colors, there is no clear-cut link between access to pigments from the diet and color expression (Chapter 12). Below I consider the intrinsic and extrinsic factors that have received either theoretical or empirical attention as potential honesty-reinforcing mechanisms of ornamental melanin colors in birds.

Substrate Availability

As melanins are synthesized from amino acid precursors (directly from tyrosine and indirectly from phenylalanine, cysteine, and methionine), it is conceivable that dietary availability of amino acids could restrict or enhance an individual's ability to produce and display melanins in the integument. The idea that amino acids affect melanin production and accumulation in the integumentary tissues of any animal remained untested until very recently, however. Human skin melanocytes cultured in vitro show increased production of melanin when tyrosine is in high supply (Smit et al. 1997). In the most intensive studies of which I am aware on live animals, Yu et al. (2001), Anderson et al. (2002), and Morris et al. (2002) manipulated dietary levels of tyrosine and phenylalanine in Domestic Cats (*Felis domestica*) and measured changes in coat color and pigment composition. Adult cats and kittens that typically have black coats were found to grow reddish-brown, melanin-depleted hair when fed tyrosine-deficient diets (Yu et al. 2001; Anderson et al. 2002; Morris et al. 2002). These animals regrew their usual black, melanin-replete hair when returned to a phenylalanine- or tyrosine-rich diet. This dietary component to melanization can be a maternally transmitted phenomenon as well, as black-coated mothers fed tyrosine-deficient diets during pregnancy gave birth to kittens with reddish-brown fur (Yu et al. 2001). In a recent study of amino acids relative to melanin color in birds, Poston et al. (in press) manipulated dietary tyrosine and phenylalanine in the diets of molting male House Sparrows (Plate 23) and found an effect of both amino acids on the blackness of the melanin-based throat badge. They did not measure melanin concentration in feathers, however, and this diet did not influence the area of the black badge, which is the variable and sexually relevant trait of interest in this species. Interestingly, there are reports of deficiencies in dietary lysine (an essential amino acid) also inhibiting normal dark feather pigmentation in such domestic birds as chickens, quail, and turkeys (reviewed in Grau and Roudybush 1986; Grau et al. 1989). Clearly, we are just now beginning to understand the specific dietary contributions to melanin coloration in animals.

Mineral Concentration

Other work suggests that there may be additional nutritional components that govern integumentary melanization. Data began to accumulate over a half century ago on the role that dietary minerals may play in maintaining mammalian coat color (Lerner and Fitzpatrick 1950). Such animals as rabbits and cattle tended to show reduced pelage coloration when given copper-deficient diets. Interestingly, copper is bound by tyrosinase and serves as a co-factor in tyrosinase-catalyzed melanogenesis (Sanchez-Ferrer et al. 1995), among many other enzyme-driven reactions in the body. Thus there may be a direct role for this trace metal in melanin pigment production. However, since the middle of the twentieth century, other minerals, such as calcium, iron, and zinc, have been shown to facilitate melanin formation; specifically, by their effects on the rate at which intermediates in the melanin pathway (e.g., dopaquinone [DOPA]) are formed (reviewed in McGraw 2003; discussed in more detail below). Thus it has been hypothesized that the degree to which animals accumulate dietary minerals affects their ability to produce ornamental melanin-based traits (McGraw 2003; Niecke et al. 2003; West and Packer 2003). In the only published empirical study on this topic, Niecke et al. (2003) found a significant positive correlation between the area of (presumed) melanized plumage in Barn Owls (*Tyto alba;* Plate 29) and the concentration of calcium in feathers (Figure 6.3). K. J. McGraw (unpubl. data) manipulated levels of calcium in the diets of fledgling and adult male Zebra Finches and showed that birds grew larger melanin breast patches when fed calcium-rich diets. To determine whether this hypothesis satisfies the assumptions of honest-advertisement models of sexual selection, more work is needed on how limiting various minerals are in the diets of wild birds (e.g., Klasing 1998) and how they may be traded off with physiological and sexual functions (McGraw 2003).

Hormonal Stimulation

Hormonal control of plumage and bare-part coloration in birds is reviewed in full in Chapter 11. Here I consider specific endocrinological pathways underlying melanin production and incorporation into the integument (also see the excellent reviews by Witschi 1961; Ralph 1969). Among the control mechanisms for avian melanic colors, hormonal processes rank as the most studied. Both steroid and nonsteroid hormones have been shown to influence melanization in birds. Most of this work was conducted in the middle part of the twentieth century, and considerably less attention has been given to this area of late (Owens and Short 1995; Kimball and Ligon 1999). With the surge of interest in the honesty-reinforcing mechanisms of coloration and in how known hormonally mediated patterns of coloration may be costly to maintain (e.g., via hormonal effects on metabolism and immunocompetence; Folstad and

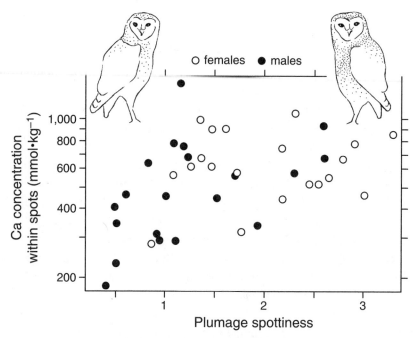

Figure 6.3. Relation between the concentration of calcium ions and plumage spottiness in black feathers of Barn Owls (n = 20 males, 20 females) in mated pairs. Plumage spottiness is measured here as the percentage of one flank that was covered by black spots. Calcium content of feathers was determined using proton induced X-ray emission. Adapted from Niecke et al. (2003).

Karter 1992), it is due time to revisit these pathways using modern methods and to further probe their causes, costs, and consequences. There are four general classes of hormones that affect melanin pigmentation in birds: (1) androgens, (2) estrogens, (3) pituitary hormones (e.g., luteinizing hormone [LH]), and (4) thyroid hormones (e.g., thyroxin). The role of melanocyte-stimulating hormone (MSH), which regulates integumentary melanosome dispersion/aggregation and hence melanin coloration in the melanophores and melanocytes of other vertebrates (Rasmussen et al. 1999), has traditionally been ignored in avian studies because birds lack the intermediate lobe of the pituitary where this hormone is made. Teshigawara et al. (2001) and Takeuchi et al. (2003), however, recently provided evidence for local production and action of α-MSH in the chicken brain and eye, which should stimulate renewed interest in the potential link between MSH and avian integumentary melanization.

Androgens

Gonadally secreted sex steroids, such as testosterone (T), have been argued to control melanin colors in members of four avian orders—songbirds (order Passeriformes), shorebirds (Charadriiformes), ducks (Anseriformes), and gamebirds (Galliformes). Unfortunately, there are only one or two study species in each order for which androgenic effects on either the development of melanin-containing plumage or variable expression of a melanin ornament have been considered. From the outset, it is important to note for these studies that T can metabolized both into estrogen (via aromatase) and other androgens (5α-dihydrotestosterone [DHT] by 5α-reductase), often peripherally at target cells (e.g., feather follicles), so unless enzyme inhibitors, receptor blockers, or these metabolites are administered experimentally, none of these studies can resolve the effects of specific sex steroids on melanin pigmentation.

Keck (1933, 1934) showed that the black beak of breeding male House Sparrows fades to horn-colored when the birds are castrated; coloration is restored by T injections (see also Witschi 1935, 1955 for similar evidence in African weaver finches [*Euplectes* sp.]). There now are several recent experimental studies that demonstrate the controlling action of T on development of the male House Sparrow's black throat patch as well. Captive adult males implanted with physiologically high levels of T during the plumage molt grew larger badges and badges that increased more in size than those of controls, castrates, and males implanted with a low-T dose (Evans et al. 2000; Buchanan et al. 2001; Figure 10.3). Gonzalez et al. (2001) also showed that natural T levels are positively correlated with badge size in field-caught males and that experimental treatment with an anti-androgen (cyproterone acetate) reduced badge area. More recently, Strasser and Schwabl (2004) found that T injections into egg yolk induced males hatching from these eggs to develop a larger badge at sexual maturity than did controls, elucidating an interesting organizational effect of maternal biochemicals on offspring development of melanin color later in life. Unlike inter-male variation in color, however, plumage difference between male and female sparrows cannot be reversed by simple hormone administration or even tissue grafts, suggesting a genetic basis for dichromatism (Keck 1934; Witschi 1961).

The physiological cost of using elevated T to develop a more melanic appearance has only been considered in passerines. The immunocompetence handicap hypothesis and the immunosuppressive effects of T have been trumpeted as a major honesty-reinforcing mechanism for androgen-dependent secondary sexual characteristics (Folstad and Karter 1992), but two studies have failed to support the notion that immunosuppression is the major cost of becoming colorful in House Sparrows. T manipulations have been found to be both immunosuppressive and immunoenhancing and exert complex effects on other hormone systems (e.g., corticosterone; Evans et al. 2000; Poiani et al. 2000). Buchanan et al. (2001) instead considered the metabolic consequences of elevated T and color expression and found that metabolic rates were higher for birds implanted with T. Whether metabolic costs are the prevailing pressure that keeps T-controlled melanin-based ornamental colors honest remains to be seen.

Among gamebirds, as in passerines, androgens also do not appear to govern sex differences in melanin coloration (Kimball and Ligon 1999), but there are instances when T manipulations (e.g., in male Willow Ptarmigan [*Lagopus lagopus lagopus*]; Stokkan 1979) induce growth of the more melanic breeding plumage. In this study, however, castration had the same effect as did T implants, and the author suggested that T as well as LH (which rises upon castration) may equally influence the process of melanization (discussed in more detail below). Hormonally based intrasexual variation in the extent of melanization has been less commonly studied, but in the two galliform species considered, Gambel's Quail (*Callipepla gambelii*; Plate 24, Volume 2) and Scaled Quail (*Callipepla squamata*), castration had no effect on the size of melanin-containing forehead plumes or belly plumage patches (Hagelin 2001).

In shorebirds, exogenous T administration induces acquisition of the more melanized nuptial plumage in adults from both sexes of three species—the Ruff (*Philomachus pugnax*; Plate 7, Volume 2; Lank et al. 1999), the Wilson's Phalarope (*Phalaropus tricolor*), and Northern Phalarope (*Lobipes lobatus*; Johns 1964)—the latter two of which show reversed sexual dichromatism (with females exhibiting the more melanic plumage). Estrogen or prolactin treatment in phalaropes did not, however, exert the same effects. Supplemental T delivery to young Black-headed Gulls (*Larus ridibundus*) induced the expression of more adultlike black masks, but also reduced melanization in some body regions (e.g., back, wing, tail; Ros 1999), which is also characteristic of adult, but not juvenile, plumage. This example nicely illustrates how such hormones as T can exert very specific effects on different plumage regions, and this likely is regulated by hormone-receptor targets or local enzymes that are differentially scattered across feather tracts and follicles (Peczely 1992). In line with this idea, Schlinger et al. (1989) showed that skin beneath colorful, melanic plumage in female Wilson's Phalaropes converts more T to DHT than that of males, due to higher reductase activity.

Androgenic effects on the ratio of melanin types that accumulate in the integument have been studied best in Mallard drakes (Plate 15, Volume 2). T not only accelerates the rate at which the melanin-containing male nuptial plumage is acquired (Haase and Schmedemann 1992), but also alters the phaeomelanin:eumelanin content of different ornamental feather patches. T administration reduced eumelanin and increased phaeomelanin in iridescent head and black undertail-covert plumage of male Mallards, but increased eumelanin and decreased phaeomelanin in brown breast plumage (Haase et al. 1995). It is still not clear how these plumage regions were differentially affected by T or what precise consequences this differentiation had for coloration. In fact, DHT treatment in this same study had far less potent effects on melanin distribution, so the authors hypothesized that estrogen, metabolized from T, is the actual molecule that drives melanin synthesis and coloration in these cases (Haase et al. 1995). Additional studies that couple this hormonal perspective with detailed analyses of coloration and melanin content may reveal how such control mechanisms act and why they exist.

Estrogens

Although I am aware of no studies that examine intrasexual variation in melanin-based color expression in relation to estrogen levels, there is a tremendous literature on estrogenic regulation of sexual plumage dichromatism, most of which is melanin-based. Kimball (Chapter 10; see also Kimball and Ligon 1999) reviewed those species, mostly in gamebird and waterfowl lineages, that exhibit estrogen-mediated melanin-based plumage dichromatism. In all of these cases, it is the presence of estrogen (thought to be largely ovary-derived) that prevents the formation of nuptial melanic plumage in females; in the absence of ovaries and estrogen, males attain the default ornamental state (see also Witschi 1936 for evidence in a passerine, the Orange Weaver [*Euplectes franciscana*]). Elegant in vivo and in vitro studies by Trinkaus (1947, 1948, 1953) on brown leghorn chickens show the cellular basis for estrogen control of plumage type. Adult males normally acquire a glossy black body plumage, whereas females are reddish-brown. As predicted, estrogen administration in adult males results in regeneration of femalelike reddish-brown plumage. However, no plumage change occurs when estrogen is delivered to male embryos or young chicks. Trinkaus even followed the differentiation of melanoblasts from tissue explants of 8-day-old male embryos and adults, and these experiments yielded similar estrogen-dependent patterns, with no shift in embryonic production of reddish pigment granules when cultured in the presence of estrogen, but notable accumulation of red pigments in adult tissue. These results indicate that feather melanization is only sensitive to estrogen in sexually mature birds, which may be driven by a receptor- or enzyme-mediated process, as both estrogen-receptor and aromatase expression can be absent in immature individuals and upregulated during sexual maturity (e.g., Balthazart et al. 2000; Nishizawa et al. 2002). In fact, T aromatization to estrogen in the skin is what controls the development of female feathering in a different strain of chicken (Sebright bantam; George et al. 1981). We might also hypothesize from Trinkaus's experiments that, based on estrogen-mediated color changes, this sex-steroid tilts the melanin-production pathway toward phaeomelanogenesis, which would be quite similar to the relative increase in phaeomelanization seen in T-treated Mallards (as above), where estrogen was thought to be the acting hormone. In fact, Hall (1970) has shown that estrogen acts in this system by facilitating the addition of cysteine into melanogenesis, which results in increased phaeomelanin synthesis and subsequent incorporation into feathers (Land and Riley 2000). In all of these studies, little consideration has been given to the adaptive significance or costliness of using estrogen to control feather coloration in this fashion.

LH

LH is a pituitary (hypophysial) protein hormone that functions as a gonadotropin by stimulating gonadal steroidogenesis and gamete maturation in response to hypothalamic input from gonadotropin-releasing hormone (GnRH). LH has been shown to exert stimulatory effects on the acquisition of melanin plumage in several songbirds, mostly the weavers (Family Ploceidae; Ralph 1969). Early studies of Red-billed Queleas (*Quelea quelea;* Plate 6, Volume 2), Yellow-crowned Bishops (*Euplectes afer*), Orange Weavers, and Paradise Whydahs showed that they were able to grow nuptial melanic plumage even after castration and that both sexes could be induced to grow dark nuptial plumage, either by plucking or from the basic plumage, by injecting them with extracts of cow pituitary (Witschi 1936, 1937). Follow-up studies proved that it was pituitary-derived LH, not follicle-stimulating hormone (FSH), that controlled plumage color (Witschi 1940; Witschi and Segal 1955; Segal 1957). LH treatment produced such predictable effects on feather melanization in these birds and at low doses (5 µg) that it was given the name the "weaver finch test" and was used as an early form of an LH bioassay in other ploceids (e.g., Baya Weaver [*Ploceus philippinus*]; Segal 1957; Thapliyal and Saxena 1961; Ortman 1967).

With the knowledge of LH-mediated melanin-based plumage expression, Hall and colleagues explored the cellular processes accompanying plumage darkening. They showed that LH effectively blackens male feathers by stimulating the activity of tyrosinase and increasing melanin-granule production (Okazaki and Hall 1965; Hall 1966, 1969a,b; Hall and Okazaki 1966). It was thought that LH exerted its effects on plumage directly at peripheral feather follicles, but these authors could never demonstrate that its actions were local (Hall et al. 1965). They employed four intricate experimental LH manipulations (injections into individual growing feathers, pellets implanted into the skin from which feathers were growing, bathing skin flaps containing growing feathers in hormone solution, and applying hormone cream to skin), and all either had no effect on feather melanization or had systemic effects on all feathers grown at that time. Remarkably, despite this being one of the most detailed links between melanin production and integumentary coloration in birds, this topic has not been revisited in over 35 years. We still do not know, for example, what factors regulate the elevation of LH levels.

Thyroxin

Another potential nongonadal source of hormones that may have complementary or independent effects on avian melanin pigmentation comes from the thyroid gland; namely, thyroxin (T_4; Markert 1948). Studies of T_4 and plumage also have produced some of the more complicated results among those that link hormones to melanization, perhaps due to the different modes and dosages of hormone administration, as well as to the important effects that thyroid/T_4 manipulations have on metabolism. Miller (1935) summarized much of the early work on T_4 and plumage pigmentation. Moderate physiological doses were found to increase melanism in male Bullfinches (*Pyrrhula pyrrhula;* Schereschewsky 1929), Domestic Chickens, and Mallards, whereas pharmacological doses (>1 g/kg) generally caused lightening in chickens, Japanese Quail, and House Sparrows. However, not all body regions of plumage were affected equally. Black and chestnut features in male House Sparrows faded to a more gray appearance, but normally light-gray feathers on the breast became darker (Miller 1935). It should be pointed out that, in virtually all of these studies, feathers were plucked from nonmolting birds, and the effects of T_4 manipulation were examined, so it is conceivable that studies conducted under normal hormonal backgrounds and with locally activated enzymes and receptors might yield different results. Unlike most other hormones, T_4 levels are usually at their annual peak during molt (e.g., in House Sparrows; Smith 1978).

Colorful melanin traits found to come under the control of other previously described hormones also respond to thyroid-hormone manipulations. Thyroidectomized male House Sparrows and Tree Sparrows (*Passer montanus*) lose their black beak color, which can be restored with T_4 treatment (Lal and Thapliyal 1982; Lal and Pathak 1987). Paralleling his studies on estrogen and plumage colors in leghorn chickens, Trinkaus (1950, 1953) considered the role of T_4 in melanoblast differentiation in both in vivo and in vitro experiments. Thryoidectomy in both sexes leads to a more reddish-brown, presumably more phaeomelanized appearance; T_4 administration ameliorates this effect. Interestingly, not all plumage regions show equal responses to T_4 treatment (e.g., the black color of ventral feathers is not as deep), suggesting again that local cellular phenomena (as discussed above) underlie feather-tract sensitivities to T_4. As found for estrogen, hypo- or hyperthryoidism early in life had no effects on plumage development in chicks, suggesting upregulation of T_4 sensitivity at sexual maturity. Although the interaction between estrogen and T_4 in this system is still unclear, some researchers have suspected that thyroid hormones regulate sex-steroid synthesis and thus potentiate their direct effects on melanic colors (Lal and Pathak 1987). That such steroids as T interact with T_4-driven metabolic processes (Buchanan et al. 2001) also makes this proposed interaction an enticing honesty-reinforcing mechanism to pursue.

Genetic Control

The genetic mechanisms underlying avian coloration are covered in Chapter 11 of this volume, but here I touch on the specific pathways linked to melanin-pigment formation and deposition. Dark melanin colors have long been known to come under genetic control. In fact, Darwin (1859) focused a great deal of his attention on the variation in and ability to artificially select for different melanic plumage morphs in Domestic Pigeons (*Columba livia*), without even knowing the units of inheritance or color basis. Perhaps the greatest historical sources of information on the genetics of melanization in birds come from pet-bird fanciers and breeders. The opportunity to breed such domesticated species as chickens (Hollander 1989), quail (Lauber 1964; Homma et al. 1969; Minezawa and Wakasugi 1977; Roberts et al. 1978; Truax et al. 1979;

Roberts and Fulton 1980; Fulton et al. 1982), Mallards (Jaap 1934; Lancaster 1963), pigeons (Sell 1994), and Zebra Finches (Landry 1997) for albinos, melanics, and schizochromatics (partial pigment loss or dilution) has revealed the single- or multi-locus, dominant or recessive, autosomal or sex-linked nature of avian melanism. Virtually every combination of these genetic modifications has been reported for melanin patterning in birds, from single-locus dominant point mutations in chickens (Takeuchi et al. 1996a,b) to multi-locus recessive control in pigeons (Sell 1994). Although these examples of artificial selection do not necessarily depict natural genetic agents of selection on coloration, they clearly show the tremendous lability in the genetic architecture behind melanization.

Genetic control of melanization has also been studied in a handful of free-ranging bird populations (reviewed in Berthold et al. 1996). These studies typically center on aberrant individuals observed in the wild (Hudon 1997; Literak et al. 1999; Rutz et al. 2004), geographically distinct populations exhibiting unusual plumage patterns for that species (Berthold et al. 1996; Bensch et al. 2000; Theron et al. 2001), or stable plumage polymorphisms that exist within a population (Cooke et al. 1995; Lank et al. 1995). In most instances, even the gross genetic determinants remain undescribed, because of the obvious difficulties with conducting controlled breeding experiments on these free-ranging animals. In those that have been well studied, however, it is clear that melanism is consistently an autosomal trait (with one exception, sex-linked plumage spottiness in Barn Owls; Roulin and Dijkstra 2003), but can be either dominant (e.g., plumage phases in Lesser Snow Geese [*Anser caerulescens caerulescens*]; Plate 28; Cooke et al. 1995) or recessive (e.g., single-locus melanistic mutants in Blackcaps [*Sylvia atricapilla*]; Berthold et al. 1996).

Recent advances in molecular techniques, stemming largely from work on mammals, have paved the way for a new era in genetic studies of melanin coloration in birds. Melanin-containing hair and skin from mice to humans vary tremendously in melanin content and appearance, and in the past decade, remarkable progress has been made in isolating the chromosomal and receptor-encoding loci for melanism in mammals (Barsh 1996; Jackson 1997). Although it is believed that as many as 100 different loci can interact or contribute to melanin patterns in animals (Urabe et al. 1993), much recent attention has been centered on the locus that encodes the melanocortin-1 receptor (*MC1R;* Rees 2000; Horth 2005). This seven-pass transmembrane G-coupled receptor is expressed on maturing melanocytes (Mountjoy et al. 1992), and in the mouse is encoded at the *extension* locus on chromosome 8 (Robbins et al.

1993). This receptor is known to modulate melanin synthesis in mammalian melanocytes by binding α-MSH, which subsequently activates tyrosinase (Hunt et al. 1994).

In birds, however, the intermediate lobe of the pituitary—the site of α-MSH synthesis—is lacking. Moreover, some of the early work with chickens showed that the *Extended black* (*E*) locus on chromosome 1 regulated eumelanin and phaeomelanin formation in melanocytes (Carefoot 1993). This observation begs the question of whether *MC1R* should make similar genetic contributions to avian melanization as it does in other vertebrates. Takeuchi et al. (1996a) and colleagues undertook the first studies of *MC1R* in relation to plumage coloration in birds and first identified the *MC1R* gene in chickens, which shares 64% homology with the mammalian form. They proceeded to demonstrate a link between the structure of the *MC1R* and the *E* locus, with the very same mutation (Glu92Lys) that induces black coat color in mice (Robbins et al. 1993) associated with black feather color in chickens (Takeuchi et al. 1996b; also see Andersson 2003 and Schioth et al. 2003 for a review of *MC1R* homology in vertebrates).

Four studies have since carried these methods into the field to investigate *MC1R*-driven patterns of melanization in wild birds. Bananaquits (*Coereba flaveola*) typically show predominantly yellow body plumage color, but they display remarkable geographic variation throughout the Caribbean islands, with the most unusual melanistic variant found on the islands of Grenada and St. Vincent (Price and Bontrager 2001). Theron et al. (2001) sequenced *MC1R* genes from yellow and black Bananaquits and found that all of the melanistic individuals, but none of the yellow birds, were homo- or heterozygous for the very same Glu92Lys mutation responsible for black color in mice and chickens. Mundy et al. (2004) revisited the classic examples of genetic melanism in Arctic Skuas (*Stercorarius parasiticus;* Plate 28) and Lesser Snow Geese and showed that different nonsynonymous point substitutions in the *MC1R* gene (Arg^{230}–His^{230} in skuas; Val^{85}–Met^{85} in geese) were perfectly associated with the more melanic plumage condition (Figure 11.3). Doucet et al. (2004) examined plumage color differences among populations of White-winged Fairywrens (*Malurus leucopterus*) and found that five nonsynonymous amino acid substitutions in *MC1R* were responsible for a change in plumage appearance between black- and blue-plumaged birds (Plate 21). However, one study showed a lack of correspondence between natural melanin plumage variation and *MC1R*; *Phylloscopus* warblers (Plate 29, Volume 2) exhibit marked differences in plumage darkening among different body regions, but interspecific variation in the *MC1R* sequence did not predict species differences in coloration and was generally conserved (MacDougall-Shackleton et al. 2003). Clearly, adaptive selection for different plumage patterns and on *MC1R* variation can occur at different rates and extents in different lineages. Once we have a firmer grasp on *MC1R* variation across many bird groups, we can use comparative approaches to understand how and why melanin colors and the *MC1R* may or may not co-vary.

Despite these dramatic advances in genetic studies of avian melanism, we still lack a good understanding for how *MC1R* is regulated hormonally. What is the source of α-MSH if it truly upregulates this receptor system? Studies have detected α-MSH in the anterior lobe of the pituitary in ducks but have established no effect on pigmentation (Iturriza et al. 1980, 1992). In vitro work on quail melanocytes, however, demonstrates notable stimulation by α-MSH (Satoh and Ide 1987). Perhaps birds produce this hormone peripherally in the skin (Takeuchi et al. 2003). We also have yet to elucidate the specific melanogenic events that are altered by *MC1R*. *MC1R* sequence variability can affect both the color (e.g., in chickens; Kerje et al. 2003; Ling et al. 2003) and extent (e.g., in Bananaquits) of melanin patches, and to what degree this linkage is due to effects on melanocyte activity versus feather-follicle receptors or sensitivity remains to be seen. In only one case have the genetic consequences for variation in phaeomelanin/eumelanin composition been uncovered; Haase et al. (1992) used chemical degradation methods (Box 6.1) to quantify the amounts of phaeomelanin and eumelanin from the feathers of various pigeon breeds. We await studies of *MC1R* in relation to plumage color in this species, however. Ultimately, dovetailing such biochemical techniques with sophisticated molecular-genetic approaches will take us one step closer to uncovering the optimal pathways for becoming colorful and will help to better frame broader mechanistic and functional studies of adaptive melanization in birds.

General Condition-Dependent Factors

Chapter 12 comprehensively covers the suite of environmental and condition-dependent control mechanisms for the expression of melanin coloration in birds. Here I briefly reiterate that, at present, the evidence more often points to the lack of a relationship between condition and melanization (Hill and McGraw 2003). Experimental manipulations of protein content in the diet (in House Sparrows; Gonzalez et al. 1999; Buchanan et al. 2001), overall food intake (in House Sparrows and Brown-headed Cowbirds [*Molothrus ater*];

Plate 7; McGraw et al. 2002), and endoparasite loads (in House Finches [*Carpodacus mexicanus*]; Plate 31; and American Goldfinches [*Carduelis tristis*]; Plate 30; Hill and Brawner 1998; McGraw and Hill 2000) consistently do not seem to alter the intensity or size of melanin-based color patches. Senar et al. (2003) reported no association between the rate of feather growth and extent of melanin coloration in Great Tits (*Parus major;* Plate 11, Volume 2), but a significant relationship between feather growth and carotenoid coloration. In addition, there is no link between melanin coloration and immunological variables in Barn Swallows (*Hirundo rustica;* Plate 7; Camplani et al. 1999; Saino et al. 1999). Results from five comparative studies in cardueline finches also show that evolutionary changes in sexual dichromatism or condition-dependence are more often associated with carotenoid coloration, not melanin coloration (Gray 1996; Badyaev and Hill 2000, 2003; Badyaev et al. 2002; Badyaev and Young 2004). However, there still are a handful of correlational links between measures of condition and melanin pigmentation in songbirds (Slagsvold and Lifjeld 1992; Veiga and Puerta 1996; Roulin et al. 2001, 2003; Parker et al. 2003). Two experiments also demonstrate the negative effects of ectoparasites and parental effort, respectively, on the size of melanic plumage patches (in male Great Tits; Fitze and Richner 2002; in House Sparrows; Griffith 2000), but because these variables were manipulated in advance of molt in these studies, it is unclear how they, as opposed to other factors, had direct effects on the development of coloration. Clearly, in future studies of condition-dependence of melanin coloration, we need to first define our measure of condition, develop an appropriate means of manipulating it, and then target its specific physiological effect on melanin production/accumulation, and do this in more and more species, so that we can continue to learn whether in fact melanin colors are universally less sensitive to changes in an individual's condition than other color types (Chapter 12).

Other Biochemical and Physiological Actions of Melanins

Melanins are not necessarily found only in the integument of animals (Muroya et al. 2000). They occur in high quantities in the eye (Imesch et al. 1997), esophagus, thyroid, brain (neuromelanin; Zecca et al. 2003), and in tumor cells (melanomas; Jurado et al. 1998). They can exist in odd forms, such as in the defensive black ink of cephalopods (Palumbo 2003). Even when they are present in integumentary tissue, they may serve valuable functions beyond chromatic signaling to conspecifics. In such vertebrates as birds, there are several noteworthy physiological properties of melanin pigments that warrant coverage here, as we consider the signal value of melanin-based colors.

Antioxidants

It is clear that the antioxidant potential of carotenoid pigments plays an important role in shaping the information content of carotenoid-based coloration (Chapter 5). The same may be true of melanins and of several other groups of pigments in animals (McGraw 2005). Melanin polymers, with their highly conjugated system of double bonds, are suggested to act as valuable free-radical scavengers in many systems (McGinness et al. 1970; Borovansky 1996), from fungi (Shcherba et al. 2000) to amphibians (Geremia et al. 1984) to humans (Rozanowska et al. 1999; Kasraee et al. 2003). In vitro work with mammalian retinal epithelial cells, for example, shows that melanotic cell lines are more resistant to cytotoxic effects at high oxygen tensions than amelanotic lines (Magomedov et al. 1990; Akeo et al. 2000). Thus, as Lozano (1994) argued for carotenoids, it is conceivable that melanin colors signal individuals' abilities to defend themselves from oxidative damage, and hence to maintain effectively functioning cells and tissues, such as those of the immune system. To date, this idea is untested in birds. Melanins are relegated to melanocytes and, unlike carotenoids, are not thought of as a mobile pool of antioxidants that can be shunted to needy body tissues. To work in this fashion, they would have to protect cells locally from free-radical attack and do so only during a very short period of time, as granules are synthesized only during feather maturation (Ralph 1969) and subsequently rapidly (on the order of hours to days) incorporated into dead keratinized tissue.

Cation Chelators

Another diagnostic and functional component of the molecular structure of melanins is their carboxyl substituents (Figure 6.1). These negatively charged end-groups function as cation chelators, selectively binding positively charged particles, such as free radicals and transition metals (Riley 1997). Melanized tissues in mice, cattle, horses, humans (Horcicko et al. 1973; Borovansky 1994), frogs, fishes (Bowness and Morton 1952), and birds (Niecke et al. 1999) are known to harbor high concentrations of zinc, copper, calcium, and iron, and melanin granules within the melanosomes are the confirmed sites of sequestration (Borovansky et al. 1976; reviewed in McGraw 2003). Calcium, manganese, and zinc, for example, were enriched in dark feathers of White-tailed Eagles (*Haliaeetus albicilla*) compared to white feathers (Niecke et al. 1999), and the same was true for calcium and zinc in spotted and unspotted feathers of Barn Owls (Niecke et al. 2003).

Thus not only do melanins have multiple chemostatic means by which they can mop up potentially harmful reactive oxygen species, but they can potentially protect individuals from the accumulation of toxic cations, such as the trace metals. Above, it was proposed that increased mineral intake might promote melanin production and coloration, but in fact minerals are known to have a series of detrimental effects at very high concentrations (reviewed in McGraw 2003). If selection has driven individuals to accumulate high levels of minerals and melanin for pigmentation, then the removal of these minerals from the body in dead tissues (e.g., hair, feathers) containing melanin might be the very escape mechanism that animals have evolved to avoid the chemotoxic effects of minerals (McGraw 2003). We should begin to evaluate this hypothesis by understanding variation in chemoprotection among individuals with varying melanized phenotypes and by manipulating both mineral accumulation and melanization to understand their effects on one another and on chemotoxicity.

Tissue Strengtheners

As large polymers that cross-link with proteins, melanins are thought to offer structural support and strength to tissues (Riley 1992), including insect cuticles, plant seed pods, the injured surfaces of fruits (Riley 1997), as well as bird feathers (Burtt 1986; Bonser 1995) and bare parts (e.g., the beak; Bonser and Witter 1993). For nearly a century, ornithologists have remarked that black feathers seem to be more durable and better resist abrasion than do other feathers (e.g., Averil 1923; Barrowclough and Sibley 1980). Many feathers are tipped with black melanin coloration (Plate 2, Volume 2), which likely functions to protect these most-exposed parts of feathers with the highest risk of degradation (Burtt 1979). Melanized feathers may also be thicker than unmelanized feathers (Voitkevich 1966). As perhaps the best empirical evidence for the tissue-strengthening effectiveness of melanin, Bonser (1995) subjected melanized and unmelanized portions of flight feathers from Willow Ptarmigan to an indentation hardness test to calculate their ability to resist standardized forces of mechanical indentation. Melanic keratin performed 40% better at resisting indentation than nonmelanic keratin. Thus birds with more melanic feathers may have more durable plumage and signal to conspecifics the ecological and physiological benefits of having fresher feathers (e.g., more efficient thermoregulation and flight). However, Butler and Johnson (2004) recently showed in Ospreys (*Pandion haliaetus*) that, when position on the feather (proximal versus distal) is taken into account, differences in toughness between melanized and unmelanized barbs are negated (Figure 6.4). The metabolic costs of abraded/damaged plumage should be determined to test this hypothesis for the advantages of being melanized.

Antimicrobials and Parasite Deterrents

Such macroparasites as flies, fleas, and mites have long been known to live on or find a way past feathers to various nutritional sources on a bird's integument. Microbes have also recently been shown to colonize bird feathers, and some secrete enzymes that degrade keratin (Burtt and Ichida 1999). Thus it would be adaptive for birds to endow their feathers with substances that are aversive to parasites and pathogens. The actions of uropygial-gland oil aside (Shawkey et al. 2003), melanin may be one means by which birds accomplish this. By strengthening the tissue (see above) through which microbes and other ectoparasites must chew and/or lyse, melanin may slow feather degradation. Melanins may also act in this fashion by directly binding the enzyme proteins (proteases) responsible for keratin lysis in microbes (Doering et al. 1999). There is a large literature for insects (reviewed in Wilson et al. 2001) and humans (reviewed in Mackintosh 2001) on the antimicrobial potential of melanin in the integument (e.g., cuticle, skin). In a recent study that confirms this hypothesis in birds, Goldstein et al. (2004) exposed black and white wing feathers from Domestic Chickens to feather-degrading bacteria (*Bacillus licheniformis*) and found that white feathers released more oligopeptide fragments (the outcome of keratin degradation) than did black feathers (Figure 6.5). Burtt and Ichida (2004) also considered differences in plumage melanization among Song Sparrow (*Melospiza melodia*) populations and found that feather bacteria from the more melanic population posed a greater threat to plumage degradation than did bacteria from a less melanic sparrow population. However, correlational and experimental links between feather melanization in Rock Pigeons and their feather lice (*Columbicola columbae*) were examined, and it was determined that dark feathers did not house fewer lice or resist louse damage better than did light feathers (S. E. Bush, unpubl. data).

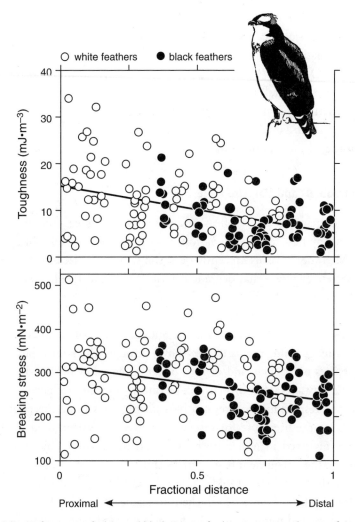

Figure 6.4. Performance of white and black Osprey feathers in two tensile tests of tissue strength in relation to feather location (proximal or distal). Both feather toughness (calculated as work per volume) and breaking stress (breaking force of the cortex wall) were unrelated to the melanin content of the feather and instead decreased distally along the rachis. Redrawn from Butler and Johnson (2004).

Perhaps melanin effectively deters only specific forms of parasites and pathogens. We now need to place these studies in the context of sexual selection to ask whether individuals with less melanized and poorer defended plumage are disfavored as mates or in mate competitions.

Mechanics of Melanin-Based Coloration 275

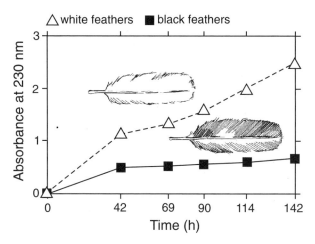

Figure 6.5. Rates of bacterial (*Bacillus licheniformis*) degradation experienced by black and white secondary feathers from leghorn chickens. Bacterial degradation was estimated by measuring light absorbance of oligopeptides released into the liquid medium surrounding the feathers as a result of bacterial lysis of β-keratin; these oligopeptides absorb light maximally at 230 nm. Media from white feathers absorbed more light at 230 nm (and thus contained more oligopeptides) than did that from black feathers. Redrawn from Goldstein et al. (2004).

Melanins as Photoprotectants

Melanin, with its large, highly conjugated structure, absorbs most wavelengths of light, thus giving it its typically dark black or brown color (Riley 1997). This property can also serve as an important line of defense from solar rays. There are several lines of evidence, in vitro and in vivo, correlational and experimental, and from many tissues, that melanins offer photoprotection to living things (reviewed in Ortonne 2002; Kadekaro et al. 2003). Daphnia, for example, use melanins to pigment their bodies and their eggs in response to depth in the water column and exposure to UV radiation (Rhode et al. 2001). Human skin color varies historically with the need for solar defenses (Robins 1991; Zeise et al. 1995). Birds and other animals in desert environments are often colored black (Ward et al. 2002; discussed below in more detail in the section on thermoregulation). That birds often use small areas of melanin coloration as signals, as opposed to covering much of their body in black, works against any hypothesis that links melanin ornaments to photoprotective benefits, unless these small patches play pivotal roles in protecting certain sun-exposed body parts. The eyes, for example, may be among the most critical and susceptible tissues to photodamage, and birds are known to deposit high concentrations of melanin in the pigment epithelium of the retina (Hudon and Oliphant 1995) and in the pecten oculi (Kiama et al. 1994). Some birds that live in open habitats, however, grow plumage that is fully pigmented with melanin (e.g., American Crows [*Corvus brachyrhynchos*]), and no studies have yet considered whether these birds might vary in coloration and melanin content or whether this variation has any functional links to photoexposure and the need to protect underlying body tissues.

Thermoregulators

As melanins absorb much of the sun's rays, they convert the photon energy into heat (McGinness and Proctor 1973). Thus heavy melanization has been thought of as a means of thermoregulating and maintaining high body temperatures (Morrison 1985). Although this strategy may prove successful for animals that inhabit polar climes (Cloudsley-Thompson 1999), it certainly would not seem to explain the long-standing observation that desert birds, faced with cooling challenges, tend to exhibit quite melanic plumages (Buxton 1923). However, careful physiological studies of radiation penetration, heat load, and plumage color in birds have shown that under simulated desert conditions, black feathers can actually reduce solar heat gain (reviewed in Wolf and Walsberg 2000; Ward et al. 2002). When black and white morphs of Domestic Pigeons were exposed to solar radiation under stagnant conditions, black plumage in fact absorbed more heat than did white plumage; however, when natural wind speeds were introduced experimentally, heat loads were lowest and best regulated in black-plumaged birds (Figure 6.6; Walsberg et al. 1978; see also Walsberg 1982 for evidence in Phainopeplas [*Phainopepla nitens;* Plate 26, Volume 2]). Heat, which is generally trapped in the layers of melanized plumage, can dissipate when feathers are ruffled by wind; in light plumage birds, solar energy penetrates most of the plumage regardless of feather position, providing little opportunity for wind-aided thermoregulation. Clearly there may be other advantages to being melanic in desert habits (e.g., UV protection); more rigorous comparative evolutionary and physiological methods are needed to confirm this thermoregulation hypothesis for dark desert birds (Ward et al. 2002). Moreover, other geographic patterns related to melanin coloration require a different explanation; namely, Gloger's rule, which states that that birds in wetter, equatorial climates tend to be darker col-

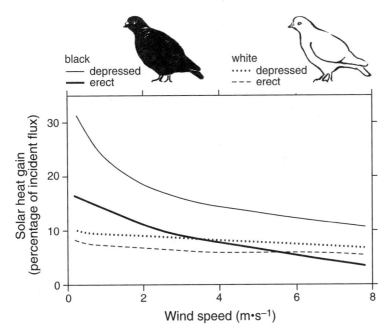

Figure 6.6. Change in solar heat load as a function of wind speed for domestic pigeons having different plumage types (black versus white) and arrangements (erect versus depressed). Although, compared to white plumage, black plumage retains more heat when depressed and at low wind speeds, it minimizes solar heat gain when erected under the highest wind speeds. Redrawn from Wolf and Walsberg (2000).

ored than conspecifics found away from the equator (Gloger 1833). Perhaps in humid habitats, where wind may not play as much of a role in thermoregulation, the light-absorptive nature of melanic factors can in fact aid in plumage drying. Given that this hypothesis was conceived over a century and a half ago for birds, it is due time that we revisit the mechanism behind it.

Summary

Melanin pigments are the most abundant and widespread pigments in animals and confer most—but not all—of the black, brown, gray, rufous, and occasionally buff colors in birds. Melanin is made up of several interlinked indole moieties that consist of a benzene and pyrrole (nitrogen-containing) ring. Its large and highly conjugated structure is responsible for its high light absorbance and thus dark color. The full chemical structure of melanin, however, has never been identified, due to its size, insolubility in most solvents, and strong binding properties with other biochemicals. There are two main forms of melanin—eumelanin and phaeomelanin—that differ in granule size and shape, chemical composition (e.g., phaeomelanin contains sulfur), and color (eumelanin tends to be black/gray whereas phaeomelanin is more brown/buff). When they have been described in bird feathers, phaeomelanin and eumelanin always co-exist, and it is the predominance of one type over the other that gives melanin-based colors their range of hues and shades. No robust procedures are available to directly analyze melanin types and amounts, but there are new techniques that generate and measure oxidative and hydrolyzed byproducts of eumelanin and phaeomelanin using HPLC.

Melanins found in the integument of animals are synthesized endogenously in specialized pigment cells (melanocytes in birds and mammals, melanophores in poikilotherms). Because melanogenesis is under local control at maturing feather follicles and keratinocytes in birds, melanin pigments can be deposited in intricate patterns (e.g., stripes, spots) within feathers and among feather tracts. In the few systems (e.g., cultured cell lines) in which melanin synthesis has been examined, the enzyme tyrosinase plays a central role. High tyrosinase concentrations and activity drive eumelanin synthesis, whereas low amounts of enzyme and high concencentrations of the amino acid cysteine drive reactions toward phaeomelanogenesis. There is strong genetic determination (e.g., melanocortin-1 receptor) of tyrosinase activity and melanin synthesis in birds and other animals. Hormones such as sex steroids, thryroxin, LH, and MSH are also capable of governing tyrosinase activity and thus extent of melanization. There may yet be other factors that affect the extent and color of melanized ornaments in birds, such as the availability of substrate (tyrosine) and minerals, but the current evidence suggests that the formation of melanin colors is less sensitive to the suite of nutritional and health parameters that influence the expression of carotenoid-based colors in birds.

Once synthesized, melanin pigments offer animals a series of biological, chemical, and physical advantages. Because of their conjugated double-bond system, these pigments can scavenge free radicals and protect cells and tissues from oxidative damage. They bind transition metals and can protect animals from the toxic effects of high metal accumulation. Melanins have strong interlinking properties and are thought to strengthen tissues and help them resist breakage, wear, and microbial degradation. Their UV-absorbing properties make them ideal photoprotectants. Future studies that target the adaptive significance and evolution of melanin colors in birds and other animals should consider both the costs and benefits of melanin synthesis and use.

References

Akeo, K., S. Amaki, T. Suzuki, and T. Hiramitsu. 2000. Melanin granules prevent the cytotoxic effects of L-DOPA on retinal pigment epithelial cells in vitro by regulation of NO and superoxide radicals. Pigment Cell Res 13: 80–88.

Anderson, P. J. B., Q. R. Rogers, and J. G. Morris. 2002. Cats require more dietary phenylalanine or tyrosine for melanin deposition in hair than for maximal growth. J Nutr 132: 2037–2042.

Andersson, L. 2003. Melanocortin receptor variants with phenotypic effects in horse, pig, and chicken. Ann NY Acad Sci 994: 313–318.

Averil, C. K. 1923. Black wing tips. Condor 25: 57–59.

Badyaev, A. V., and G. E. Hill. 2000. Evolution of sexual dichromatism: Contribution of carotenoid- versus melanin-based coloration. Biol J Linn Soc 69: 153–172.

Badyaev, A. V., and G. E. Hill. 2003. Avian sexual dichromatism in relation to phylogeny and ecology. Annu Rev Ecol Syst 34: 27–49.

Badyaev, A. V., and R. L. Young. 2004. Complexity and integration in sexual ornamentation: An example with carotenoid and melanin plumage pigmentation. J Evol Biol 17: 1317–1327.

Badyaev, A. V., G. E. Hill, and B. V. Weckworth. 2002. Species divergence in sexually selected traits: Increase in song elaboration is related to decrease in plumage ornamentation in finches. Evolution 56: 412–419.

Bagnara, J. T., and M. E. Hadley. 1973. Chromatophores and Color Change: The Comparative Physiology of Animal Pigmentation. Englewood Cliffs, NJ: Prentice-Hall.

Bagnara, J. T., J. Matsumoto, W. Ferris, S. K. Frost, W. A. Turner, Jr., et al. 1979. Common origin of pigment cells. Science 203: 410–415.

Balthazart, J., O. Tlemçani, N. Harada, and M. Baillien. 2000. Ontogeny of aromatase and tyrosine hydroxylase activity and or aromatase-immunoreactive cells in the preoptic area of male and female Japanese Quail. J Neuroendocrinol 12: 853–866.

Barral, D. C., and M. C. Seabra. 2004. The melanosome as a model to study organelle motility in mammals. Pigment Cell Res 17: 111–118.

Barrowclough, G. F., and F. C. Sibley. 1980. Feather pigmentation and abrasion—Test of a hypothesis. Auk 97: 881–883.

Barsh, G. S. 1996. The genetics of pigmentation: From fancy genes to complex traits. Trends Genet 12: 299–305.

Bensch, S., B. Hansson, D. Hasselquist, and B. Nielsen. 2000. Partial albinism in a semi-isolated population of Great Reed Warblers. Hereditas 133: 167–170.

Berthold, P., G. Mohr, and U. Querner. 1996. The legendary "Veiled Blackcap" (Aves): A melanistic mutant with single-locus autosomal recessive inheritance. Naturwissenschaften 83: 568–570.

Bilodeau, M. L., J. D. Greulich, R. L. Hullinger, C. Bertolotto, R. Ballotti, and O. M. Andrisani. 2001. BMP-2 stimulates tyrosinase gene expression and melanogenesis in differentiated melanocytes. Pigment Cell Res 14: 328–336.

Bonser, R. H. C. 1995. Melanin and the abrasion resistance of feathers. Condor 97: 590–591.

Bonser, R. H. C., and M. S. Witter. 1993. Indentation hardness of the bill keratin of the European Starling. Condor 95: 736–738.

Borovansky, J. 1994. Zinc in pigmented cells and structures—Interactions and possible roles. Sborn Lek 95: 309–320.

Borovansky, J. 1996. Free radical activity of melanins and related substances: Biochemical and pathobiochemical aspects. Sborn Lek 97: 49–70.

Borovansky, J., J. Horcicko, and J. Duchon. 1976. The hair melanosome: Another tissue reservoir of zinc. Physiol Bohemoslov 25: 87–91.

Bowness, J. M., and R. A. Morton. 1952. Distribution of copper and zinc in the eyes of fresh-water fishes and frogs—Occurrence of metals in melanin fractions from eye tissues. Biochem J 51: 530–534.

Buchanan, K. L., M. R. Evans, A. R. Goldsmith, D. M. Bryant, and L. V. Rowe. 2001. Testosterone influences basal metabolic rate in male House Sparrows: A new cost of dominance signalling? Proc R Soc Lond B 268: 1337–1344.

Burtt, E. H. 1979. Tips on wings and other things. In E. H. Burtt, ed., The Behavioral Significance of Color, 70–123. New York: Garland STPM Press.

Burtt, E. H. 1986. An analysis of physical, physiological, and optical aspects of avian colouration with emphasis on wood-warblers. Ornithol Mongr 38: 1–126.

Burtt, E. H., and J. M. Ichida. 1999. Occurrence of feather-degrading bacilli in the plumage of birds. Auk 116: 364–372.

Burtt, E. H., and J. M. Ichida. 2004. Gloger's rule, feather-degrading bacteria, and color variation among song sparrows. Condor 106: 681–686.

Butler, M., and A. S. Johnson. 2004. Are melanized feather barbs stronger? J Exp Biol 207: 285–293.

Buxton, P. A. 1923. Animal Life in Deserts. London: Edward Arnold and Co.

Camplani, A., N. Saino, and A. P. Møller. 1999. Carotenoids, sexual signals and immune function in Barn Swallows from Chernobyl. Proc R Soc Lond B 266: 1111–1116.

Carefoot, W. C. 1993. Further studies of linkage and mappings of the loci of genes in group 3 on chromosome 1 of the domestic fowl. Br Poult Sci 34: 205–209.

Cloudsley-Thompson, J. L. 1999. Multiple factors in the evolution of animal coloration. Naturwissenschaften 86: 123–132.

Cooke, F., R. F. Rockwell, and D. B. Lank. 1995. The Snow Geese of La Perouse Bay—Natural Selection in the Wild. Oxford: Oxford University Press.

Cowie, V., and L. S. Penrose. 1951. Dilution of hair colour in phenylketonuria. Ann Eugen 16: 297–301.

Darwin, C. 1859. The Origin of Species by Means of Natural Selection. London: Murray.

Doering, T. L., J. D. Nosanchuk, W. K. Roberts, and A. Casadevall. 1999. Melanin as a potential cryptococcal defence against microbicidal proteins. Med Mycol 37: 175–181.

Doucet, S. M., M. D. Shawkey, M. K. Rathburn, H. L. Mays, Jr., and R. Montgomerie. 2004. Concordant evolution of plumage colour, feather microstructure and a melanocortin receptor gene between mainland and island populations of a Fairy-wren. Proc R Soc Lond B 271: 1663–1670.

Dupin, E., and N. M. LeDouarin. 1995. Retinoic acid promotes the differentiation of adrenergic cells and melanocytes in quail neural crest cultures. Dev Biol 168: 529–548.

Dupin, E., and N. M. LeDouarin. 2002. The avian embryo, a model for the role of cellular interactions in development; the example of neural crest-derived pigment cells. Avian Poult Biol Rev 13: 155–167.

Dupin, E., and N. M. LeDouarin. 2003. Development of melanocyte precursors from the vertebrate neural crest. Oncogene 22: 3016–3023.

DuShane, G. P. 1934. The origin of pigment cells in amphibia. Science 80: 620–621.

Evans, M. R., A. R. Goldsmith, and S. R. A. Norris. 2000. The effects of testosterone on antibody production and plumage coloration in male House Sparrows (*Passer domesticus*). Behav Ecol Sociobiol 47: 156–163.

Evrard, Y. A., L. Mohammad-Zadeh, and B. Holton. 2004. Alterations in Ca^{2+}-dependent and cAMP-dependent signaling pathways affect neurogenesis and melanogenesis of quail neural crest cells in vitro. Dev Gen Evol 214: 193–199.

Fattorusso, E., L. Minale, S. DeStefano, G. Cimino, and R. A. Nicolaus. 1968. Struttura e biogenesi delle feomelanine. Nota V. Sulla struttura della gallofeomelanina-1. Gazz Chim Ital 98: 1443–1463.

Fattorusso, E., L. Minale, S. DeStefano, G. Cimino, and R. A. Nicolaus. 1969. Struttura e biogenesi delle feomelanine. Nota VI. Sulla struttura della gallofeomelanina. Gazz Chim Ital 99: 29–45.

Fattorusso, E., L. Minale, and G. Sodano. 1970. Feomelanine e eumelanine da nuove fonti naturali. Gazz Chim Ital 100: 452–460.

Finger, E., D. Burkhardt, and J. Dyck. 1992. Avian plumage colors: Origin of UV reflection in a black parrot. Naturwissenschaften 79: 187–188.

Fitze, P. S., and H. Richner. 2002. Differential effects of a parasite on ornamental structures based on melanins and carotenoids. Behav Ecol 13: 401–407.

Folstad, I., and A. Karter. 1992. Parasites, bright males, and the immunocompetence handicap. Am Nat 139: 938–946.

Fox, H. M., and G. Vevers. 1960. The Nature of Animal Colors. New York: Macmillan.

Fujii, R. 2000. The regulation of motile activity in fish chromatophores. Pigment Cell Res 13: 300–319.

Fulton, I. E., C. W. Roberts, and K. M. Cheng. 1982. Cinnamon: A mutant of Japanese Quail. Can J Genet Cytol 24: 163–166.

George, F. W., J. F. Noble, and J. D. Wilson. 1981. Female feathering in Sebright cocks is due to conversion of testosterone to estradiol in skin. Science 213: 557–559.

Geremia, E., C. Corsaro, R. Bonomo, R. Giardinelli, P. Pappalardo, A. Vanella, and G. Sichel. 1984. Eumelanins as free radical trap and superoxide dismutase activity in Amphibia. Comp Biochem Physiol B 79: 67–69.

Gerrish, G. A., and C. E. Caceres. 2003. Genetic versus environmental influence on pigment variation in the ephippia of *Daphnia pulicaria*. Freshwater Biol 48: 1971–1982.

Gloger, C. W. L. 1833. Das Abandern der Vögel durch Einfluss des Klimas. Breslau: A. Schulz.

Goldstein, G., K. R. Flory, B. A. Browne, S. Majid, J. M. Ichida, and E. H. Burtt. 2004. Bacterial degradation of black and white feathers. Auk 121: 656–659.

Gonzalez, G., G. Sorci, A. P. Møller, P. Ninni, C. Haussy, and F. de Lope. 1999. Immunocompetence and condition-dependent sexual advertisement in male House Sparrows (*Passer domesticus*). J Anim Ecol 73: 1225–1234.

Gonzalez, G., G. Sorci, L. C. Smith, and F. de Lope. 2001. Testosterone and sexual signaling in male House Sparrows (*Passer domesticus*). Behav Ecol Sociobiol 50: 557–562.

Grau, C. R., and T. E. Roudybush. 1986. Lysine requirement of Cockatiel chicks. AFA Watchbird 12: 12–14.

Grau, C. R., T. E. Roudybush, P. Vohra, F. H. Kratzer, M. Yang, and D. Nearenberg. 1989. Obscure relations of feather melanization and avian nutrition. World's Poult Sci J 45: 241–246.

Gray, D. A. 1996. Carotenoids and sexual dichromatism in North American passerine birds. Am Nat 148: 453–480.

Griffith, S. C. 2000. A trade-off between reproduction and a condition-dependent sexually selected ornament in the House Sparrow *Passer domesticus*. Proc R Soc Lond B 267: 1115–1119.

Haase, E., and R. Schmedemann. 1992. Dose dependent effect of testosterone on the induction of eclipse coloration in castrated wild Mallard drakes (*Anas platyrhynchos* L.). Can J Zool 70: 428–431.

Haase, E., S. Ito, A. Sell, and K. Wakamatsu. 1992. Melanin concentrations in feathers from wild and Domestic Pigeons. J. Hered 83: 64–67.

Haase, E., S. Ito, and K. Wakamatsu. 1995. Influences of sex, castration, and androgens on the eumelanin and phaeomelanin contents of different feathers in wild Mallards. Pigment Cell Res 8: 164–170.

Hach, P., J. Borovansky, and E. Vendralova. 1993. Melanosome—A sophisticated organelle. Sborn Lek 94: 113–123.

Hadley, M. E., and W. C. Quevedo, Jr. 1966. Vertebrate epidermal melanin unit. Nature 209: 1334–1335.

Hagelin, J. C. 2001. Castration in Gambel's and Scaled Quail: Ornate plumage and dominance persist, but courtship and threat behaviors do not. Horm Behav 39: 1–10.

Hall, P. F. 1966. Tyrosinase activity in relation to plumage color in weaver birds (*Steganura paradisaea*). Comp Biochem Physiol 18: 91–100.

Hall, P. F. 1969a. The influence of hormones on melanogenesis. Aust J Dermatol 10: 125–139.

Hall, P. F. 1969b. Hormonal control of melanin synthesis in birds. Gen Comp Endocrinol 2 (Suppl.): 451–458.

Hall, P. F. 1970. Influence of oestradiol upon incorporation of tyrosine and cysteine into phaeomelanin by particles from feather tracts of brown leghorn chickens (*Gallus domesticus*). Int J Biochem 1: 638.

Hall, P. F., and K. Okazaki. 1966. The action of interstitial cell stimulating hormone upon avian tyrosinase. Biochem 5: 1202–1208.

Hall, P. F., C. L. Ralph, and D. L. Grinwich. 1965. On the locus of action of interstitial cell-stimulating hormone (ICSH or LH) on feather pigmentation of African weaver birds. Gen Comp Endocrinol 5: 552–557.

Hearing, V. 2000. The melanosome: The perfect model for cellular responses to the environment. Pigment Cell Res 13: 23–34.

Hill, G. E., and W. R. Brawner, III. 1998. Melanin-based plumage coloration in the House Finch is unaffected by coccidial infection. Proc R Soc Lond B 265: 1105–1109.

Hill, G. E., and K. J. McGraw. 2003. Melanin, nutrition, and the Lion's mane. Science 299: 660.

Hollander, W. F. 1989. ABC's of Poultry Genetics: A Short Course. Pine River, MN: Stromberg.

Homma, K., M. Jinno, and J. Kito. 1969. Studies on silver feathered Japanese Quail. Jap J Zootech Sci 40: 129–130.

Horcicko, J., J. Borovansky, J. Duchon, and B. Prochazkova. 1973. Distribution of zinc and copper in pigmented tissues. Hopp-Seyl Z Physiol Chem 354: 203–204.

Horth, L. 2005. Melanic pigmentation and melanocortin-1 receptor mutations in animals. Proc Ind Nat Sci Acad (in press).

Hudon, J. 1997. Non-melanic schizochroism in Alberta Evening Grosbeaks, *Coccothraustes vespertinus*. Can Field Nat 111: 652–654.

Hudon, J., and L. W. Oliphant. 1995. Reflective organelles in the anterior pigment epithelium of the iris of the European Starling *Sturnus vulgaris*. Cell Tissue Res 280: 383–389.

Hunt, G., C. Todd, J. E. Cresswell, and A. J. Thody. 1994. Alpha-melanocyte stimulating hormone and its analogue Nle4DPhe7 alpha-MSH affect morphology, tyrosinase activity and melanogenesis in cultured human melanocytes. J Cell Sci 107: 205–211.

Imesch, P. D., I. H. L. Wallow, and D. M. Albert. 1997. The color of the human eye: A review of morphologic correlates and of some conditions that affect iridial pigmentation. Surv Ophthalmol 41: S117–S123.

Ito, S., and K. Fujita. 1985. Microanalysis of eumelanin and phaeomelanin in hair and melanomas by chemical degradation and liquid chromatography. Analyt Biochem 144: 527–536.

Ito, S., and K. Wakamatsu. 1994. An improved modification of permanganate oxidation that gives constant yield of pyrrole-2,3,5-tricarboxylic acid. Pigment Cell Res 1: 141–144.

Ito, S., and K. Wakamatsu. 2003. Quantitative analysis of eumelanin and phaeomelanin in humans, mice, and other animals: A comparative review. Pigment Cell Res 16: 523–531.

Ito, S., K. Wakamatsu, and H. Ozeki. 2000. Chemical analysis of melanins and its application to the study of regulation of melanogenesis. Pigment Cell Res 13: 103S–109S.

Iturriza, F. C., F. E. Estivariz, and H. P. Levitin. 1980. Coexistence of alpha-melanocyte–stimulating hormone and adrenocorticotropin in all cells containing either of the 2 hormones in the duck pituitary. Gen Comp Endocrinol 42: 110–115.

Iturriza, F. C., R. A. Venosa, M. G. Pujol, and N. B. Quintas. 1992. Alpha-melanocyte–stimulating hormone stimulates sodium excretion in the salt gland of the duck. Gen Comp Endocrinol 87: 369–374.

Jaap, R. G. 1934. Alleles of the Mallard plumage pattern in ducks. Genetics 19: 310–322.

Jackson, I. J. 1997. Homologous pigmentation mutations in humans, mouse, and other model organisms. Hum Molec Genet 6: 1613–1624.

Jawor, J. M., and R. Breitwisch. 2003. Melanin ornaments, honesty, and sexual selection. Auk 120: 249–265.

Jimbow, K., C. Hua, P. F. Gomez, K. Hirosaki, K. Shinoda, et al. 2000. Intracellular vesicular trafficking of tyrosinase gene family protein in eu- and pheomelanosome biogenesis. Pigment Cell Res 13: 110S–117S.

Johns, J. E. 1964. Testosterone-induced nuptial feathers in phalaropes. Condor 66: 449–455.

Jurado, I., A. Saez, J. Luelmo, J. Diaz, I. Mendez, and M. Rey. 1998. Pigmented squamous cell carcinoma of the skin—Report of two cases and review of the literature. Am J Dermatopathol 20: 578–581.

Kadekaro, A. L., R. J. Kavanagh, K. Wakamatsu, S. Ito, M. A. Pipitone, and Z. A. Abdel-Malek. 2003. Cutaneous photobiology. The melanocyte vs. the sun: Who will win the final round? Pigment Cell Res 16: 434–447.

Kasraee, B., O. Sorg, and J. H. Saurat. 2003. Hydrogen peroxide in the presence of cellular antioxidants mediates the first and key step of melanogenesis: A new concept introducing melanin production as a cellular defence mechanism against oxidative stress. Pigment Cell Res 16: 571.

Keck, W. N. 1933. Control of the bill color of the English Sparrow by injection of male hormone. Proc Soc Exp Biol Med 30: 1140–1141.

Keck, W. N. 1934. The control of secondary sex characters in the English Sparrow, *Passer domesticus* (Linnaeus). J Exp Biol 67: 315–347.

Kerje, S., J. Lind, K. Schutz, P. Jensen, and L. Andersson. 2003. Melanocortin-1 receptor (*MC1R*) mutations are associated with plumage colour in chicken. Anim Gen 34: 241–248.

Kiama, S. G., J. Bhattacharjee, J. N. Maina, and K. D. Weyrauch. 1994. A scanning electron-microscope study of the pecten oculi of the Black Kite (*Milvus migrans*) —Possible involvement of the melanosomes in protecting the pecten against damage by ultraviolet light. J Anat 185: 637–642.

Kimball, R. T., and J. D. Ligon. 1999. Evolution of avian plumage dichromatism from a proximate perspective. Am Nat 14: 182–193.

Klasing, K. C. 1998. Comparative Avian Nutrition. Wallingford, UK: CAB International.

Kobayashi, T., W. D. Vieira, B. Potterf, C. Sakai, G. Imokawa, and V. J. Hearing. 1995. Modulation of melanogenic protein expression during the switch from eu- to phaeomelanogenesis. J Cell Sci 108: 2301–2309.

Krishnaswamy, A., and G. V. G. Baranoski. 2004. A biophysically-based spectral model of light interaction with human skin. Eurographics 23: 331–340.

Lahav, R., L. Lecoin, C. Ziller, V. Nataf, J. F. Carnahan, et al. 1994. Effect of the Steel gene product on melanogenesis in avian neural crest cell cultures. Differentiation 58: 133–139.

Lal, P., and V. K. Pathak. 1987. Effect of thyroidectomy and L-thyroxine on testes, body weight and bill color of the Tree Sparrow *Passer montanus*. Ind J Exp Biol 25: 660–663.

Lal, P., and J. P. Thapliyal. 1982. Role of thyroid in the response of bill pigmentation to male hormone of the House Sparrow, *Passer domesticus*. Gen Comp Endocrinol 48: 135–142.

Lancaster, F. M. 1963. The inheritance of plumage color in the Common Duck. Biblio Genet 19: 317–404.

Land, E. J., and P. A. Riley. 2000. Spontaneous redox reactions of dopaquinone and the balance between the eumelanic and phaeomelanic pathways. Pigment Cell Res 13: 273–277.

Landry, G. P. 1997. The varieties and genetics of the Zebra Finch. Franklin, LA: Poule d'eau Publishing.

Lank, D. B., C. M. Smith, O. Hanotte, T. Burke, and F. Cooke. 1995. Genetic polymorphism for alternative mating behaviour in lekking male Ruff *Philomachus pugnax*. Nature 378: 59–62.

Lank, D. B., M. Coupe, and K. E. Wynne-Edwards. 1999. Testosterone-induced male traits in female Ruffs (*Philomachus pugnax*): Autosomal inheritance and gender differentiation. Proc R Soc Lond B 266: 2323–2330.

Lauber, J. K. 1964. Sex-linked albinism in the Japanese Quail. Science 146: 948–950.

Lerner, A. B., and T. B. Fitzpatrick. 1950. Biochemistry of melanin formation. Physiol Rev 30: 91–126.

Lesser, M. P., S. L. Turtle, J. H. Farrell, and C. W. Walker. 2001. Exposure to ultraviolet radiation (290–400 nm) causes oxidative stress, DNA damage, and expression of *p53/p73* in laboratory experiments on embryos of the Spotted Salamander, *Ambystoma maculatum*. Physiol Biochem Zool 74: 733–741.

Ling, M. K., M. C. Lagerstrom, R. Fredriksson, R. Okimoto, N. I. Mundy, et al. 2003. Association of feather colour with constitutively active melanocortin 1 receptors in chicken. Eur J Biochem 270: 1441–1449.

Literak, I., A. Roulin, and K. Janda. 1999. Close inbreeding and unusual melanin distribution in Barn Owls (*Tyto alba*). Folia Zool 48: 227–231.

Lozano, G. A. 1994. Carotenoids, parasites, and sexual selection. Oikos 70: 309–311.

Lucas, A. M., and P. R. Stettenheim. 1972. Avian Anatomy: Integument. Agriculture Handbook 362. Washington, DC: U.S. Government Printing Office.

MacDougall-Shackleton, E. A., L. Blanchard, and H. L. Gibbs. 2003. Unmelanized plumage patterns in Old World Leaf Warblers do not correspond to sequence variation at the melanocortin-1 receptor locus (*MC1R*). Molec Biol Evol 20: 1675–1681.

Mackintosh, J. A. 2001. The antimicrobial properties of melanocytes, melanosomes and melanin and the evolution of black skin. J Theor Biol 211: 101–113.

Magomedov, N. M., A. I. Dzharfarov, and E. J. Yusifov. 1990. On melanin role in regulation of free radical processes in the eye pigmental epithelium under the acute hypoxia. Biofizika 35: 977–980.

Markert, C. L. 1948. The effects of thyroxine and antithyroid compounds on the synthesis of pigment granules in chick melanoblasts cultured in vitro. Physiol Zool 21: 309–327.

Marks, M. S., and M. C. Seabra. 2001. The melanosome: Membrane dynamics in black and white. Nat Rev Molec Cell Biol 2: 738–748.

Mason, K. A., and S. K. Frost-Mason. 2000. Evolution and development of pigment cells: At the crossroads of the discipline. Pigment Cell Res 13: 150–155.

McGinness, J., and P. H. Proctor. 1973. The importance of the fact that melanin is black. J Theor Biol 39: 677–678.

McGinness, J. E., R. Kono, and W. D. Moorhead. 1970. The melanosome: Cytoprotective or cytotoxic? Pigment Cell 4: 270–276.

McGraw, K. J. 2003. Melanins, metals, and mate quality. Oikos 102: 402–406.

McGraw, K. J. 2005. The antioxidant functions of many animal pigments: Are there consistent health benefits of sexually selected colorants? Anim Behav 69: 757–764.

McGraw, K. J., and G. E. Hill. 2000. Differential effects of endoparasitism on the expression of carotenoid- and melanin-based ornamental coloration. Proc R Soc Lond B 267: 1525–1531.

McGraw, K. J. and K. Wakamatsu. 2004. Melanin basis of ornamental feather colors in male Zebra Finches. Condor 106: 686–690.

McGraw, K. J., E. A. Mackillop, J. Dale, and M. E. Hauber. 2002. Different colors reveal different information: How nutritional stress affects the expression of melanin- and structurally based ornamental plumage. J Exp Biol 205: 3747–3755.

McGraw, K. J., K. Wakamatsu, S. Ito, P. M. Nolan, P. Jouventin, et al. 2004a. You can't judge a pigment by its color: Carotenoid and melanin content of yellow and brown feathers in swallows, bluebirds, penguins, and Domestic Chickens. Condor 106: 390–395.

McGraw, K. J., K. Wakamatsu, A. B. Clark, and K. Yasukawa. 2004b. Red-winged Blackbirds *Agelaius phoeniceus* use carotenoid and melanin pigments to color their epaulets. J Avian Biol 35: 543–550.

McGraw, K. J., R. J. Safran, M. R. Evans, and K. Wakamatsu. 2004c. European Barn Swallows use melanin pigments to color their feathers brown. Behav Ecol 15: 889–891.

McGraw, K. J., J. Hudon, G. E. Hill, and R. S. Parker. 2005. A simple and inexpensive chemical test to determine the presence of carotenoid pigments in animal tissues. Behav Ecol Sociobiol 57: 391–397.

Miller, D. S. 1935. Effects of thyroxin on plumage of the English Sparrow, *Passer domesticus*. J Exp Zool 71: 293–309.

Minale, L., E. Fattorusso, G. Cimino, S. DeStefano, and R. A. Nicolaus. 1967. Struttura e biogenesi delle feomelanine. Nota III. Prodotti di degradazione. Gazz Chim Ital 97: 1636–1663.

Minale, L., E. Fattorusso, G. Cimino, S. DeStefano, and R. A. Nicolaus. 1969. Struttura e biogenesi delle feomelanine. Nota VIII. Sulla struttura della gallofeomelanina-1. Gazz Chim Ital 100: 461–466.

Minezawa, M., and K. Wakasugi. 1977. Studies on a plumage mutant (black at hatch) in the Japanese Quail. Jap J Genet 52: 183–195.

Moreau, R. E. 1958. Some aspects of the Musophagidae. Part 3. Ibis 100: 238–270.

Morris, J. G., S. Yu, and Q. R. Rogers. 2002. Red hair in black cats is reversed by addition of tyrosine to the diet. J Nutr 132: 1646S–1648S.

Morrison, W. L. 1985. What is the function of melanin? Arch Dermatol 121: 1160–1163.

Mountjoy, K. G., L. S. Robbins, M. T. Nortrud, and R. D. Cone. 1992. The cloning of a family of genes that encode the melanocortin receptors. Science 257: 1248–1251.

Mundy, N. I., N. Badcock, T. Hart, K. Scribner, K. Janssen, and N. J. Nadeau. 2004. Conserved genetic basis of a quantitative plumage trait involved in mate choice. Science 303: 1870–1873.

Muroya, S., R. Tanabe, I. Nakajima, and K. Chikuni. 2000. Molecular characteristics and site-specific distribution of the pigment of the silky fowl. J Vet Med Sci 62: 391–395.

Nataf, V., and N. M. LeDouarin. 2000. Induction of melanogenesis by tetradecanoylphorbol-13 acetate and endothelin 3 in embryonic avian peripheral nerve cultures. Pigment Cell Res 13: 172–178.

Needham, A. E. 1974. The Significance of Zoochromes. New York: Springer-Verlag.

Nickerson, M. 1941. Relation between black and red melanin pigments in feathers. Physiol Zool 19: 66–77.

Niecke, M., M. Heid, and A. Kruger. 1999. Correlations between melanin pigmentation and element concentration in feathers of White-tailed Eagles (*Haliaeetus albicilla*). J Ornithol 140: 355–362.

Niecke, M., S. Rothlaender, and A. Roulin. 2003. Why do melanin ornaments signal individual quality? Insights from metal element analysis of Barn Owl feathers. Oecologia 137: 153–158.

Nishizawa, H., T. Okamoto, and Y. Yoshimura. 2002. Immunolocalization of sex steroid receptors in the epididymis and ductus deferens of immature and mature Japanese Quail, *Coturnix japonica*. Anim Sci J 73: 339–346.

Niwa, T., M. Mochii, A. Nakamura, and N. Shiojiri. 2002. Plumage pigmentation and expression of its regulatory genes during quail development—Histochemical analysis using Bh (black at hatch) mutants. Mech Dev 118: 139–146.

Okazaki, K., and P. F. Hall. 1965. The action of interstitial cell-stimulating hormone upon tyrosinase activity in the weaver bird. Biochem Biophys Res Comm 20: 667–673.

Ortman, R. 1967. The performance of the Napolean Weaver, the Orange Weaver, and the Paradise Whydah in the weaver finch test for luteinizing hormone. Gen Comp Endocrinol 9: 368–373.

Ortonne, J. P. 2002. Photoprotective properties of skin melanin. Brit J Dermatol 146: 7–10.

Owens, I. P. F., and R. V. Short. 1995. Hormonal basis of sexual dimorphism in birds: Implications for new theories of sexual selection. Trends Ecol Evol 10: 44–47.

Ozeki, H., S. Ito, K. Wakamatsu, and I. Ishiguro. 1997. Chemical characterization of phaeomelanogenesis starting from dihydroxyphenylalanine or tyrosine and cysteine. Effects of tyrosinase and cysteine concentrations and reaction time. Biochim Biophys Acta Gen Subj 1336: 539–548.

Palumbo, A. 2003. Melanogenesis in the ink gland of *Sepia officinalis*. Pigment Cell Res 16: 517–522.

Parichy, D. M. 2003. Pigment patterns: Fish in stripes and spots. Curr Biol 13: R947–R950.

Parker, T. H., B. M. Stansberry, C. D. Becker, and P. S. Gipson. 2003. Do melanin- or carotenoid-pigmented plumage ornaments signal condition and predict pairing success in the Kentucky Warbler? Condor 105: 663–671.

Peczely, P. 1992. Hormonal regulation of feather development and molt on the level of feather follicles. Orn Scand 23: 346–354.

Poiani, A., A. R. Goldsmith, and M. R. Evans. 2000. Ectoparasites of House Sparrows (*Passer domesticus*): An experimental test of the immunocompetence handicap hypothesis and a new model. Behav Ecol Sociobiol 47: 230–242.

Pollock, R. T. 1967. Factors Affecting Melanin Synthesis in the Pigment Cells of Chick Embryo Skin. Ph.D. diss., University of California, Davis, Davis.

Poston, J. P., D. Hasselquist, I. R. K. Stewart, and D. F. Westneat. Dietary amino acids influence plumage traits and immune responses of male House Sparrows (*Passer domesticus*), but not as expected. Anim Behav (in press).

Potterf, S. B., J. Muller, I. Bernardini, F. Tietzej, T. Kobayashi, et al. 1996. Characterization of a melanosomal transport system in murine melanocytes mediating entry of the melanogenic substrate tyrosine. J Biol Chem 271: 4002–4008.

Potterf, S. B., V. Virador, K. Wakamatsu, M. Furumura, C. Santis, et al. 1999. Cysteine transport in melanosomes from murine melanocytes. Pigment Cell Res 12: 4–12.

Price, T., and A. Bontrager. 2001. The evolution of plumage patterns. Curr Biol 11: 405–408.

Price, T., and M. Pavelka. 1996. Evolution of a colour pattern: History, development, and selection. J Evol Biol 9: 451–470.

Prota, G. 1992. Melanins and Melanogenesis. San Diego: Academic Press.

Prota, G., and R. A. Nicolaus. 1967. Struttura e biogenesi delle feomelanine. Nota I. Isolamento e proprieta dei pigmenti delle piume. Gazz Chim Ital 97: 666–684.

Prota, G., M. Dischia, and D. Mascagna. 1994. Melanogenesis as a targeting strategy against metastatic melanoma—A reassessment. Melanoma Res 4: 351–358.

Prum, R. O., and S. Williamson. 2002. Reaction-diffusion models of within-feather pigmentation patterning. Proc R Soc Lond B 269: 781–792.

Ralph, C. L. 1969. The control of color in birds. Am Zool 9: 521–530.

Rasmussen, N., F. Nelson, P. Govitrapong, and M. Ebadi. 1999. The actions of melanin and melanocyte stimulating hormone (MSH). Neuroendocrinol Lett 20: 265–282.

Rawles, M. E. 1944. The migration of melanoblasts after hatching into pigment-free skin grafts of the common fowl. Physiol Zool 17: 167–183.

Rees, J. L. 2000. The melanocortin-1 receptor (*MC1R*): More than just red hair. Pigment Cell Res 13: 135–140.

Rhode, S. C., M. Pawlowski, and R. Tollrian. 2001. The impact of ultraviolet radiation on the vertical distribution of zooplankton of the genus *Daphnia*. Nature 412: 69–72.

Riley, P. A. 1992. Materia melanica: Further dark thoughts. Pigment Cell Res 5: 101–106.

Riley, P. A. 1997. Melanin. Int J Biochem Cell Biol 29: 1235–1239.

Robbins, L. S., J. H. Nadeau, K. R. Johnson, M. A. Kelly, L. Roselli-Rehfuss, et al. 1993. Pigmentation phenotypes of the variant extension locus alleles result from point mutations that alter MSH receptor function. Cell 72: 827–834.

Roberts, C. W., and I. E. Fulton. 1980. Yellow: A mutant plumage colour, segregating independently from brown, in Japanese Quail. Can J Genet Cytol 22: 411–416.

Roberts, C. W., I. E. Fulton, and C. R. Barnes. 1978. Genetics of white-breasted, white and brown colors and descriptions of feather patterns in Japanese Quail. Can J Genet Cytol 20: 1–8.

Robins, A. H. 1991. Biological Perspectives on Human Pigmentation. Cambridge, UK: Cambridge University Press.

Ros, A. F. H. 1999. Effects of testosterone on growth, plumage pigmentation, and mortality in Black-headed Gull chicks. Ibis 141: 451–459.

Roulin, A., and C. Dijkstra. 2003. Genetic and environmental components of variation in eumelanin and phaeomelanin sex-traits in the Barn Owl. Heredity 90: 359–364.

Roulin, A., C. Riols, C. Dijkstra, and A. L. Ducrest. 2001. Female plumage spottiness signals parasite resistance in the Barn Owl (*Tyto alba*). Behav Ecol 12: 103–110.

Roulin, A., A. L. Ducrest, F. Balloux, C. Dijkstra, and C. Riols. 2003. A female melanin ornament signals offspring fluctuating asymmetry in the Barn Owl. Proc R Soc Lond B 270: 167–171.

Rozanowska, M., T. Sarna, E. J. Land, and T. G. Truscott. 1999. Free radical scavenging properties of melanins: Interaction of eu- and pheo-melanin models with reducing and oxidising radicals. Free Rad Biol Med 26: 518–525.

Rutz, C., A. Zinke, T. Bartels, and P. Wohlsein. 2004. Congenital neuropathy and dilution of feather melanin in nestlings of urban-breeding Northern Goshawks (*Accipiter gentilis*). J Zoo Wildlife Med 35: 97–103.

Saino, N., R. Stradi, P. Ninni, and A. P. Møller. 1999. Carotenoid plasma concentration, immune profile and plumage ornamentation of male Barn Swallows (*Hirundo rustica*). Am Nat 154: 441–448.

Sanchez-Ferrer, A., J. N. Rodriguez-Lopez, F. Garcia-Canovas, and F. Garcia-Carmona. 1995. Tyrosinase—A comprehensive review of its mechanism. BBA Protein Struct Molec Enzymol 1247: 1–11.

Sarna, T., and H. M. Swartz. 1998. The physical properties of melanins. In J. J. Nordlund, R. E. Boissy, V. J. Hearing, R. A. King, and J. P. Ortonne, ed., The Pigmentary System: Physiology and Pathophysiology, 333–358. New York: Oxford University Press.

Sato, S., M. Tanaka, H. Miura, K. Ikeo, T. Gojobori, et al. 2001. Functional conservation of the promoter regions of vertebrate tyrosinase genes. J Inv Derm Sym Proc 6: 10–18.

Satoh, M., and H. Ide. 1987. Melanocyte-stimulating hormone affects melanogenic differentiation of quail neural crest cells in vitro. Dev Biol 119: 579–586.

Schallreueter, K. U. 1999. A review of recent advances on the regulation of pigmentation in the human epidermis. Cell Molec Biol 45: 943–949.

Schereschewsky, H. 1929. Einige Beiträge zum Problem der Verfarbung des Gefieders beim Gimpel. Arch Entwick Org 115: 110–153.

Schioth, H. B., T. Raudsepp, A. Ringholm, R. Fredriksson, S. Takeuchi, et al. 2003. Remarkable synteny conservation of melanocortin receptors in chicken, human, and other vertebrates. Genomics 81: 504–509.

Schlinger, B. A., A. J. Fivizzani, and G. V. Callard. 1989. Aromatase, 5α-, and 5β-reductase in brain, pituitary and skin of the sex-role reversed Wilson's phalarope. J Endocrinol 122: 573–581.

Schofer, C., K. Frei, K. Weipoltshammer, and F. Wachtler. 2001. The apical ectodermal ridge, fibroblast growth factors (FGF-2 and FGF-4) and insulin-like growth factor I (IGF-I) control the migration of epidermal melanoblasts in chicken wing buds. Anat Embryol 203: 137–146.

Schraermeyer, U. 1996. The intracellular origin of the melanosome in pigment cells. A review of ultrastructural data. Histol Histopathol 11: 445–462.

Sealy, R. C., J. S. Hyde, C. C. Felix, I. A. Menon, and G. Prota. 1982. Eumelanins and phaeomelanins: Characterization by electron spin resonance spectroscopy. Science 217: 545–547.

Segal, S. J. 1957. Response of weaver finch to chorionic gonadotropin and hypophysial luteinizing hormone. Science 126: 1242–1243.

Sell, A. 1994. Breeding and Inheritance in Pigeons. Hengersberg: Schober.

Senar, J. C., J. Figuerola, and J. Domenech. 2003. Plumage coloration and nutritional condition in the Great Tit *Parus major:* The roles of carotenoids and melanins differ. Naturwissenschaften 90: 234–237.

Shawkey, M. D., S. R. Pillai, and G. E. Hill. 2003. Chemical warfare? Effects of uropygial oil on feather-degrading bacteria. J Avian Biol 34: 345–349.

Shcherba, V. V., V. G. Babitskaya, V. P. Kurchenko, N. V. Ikonnikova, and T. A. Kukulyanskaya. 2000. Antioxidant properties of fungal melanin pigments. Appl Biochem Microbiol 36: 491–495.

Shiojiri, N., T. Niwa, K. Wakamatsu, S. Ito, and A. Nakamura. 1999. Chemical analysis of melanin pigments in feather germs of Japanese Quail Bh (black at hatch) mutants. Pigment Cell Res 12: 259–265.

Slagsvold, T., and J. T. Lifjeld. 1992. Plumage color is a condition-dependent sexual trait in male Pied Flycatchers. Evolution 46: 825–828.

Smit, N. P. M., H. van der Meulen, H. K. Koerten, R. M. Kolb, A. M. Mommaas, et al. 1997. Melanogenesis in cultured melanocytes can be substantially influenced by L-tyrosine and L-cysteine. J Invest Dermatol 109: 796–800.

Smith, J. P. 1978. Annual cycles of thyroid-hormones in plasma of White-crowned Sparrows (*Zonotrichia leucophrys gambelii*) and House Sparrows (*Passer domesticus*). Am Zool 18: 591.

Somes, Jr., R. G., and J. R. Smyth, Jr. 1965. Feather phaeomelanin intensity in Buff Orpington, New Hampshire and Rhode Island red breeds of fowl. Poult Sci 154: 40–46.

Stinchcombe, J., G. Bossi, and G. M. Griffiths. 2004. Linking albinism and immunity: The secrets of secretory lysosomes. Science 305: 55–59.

Stokkan, K. A. 1979. Testosterone and daylength-dependent development of comb size and breeding plumage of male Willow Ptarmigan (*Lagopus lagopus lagopus*). Auk 96: 106–115.

Strasser, R., and H. Schwabl. 2004. Yolk testosterone organizes behavior and male plumage coloration in House Sparrows (*Passer domesticus*). Behav Ecol Sociobiol 56: 491–497.

Sugumaran, M. 2002. Comparative biochemistry of eumelanogenesis and the protective roles of phenoloxidase and melanin in insects. Pigment Cell Res 15: 2–9.

Sulaimon, S. S., and B. E. Kitchell. 2003. The biology of melanocytes. Vet Dermatol 14: 57–65.

Takeuchi, S., H. Suzuki, S. Hirose, M. Yabuuchi, C. Sato, et al. 1996a. Molecular cloning and sequence analysis of the chick melanocortin-1 receptor gene. Biochim Biophys Acta 1306: 122–126.

Takeuchi, S., H. Suzuki, M. Yabuuchi, and S. Takahashi. 1996b. A possible involvement of melanocortin-1 receptor in regulating feather color pigmentation in the chicken. Biochim Biophys Acta 1308: 164–168.

Takeuchi, S., K. Teshigawara, and S. Takahashi. 2000. Widespread expression of agouti-related protein (AGRP) in the chicken: A possible involvement of AGRP in regulating peripheral melanocortin systems in the chicken. Biochim Biophys Acta—Molec Cell Res 1496: 261–269.

Takeuchi, S., S. Takahashi, S. Okimoto, H. B. Schioth, and T. Boswell. 2003. Avian melanocortin system: Alpha-MSH may act as an autocrine/paracrine hormone—A minireview. Ann NY Acad Sci 994: 366–372.

Teshigawara, K., S. Takahashi, T. Boswell, Q. Li, S. Tanaka, and S. Takeuchi. 2001. Identification of avian alpha-melanocyte–stimulating hormone in the eye: Temporal and spatial regulation of expression in the developing chicken. J Endocrinol 168: 527–537.

Thapliyal, J. P., and R. N. Saxena. 1961. Plumage control in Indian Weaver bird (*Ploceus philippinus*). Naturwissenschaften 48: 741–742.

Theron, E., K. Hawkins, E. Bermingham, R. Ricklefs, and N. I. Mundy. 2001. The molecular basis of an avian plumage polymorphism in the wild: A point mutation in the melanocortin-1 receptor is perfectly associated with melanism in the Bananaquit (*Coereba flaveola*). Curr Biol 11: 550–557.

Trinkaus, J. P. 1947. An analysis of the effects of estradiol on melanoblast differentiation in the brown leghorn fowl. Anat Rec 99: 588–599.

Trinkaus, J. P. 1948. Factors concerned in the response of melanoblasts to estrogen in the brown leghorn fowl. J Exp Zool 109: 135–170.

Trinkaus, J. P. 1950. The role of thyroid hormone in melanoblast differentiation in the brown leghorn. J Exp Zool 113: 149–178.

Trinkaus, J. P. 1953. Estrogen, thyroid hormone, and the differentiation of pigment cells in the brown leghorn. In M. Gordon, ed., Pigment Cell Growth, 73–91. New York: Academic Press.

Truax, R. E., P. B. Siegel, and W. A. Johnson. 1979. Redhead, a plumage color mutant in Japanese Quail. J Hered 70: 413–415.

Urabe, K., P. Aroca, and V. J. Hearing. 1993. From gene to protein—Determination of melanin synthesis. Pigment Cell Res 6: 186–192.

Veiga, J. P., and P. Puerta. 1996. Nutritional constraints determine the expression of a sexual trait in the House Sparrow, *Passer domesticus*. Proc R Soc Lond B 263: 229–234.

Verne, J. 1926. Les Pigments Dans L'organisme Animal. Paris: Gaston Doin et Cie.

Voitkevich, A. A. 1966. The Feathers and Plumage of Birds. London: Sidgwick and Jackson.

Wakamatsu, K., and S. Ito. 2002. Advanced chemical methods in melanin determination. Pigment Cell Res 15: 174–183.

Wakamatsu, K., S. Ito, and J. L. Rees. 2002. The usefulness of 4-amino-3-hydroxyphenylalanine as a specific marker of phaeomelanin. Pigment Cell Res 15: 225–232.

Walsberg, G. E. 1982. Coat color, solar heat gain, and conspicuousness in the Phainopepla. Auk 99: 495–502.

Walsberg, G. E., G. S. Campbell, and J. R. King. 1978. Animal coat color and radiative heat gain: A re-evaluation. J Comp Physiol 126: 211–222.

Ward, J. M., J. D. Blount, G. D. Ruxton, and D. C. Houston. 2002. The adaptive significance of dark plumage for birds in desert environments. Ardea 90: 311–323.

West, P. M., and C. Packer. 2003. Response. Science 299: 660.

Wilson, K., S. C. Cotter, A. F. Reeson, and J. K. Pell. 2001. Melanism and disease resistance in insects. Ecol Lett 4: 637–649.

Witschi, E. 1935. Seasonal sex characters in birds and their hormonal control. Wilson Bull 47: 177–188.
Witschi, E. 1936. Effect of gonadotropic and oestrogenic hormones on regenerating feathers of weaver finches (*Pyromelana franciscana*). Proc Soc Exptl Biol Med 35: 484–489.
Witschi, E. 1937. Comparative physiology of the vertebrate hypophysis (anterior and intermediate lobes). Cold Spring Harbor Symp Quant Biol 5: 180–190.
Witschi, E. 1940. The quantitative determination of follicle stimulating and luteinizing hormones in mammalian pituitaries and a discussion of the gonadotropic quotient, F/L. Endocrinol 27: 437–446.
Witschi, E. 1955. Vertebrate gonadotrophins. Memb Soc Endocrinol 4: 149–165.
Witschi, E. 1961. Sex and secondary sexual characters. In A. L. Marshall, ed., Biology and Comparative Physiology of Birds, Volume 2, 115–168. New York: Academic Press.
Witschi, E., and S. J. Segal. 1955. Color phases and pigmentary reactions in African weaver finches of the genus *Euplectes*. Anat Rec 122: 465.
Wolf, B. O., and G. E. Walsberg. 2000. The role of plumage in heat transfer processes in birds. Am Zool 40: 575–584.
Wolff, G. L. 2003. Regulation of yellow pigment formation in mice: A historical perspective. Pigment Cell Res 16: 2–15.
Yu, M., Z. Yue, P. Wu, W. Da-Yu, J. A. Mayer, et al. 2004. The developmental biology of feather follicles. Int J Dev Biol 48: 181–191.
Yu, S., Q. R. Rogers, and J. G. Morris. 2001. Effect of low levels of dietary tyrosine on the hair colour of cats. J Small Anim Pract 42: 176–180.
Zang, L. Y., O. Sommerburg, and F. J. G. M. van Kuijk. 1997. Absorbance changes of carotenoids in different solvents. Free Rad Biol Med 23: 1086–1089.
Zecca, L., F. A. Zucca, P. Costi, D. Tampellini, A. Gatti, et al. 2003. The neuromelanin of human substantia nigra: Structure, synthesis and molecular behaviour. J Neural Transm 65 (Suppl.): 145–155.
Zeise, L., M. R. Chedekel, and T. B. Fitzpatrick. 1995. Melanin: Its role in human photoprotection. Overland Park, Kansas: Valdenmar.

7

Anatomy, Physics, and Evolution of Structural Colors

RICHARD O. PRUM

The coloration of birds is produced by chemical pigments and by the physical interactions of light waves with biological structures. Pigments are molecules that differentially absorb and emit wavelengths of visible light. The color produced by a pigment is determined by the molecular structure of the pigment molecule and its concentration. Pigmentary colors are common in the plumage, eggshells, dermis, and eyes of birds (Chapters 5, 6, and 8). In contrast, structural colors of birds are produced by light interacting physically with nanometer-scale variation in the structure of avian integument and eye tissues. Structural colors are an important component of the appearance of plumage, skin, and irides of many bird species and are broadly distributed throughout Aves (Auber 1957; Prum and Torres 2003b).

The production mechanisms of structural colors of bird feathers have been researched extensively over the past century (reviewed in Dyck 1971b, 1976; Fox 1976; Durrer 1986). A complete list of the 164 bird species with integumentary structural colors that have been examined by electron microscopy is listed in Table 7.1. Despite this body of historical research, there have been many recent advances in our understanding of the physics and evolution of avian structural colors. Furthermore, relatively few papers have investigated the anatomy and physics of structural colors found in the avian skin and eye. Recent developments in these areas indicate new frontiers in the study of avian structural coloration.

Table 7.1 List of All Published Observations of Color-Producing Nanostructures in Avian Integument Using Electron Microscopy

Family	Taxon	Feather barbules	Feather barbs	Dermal collagen	Reference
Casuariidae	*Dromaius novahollandiae*			x	32
Anatidae	*Anas crecca*	x			18
Anatidae	*Anas platrhynchos*	x			37
Anatidae	*Anas brasiliensis*	x			36
Anatidae	*Anas specularis*	x			36
Anatidae	*Anas gibberifrons*	x			36
Anatidae	*Aix galericulata*	x			36
Anatidae	*Aix sponsa*	x			8
Anatidae	*Oxyura jamaicensis*			x	1, 32
Anatidae	*Tadorna tadorna*	x			8
Cracidae	*Chaemapetes unicolor*			x	32
Cracidae	*Oreophasis derbianus*	x			36
Megapodidae	*Megapodius freycinet*	x			8
Numididae	*Numida meleagris*			x	1, 32
Numididae	*Acryllium vulturinum*	x			8
Phasianidae	*Afropavo congensis*	x			8, 14
Phasianidae	*Agriocharis ocellata*	x			8
Phasianidae	*Caloperdix oculea*	x			8
Phasianidae	*Chrysolophus pictus*	x			8
Phasianidae	*Francolinus francolinus*	x			8
Phasianidae	*Gallus gallus*	x			8
Phasianidae	*Lagopus mutus*	x			19
Phasianidae	*Lophura bulweri*			x	32
Phasianidae	*Lophura erythropthalmus*	x			8
Phasianidae	*Lophura ignata*	x			8
Phasianidae	*Lophura leucomelanos*	x			8
Phasianidae	*Lophophorus impejanus*	x		x	32, 40
Phasianidae	*Melanoperdix nigra*	x			8
Phasianidae	*Meleagris gallopavo*	x		x	8, 32
Phasianidae	*Pavo cristatus*	x			7, 8, 18, 28
Phasianidae	*Phasianus colchicus*	x			8
Phasianidae	*Polyplectron bicaratum*	x			8
Phasianidae	*Rollulus roulroul*	x			8
Phasianidae	*Syrmaticus ellioti*	x			8
Phasianidae	*Tetrao tetrix*	x			8
Phasianidae	*Tragopan temmincki*			x	32
Phasianidae	*Tragopan caboti*			x	1, 32
Phasianidae	*Tragopan satyra*			x	1, 32
Phasianidae	*Tragopan melanocephalus*	x			8
Sulidae	*Sula nebouxii*			x	32
Ardeidae	*Butorides striatus*	x			8
Ardeidae	*Pilherodius pileatus*			x	32
Ardeidae	*Syrigma sibalatrix*			x	1, 32

Table 7.1 (continued)

Family	Taxon	Feather barbules	Feather barbs	Dermal collagen	Reference
Threskiornithidae	*Bostrichia hagedash*	x			3
Threskiornithidae	*Carphibis spinicollis*	x			8
Charadridae	*Vanellus vanellus*	x			8, 18
Columbidae	*Caloenas nicobarica*	x			8
Columbidae	*Phaps chocoptera*	x			8
Columbidae	*Ptilinopus cincta*	x			20
Columbidae	*Ptilinopus jambu*	x			20
Columbidae	*Ptilinopus rivoli*	x			1, 20, 21
Columbidae	*Ptilinopus superbus*	x			20
Columbidae	*Ptilinopus viridis*	x			20
Columbidae	*Ptilinopus victor*	x			20
Columbidae	*Ptilinopus luteovirens*	x			8, 20
Columbidae	*Ducula concinna*	x			8, 20
Columbidae	*Chalcophaps indica*	x			20
Columbidae	*Columba trocaz*	x			38
Columbidae	*Columba livia*	x			8
Columbidae	*Columba fasciata*	x			8
Columbidae	*Zenaida asciatica*	x			8
Opisthocomidae	*Opisthocomus hoazin*			x	32
Cuculidae	*Chrysococcyx cupreus*	x			8, 13
Cuculidae	*Coua caerulea*			x	32
Cuculidae	*Coua reynaudii*			x	32
Cuculidae	*Ceuthmochares aereus*			x	32
Cuculidae	*Crotophaga sp.*	x			18
Musophagidae	*Tauraco corythaix*	x			8
Psittacidae	*Ara ararauna*		x		39
Psittacidae	*Agapornis roseicollis*		x		1, 15, 16, 18, 22, 35
Psittacidae	*Chalcopsitta atra*		x		22
Psittacidae	*Trichoglossus haematodus*		x		18
Psittacidae	*Melopsittacus undulatus*		x		35, 22
Apodidae	*Collocalia esculenta*	x			8
Hemiprocnidae	*Hemiprocne comata*	x			8
Trochilidae	*Amazilia cyanura*	x			8
Trochilidae	*Chlorestes notatus*	x			18
Trochilidae	*Heliangelus strophianus*	x			40
Trochilidae	*Heliangelus clemenciae*	x			25, 26
Trochilidae	*Heliangelus viola*	x			25, 26
Trochilidae	*Thalurania furcata*	x			25, 26
Trochilidae	*Clytolaema rubicauda*	x			25, 26
Trochilidae	*Chrysolampis mosquitos*	x			25, 26
Trogonidae	*Pharomachrus mocino*	x			1, 8, 9, 27
Trogonidae	*Pharomachrus pavoninus*	x			8, 9
Trogonidae	*Trogon rufus*	x			8, 9

Table 7.1 (continued)

Family	Taxon	Feather barbules	Feather barbs	Dermal collagen	Reference
Trogonidae	*Trogon violaceous*	x			8, 9, 18
Trogonidae	*Trogon massena*	x			1
Trogonidae	*Priotelus temnurus*	x			8, 9
Trogonidae	*Apaloderma narina*	x			8, 9
Trogonidae	*Apaloderma equatoriale*			x	32
Trogonidae	*Harpactes reinwardtii*	x			8, 9
Alcenidae	*Alcedo atthis*		x		18, 22
Alcenidae	*Alcedo ispida*		x		37
Alcenidae	*Chloroceryle americana*	x			8
Alcenidae	*Halcyon pileata*		x		1
Meropidae	*Nyctyornis amicta*		x		1
Motmotidae	*Motmotus motmota*		x		1
Phoeniculidae	*Phoeniculus purpureus*	x			8
Galbulidae	*Galbula ruficauda*	x			8, 9
Ramphastidae	*Megalaima chrysopogon*		x		1
Ramphastidae	*Selenidera culik*			x	32
Ramphastidae	*Ramphastos vitellinus*			x	32
Ramphastidae	*Ramphastos toco*			x	1, 32
Pittidae	*Pitta maxima*		x		23, 24
Eurylaimidae	*Calyptomena viridis*		x		1
Eurylaimidae	*Neodrepanis coruscans*			x	32, 33
Eurylaimidae	*Neodrepanis hypoxantha*			x	32, 33
Eurylaimidae	*Philepitta castanea*			x	1, 30, 32, 33
Thamnophilidae	*Gymnopithys leucapsis*			x	1, 32
Thamnophilidae	*Rhegmatorhina melanosticta*			x	32
Thamnophilidae	*Myrmeciza ferruginea*			x	32
Cotingidae	*Cotinga cayana*		x		39
Cotingidae	*Cotinga cotinga*		x		1
Cotingidae	*Cotinga maynana*		x		15, 34
Cotingidae	*Perissocephalus tricolor*			x	32
Cotingidae	*Procnias nudicollis*			x	31, 32
Pipridae	*Chiroxiphia caudata*		x		17
Pipridae	*Lepidothrix coronta*		x		1
Pipridae	*Lepidothrix natereri*		x		1
Paradiseadiae	*Astrapia rothschildi*	x			8
Paradiseadiae	*Astrapia splendissima*	x			8
Paradiseadiae	*Epimachus fastuosus*	x			8
Paradiseadiae	*Lophorina superba*	x			8
Paradiseadiae	*Paradisea raggiana*	x			1
Paradiseadiae	*Paradisea rubra*	x			8
Paradiseadiae	*Parotia sefilata*	x			5
Paradiseadiae	*Ptiloris paradiseus*	x			5
Paradiseadiae	*Ptiloris magnificus*	x			1, 8
Paradiseadiae	*Cicinnurus magnificus*	x			5

Table 7.1 (continued)

Family	Taxon	Feather barbules	Feather barbs	Dermal collagen	Reference
Paradiseadiae	*Cicinnurus regius*	x			8
Meliphagidae	*Prosthemadura novaeseelandiae*	x			8
Maluridae	*Malurus leucopterus*		x		6
Irenidae	*Irena puella*		x		1
Irenidae	*Aegithina tiphia*	x			8
Corvidae	*Pica pica*	x			18
Corvidae	*Garrulus glandularis*		x		39
Corvidae	*Cyanocorax beechi*		x		1
Hirundinidae	*Tachycineta albiventer*	x			8
Pycnonotidae	*Pycnonotus atriceps*	x			18
Platysteridae	*Dyaphorophyia concreta*			x	32
Monarchidae	*Terpsiphone mutata*			x	1, 32
Sturnidae	*Acridotheres cristatellus*	x			8, 12
Sturnidae	*Ampeliceps coronatus*	x			8, 12
Sturnidae	*Aplonis panayensis*	x			8, 12
Sturnidae	*Basilornis celebensis*	x			8, 12
Sturnidae	*Buphagus erythrorhynchus*	x			8
Sturnidae	*Cinnyricinclus femoralis*	x			4
Sturnidae	*Cinnyricinclus leucogaster*	x			1, 4, 8, 12
Sturnidae	*Coccycolius iris*	x			4, 8, 12
Sturnidae	*Cosmopsarus regius*	x			4, 8, 12
Sturnidae	*Cosmopsarus unicolor*	x			4
Sturnidae	*Creatophora cinerea*	x			8, 12
Sturnidae	*Enodes erythrophris*	x			8, 12
Sturnidae	*Gracula religiosa*	x			8, 12
Sturnidae	*Grafisia torquata*	x			4, 8, 12
Sturnidae	*Lamprotornis australis*	x			4
Sturnidae	*Lamprotornis caudatus*	x			4, 8, 12
Sturnidae	*Lamprotornis chalcurus*	x			4
Sturnidae	*Lamprotornis chalybaeus*	x			4
Sturnidae	*Lamprotornis chloropterus*	x			4, 8, 10, 12
Sturnidae	*Lamprotornis corruscus*	x			4
Sturnidae	*Lamprotornis cupreocauda*	x			4
Sturnidae	*Lamprotornis mevesii*	x			4
Sturnidae	*Lamprotornis nitens*	x			4
Sturnidae	*Lamprotornis ornatus*	x			4
Sturnidae	*Lamprotornis purpureiceps*	x			4
Sturnidae	*Lamprotornis purpureus*	x			4
Sturnidae	*Lamprotornis splendidus*	x			4
Sturnidae	*Leucopsar rothschildi*	x		x	8, 12, 32
Sturnidae	*Mino dumontii*	x			8, 12
Sturnidae	*Neocichla gutturalis*	x			4, 8, 12
Sturnidae	*Onychognathus albirostris*	x			4
Sturnidae	*Onychognathus blythii*	x			4

Table 7.1 (continued)

Family	Taxon	Feather barbules	Feather barbs	Dermal collagen	Reference
Sturnidae	*Onychognathus frater*	x			4
Sturnidae	*Onychognathus fulgidus*	x			4
Sturnidae	*Onychognathus morio*	x			4
Sturnidae	*Onychognathus nabouroup*	x			4
Sturnidae	*Onychognathus salvadorii*	x			4, 8, 12
Sturnidae	*Onychognathus tenuirostris*	x			4, 8, 12
Sturnidae	*Onychognathus tristamii*	x			4
Sturnidae	*Poeptera kenricki*	x			4
Sturnidae	*Poeptera lugubris*	x			8, 12
Sturnidae	*Poeptera stuhlmanni*	x			4
Sturnidae	*Sarcops calvus*	x			8, 12
Sturnidae	*Saroglossa aurata*	x			8, 12
Sturnidae	*Scissirostrum dubium*	x			8, 12
Sturnidae	*Speculipastor bicolor*	x			4, 8, 12
Sturnidae	*Spreo albicapillus*	x			4
Sturnidae	*Spreo bicolor*	x			4
Sturnidae	*Spreo fischeri*	x			4
Sturnidae	*Spreo hildebrandti*	x			4
Sturnidae	*Spreo pulcher*	x			4
Sturnidae	*Spreo shelleyi*	x			4
Sturnidae	*Spreo superbus*	x			4, 8, 12, 18
Sturnidae	*Streptocitta albicollis*	x			8
Sturnidae	*Sturnus vulgaris*	x			1, 8, 12
Turdidae	*Myiophoneus caeruleus*		x		2, 29
Turdidae	*Sialia sialis*		x		41
Nectariniidae	*Nectarinia coccinigaster*	x			1, 18, 31
Nectariniidae	*Nectarinia cuprea*	x			8, 11
Nectariniidae	*Nectarinia sperata*	x			8, 11
Estrildidae	*Taeniopygia guttata*		x		35
Fringillidae	*Euphonia affinis*	x			1
Cardinalidae	*Passerina cyanea*		x		1
Thraupidae	*Tangara mexicana*		x		23
Thraupidae	*Tangara cyanocollis*		x		22
Thraupidae	*Cyanerpes cyaneus*		x		1
Icteridae	*Molothrus bonariensis*	x			8

References: 1, This chapter; 2, Andersson (1999); 3, Brink and van der Berg (2004); 4, Craig and Hartley (1985); 5, Dorst et al. (1974); 6, Doucet et al. (2004); 7, Durrer (1962, 1965); 8, Durrer (1977); 9, Durrer and Villiger (1966); 10, Durrer and Villiger (1967); 11, Durrer and Villiger (1968); 12, Durrer and Villiger (1970a); 13, Durrer and Villiger (1970b); 14, Durrer and Villiger (1975); 15, Dyck (1971a); 16, Dyck (1971b); 17, Dyck (1973); 18, Dyck (1976); 19, Dyck (1979); 20, Dyck (1987); 21, Dyck (1999); 22, Finger (1995); 23, Frank (1939); 24, Frank and Ruska (1939); 25, Greenewalt et al. (1960a); 26, Greenewalt et al. (1960b); 27, LaBastille et al. (1972); 28, Li et al. (2003); 29, Prum et al. (2003); 30, Prum et al. (1994); 31, Prum and Torres (2003a); 32, Prum and Torres (2003b); 33, Prum et al. (1999a); 34, Prum et al. (1998); 35, Prum et al. (1999b); 36, Rutschke (1966a); 37, Rutschke (1966b); 38, Schmidt and Ruska (1961); 39, Schmidt and Ruska (1962b); 40, Schmidt and Ruska (1962a); 41, Shawkey et al. (2003).

Because structural colors comprise an important part of the phenotype of many birds, a thorough understanding of the physics of structural color production is essential for an analysis of the function, development, and evolution of avian coloration. An understanding of the physics and anatomy of avian structural colors provides a corroborated biophysical basis for investigations of the function of structural colors in the lives of birds.

Ornithologists have recently become interested in the consequences of avian ultraviolet (UV) vision for avian sensory biology, social signaling, and mate choice (Chapter 1). Although some pigments may produce additional UV reflectance peaks, purely UV-colored pigments are unknown in the vertebrate integument. To focus on the importance of UV stimuli in avian communication, many researchers have turned to research on structural colors that are predominantly expressed in the UV range of avian vision. This new interest in the biology of UV structural colors underscores the importance of having an accurate anatomical and physical understanding of these phenomena.

In this chapter, I review the anatomy of avian structurally colored tissues and the evidence concerning the physical mechanisms by which these tissues produce colors. First, I briefly discuss the distinction between pigmentary and structural colors. Then I review and categorize the various physical models of light-scattering physics and their applications for structural color production in birds. Next I present the available data concerning the anatomy and physics of the structural colors of avian feathers, skin, and eyes. I conclude with a discussion of topics in the development and evolution of structural coloration. A discussion of the mechanisms of structural coloration necessitates the use of terms that may not be familiar to some readers; Box 7.1 provides a glossary of terms.

Structural versus Pigmentary Colors

Pigmentary and structural colors both contribute to the phenotype of birds. Many of the most striking colors are created by a combination of pigmentary and structural colors. For example, many noniridescent green plumages are hypothesized to be produced by a combination of a structural blue component and either a yellow carotenoid pigment or a yellow psittacofulvin (in the case of parrots; Chapter 8) in the feather barbs (Lucas and Stettenheim 1972; Dyck 1976, 1978; Durrer 1986). In addition, saturated yellow colors in bird skin (e.g., Cabot's Tragopan [*Tragopan caboti*] and the Toco Toucan [*Ramphastos toco;* Plate 11]) can be produced by a combination of carotenoid pigments and an underlying structural yellow color (Prum and Torres 2003b). But what about

Box 7.1. Glossary of Terms

Coherent scattering—A form of light scattering by multiple objects in which the phases of waves scattered from different objects are nonrandom. Physical models of coherent scattering account for the phase relationships among waves scattered by different objects. Differential reinforcement of in-phase waves and interference of out-of-phase waves produces coherent scattering of a subset of light wavelengths. Coherent scattering from different physical structures can been called "constructive interference," "reinforcement," "diffraction," "Bragg scattering," and "thin-film reflection." Coherent scattering is the main mechanism of structural color production in birds.

Fourier transform—A mathematical technique that decomposes a function, signal, or image into an equivalent sum of sine waves of different wavelengths and amplitudes.

Fourier power spectrum—A distribution of the relative amplitudes of the component sine waves of the Fourier transform of a function. The Fourier power spectrum reveals the relative contribution of different frequencies of periodicity to the total energy of the original function, signal, or image.

Incoherent scattering—A form of light scattering by individual objects in which the phases of the waves scattered from different objects are random. Physical models of incoherent scattering ignore the phase relationships among waves from multiple objects. There is no evidence that any avian tissues get their color from this mechanism.

In phase—The relationship of the phases of two waves of the same wavelength in which the maxima and minima (i.e., peaks and troughs) of the waves are coincident. In-phase waves scattered by different objects reinforce one another and sum to form a wave of the same wavelength with a greater amplitude than that of the original waves.

Iridescence—The optical phenomena of change in color with change in angle of observation or angle of illumination. Iridescence is commonly produced by coherent scattering from laminar and crystal-like nanostructures, but it is not produced by quasi-ordered nanostructures under general light conditions.

Light scattering—The propagation of a light wave in all directions from a point other than the forward direction of transmission. Light scattering increases as a light wave passes from one medium to another with a different refractive index.

Nanostructure—A nanometer-scale ordered array. Bio-optical nanostructures are usually composed of alternating structures with different refractive indices.

Optical distance—The path length of light traveling a given distance through a given material. Calculated as the width of the material times its refractive index.

Out of phase—The relationship of the phases of two waves of the same wavelength in which the maxima and minima (i.e., peaks and troughs) of the waves are not coincident. Out-of-phase waves scattered by different objects will cancel one another and sum to zero scattered reflections.

Path-length addition—The difference in optical distance traveled by light waves scattered by two different objects. Random path-length additions create incoherent scattering. Nonrandom path-length additions create coherent scattering if the size of the wavelengths are close to the size of path-length additions.

Pigment—A molecule that differentially absorbs and emits wavelengths of incident light.

Refractive index (n)—A physical characteristic of a material calculated as the ratio of the speed of light in a vacuum to the speed (or phase velocity) of light in the medium. The difference in indices of refraction of two materials determines the amount of light scattering when light waves pass from one material to another. Most color-producing nanostructures are composed of a combination of high- and low-refractive–index substances.

Spatial frequency—Cycles per unit length (usually cycles/nm); a unit used to describe the periodicity of variation in refractive index in a color-producing nanostructure. Spatial frequency is the inverse of size of the periodic structures.

Structural color—A color produced by the physical, optical interaction of light waves with the structure of an organism. Most structural colors are produced by light scattering by nanometer-scale variations in the refractive index of a substance.

the brilliantly iridescent hummingbird feather barbs that are composed of a matrix of feather β-keratin, melanin granules, and air? Are hummingbird colors pigmentary or structural?

The complex relationship between structural and pigmentary colors often creates confusion about the distinction between the two. To reiterate the definitions above, pigmentary colors are those created solely as a result of molecular absorbance and emission of light. The hue of a pigmentary color is determined by the molecular structure of the pigment and its density in the tissue (Chapters 2 and 5). In contrast, structural colors result from the physical interactions of light waves that are scattered at the interfaces of biological materials of different refractive indices. Although light scattering may produce colors by various physical mechanisms (discussed below), all mechanisms depend on the nanoscale physical properties of the structures and their refractive indices. The relative index of refraction is defined as the ratio of the speed of light in a medium to the speed of light in a vacuum.

Because back scattering increases dramatically when light waves travel through material of one refractive index into a material with a different refractive index, structural-color–producing biological nanostructures are usually composed of a combination of different substances with higher and lower refractive indices. For example, in birds, structural-color–producing arrays are composed of combinations of keratin and air, melanin and keratin, melanin and air, collagen and mucopolysaccharide, purines and cytoplasm, or pterins and cytoplasm (see below). Confusion arises because some the components of these structurally colored arrays are themselves molecular pigments; for example, melanin (Chapter 6), pterins, or purines (Chapter 8). Nevertheless, the colors produced by these nanometer-scale arrays of pigment granules are structural, because the colors produced are a consequence of the size, spatial distribution, and refractive indices of the pigment granules, not merely their molecular properties. Colors produced by nanometer-scale physical structures are structural colors even if those nanostructures are partially composed of pigments. The same molecular properties that make these molecules pigments can also mean that they have higher refractive indices than most biomolecules. This property has created an interesting interface between the evolution of certain classes of pigmentary and structural coloration (see below). In conclusion, some structural colors are not strictly "nonpigmentary" colors; they are produced by nanometer-scale physical structures that can be made of pigments.

Thus the iridescent color of a hummingbird feather is a structural color, even though it is created by a matrix of melanin, keratin, and air in the feather barbule (see below). In complex systems in which a combination of structural and pigmentary colors interact to create the observed hue (e.g., most non-iridescent green birds), it is best to analyze and discuss the structural and pigmentary components of the color separately.

Physics of Light Scattering and Mechanisms of Structural Color Production

The physical mechanisms for the production of biological structural colors, including those of birds, are often described as being diverse (Fox 1976; Herring 1994; Parker 1999; Srinivasarao 1999). An exhaustive list of previously proposed mechanisms would include Rayleigh scattering, Tyndall scattering, Mie scattering, constructive interference, thin-film reflection, Bragg reflection, and diffraction. However, much of this apparent diversity refers to the diversity of the specific mathematical tools that are used to analyze the optical properties of different anatomical situations, rather than to actual differences in the physical mechanisms of color production.

The most fundamental physical classification of structural color mechanisms is to describe them as variations of light scattering. Whenever a light wave travels from a medium of one refractive index into a medium with a different refractive index, some of the total energy of the light wave is absorbed as heat, some of the scattered light waves travel forward and propagate the wave farther, and the remainder is scattered in other directions. For the materials relevant to structural coloration in birds, the refractive indices vary between approximately 1 for air, to 1.56–1.58 for β-keratin (Brink and van der Berg 2004), and 2.0 for eumelanin (Durrer and Villiger 1966, 1970a,b; Brink and van der Berg 2004). The amount of light scattered at the interface between two materials is proportional to the difference in the refractive index of the two materials (see the Fresnel equation below; Land 1972). The scattering behavior of light waves in an optically heterogeneous tissue is determined by the size, shape, and spatial distribution of the scattering objects and their refractive indices.

Mechanisms of biological color production can be classified according to a fundamental physical dichotomy as either incoherent or coherent scattering (van de Hulst 1981; Bohren and Huffman 1983). Incoherent and coherent scattering differ in whether the behavior of scattered light is independent of or dependent on the phases of scattered light waves, and in whether the phases of scattered wave can be ignored or must be taken into account to describe the behavior of scattered light waves.

Incoherent Scattering

Incoherent scattering occurs when individual light-scattering objects are responsible for differentially scattering visible wavelengths (Figure 7.1). Incoherent scattering models assume that light-scattering objects in a biological material are spatially independent. In other words, the scatterers are assumed to be randomly distributed on a spatial scale of the same order of magnitude as the wavelengths of visible light. If the scatterers are randomly distributed in space, then the phase relationships among the scattered waves will also be random and can be ignored in any description of light-scattering behavior of the medium. Consequently, incoherent scattering models ignore the phase relationships among the scattered waves and describe color production as the result solely of differential scattering of wavelengths by the individual scatterers themselves (van de Hulst 1981; Bohren and Huffman 1983). Incoherent scattering mechanisms include Rayleigh scattering, Tyndall scattering, and Mie scattering.

Historically, Tyndall scattering refers to light scattering by small (microscopic) particles that are larger than molecules, whereas Rayleigh scattering refers to scattering by all small particles down to the size of individual molecules (Young 1982). Thus Rayleigh scattering includes the phenomenon described as Tyndall scattering. Some researchers have argued that, because the biological objects that scatter light are much larger than individual molecules, the term "Tyndall scattering" should be used in biological contexts (Huxley 1975). However, because most of the explicit, testable predictions concerning small-particle scattering—including polarization and the inverse relationship between scattering magnitude and the fourth power of the wavelength—are derived from the work of Lord Rayleigh and because Rayleigh scattering includes Tyndall scattering (Young 1982), it is more informative to refer to biological color production by incoherent scattering from small particles as "Rayleigh scattering" (Dyck 1971a,b; Prum et al. 1998, 1999a, 2003; Prum and Torres 2003b; Brink and van der Berg 2004). Mie scattering is a mathematically exact description of incoherent scattering by single particles of small or large size that simplifies to the Rayleigh scattering conditions for small particles (van de Hulst 1981; Bohren and Huffman 1983). Thus Mie scattering is a more precise model that includes the phenomena described previously as Rayleigh and Tyndall scattering.

Examples of blue incoherent scattering include the blue sky, skim milk, blue smoke, and blue ice and snow. Although many biological examples of incoherent scattering have been proposed (Fox 1976), none have been rigorously confirmed by testing the congruence with (1) Rayleigh's prediction that incoherent scattering by small particles should be proportional to the inverse of the fourth power of the wavelength, and (2) the spatial independence of scattering objects over the spatial scale of visible light. Incoherent scattering

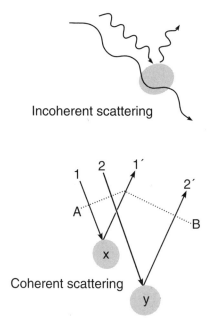

Figure 7.1. Comparison of incoherent and coherent scattering mechanisms of biological structural color production. Incoherent scattering is differential scattering of wavelengths by individual light-scattering objects. In Rayleigh (or Tyndall) scattering, smaller wavelengths are preferentially scattered. The phase relationships among light waves scattered from different objects are ignored and assumed to be random. Coherent scattering is differential interference or reinforcement of wavelengths scattered by multiple light-scattering objects. Coherent scattering of specific wavelengths is determined by the phase relationships among the scattered waves. Scattered wavelengths that are out of phase cancel out, but scattered wavelengths that are in phase are constructively reinforced and coherently scattered. Phase relationships of wavelengths scattered by two different objects (x, y) are given by the differences in the path lengths of light scattered by the first object (x; 1–1′) and a second object (y; 2–2′), as measured from planes (A, B) perpendicular to the incident and reflected waves in the average refractive index of the media.

by a random aggregation of large scatterers can produce white, as in the cloudy sky, fog, or a white piece of paper.

Coherent Scattering

Coherent scattering occurs when the spatial distribution of light scatterers is *not* random with respect to the wavelengths of visible light, so that the phases of scattered waves are also nonrandom (Bohren and Huffman 1983). Consequently, coherent-scattering models describe color production in terms of the phase interactions among light waves scattered by multiple scatterers (Figure 7.1). Light waves scattered by independent objects that are out of phase will destructively interfere and cancel out, whereas scattered waves that are in phase will constructively reinforce one another and will be coherently scattered by the material.

The phase relationships among the wavelengths scattered by two objects are determined by the differences between the path lengths of the waves scattered by the two objects and the average refractive index of the medium (Huxley 1968; Benedek 1971; Land 1972). Light waves scattered by different objects will travel different distances (or path lengths) before they are observed, which creates the potential for light waves to shift phase relative to one another. The difference in the distance traveled by waves scattered by adjacent scatterers is called the "path-length addition." In general, the path-length addition depends on the distance between the scatterers, the angles of observation and illumination, and the average refractive index. For a given path-length addition, most wavelengths will be substantially out of phase with one another and will cancel out. In contrast, scattered wavelengths that are near the path-length addition in size (or that travel an integer multiple of their wavelength in the same distance) will be in-phase after scattering. This limited subset of wavelengths constructively reinforce one another, producing coherently scattered waves of a larger amplitude.

Coherent scattering encompasses optical phenomena that have been referred to variously as "constructive interference," "reinforcement," "diffraction," and "Bragg scattering." Nanostructures that coherently scatter have been referred to as "thin-film reflectors," "multilayer reflectors," "quarter-wave stacks," "Bragg scatterers," and "diffraction gratings." Coherent scattering produces the colors of an oil slick or a soap bubble. Well-known biological examples of coherent scattering include brilliant iridescent butterfly wing scales and such bird feathers as the peacock's tail (Dyck 1976; Fox 1976; Durrer 1986; Ghiradella 1991; Parker 1999).

In contrast to incoherent scattering, coherent scattering models account for the phase relationships of waves scattered by the array. A coherently scattering nanostructure is characterized by variations in refractive index that are periodically distributed in space, so that the light waves scattered from nearby scatterers are nonrandom in phase. Predictable interactions of the light waves scattered by adjacent objects produce coherent scattering or selective reinforcement and interference of different portions of the visible spectrum. Thus the phase relationships among scattered waves are determined by the size and spatial distribution of the scatterers and the average refractive index of the medium. Periodic spatial relationships among scatterers will produce predictable path-length differences among scattered waves and reinforcement of a limited set of wavelengths. Genuinely random spatial relationships among light scatterers does not result in reinforcement of a specific subset of wavelengths and creates opportunities for incoherent scattering.

Classes of Coherently Scattering Nanostructures

Coherent scattering requires spatially periodic variations in the refractive index. There are three fundamental classes of coherently scattering biological nanostructures: laminar, crystal-like, and quasi-ordered (Prum and Torres 2003a). In addition, spatial periodicity in the refractive index can exist in one, two, or three dimensions.

Laminar nanostructures are composed of a series of alternating layers of two materials of different refractive indices (Figure 7.2a). Essentially, by definition, the spatial variation in refractive index in a laminar system is one dimensional. In the simplest "ideal" arrays, the thickness of each layer multiplied by its refractive index is approximately one quarter of the wavelength of the peak reflectance. Such structures are frequently referred to as "quarter-wave stacks" (Huxley 1968; Land 1972; Macleod 1986).

A single layer with a refractive index distinct from the underlying substrate (e.g., oil slick, soap bubble) can function as the simplest form of laminar optical nanostructure. These single-film arrays are often quite thick (>400 nm), so that they can coherently scatter a fundamental frequency in the infrared and also produce a broad series of smaller harmonic peaks at visible frequencies, producing an oily iridescent appearance (Macleod 1986; Brink and van der Berg 2004). Such single, thick layers can be referred to as "thick films."

Crystal-like nanostructures are composed of a square or hexagonal array of light-scattering objects of a refractive index that is distinct from the surrounding medium (Figure 7.2b). A crystal-like array is considered two dimensional if it is composed of parallel rods, bars, or nanofibers that are equivalently spaced in all perpendicular planes (i.e., periodic refractive index variation in only two dimensions). Alternatively, a crystal-like nanostructure is three dimensional if it is composed of objects in a true three-dimensional lattice of spherical objects. Color production from many superficial diffraction gratings

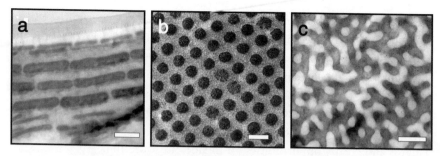

Figure 7.2. Examples of the three major classes of coherently scattering nanostructures. (a) Laminar array of plate-shaped melanosomes in the barbules of the iridescent green back feathers of the Superb Sunbird (*Nectarinia coccinigastra*; transmission electron micrograph courtesy of Jan Dyck). (b) Crystal-like, hexagonal array of parallel collagen fibers from the green facial caruncle of the Velvet Asity (Prum et al. 1994, 1999a). (c) Quasi-ordered arrays of β-keratin and air vacuoles from the medullary layer of blue feather barbs of the Peach-faced Lovebird (Dyck 1971b; Prum et al. 1999b). All scale bars are 200 nm.

can also be understood as coherent scattering from a laminar or crystal-like nanostructure restricted to the surface of the organism (Kinoshita et al. 2002), but these anatomical systems are not found in birds.

Quasi-ordered nanostructures are characterized by arrays of light-scattering objects that are unimodal in size and interscatterer spacing, but that lack the spatial ordering at larger spatial scales (Figure 7.2c; see also Figures 7.10–7.12). Quasi-ordered arrays can be either a two-dimensional array of rods or fibers or a three-dimensional array of objects. The possibility of coherent scattering by quasi-ordered nanostructures has only recently become appreciated, and research on structural coloration of birds has been central to this development (reviewed in Prum and Torres 2003a). Dyck (1971a,b) was the first to propose explicitly that the spongy medullary keratin of structurally colored feather barbs would produce color by coherent scattering. However, Dyck was unable to demonstrate that the light-scattering air bubbles in the spongy medullary keratin are sufficiently ordered in spatial distribution to selectively reinforce specific visible wavelengths. Benedek (1971) presented a model of coherent scattering by the two-dimensional quasi-ordered arrays of collagen fibers in the transparent vertebrate cornea. Benedek documented that the very short path-length additions produced by the small corneal collagen fibers and spacing create coherent scattering of far UV light and destructive interference of all visible wavelengths, resulting in optical transparency. Quasi-ordered biological nanostructures exploit the same physical mechanism for the production of structural colors or optical transparency; they differ only in size. In recent years, my colleagues and I have documented coherent scattering in the quasi-ordered spongy medullary keratin of bird feather barbs (Prum et al. 1998, 1999b, 2003), the quasi-ordered collagen fiber arrays in avian skin (Prum et al. 1999a; Prum and Torres 2003b), mammalian skin (Prum and Torres 2004), and the dragonfly integument (Prum et al. 2004).

Methods of Analysis of Coherent Scattering

Coherent scattering by laminar systems can be analyzed using traditional thin-film optics (Huxley 1968; Land 1972; Macleod 1986). The calculations of hue, or the wavelength at peak reflectance, can be very simple when the array is ideal. An ideal array has alternating layers of different materials that maintain the same optical distance (i.e., the product of the refractive index and the film thickness; Land 1972). For nonideal arrays, in which the optical distance of all layers is not uniform, there are well-established matrix methods for calculating the reflectance by a system with great precision, including reflectance spectrum and polarization (Huxley 1968; Macleod 1986).

Color production by crystal-like nanostructures, sometimes referred to as "Bragg scatterers," has traditionally been analyzed using Bragg's Law:

$$\lambda_{max} = n\, 2d \sin \alpha,$$

where λ_{max} is the peak wavelength of reflection, n is the average refractive index of the array, d is the space lattice dimension (the distance between parallel planes of reflection), and α is the angle of incidence (Bragg and Bragg 1915; Hecht 1987).

There has not been a traditional method of analysis for quasi-ordered arrays, because they were usually hypothesized to be incoherently scattering and were not recognized as coherently scattering nanostructures (Prum and Torres 2003a). Following the transparency theory of Benedek (1971), Rodolfo Torres and I developed a new method for the analysis of coherent scattering by quasi-ordered arrays—a two-dimensional Fourier analysis tool for the study of coherent light scattering by biological nanostructures (Prum et al. 1998, 1999a,b, 2003; Prum and Torres 2003b, 2004; see Box 7.2). Discrete Fourier analysis is a well-known analytical tool that transforms a sample of data points into a mathematically equivalent sum of component sine waves of different

Box 7.2. Fourier Analysis of Color-Producing Nanostructures

Traditional optics provides many different mathematical methods for the analysis of structural color production by optical nanostructures. However, most traditional methods are applicable only to physical systems that can be assumed to be periodic—that is, laminar or crystal-like nanostructures. These mathematical methods do not apply to materials that are quasi-ordered, in which the scatterers are all roughly the same size and distance apart, but vary substantially from a perfectly laminar or crystal-like distribution. It was not until new mathematical methods using Fourier analysis were developed to analyze coherent scattering by such structures that quasi-ordered nanostructures were shown definitively to produce structural colors by interference among scattered waves.

Fourier analysis is a classical mathematical method that decomposes a function, signal, or image into a mathematically equivalent sum of sine waves. With computers, calculating the discrete form of the Fourier transform for a digital data set has become very efficient, leading to many applications in signal processing, filtering, and data compression. Most ornithologists are familiar with the common "sonogram" of a bird vocalization, which is a Fourier transform of the function of sound waves over a series of time periods. The result is an analysis of the frequency composition of the original sound waves and its change over time.

The two-dimensional Fourier transform of an image provides a mathematically precise way to quantify periodicity in all directions in the image. If we imagine a digital, black-and-white transmission electron micrograph of a color-producing nanostructure as a mathematical function mapping each pixel to a numerical value that varies from pure black to pure white, the two-dimensional Fourier transform compartmentalizes the variation of that function into component sine waves of different wavelengths and spatial frequencies (cycles/nm) in all different directions. The easiest way to visualize the Fourier transform is with the Fourier power spectrum, which depicts the relative amplitude, or contribution, of each of the component sine waves. The two-dimensional Fourier power spectrum of a color-producing nanostructure shows the relative contributions of each period (or spatial frequency) to the total spatial variation in the material. Because light is a wave

Anatomy, Physics, and Evolution of Structural Colors 313

> (just like the component sine waves of the Fourier transform), the Fourier power spectrum is actually a physically realistic description of the periodicity of scattering opportunities of light waves incident on a nanostructure.
>
> The Fourier Tool for Biological Nano-optics (available free at http://www.yale.edu/eeb/prum/fourier.htm) calculates the Fourier transform of an electron micrograph of a biological structure, presents its Fourier power spectrum, and predicts the shape of the reflectance spectrum from the nanostructure, based on coherent scattering.

frequencies and amplitudes (Briggs and Henson 1995). The amplitudes of each of the component sine waves in the Fourier transform express the relative contributions of each frequency of variation to the periodicity of the original data. The plot of the variation in the modulus (or complex norm) of the coefficients, or amplitudes, of the Fourier components is called the "Fourier power spectrum." The relative values of the different Fourier components in the power spectrum express the comparative contribution of those frequencies of variation to the total energy of the original function.

With the nano-optic Fourier tool (see Box 7.2), coherent scattering by digital or digitized two-dimensional transmission electron micrographs (TEMs) can be analyzed using the matrix algebra program MATLAB (http://www.mathworks.com). The numerical computation of the Fourier transform is done with the well-established two-dimensional Fast Fourier Transform (FFT2) algorithm (Briggs and Henson 1995). The Fourier tool is described in detail in Prum and Torres (2003a), and its application to the study of avian structural colors in Prum et al. (1998, 1999a,b, 2003) and Prum and Torres (2003b, 2004).

The Fourier transformed data are visualized with the two-dimensional Fourier power spectrum (Plate 11), which resolves the spatial variation of the refractive index in the tissue into its periodic components in any direction from a given point (Figure 7.3a,b). Each value in the two-dimensional power spectrum reports the magnitude of the periodicity in the original data of a specific spatial frequency in a given direction from all points in the original image (Figure 7.3b). The spatial frequency and direction of any component in the power spectrum are given by the length and direction, respectively, of a vector from the origin to that point. The magnitude is depicted by a gray

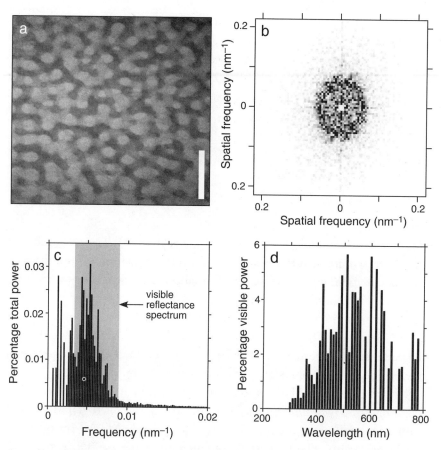

Figure 7.3. Example of two-dimensional Fourier analysis of the spongy medullary keratin from a turquoise brow feather from the Turquoise-browed Motmot (*Motmotus motmota*; Motmotidae; KU 40579). (a) Transmission electron micrograph of the spongy β-keratin nanostructure at 50,000 magnification. Scale bar is 500 nm. (b) Two-dimensional Fourier power spectrum. (c) Radial average of the Fourier power spectrum based on 100 radial frequency bins, showing that the predominant Fourier power is concentrated at spatial frequencies that are likely to coherently scatter visible reflections (shaded region). (d) Predicted reflectance using the two-dimensional Fourier transform based on 50 radial wavelength bins and showing a unimodal peak at ~500 nm.

scale or color scale, but the units are dimensionless values related to the total lightness and darkness of the original images. Radial averages of the power spectra can be used to summarize the distribution of Fourier power among the different spatial frequencies of variation in refractive index over all directions within an image of a tissue (Figure 7.3c). The nano-optical Fourier tool can also predict reflectance spectra for coherently back-scattered light based on the two-dimensional Fourier power spectra, image scales, and average refractive indices (Figure 7.3d). The result is a theoretical prediction of the relative magnitude of coherent light scattering by the tissue, expressed in percentage of Fourier power, that is based entirely on the spatial variation in refractive index of the tissue.

Coherent Scattering and Iridescence

Unlike incoherent scattering, coherent scattering can produce the phenomenon of iridescence—a prominent change in color with angle of observation or illumination. Iridescence occurs when changes in the angle of observation or illumination affects the average path length of scattered waves. Alterations in the distribution of path-length additions to scattered waves change the phase relationships among scattered wavelengths and alter which wavelengths are constructively reinforced. Strong iridescence conditions are met when the light-scattering objects are arranged in laminar or crystal-like nanostructure.

In the biological literature, since at least Mason (1923a,b), iridescence has frequently been synonymized with mechanisms of coherent scattering (e.g., Fox 1976; Nassau 1983; Herring 1994; Lee 1997). Accordingly, all iridescent structural colors were hypothesized (correctly) to be produced by coherent scattering, but all noniridescent structural colors were erroneously hypothesized to be exclusively produced by incoherent scattering (Fox 1976; Herring 1994). The absence of iridescence was considered prima facie evidence of incoherent scattering.

Recently, however, it has been demonstrated that coherent light scattering by quasi-ordered arrays of light scatterers can produce biological structural colors that are non- or only weakly iridescent (Prum et al. 1998, 1999a,b, 2003; Prum and Torres 2003b). Quasi-ordered arrays lack the organization at larger spatial scales that produce strong iridescence. Quasi-ordered arrays have an organization that is similar to a bowl of grapes or popcorn; each grape or kernel is similar in size to its neighbor and center-to-center distances are nearly equal, but beyond the spatial scale of a single kernel, there is no organization. An example of a color-producing, quasi-ordered nanostructure is the light scattering air bubble/keratin interfaces in the layer of structurally colored avian feather barbs (Prum et al. 1998, 1999b, 2003; Figure 7.2c; see also Figures 7.11, 7.12); these air-keratin matrices are sufficiently spatially ordered at the appropriate nanoscale to produce the observed hues by coherent scattering, but are not ordered at larger spatial scales, so these colors are not iridescent or only weakly iridescent. Because they are equivalently nanostructured in all directions, quasi-ordered nanostructures have the capacity to back-scatter light when viewed from any direction.

Thus iridescence can be created by both variation in the differential capacity of a nanostructure to back-scatter ambient light in different directions and by variation in the directionality of light. Laminar and crystal-like nanostructures produce iridescence under many conditions, because they back-scatter different colors when viewed from different directions. Even quasi-ordered tissues produce iridescence under highly directional, artificial light. Recently, Osorio and Ham (2002) documented that the quasi-ordered spongy medullary keratin of avian feather barbs displays prominent iridescence when artificially illuminated and observed from independent angles. However, under natural lighting conditions, these quasi-ordered nanostructures generally lack iridescence, because they can coherently back-scatter the same color from any direction of observation, and these back-scattered spectra predominate under the generally omni-directional illumination that exists in nature.

Role of Basal Pigment Layers in the Production of Structural Colors

All color-producing nanostructures must be connected to and integrated with the rest of the organism. Except for the extraordinary instances of entirely transparent aquatic organisms (Johnsen 2001), the entire organism cannot be nanostructured to produce a single optical property. Typically, the tissue below a color-producing nanostructure in the integument, which cannot itself be engineered for color production, produces incoherent scattering of all visible wavelengths (i.e., white light scattering up toward the observer). Incoherent, back-scattered white light greatly reduces the saturation (or purity) of any wavelength-specific, structurally colored tissue. By analogy, it is easy to imagine that a flashlight shining bright white light toward the viewer from inside a television or computer screen would wash out any color produced by the screen itself.

If the color-producing nanostructure is very thick and has many arrays, then it may be essentially infinitely optically thick, and no light can pass through to scatter from the tissue below (e.g., Prum and Torres 2003b). However, in most cases, the number of scattering layers or arrays and their combined optical efficiency is limited. Thus color-producing nanostructures need to be isolated, or "optically insulated," from the underlying tissues of the organism to produce saturated colors.

To accomplish this function, most structurally colored tissues are anatomically underlain by a layer of pigment granules that can absorb incident light that has propagated through the entire color-producing nanostructure. These structures prevent incoherent scattering from the tissue below. Most avian structural color systems are underlain by a layer of melanin granules, but in some cases, the absorbing pigment is a carotenoid (e.g., feather barbs of some but not all species of *Cotinga;* Dyck 1971a).

Structural Colors of Feathers

There are three main anatomical classes of color-producing structures in feathers: (1) unspecialized, unpigmented feather keratin, which produces white by incoherent scattering of all visible light waves; (2) structurally colored feather barbules, which typically produce iridescent structural colors by scattering from arrays of melanin granules, or melanosomes (which sometimes include internal air vacuoles), suspended in the β-keratin of the feather barbules; and (3) the spongy medullary layer of the barb rami, or the main shaft of the feather barbs. Structural colors of feather barbs include the generally noniridescent blue, violet, green, and UV hues, which are broadly distributed among bird species and must have arisen numerous times independently in the phylogenetic history of birds.

White of Unpigmented Feathers

The white appearance of unpigmented feathers is a structural color that is produced by incoherent scattering of all visible wavelengths from unpigmented feather keratin. In typical unpigmented bird feathers, light scatters at the interfaces of the feather β-keratin and the large air-filled cavity that lies at the center of many feather keratinocytes. The magnitude of light scattering is increased by the surfaces of the feather structures and air vacuoles within unpigmented feather keratin. Because the light-scattering objects are of many different sizes (and not just smaller than visible light wavelengths), the result is incoherent scattering of all wavelengths, or white light. All parts of the feather can be white, although the barbules are composed of a single string of interconnected cells, and may be so thin that they appear transparent.

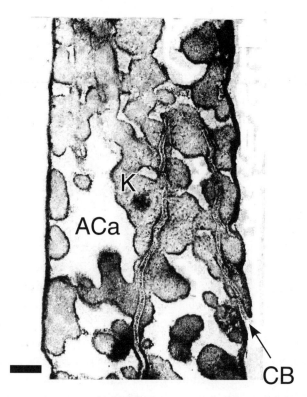

Figure 7.4. Cross-section of a white feather barbule from the winter plumage of a Rock Ptarmigan (from Dyck 1979). ACa = air-filled channels that extend down the length of the feather barbule; CB = cell boundary between neighboring barbule cells; K = keratin. Scale bar is 200 nm.

In a detailed analysis of the brilliantly white, basic (winter) plumage of Rock Ptarmigan (*Lagopus mutus;* Plate 19), Dyck (1976) demonstrated that white ptarmigan feathers are a brighter white than are unpigmented feathers of other galliforms. Single feathers reflect almost 50% of visible light of all wavelengths, compared to 15–18% reflectance of visible wavelengths for white feathers of Domestic Chickens (*Gallus gallus;* Dyck 1976). Using TEM, Dyck demonstrated that this bright white color is produced by large, randomly organized air vacuoles in the barbules of the ptarmigan contour feathers (Figure 7.4), and that these vacuoles are absent from white feathers of other species. Thus white ptarmigan feathers are structured to create a brighter white color. Dyck (1976) hypothesized that many feathers overlapping in the plumage of the bird would eventually approach the highly efficient 80% reflectance of fallen snow. Furthermore, the reflectance spectra of white ptarmigan feathers are slightly bluish (scattering smaller wavelengths more efficiently), even further approximating the genuinely bluish reflectance spectrum of snow. This white color and its slightly bluish hue may be produced by incoherent (Rayleigh) scattering by the irregularly shaped, small, randomly distributed air vacuoles in the ptarmigan feather barbules.

In several avian clades, structurally white feathers have evolved from plesiomorphic (i.e., primitive), homologous, structurally colored blue feathers created by the spongy medullary layer of feather barbs (see below). For example, in the seven biological species of the manakin genus *Lepidothrix* (Pipridae), blue and violet crowns and rump patches are broadly distributed, and white crown and rump patches have convergently evolved three times (Prum 1988). The blue crown feathers of the Blue-crowned Manakin (*Lepidothrix coronata*) have typical, quasi-ordered spongy medullary keratin, whereas the white crown feathers of the Snowy-capped (*L. nattereri*) and White-fronted (*L. serena*) Manakins have entirely unpigmented feathers with a virtually identical spongy medullary keratin in the barb rami. These brilliantly white feather colors appear to have evolved from structural blue as a result of the derived loss of the underlying melanin-granule deposition.

Structural Colors of Feather Barbules

Structural colors of avian feather barbules are produced by coherent scattering from arrays of granules of melanin, called "melanosomes," that are distributed in the keratin of the feather barbules (Dyck 1976; Fox 1976; Durrer 1986). These color-producing structures are found in many avian orders and families and have evolved numerous times independently in groups of birds that vary widely in size, behavior, and ecology (Greenewalt et al. 1960a; Schmidt and Ruska 1961, 1962a; Durrer 1962, 1965, 1986; Durrer and Villiger 1966, 1968, 1970a,b; Rutschke 1966a; Dorst et al. 1974; Dyck 1976, 1987; Fox 1976; Craig and Hartley 1985).

Given the diversity of avian taxa in which barbule structural colors have evolved, it is not surprising that there is a tremendous diversity in the composition, size, shape, and spatial distribution of the melanosomes in these color-producing arrays. The melanosomes can be spherical, rod-, or disk-shaped, and entirely solid or with a hollow air-filled cavity in the center. In most structurally colored barbules, the melanosomes are composed of eumelanin,

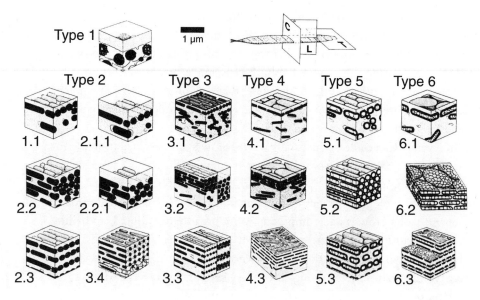

Figure 7.5. Types of iridescent arrays of β-keratin, melanin, and air from avian feather barbules (redrawn from Durrer 1977, 1986). See Durrer (1986) for complete list of taxa in each type. Type 1: a thin layer of keratin above a layer of large spherical melanocytes and air vacuoles (e.g., Rock Pigeon [*Columba livia*], *Zenaida* sp.; Schmidt and Ruska 1961). Type 2: single or multiple layers of adjacent rod-like melanin granules (1–2 μm long; 0.2 μm in diameter). Type 3: single or multiple layers of thinner rods of melanin (1 μm long; 0.1 μm in diameter) that are arranged in nearly square or hexagonal arrays, or in tight packed single or double layers. Type 4: melanin platelets or lozenges (1.5–2.5 μm long; 0.25–0.4 μm wide; 0.25–0.4 μm thick) arranged in single or multiple layers. Type 5: air-filled melanin tubes (1.2–1.6 μm long; 0.12–0.27 μm in diameter) arranged in single or multiple layers or in a hexagonal array. Type 6: air-filled melanin platelets (2.5 × 1.0 μm in area; 0.3 μm thick) arranged in single or multiple layers, which may be immediately adjacent or separated by an additional keratin layer. Inset at top right shows locations of cross (C), longitudinal (L) and transverse (T) sections of a barbule.

but in the Trocaz Pigeon (*Columba trocaz*) they are composed of phaeomelanin (Schmidt and Ruska 1961). Durrer (1986) classified the diversity of light-scattering arrays in avian barbules into six basic types (Figure 7.5). Each of these six types of light-scattering structures has convergently evolved in multiple families and orders of birds (e.g., they are found in both passerine and nonpasserine families).

The color-producing melanosome/keratin arrays of avian barbules can be organized into one-dimensional laminar nanostructures or two-dimensional crystal-like nanostructures, which can have either a square or hexagonal unit cell (Figure 7.6). For example, *Pavo* and *Afropavo* peacocks (Phasianidae) have a square array of solid, rod-shaped melanosomes (Durrer 1962, 1986; Durrer and Villiger 1975; Li et al. 2003), whereas *Trogon* spp. (Trogonidae) have a hexagonal closely packed array of hollow melanosomes (Durrer and Villiger 1966; Dyck 1976; Durrer 1986; Figure 7.6f). In some species, the hollow melanosomes can be quite irregularly packed, bordering on a quasi-order; for example, the dark blue back feathers of the Scrub Euphonia (*Euphonia affinis;* Fringillidae, Carduelinae) reveal an array of hollow melanosomes that lack a laminar or strictly packed, crystal-like spatial organization (Figure 7.6e). The capsule or rod-shaped melanosomes may be organized into laminar, crystal-like, or nearly quasi-ordered (i.e., *Euphonia*) nanostructures, but, not surprisingly, the pancake-shaped melanosomes, which are found exclusively in laminar nanostructures (e.g., *Trochilidae;* Greenewalt et al. 1960a). The air spaces within the melanosomes may be single cavities (e.g., *Trogon, Euphonia*) or composed of multiple smaller compartments separated by bars of melanin (e.g., *Pharomachrus,* hummingbirds; Figure 7.6).

Structurally colored barbules typically show some specialization in morphology for presentation of the color-producing nanostructures to the surfaces of the feather. Usually the distal barbules, which extend from the barb ramus toward the distal tip of the feather, lie on top of proximal barbules (Lucas and Stettenheim 1972). Consequently, on a pennaceous feather with a closed vane, the color-producing nanostructures are usually restricted to the distal barbules only. In the upper-tail–covert display plumes of the male *Pavo* peacock (Phasianidae), however, the structural color arrays are found on both proximal and distal barbules (R. O. Prum, unpubl. data). In some feathers, the color-producing arrays are restricted to the obverse (upper or outer) surface of the barbule, whereas in others, both the obverse and reverse surfaces of the feather are structurally colored. Finally, many structurally colored barbules are elongate, flattened, and rotated to create a planar surface for the efficient presentation of the color-producing arrays (Durrer and Villiger 1970b; Lucas and Stettenheim 1972; Durrer 1986).

Mechanisms of Color Production

The structural colors of feather barbules are produced by coherent scattering of light from the interfaces between the keratin, melanin, and air. Many of these systems have been successfully modeled using thin-film optics or Bragg's Law (Greenewalt et al. 1960a; Durrer 1962, 1986; Durrer and Villiger 1966, 1970a,b; Dyck 1976). Nanometer-scale variations in the size and spatial

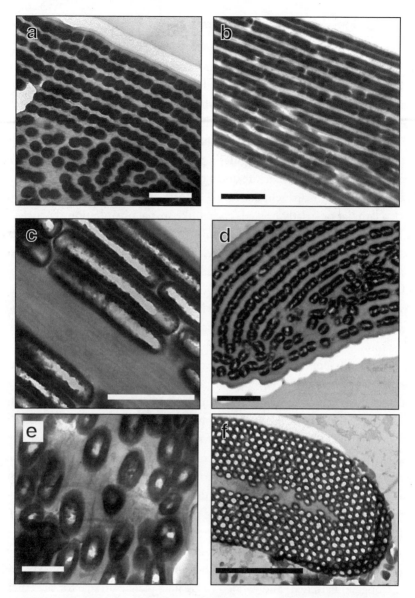

Figure 7.6. Transmission electron micrographs of iridescent arrays of β-keratin, melanin, and air in the distal barbules of feather barbs. (a) Male Raggiana Bird of Paradise (*Paradisea raggiana*, Paradiseaidae; KU 108321) green throat feather. Magnification, 25,000. (b) Male Magnificent Riflebird (*Ptiloris magnifica*, Paradiseaidae; KU 43615) green throat feather. Magnification, 25,000. (c) Plum-colored Starling (*Cinnyricinclus leucogaster*, Sturnidae; KU 29536) purple back feather. Magnification, 50,000. (d) Resplendent Quetzal (*Pharomachus moccino*, Trogonidae; KU 17159) green back feather. Magnification, 25,000. (e) Scrub Euphonia (Fringillidae; KU 101212) blue back feather. Magnification, 50,000. (f) Slaty-tailed Trogon (*Trogon massena*, Trogonidae; KU 89364) green back feather. Magnification, 20,000. All scale bars are 1 μm.

distribution of these structures have been accurately correlated with the observed hue or the reflectance spectrum of the barbules. Functionally, both laminar and crystal-like arrays create opportunities for iridescence and strongly directional color production.

This arrangement creates iridescence, or potential changes in color with the angle of observation, which is such a striking feature of most structural colors of barbules. The change in hue occurs because the path-length addition (or difference in distance traveled) between light waves scattered by adjacent scatterers or layers decreases with decreasing angle of observation. Usually the hue of iridescent feathers should shift toward shorter visible wavelengths with decreasing angles of observation.

Scattering theory predicts a strong positive relationship between the number of scattering nanostructures in a cross-section of an array and the magnitude of reflectance, or brilliance, of the color produced. Dyck (1987) empirically documented this relationship very well for *Ptilinopus* pigeons and their relatives (Columbidae). Furthermore, the efficiency of scattering increases with the magnitude of the difference in refractive index between the scattering media. For directly perpendicular light, the reflectance, R (the portion of ambient light scattered at a single interface between two materials of different refractive indices), is calculated by the Fresnel equation to be:

$$R = ([n_D - n_A]/[n_D + n_A])^2,$$

where n_D and n_A are the refractive indices of the two materials (Huxley 1968; Land 1972; Macleod 1986). For feather keratin ($n = 1.55$–1.58) and melanin ($n = 2$; Land 1972; Brink and van der Berg 2004), R is 1.61%. But for melanin and air ($n = 1$), R is 11%. By including hollow air cavities in the melanosomes, these nanostructures can increase their optical efficiency (or percentage reflectance per layer) by an entire order of magnitude. Thus hummingbirds can produce their brilliant structural colors with only a minimal number of layers (Greenewalt et al. 1960a,b). Structural colors produced by air-filled melanosomes (Type 6 of Durrer 1986; Figure 7.5) should be much more brilliant than solid melanin arrays having similar dimensions and numbers of layers.

Coherent scattering predicts the existence of higher-order reflectance peaks. These peaks are due to reinforcement among wavelengths that travel some higher integer number of wavelengths in the predominant path-length addition. Thus, the "harmonic" wavelengths are approximately one-half, one-third, one-fourth, and so forth, of the wavelength of the fundamental (or first-order) wavelength coherently scattered by the array. If the first-order reflectance peak is at the long-wavelength extreme of the visual spectrum, then it is possible for the secondary peak to be at the short-wavelength end of the visual spectrum, producing complex, combined hues (e.g., purple). For example, the Plum-colored Starling (*Cinnyricinclus leucogaster*) has brilliant, rich purple plumage that is produced by a laminar array of air filled melanosomes (Figure 7.7a). Its reflectance spectrum reveals a primary peak reflectance at 800 nm and a second-order harmonic peak at approximately 380 nm (Figure 7.7b). A two-dimensional Fourier analysis of these arrays documents the laminar nanostructure and predicts a similar reflectance spectrum with a primary red peak and a secondary shorter-wavelength harmonic peak (Figure 7.7c,d).

Thin-film scattering theory predicts that higher peaks will be closer in wavelength to λ_{max} in nonideal systems, in which the optical distance of the layers of media are not equal, than in ideal systems (Land 1972). Multiple peaks in the visible spectrum can be produced by laminar nanostructures with a single, very thick superficial layer, so that the first-order reflectance peaks are in the infrared spectrum, placing multiple higher-order peaks in the visible spectrum (e.g., Cuthill et al. 1999; Osorio and Ham 2002; McGraw 2004; see below).

Single versus Multilayer Arrays

A fundamental and underappreciated mechanistic distinction in structural color production by feather barbules is that between nanostructures with single or multiple layers. The single-layer structures are characterized by a superficial layer of keratin overlying a single layer or a solid body of melanosomes (e.g., the iridescent plumage of European Starlings [*Sturnus vulgaris*]; Plate 23; Figure 7.8a). Although they produce color by coherent scattering, single-layer systems display some optical properties that are distinct from multilayer arrays. Surprisingly, the physical details of this situation have not been worked out until recently. In a detailed thin-film optical analysis of structural color production in the Hadeda Ibis (*Bostrychia hagedash*), Brink and van der Berg (2004) established that the iridescent color was produced by coherent reinforcement between light waves scattering by the surface of the barbule and at the upper surface of a layer of hollow, plate-shaped melanosomes. Interestingly, this single superficial layer of β-keratin is *much* thicker than the traditional "quarter-wave stack," in which the optical distance of the layer (thickness [nm] × refractive index) is equal to one-fourth of the wavelength of the peak hue (λ_{max}; Land 1972). In the feathers of the Hadeda Ibis, Brink and van der

Anatomy, Physics, and Evolution of Structural Colors

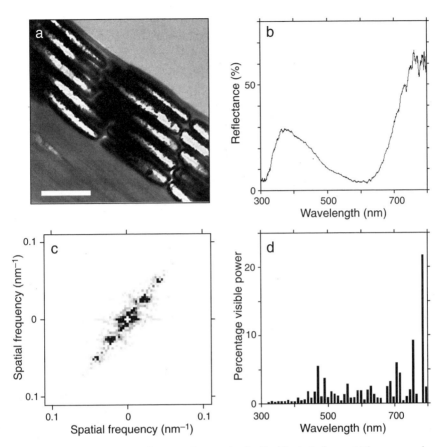

Figure 7.7. (a) Transmission electron micrograph of a distal barbule from a iridescent purple back feather from the Plum-colored Starling (KU 29536), showing the laminar array of air-filled melanosomes. Scale bar is 1 μm. (b) Reflectance spectra of these purple feathers reveals a primary peak reflectance at ~800 nm and a first-order harmonic reflectance peak at ~380 nm. To human vision, the result is a strong purple hue. (c) The two-dimensional Fourier power spectrum of the melanin-air array in panel A, showing peaks above and below the origin in the same direction as the primary periodicity of the array. (d) The predicted reflectance spectrum based on the Fourier transform, showing a primary reflectance peak at 800 nm and a short-wavelength harmonic peak at 480 nm.

Berg (2004) demonstrated that the first-order reflectance was in the non-visible infrared, but that the visible reflectance spectrum includes five or six reflectance peaks that are the higher-order harmonics of the fundamental wavelengths (integer multiples of the fundamental frequency). The result is a complex reflectance spectrum with a series of visible peaks that are more

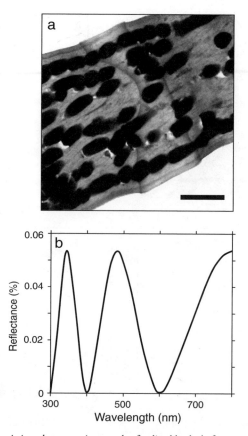

Figure 7.8. (a) Transmission electron micrograph of a distal barbule from an iridescent greenish breast feather of the European Starling (KU 46688). The melanin granules create a coherently scattering, single, superficial layer of β-keratin. Scale bar is 1 μm. (b) A thin-film optical prediction of the reflectance spectrum of a "thick-film" system made of single β-keratin and melanin layers each 300-nm thick, using the matrix method of Huxley (1968).

---

closely packed toward the short-wavelength end of the spectrum. The visible effect from many barbules of many feathers is a dark blackish plumage with a glossy or oily spectrum of highlights. This appearance is very different from the highly saturated peak reflectance from a hummingbird, for example.

Reflectance spectra similar to the Hadeda Ibis have recently been published for iridescent plumage in Rock Pigeons (*Columbia livia;* Osorio and Ham 2002; McGraw 2004) and European Starlings (Cuthill et al. 1999). Schmidt and Ruska (1961) documented that a single keratin layer above a melanin layer produced the structural colors of the Trocaz Pigeon. Durrer (1986) assigned the structural coloration of numerous genera—including *Gallus* (Phasianidae), *Carphibis* (Threskiornithidae), *Chloroceryle* (Alcenidae), *Prosthemadura* (Meliphagidae), *Tachycineta* (Hirundinidae), *Sturnus* and many other starlings (Sturnidae), and *Molothorus* (Icteridae)—to a single-keratin-layer nanostructure. For example, a thin-film optical analysis of the single pair of β-keratin and melanin layers of 300-nm thickness, as found in oily greenish-black breast feathers of European Starlings, shows a series of reflectance peaks at approximately 800, 480, and approximately 350 nm. These predicted reflectance peaks are quite similar to those previously reported for starling feathers by Cuthill et al. (1999).

In future discussions of the barbule structural colors, it will be important to distinguish between single and multilayer arrays. Single layers are much simpler in nanostructure, much thicker in dimensions, and produce much less saturated structural color. However, because they produce visible reflectance spectra with multiple reflectance peaks, they may actually be brighter (i.e., have greater total reflectance) than some simple multilayer systems. Single-layer arrays that produce oily red, green, blue, and purple iridescent colors are likely to evolve relatively easily in birds with black plumage, because this mechanism of structural color production requires only the developmental specification of a single parameter—the superficial β-keratin layer thickness—above a layer of melanosomes. Black feathers are already melanized, so that heritable variations in superficial β-keratin layer thickness could easily have optical consequences that would lead to the evolution of structural coloration. Comparative analyses are necessary to investigate whether oily iridescent black plumages produced by single "thick-films" have evolved more frequently than the multilayer thin-film arrays that produce saturated colors and that require the much more rigid specification of spatial distribution to function. Comparative analyses would also be able to identify whether these glossy black plumages evolve more frequently in clades of birds in which black plumage is common.

Noniridescent Structural Barbule Colors

Laminar and crystal-like melanosome arrays found in feather barbules create inherent opportunities for iridescence, because changes in the angle of observation produce changes in the average path-length additions to light waves scattered from different surfaces. However, several lineages of birds have evolved vivid, structural, noniridescent barbule colors. Dyck (1987) elegantly demonstrated that the noniridescent structural greens of the barbules of *Ptilinopus* fruit pigeons and related genera (Columbidae) are produced by arranging the layers of contiguous melanin granules in concentric circular arrays in specialized rounded ridges on the barbule surface (Figure 7.9a). Because the array is nearly circular, light from any angle is essentially directly incident on the array and produces equivalent, back-scattered reflections from any angle of observation. Thus the hue of the color does not change with angle of observation under omnidirectional light. The uniformity of the spatial periodicity over multiple directions is clearly indicated by the two-dimensional Fourier power spectrum of a curved array, which shows nearly a complete ring of highest-power values at a single intermediate spatial frequency (Figure 7.9b).

A nearly identical arrangement has evolved convergently in African Emerald Cuckoos (*Chrysococcyx cupreus;* Durrer and Villiger 1970b). In the feathers of both fruit pigeons and emerald cuckoos, it appears that the noniridescent, curved laminar arrays evolved from plesiomorphic iridescent, laminar arrays in these genera through natural or sexual selection for reduced iridescence in structural color production (Durrer and Villiger 1970b; Dyck 1987).

Structural Colors of Feather Barbs

Structural colors of avian feather barbs are created by coherent light scattering from a layer of typically box-shaped, medullary cells that are filled with a spongy matrix of the feather β-keratin and air. These color-producing, spongy medullary keratin nanostructures are three-dimensional quasi-ordered arrays. The spongy keratin nanostructures come in two fundamental nanostructural classes, each of which has evolved convergently many times in birds. The first class of spongy barb nanostructure is characterized by an array of almost perfectly spherical air bubbles that are connected by tiny openings in a matrix of β-keratin (Figure 7.10). This organization roughly approximates Swiss cheese, in which the volume of air is larger than the volume of the cheese. Each of the air bubbles is similar in size and shape, but the keratin structures between the air bubbles vary tremendously in thickness at different points between neighboring air bubbles. Avian genera known to demonstrate this type of structure include *Megalaima* (Ramphastidae), *Cotinga* (Cotingidae), *Lepidothrix* (=*Pipra*, Pipridae), *Cyanerpes* (Thraupidae), and *Poephila* (Estrildidae).

The second nanostructural class is characterized by air-filled channels and keratin bars in which both the air and keratin channels are approximately equivalent in shape and width, and form a complex network of interconnecting

Anatomy, Physics, and Evolution of Structural Colors

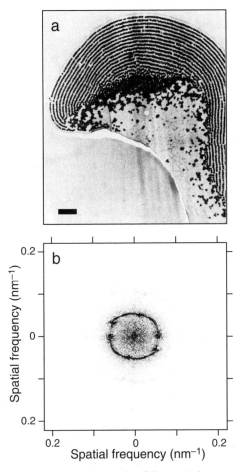

Figure 7.9. (a) Transmission electron micrographs of the noniridescent structural color-producing arrays of β-keratin and melanin in the distal barbules of the White-breasted Fruit Pigeon *Ptilinopus rivoli* (Columbidae; from Dyck 1987). Scale bar is 1 μm. (b) Two-dimensional Fourier transform of the arrays in (a), showing the ringlike Fourier power distribution, which documents the equivalent nanostructure in almost all directions. *Ptilinopus* fruit pigeons are not iridescent, because observers from any angle see back-scattered light produced by an array with uniform nanostructure. The curvature of the laminar arrays eliminates the opportunity for iridescence produced by the plesiomorphic laminar arrays (Dyck 1987).

channels (Figure 7.11). This type of nanostructure is known from many parrots (Psittacidae), *Motmotus* (Motmotidae), *Nyctyornis* (Meropidae), *Halcyon* (Alcenidae), *Coracias* (Coraciidae), *Irena* (Irenidae), *Cyanocorax* (Corvidae), *Sialis* (Turdidae), and others.

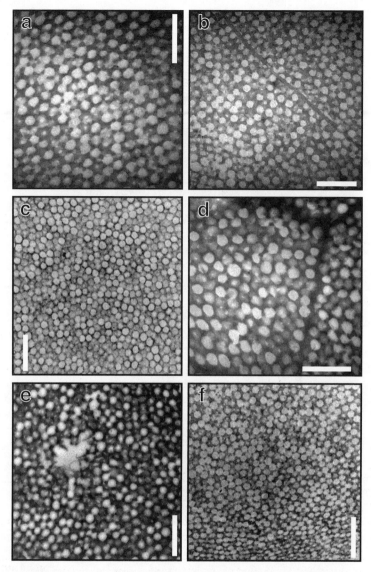

Figure 7.10. Transmission electron micrographs of the spongy β-keratin and air matrices of the structurally colored feather barbs from six species with arrays of spherical air bubbles. (a) Gold-whiskered Barbet (*Megalaima chrysopogon,* Ramphastidae; KU 43182) green breast feather. Magnification, 30,000. (b) Male Purple-breasted Cotinga (*Cotinga cotinga,* Cotingidae; KU 89001) blue back feather. Magnification, 25,000. (c) Male Blue-crowned Manakin (KU 87685) blue crown feather. Magnification, 20,000. (d) Male Snowy-capped Manakin (MPEG 7050) white crown feather. Magnification, 30,000. (e) Indigo Bunting (*Passerina cyanea,* Cardinalidae; KU 33693) blue back feather. Magnification, 25,000. (f) Red-legged Honeycreeper (*Cyanerpes cyaneus,* Thraupidae; KU 88256) blue breast feather. Magnification, 25,000. All scale bars are 1 μm.

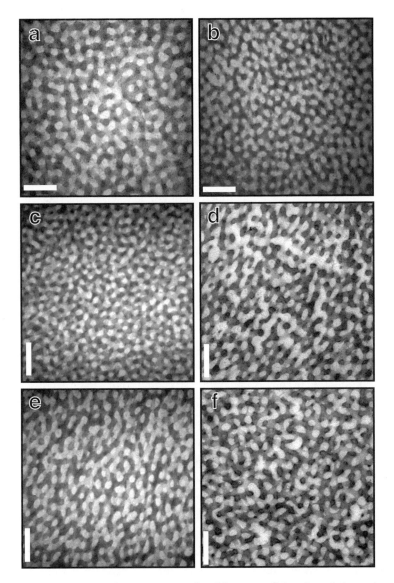

Figure 7.11. Transmission electron micrographs of the spongy β-keratin and air matrices of the structurally colored feather barbs from six species with arrays of complex networks of air and keratin channels. (a) Red-bearded Bee-eater (*Nyctyornis amicta,* Meropidae, KU 48813) green back feather. (b) Black-capped Kingfisher (*Halcyon pileata,* Alcenidae; KU 48795) blue back feather. (c) Green Broadbill (*Calyptomena viridis,* Eurylaimidae; KU 43200) green back feather. (d) Purplish-backed Jay (*Cyanocorax beechii,* Corvidae; KU 101842) blue back feather. (e) Asian Fairy-bluebird (*Irena puella,* Irenidae; KU 101420) blue back feather. (f) Eastern Bluebird (KU 107355) blue lower back feather. All magnifications are 40,000. All scale bars are 500 nm.

Despite more than a century of research (reviewed in Dyck 1971b; Fox 1976), the physical mechanism of color production by feather barbs has only recently been understood. Traditionally, incoherent Rayleigh-Tyndall scattering was the predominant explanation of structural colors of feather barbs, because they are noniridescent, mostly blue colors. The support for this hypothesis first came from reflectance spectra of blue feather barbs by Häcker and Meyer (1902), which appeared to show adherence to Rayleigh's predicted inverse fourth-power law for a limited set of visible wavelengths (500–650 nm). However, Häcker and Meyer (1902) did not measure a broad enough section of the visible spectrum to detect the peak hue in the lower portion of the human-visible spectrum below 500 nm, and erroneously interpreted their data as support for a critical Rayleigh scattering prediction.

Subsequently, Dyck (1971a, 1976) demonstrated that many structurally colored feather barbs have distinct peaks in the visible spectrum. These distinct peaks falsified a major prediction of Rayleigh scattering theory—that reflectance should continue to increase in value into the UV, following the inverse fourth-power law. Dyck (1971a) proposed an alternative hypothesis for the production of spongy barb ramus structural colors: reinforcement among light waves scattered by the keratin and air interfaces of the "hollow cylinders" in the keratin matrix (i.e., coherent scattering). The hollow-cylinder model hypothesized that the sizes of the keratin bars and air vacuoles in the spongy medullary keratin creates spatial periodicities in the light-scattering interfaces and produces constructive interference and vivid structural color production. Although Dyck's hypothesis was apparently accepted by Auber (1971/1972) and Durrer (1986), it received limited general acceptance, perhaps because of the lack of evidence demonstrating that the spongy medullary keratin was sufficiently ordered to produce coherent phase relationships among scattered waves. Subsequently, Finger (1995) hypothesized that incoherent Mie scattering theory, combined with cortical absorption of shorter light wavelengths, could accurately explain structural color production by feather barbs. In a study of the UV structural color of feather barbs of the Blue Whistling-thrush (*Myiophonus caeruleus:* Turdidae), Andersson (1999) found a positive correlation between the number of scattering air bubbles (i.e., array thickness) and percentage reflectance, but he could not distinguish between the coherent and incoherent mechanisms of structural color production.

Most recently, Prum et al. (1998, 1999b, 2003) used two-dimensional Fourier analysis to examine the periodicity of the spongy medullary keratin matrix and test Dyck's (1971b) coherent hollow-cylinder hypothesis. These results demonstrated that the spongy medullary keratin is highly nanostructured at the appropriate spatial scale to produce the observed hues by coherent scattering, or constructive interference, alone. Furthermore, these Fourier analyses demonstrate that the spongy medullary keratin is substantially nanostructured, directly falsifying a fundamental assumption of all incoherent scattering mechanisms (including Rayleigh, Tyndall, and Mie scattering) that the light-scattering objects must be spatially independent or randomly distributed. After more than a century of near unanimous support for various incoherent scattering mechanisms (Häcker and Meyer 1902; Fox 1976; Gill 1995), two-dimensional Fourier analyses have falsified these hypotheses and provided strong support that feather barb structural colors are produced by coherent scattering.

Frank (1939) and Auber (1957) described and classified the variations in shape, dimensions, position, and pigmentary composition of the spongy medullary tissue of many families of birds with structurally colored barbs. The medullary cells may form a ring around the surface of the barb ramus, a complete obverse (or upper) layer, or a even a divided pair of obverse and reverse (lower) cell bundles to produce structural colors on both surfaces of the wing. Auber (1971/1972) also described the development of the medullary cells using light-microscope histology. He documented that, during the development of the feather barbs, the medullary keratinocytes first expand to their final size to create the boxlike, polygonal, keratinized cell boundaries. Subsequently the medullary cells gradually create the spongy medullary keratin matrix on the inner surface of the original polygonal cell boundaries. Eventually, when the volume of the cell is largely filled with the spongy keratin matrix, the keratinocytes die.

Structural Colors of Avian Skin

Given the intricate variation and striking beauty of the structural colors of bird feathers, it is perhaps understandable that ornithologists have almost completely neglected to investigate the anatomy and physics of the structural colors of the avian skin. However, structural colors of the skin comprise an important component of the phenotype of many bird species (Table 7.1), and they likely play a fundamental role in social communication of many them (Prum and Torres 2003b).

Noniridescent structural colors occur in the skin, bill (ramphotheca), legs, and feet (podotheca) of a broad diversity of birds from many avian orders and families. Auber (1957) reported structurally colored skin in 19 avian families from 11 avian orders. Auber assumed that all blue or green skin colors are structural rather than pigmentary, because blue and green pigments are unknown or very rare in the avian integument (Fox 1976). Prum and Torres (2003b) identified structurally colored skin, ramphotheca, and podotheca in 129 avian genera in 50 families from 16 avian orders. Structurally colored skin is apparently present in more than 250 bird species, or approximately 2.5% of avian biological species.

Because they are noniridescent, structural colors of avian skin were traditionally hypothesized to be produced by incoherent Rayleigh or Tyndall scattering (Camichel and Mandoul 1901; Mandoul 1903; Tièche 1906; Auber 1957; Rawles 1960; Lucas and Stettenheim 1972; Fox 1976). The light-scattering structures were variously hypothesized to be melanin granules, biological colloids, or a turbid media of proteins and lipids in the dermis. The Rayleigh scattering hypothesis was never actually tested with either spectrophotometry (to examine whether these structural colors conform to the prediction of Rayleigh's inverse fourth-power law) or with electron microscopy (to examine whether the hypothesized light-scattering objects were spatially independent). Green skin colors were further hypothesized to be a combination of Rayleigh scattering blue and carotenoid yellow (e.g., Fox 1976), but this hypothesis was never confirmed.

Prum et al. (1994) were the first to examine structurally colored avian skin with electron microscopy. We documented that the green and blue colors of the supraorbital caruncles of the male Velvet Asity (*Philepitta castanea;* Plate 10*)* of Madagascar are produced by coherent scattering from hexagonally organized arrays of parallel collagen fibers in the dermis. Subsequently, Prum et al. (1999a) performed a comparative analysis of the anatomy, nanostructure, and structural coloration of three species of the Malagasy asities. We applied the two-dimensional Fourier method to asity caruncle collagen arrays to confirm the coherent scattering mechanisms and describe the variations in hue produced by variations in the dimensions of the dermal collagen nanostructure. We also documented that the crystal-like hexagonal nanostructure of the Velvet Asity is derived from the plesiomorphic, quasi-ordered nanostructure found in the sunbird asities *Neodrepanis.*

More recently, Prum and Torres (2003b) investigated the anatomy and physics of structurally colored skin, ramphotheca, and podotheca from 31 species of birds from 17 families in 10 different orders. The sample included an ecologically and phylogenetically (from paleognathes to passerines) diverse collection of birds with a wide variety of colors. Prum and Torres (2003b) confirmed that the structural colors of avian skin, ramphotheca, and podotheca are produced by coherent scattering of light from arrays of parallel collagen fibers in the dermis. Variation in hue is created by variations in collagen fiber size and spacing. These integumentary color-producing collagen arrays have convergently evolved in more than 50 independent lineages of birds (Prum and Torres 2003b). The collagen arrays in all but one species observed were quasi-ordered. The hexagonally ordered collagen arrays of Velvet Asity are unique among animals (previously described by Prum et al. 1994, 1999a; Figure 7.12). The quasi-ordered nanostructure of these collagen arrays is equivalent in all directions in the tissue perpendicular to the collagen fibers, which explains why these colors are not iridescent. No differences were found in the anatomy or nanostructure among the structurally colored ramphotheca (e.g., Ruddy Duck [*Oxyura jamaicensis*] and Channel-billed Toucan [*Ramphastos vitellinus*]), podotheca (e.g., Blue-footed Booby [*Sula nebouxii;* Plate 16, Volume 2]), and skin.

The reflectance spectra of structurally colored bird skin falsify the inverse fourth-power prediction of the incoherent, Rayleigh scattering hypothesis. Furthermore, the ring-shaped maxima of the Fourier power spectra demonstrate that spatial variations in refractive index in the collagen arrays are not spatially independent, as required by the incoherent scattering mechanisms. After more than a century of unquestioned support, the incoherent Rayleigh, or Tyndall, scattering hypothesis was falsified for a wide diversity of structurally colored bird skin (Prum and Torres 2003b).

In at least two avian lineages, *Oxyura* (Anatidae) and the Malagasy asities *Philepitta* and *Neodrepanis* (Eurylaimidae), structurally colored integumentary ornaments are seasonally variable. In *Oxyura,* males develop blue ramphotheca coloration during the breeding season (April–July), but have black bills during the rest of the year (Wetmore 1917; Hays and Habermann 1969). In asities, the structurally colored facial caruncles develop in the breeding season and atrophy completely (including the dermal melanosomes) during the rest of the year (Prum and Razafindratsita 1997; Prum et al. 1999a). All other structural colors in avian skin are apparently permanent once developed. One bird not examined by Prum and Torres (2003b) shows a change in the color of the facial caruncle during ontogeny. In the Blue-faced Honeyeater (*Entomyzon cyanotis;* Meliphagidae), the adults have deep blue facial skin, but immature individuals have greenish facial skin. Apparently large portions of the head and neck of Wild Turkeys (*Meleagris gallipavo;* Plate 32) can change rapidly from

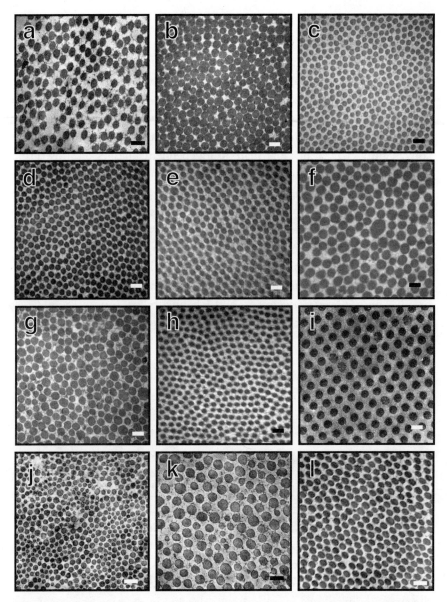

Figure 7.12. Transmission electron micrographs of nanostructured arrays of dermal collagen from (a) Ruddy Duck, light blue; (b) Helmeted Guineafowl (*Numida meleagris*), dark blue; (c) Satyr Tragopan (*Tragopan satyra*), dark blue; (d) Cabot's Tragopan, dark blue; (e) Cabot's Tragopan, light blue; (f) Cabot's Tragopan, orange; (g) Whistling Heron (*Syrigma sibilatrix*), blue; (h) Toco Toucan, dark blue; (i) Velvet Asity, light blue; (j) Bicolored Antbird (*Gymnopithys leucapsis*), light blue; (k) Bare-throated Bellbird, green; (l) Madagascar Paradise Flycatcher (*Terpsiphone mutata*), dark blue. All magnifications are 30,000. All scale bars are 200 nm.

white to blue (Schorger 1966). The rapid color change in turkeys may occur by mobilization of melanosomes in melanocytes in response to hormonal cues, as in amphibians and other reptiles (Bagnara and Hadley 1973; Bagnara 1998), but this hypothesis has not been examined. It is possible that other birds may have variable dermal structural coloration mediated by similar mechanisms.

Evolution of Structurally Colored Skin in Birds

How have the structural colors in bird skin evolved? Why has convergent evolution of these arrays among different lineages of bird been so frequent? Collagen is a ubiquitous and abundant extracellular matrix molecule in connective tissues of all metazoan animals. Collagen forms self-assembled, triple-helical fibers that are composed of collagen polypeptides and are surrounded by a mucopolysaccharide matrix. Production of structural coloration by arrays of collagen fibers requires the appropriate specification of two components of collagen nanostructure that are already intrinsic to collagen itself—fiber diameter and interfiber spacing. Thus the collagenous extracellular matrix of the skin provides an inherent nanostructure that is very near to the appropriate spatial frequency to produce visible hues. Selection on integumentary collagen for a color-production function requires more rigid specification of pre-existing features of this extracellular matrix. Genetic variation in the nanostructure of integumentary collagen may occasionally create visible variations in reflectance that could become subject to subsequent natural, sexual, or social selection for structural color production. The frequent convergent exaptation of integumentary collagen for a novel color-production function in birds has apparently been fostered by the nature of collagen itself. Interestingly, the broad visual sensitivity of birds to near UV light permits them to observe optical consequences of a broader class of latent variations in integumentary collagen nanostructure, and may create a broader set of opportunities for the evolution of nanostructured, color-producing collagen from plesiomorphic collagen in the skin.

Supporting this notion is the observation that color-producing nanostructures in the dermal collagen have convergently evolved in mammals only twice, and only in the two groups of mammals that are known to have trichromatic color vision—primates and marsupials (Prum and Torres 2004). Elsewhere in animals, color-producing, quasi-ordered collagen arrays have only been described in the tapetum lucidum of the sheep eye (Bellairs et al. 1975) and in the cornea of the eyes of certain fishes (Lythgoe 1974).

At larger, anatomical scales, the evolution of structural color production by integumentary collagen also requires the development of a sufficient number of light-scattering arrays to produce an observable color and some physical mechanism to limit incoherent scattering from underlying tissue. Percentage reflectance, R, is given by the Fresnel equation (see above; Huxley 1968; Land 1972; Hecht 1987). For other structural-color–producing, composite biological media, R varies widely: from 4.5% (keratin-air) to 11.1% (melanin-air). Theoretically, these high-R biological arrays can produce nearly total reflection of all ambient light with 10–20 layers of light scattering objects (Land 1972). However, the difference in refractive index between collagen ($n = 1.42$) and the mucopolysaccharide ($n = 1.35$) is much smaller, yielding $R \sim 0.05\%$. Thus collagen arrays need two orders of magnitude more scattering objects to produce the same magnitude of reflectance as made by other color-producing avian nanostructures. The evolution of structurally colored bird skin requires the proliferation of nanostructured dermal collagen arrays to produce a thicker dermis than found in normal skin. Structurally colored bird skin typically includes 500–2,000 light-scattering collagen fibers in any dermal cross-section (Prum and Torres 2003b). Exceptionally, the color-producing dermis of the Bare-throated Bellbird (*Procnias nudicollis;* Cotingidae; Plate 9) includes more than 5,000 fibers in a single cross-section (Prum and Torres 2003b).

As described above, the evolution of integumentary structural coloration also requires the development of a physical mechanism to prevent incoherent scattering of white light from the deeper tissues that underlie the superficial color-producing nanostructures. In almost all species examined, color-producing collagen arrays are underlain by a thick layer of melanin granules (Prum and Torres 2003b). The anatomical association between integumentary structural coloration and melanin deposition is so strong that the melanosomes were frequently hypothesized to produce the color themselves by incoherent scattering (Mandoul 1903; Rawles 1960; Hays and Habermann 1969; Fox 1976). Several authors have remarked that the disappearance of the structural color upon removal of the melanin layer supports the conclusion that melanin produces the color (e.g., Hays and Habermann 1969). Actually, the functional role of underlying melanin is to absorb any light transmitted completely through the array and to prevent incoherent scattering of white light from the deeper tissues that are not nanostructured for color production. A functional alternative to an underlying melanin barrier is to have enough light-scattering objects in the array that virtually all the incident light is coherently scattered to generate the appropriate hue. The approximately 5,000 light-scattering collagen fibers in a dermal cross-section in bellbirds, which lack dermal melanin, apparently exceeds the level of total reflection of incident light that would render unnecessary any underlying melanin.

Pre-existing melanin deposition in the skin may enhance the likelihood of subsequent evolution of structural coloration within a lineage by making any optical effects of chance variation in superficial collagen nanostructure immediately more visible (Prum and Torres 2003b). Many bird genera with structurally colored skin have close relatives with melanin-pigmented facial skin; for example, cormorants (Phalacrocoracidae), ducks (Anatidae), avocets (Recurvirostridae), and honeyeaters (Meliphagidae). Likewise, plesiomorphic bare skin may also foster the evolution of structurally colored skin, because variations in integumentary nanostructure would be immediately observable and potentially subject to selection. The featherless eye ring is a good example of broadly distributed bare skin, and not surprisingly, the most common position for structural colors is around the eyes (Plate 9). There is at least one instance of evolution of integumentary structural color by artificial selection in silkie chickens, an ancient domestic breed of the Domestic Chicken. Silkies have many associated novel phenotypic features that evolved by artificial selection during domestication, including feather abnormalities, polydactyly, highly melanized skin, and earlobes that are deep blue to turquoise. It is likely that the original integumentary mutation in silkies that produced extreme skin melanization, unusual in *Gallus,* created the opportunity for subsequent selection for structurally colored blue earlobes.

Fishes, amphibians, and reptiles produce primitive integumentary structural colors with iridophores—specialized pigment cells in the dermis that contain guanine or pterin crystals (Bagnara and Hadley 1973; Bagnara 1998), but iridophores are absent from structurally colored bird and mammal skin. These observations confirm the conclusion that birds and mammals have evolutionarily lost integumentary iridophores, because they evolved integumentary appendages that entirely cover the skin (i.e., feathers, hair; Oliphant et al. 1992; Bagnara 1998), although birds retain iridophores as an important mechanism of structural color production in the iris (see below; Oehme 1969; Ferris and Bagnara 1972; Oliphant 1981, 1987a,b; Oliphant et al. 1992; Oliphant and Hudon 1993).

Structural Colors of the Avian Iris

Little research has been done on the nature of avian iris colors. Oehme (1969) produced an extensive treatment of the pigments of avian iris (see also Chapter 8). Although he did not separate the pigments chromatographically, Oehme identified many bird species with carotenoid, pterin, and purine pigments in their irides (Chapters 5 and 8). The carotenoids were present in intracellular lipid droplets. The pterin or purine pigments were found in crystalline form in some species, but they were not chemically identified.

The detection of crystalline pigments in the avian iris by Oehme (1969) implied the presence of iridophores, or special pigment cells that produce structural colors by constructive interference (i.e., coherent scattering) from arrays of purine or pterin pigments. Purines are only present in pigment cells in crystalline form, but color-producing pterins can be found in both crystalline and noncrystalline form.

Ferris and Bagnara (1972) confirmed that the irides of the Inca Dove (*Columbina inca*) and the Common Ground-dove (*Columbina passerina*) include two structurally distinct types of iridiphores. One type of iridiphore included melanin granules and abundant, spherical guanine crystals, and the second type included many thinner, rod-shaped guanine crystals. The two types of purine-containing pigment cells were distributed in different parts of the iris.

Subsequently, Oliphant (1981, 1987a,b) and Oliphant and Hudon (1993) identified numerous species of birds from many families with crystalline purines and pterins in the iris. Oliphant (1981) identified pterin-containing leucophores (cells producing structural white) in the iris of the Great Horned Owl (*Bubo virginianus;* Plate 14). In a sample of 28 species from 11 families, Oliphant (1987a,b) and Oliphant and Hudon (1993) identified crystalline purines in the irides of 20 species in 11 families and crystalline pterins in three species. In all species with purine crystals, noncrystalline pterins were also detected. Oliphant hypothesized that these pterin pigments were superficial to the purine crystal-containing iridiphores, and were modifying their structural color. Similar structural-pigmentary color combinations are found widely in poikilothermic vertebrates.

The colors of the avian iris appear to be the most mechanistically complex in the entire phenotype of birds, and obviously there is much more to be learned about the pigmentary and structural colors of avian iris. However, based on the small sample of current data, it is clear that the coloration of the iris is based on entirely distinct combinations of pigmentary and cellular mechanisms from the avian integument. Quantitative studies are required to determine how a complex of pigments and structural colors produce the diversity of iris colors of birds. Currently, nothing is known about the ontogeny or evolution of these structures in the avian iris.

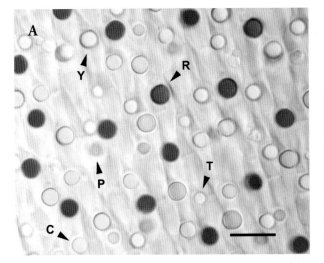

Plate 1. (A) Cone oil droplets in the periphery of the retina of (B) the Wedge-tailed Shearwater (*Puffinus pacificus*). C, P, R, T, and Y are, respectively, the oil droplets found in short-wave-sensitive (SWS) single cones, the principal member of double cones, long-wave-sensitive (LWS) single cones, violet-sensitive (VS) single cones, and medium-wave-sensitive (MWS) single cones. Scale bar, 10 μm. (C) The Red-billed Leiothrix (*Leiothrix lutea*) was the first species in which the absorptance of ultraviolet cones was measured using microspectrophotometry. *Credits:* (A) Reproduced with permission from N. S. Hart. 2004. Microspectrophotometry of visual pigments and oil droplets in a marine bird, the Wedge-tailed Shearwater *Puffinus pacificus:* Topographic variations in photoreceptor spectral characteristics. J Exp Biol 207: 1229–1240, © 2004 The Company of Biologists; (B) Don Roberson; (C) Alister Benn.

Plate 2. The "yellow" carotenoid pigments lutein and zeaxanthin can produce orange and red as well as yellow hues (see Fig. 2.12). (A) Growing rump feathers on a male Golden-backed Bishop (*Euplectes aureus*). (B) Red epaulets of a male Red-shouldered Widowbird (*Euplectes axillaris*). No "red" carotenoids ($l_{max} > 460$ nm) have been detected in these plumages, but in the epaulets, brown melanin may contribute absorptance. *Credit:* Staffan Andersson.

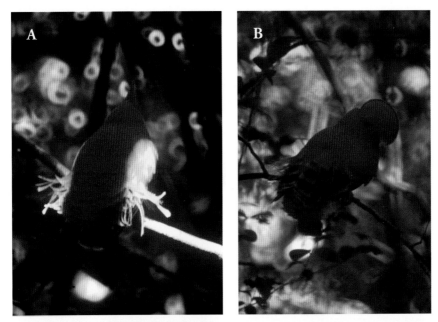

Plate 3. (A) A male Guianan Cock-of-the-Rock (*Rupicola rupicola*) displaying its vividly colored plumage in a light environment that is partly "Small Gaps" and partly "Forest Shade." (B) The same male appearing much more cryptic against the background after moving entirely into a "Forest Shade" light environment a few meters away. (C) A male Golden-headed Manakin (*Pipra erythrocephala*) in a "Small Gap" light environment, with bright head coloration contrasting against the dark background of tropical forest. *Credit:* Marc Théry, CNRS.

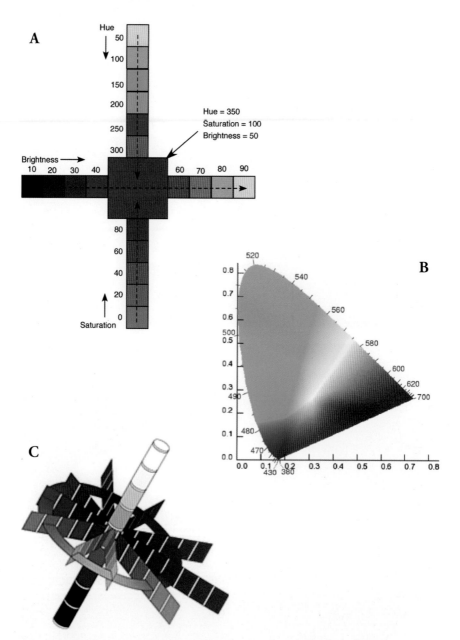

Plate 4. Color spaces. (A) Hue, saturation, and brightness (HSB) in the HSB color space. The central box shows a color that a human would perceive as scarlet, and the horizontal and vertical bars show uniform changes in brightness (horizontal), hue (vertical top), and saturation (vertical bottom) that include this color. (B) CIE chromaticity diagram, showing all of the colors (in standardized arbitrary units) that can be perceived by the average human eye. The wavelengths of all human-perceived hues are shown around the perimeter. (C) Munsell color space, showing hue on the color wheel, saturation as spokes on the wheel, and brightness on the vertical axis. *Credit:* Robert Montgomerie.

Plate 5. Carotenoid-based colors in avian plumage and bare parts. (A) Male Madagascar Fodies (*Foudia madagascariensis*) deposit red ketocarotenoids, synthesized from yellow and orange carotenoids acquired from the diet, into their feathers. (B) The yellow plumage of Yellow-breasted Chats (*Icteria virens*) results from dietary lutein deposited unaltered. (C) White Storks (*Ciconia ciconia*) get their red coloration from depositing carotenoids in the beak and legs. *Credits:* (A) Doug Jansen; (B) Geoffrey E. Hill; (C) Dan Sorchirca.

Plate 6. Examples of black melanin-based colors and patterns in bird feathers and bare parts. Bold black color displays are pigmented primarily by eumelanins, as in (A) the black bands on individual feathers of a Blue Jay (*Cyanocitta cristata*), (B) the black beak of a juvenile Zebra Finch (*Taeniopygia guttata*), and (C) the breast bands of the Killdeer (*Charadrius vociferus*). *Credits:* (A) Mark Liu; (B) Kevin McGraw; (C) William Hull.

Plate 7. Examples of brown and rufous melanin-based colors and patterns in bird feathers. Brown and rufous coloration typically results from a mix of eumelanins and phaeomelanins, as in (A) the wing bars in the male Blue Grosbeak (*Passerina caerulea*), (B) the head feathers of the male Brown-headed Cowbird (*Molothrus ater*), and (C) the breast plumage of the Barn Swallow (*Hirundo rustica*). *Credits:* (A) Mark Liu; (B) Howard Cheek; (C) Brad Lachappelle.

Plate 8. All structural plumage colors are produced by the interaction of light and feather microstructure but can differ in anatomy and arrangement of the color-producing microstructures. (A) The iridescent coloration of the male Peafowl (*Pavo cristatus*) is produced by crystal-like arrays of solid, rod-shaped melanin granules. (B) The glossy, violet-blue iridescence of the Satin Bowerbird (*Ptilonorhynchus violaceus*) plumage is produced primarily by coherent scattering from a thick keratin cortex that is underlain by a single, thick layer of melanin granules. *Credits:* (A) Jean Louis; (B) Daniel J. Mennill.

Plate 9. Structurally colored skin of birds produces ultraviolet, blue, or green hues. The greenish-yellow color of the wattles of (A) the Yellow-bellied Wattle-eye (*Dyaphorophia concreta*) is produced by a combination of structural color and carotenoid pigments, but the green color of the facial and throat skin of (B) the Bare-throated Bellbird (*Procnias nudicollis*) is produced entirely by a structural mechanism. *Credits:* (A) Kristof Zyskowski; (B) Nate Rice.

Plate 10. The structural colors of the caruncles of (A) the Velvet Asity (*Philepitta castanea*) are produced by unique, hexagonally organized arrays of parallel collagen fibers. The differences in the diameter and spacing of the collagen fibers in the blue and green areas create the differences in color. The surface of the caruncle is also covered with (B) unique cone-shaped papillae that provide a larger surface area for light to penetrate the color-producing collagen and thus produce a brighter color. *Credits:* (A) Joe Kaplan; (B) Richard Prum.

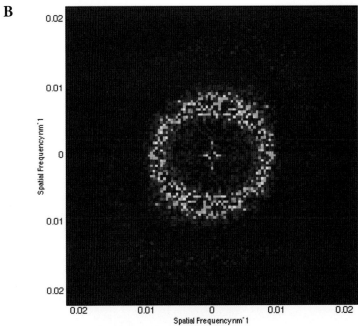

Plate 11. Fourier analysis of the parallel collagen fibers in the ultraviolet-blue eye-ring of (A) the Toco Toucan (*Ramphastos toco*) provides a description of the periodicity of light-scattering opportunities in the skin. The fibers are relatively uniform in size and spacing, but lack any crystal-like organization. (B) The two-dimensional Fourier power spectrum of the fiber distribution shows a ring of high values at intermediate spatial frequencies. These data show that the tissue will produce brilliant structural color that is not iridescent. *Credits:* (A) Marcelo Clemente; (B) Richard Prum and Tom Quinn.

Plate 12. Plumage colors in birds can be generated by pigments other than carotenoids and melanins. Porphyrin pigments create (A) the brown coloration in the plumage of the Red-crested Bustard (*Eupodotis ruficrista*) and (B) the red and green coloration (from turacin and turacoverdin pigments, respectively) in the plumage of the Fischer's Turaco (*Tauraco fischeri*). *Credits:* (A) James Dale; (B) Doug Jansen.

Plate 13. Eggs and the bare parts of nestlings primarily derive their colors from pigments. (A) The blue coloration of American Robin (*Turdus migratorius*) eggs and the brown coloration of Brown-headed Cowbird (*Molothrus ater*) eggs are both due to porphyrins. (B) The mouth coloration of Eastern Bluebird (*Sialia sialis*) chicks is at least partly due to hemoglobin in the blood; whether carotenoid pigments contribute to mouth coloration in any birds remains to be determined. *Credits:* (A) Mark Hauber; (B) Mark Liu.

Plate 14. Avian eyes vary widely in color and in the pigments responsible for such colors.
(A) Yellow pterin-based iris coloration in a Great Horned Owl (*Bubo virginianus*).
(B) Hemoglobin-based red coloration in the eye of a Phainopepla (*Phainopepla nitens*).
(C) Purine-based white coloration in the eye of an Acorn Woodpecker (*Melanerpes formicivorus*). *Credits:* (A) Mike Lanzone/Powdermill Avian Research Center; (B) Will Turner; (C) Walt Koenig.

Plate 15. Parrots synthesize their own red and yellow pigments, known as psittacofulvins, to color their feathers, as in the red plumage of (A) a female Eclectus Parrot (*Eclectus roratus*) and (B) Scarlet Macaws (*Ara macao*). *Credit:* Doug Jansen.

Plate 16. The yellow plumage of parrots often fluoresces under ultraviolet light. White-bellied Caiques (*Pionites leucogaster*) in (A) sunlight and (B) under ultraviolet light. *Credit:* Andrew Davidhazy.

Plate 17. Many sources of plumage coloration remain to be described. (A) Yellow plumage coloration in the domestic chick (*Gallus gallus*) is produced by an undescribed pterin-like pigment. (B) Orange plumage coloration in the King Penguin (*Aptenodytes patagonicus*) is produced by an undescribed yellow pterin-like pigment in association with brown melanin. (C) Green plumage coloration in the male Common Eider (*Somateria mollissima*) is produced by an undescribed pigment. *Credits:* (A) Kevin McGraw; (B) Anne-Sophie Coquel; (C) David Taylor.

Plate 18. Cosmetic colors applied to feathers. (A) Bearded Vulture (*Gypaetus barbatus*), in captivity, bathing in iron-oxide-rich muddy water. (B) Great Hornbill (*Buceros bicornis*) showing yellow stains, presumably from carotenoid pigments in uropygial gland secretions, on its head, bill, casque, wing stripes, rump, and upper tail coverts. (C) Ross's Gull (*Rhodostethia rosea*) showing a pink wash on its plumage that results from the application of uropygial gland secretions containing carotenoid pigments. *Credits:* (A) Klaus Robin; (B) Tim Laman/VIREO; (C) Jack Folkers.

Plate 19. Rock Ptarmigan (*Lagopus mutus*) use both molt and feather soiling to change from white to brown. (A) Male and female at the start of spring snow melt—the female is just beginning to molt. (B) Male in immaculate white plumage with no soiling. (C) Female during the same period as the male in (B) but showing advanced molt to brown plumage. (D) Male with heavily soiled white plumage. (E) Female on nest after replacing all her white feathers with mottled brown ones. *Credit:* Robert Montgomerie.

Plate 20. Bowerbirds use colored decorations as an extended phenotype. (A) Male Satin Bowerbird (*Ptilonorhynchus violaceus*) standing inside his bower, inspecting his blue decorations from the perspective of a potential mate. (B) Satin Bowerbird bower and platform from directly above, showing the distribution of blue decorations in dappled sunlight in front of the bower. (C) Male Spotted Bowerbird (*Chlamydera maculata*) with his bower and white decorations. *Credits:* (A, B) Robert Montgomerie; (C) Becky Coe.

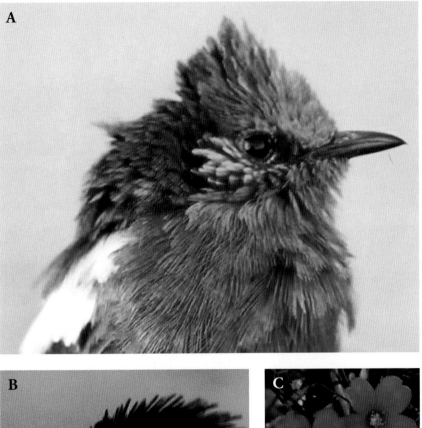
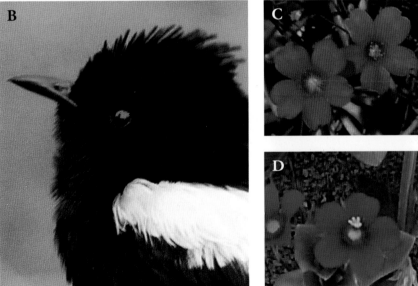

Plate 21. Male White-winged Fairy-wrens (*Malurus leucopterus*) are (A) blue on the mainland of Western Australia but (B) black on Dirk Hartog Island. The color difference is correlated with variation at a single locus, the melanocortin-1 receptor. Males present females with flowers during courtship—blue mainland males favor (C) purple flowers, whereas black island males favor (D) blue flowers; see Figure 9.3. *Credit:* Melanie Rathburn.

Plate 22. Seasonal change in plumage coloration can result from feather abrasion. A male Lapland Longspur (*Calcarius lapponicus*) in (A) nonbreeding and (B) breeding plumage; a male Snow Bunting (*Plectrophenax nivalis*) in (C) nonbreeding and (D) breeding plumage. No molt is involved in these color changes; the buffy feather tips, as on (E) the throat feather of a male Lapland Longspur, wear away, revealing bold coloration beneath. *Credits:* (A) Mike Yip; (B) Svein Bekkum; (C) Henry Ekholm; (D, E) Robert Montgomerie.

Plate 23. Seasonal change in plumage pattern can result from feather abrasion. Male House Sparrow (*Passer domesticus*) in (A) nonbreeding plumage with lower half of black bib obscured by pale feather edges and (B) breeding plumage with black bib fully displayed. European Starling (*Sturnus vulgaris*) in (C) nonbreeding plumage with light feather tips, giving a spotted appearance and (D) breeding plumage with tips worn away, revealing bright iridescent coloration beneath. *Credits:* (A) Michael Roach; (B) Janine Russell; (C, D) Christine Redgate.

Plate 24. Exposure to iron oxides in soil and water can lead to adventitious coloration. (A) Sandhill Cranes (*Grus canadensis*) with iron stains on their backs. (B) Emperor Goose (*Chen canagica*) and (C) Snow Geese (*Chen caerulescens*) with iron oxide stains on their heads. *Credits:* (A) Allan Taylor; (B) Tim Bowman; (C) Claude Nadeau.

Plate 25. Dirt can change the coloration of feathers as in (A) Evening Grosbeaks (*Coccothraustes vespertinus*), with the bird on the right having a much dirtier breast than the one on the left due to resin staining, and (B) a Pine Grosbeak (*Pinicola enucleator*) showing resin staining on its breast. Patches from the breasts of museum skins of (C, D) Evening Grosbeaks, (E, F) Pine Grosbeaks, and (G, H) House Sparrows, comparing dirty plumages (C, E, G) to plumages that had been washed after skinning (D, F, H). Each patch is 1 cm × 1 cm. *Credits:* (A) Jean-Sébastien Guénette; (B) Karl Egressy; (C–H) Robert Montgomerie.

Plate 26. A normal male Red Junglefowl (*Gallus gallus;* top), a normal hen (middle), and a hen-feathered male (bottom), exhibiting normal male comb and wattles, but having the female-type plumage. Hen feathering in males is caused by an insertion of a transposon in the promoter that increases production of aromatase, resulting in high levels of estrogen at the feather follicles, which cause the production of female-like plumage. *Credit:* J. David Ligon.

Plate 27. (A) Normal male and (C) normal female Zebra Finches flanking (B) a bilaterally gynandromorphic individual that expressed genes from both Z and W chromosomes. This expression results in female plumage on the bird's left side but male plumage on the right side, due to differential expression of the W and Z chromosomes, respectively. (D) In the male Pied Flycatcher (*Ficedula hypoleuca*), the size of the demelanized white forehead patch has been the subject of extensive quantitative genetic studies in relation to sexual selection and condition-dependence. *Credits:* (A) Roy Beckham; (B) Art Arnold, with permission from the National Academy of Sciences, USA; (C) Neil Fifer; (D) Henry Ekholm.

Plate 28. (A) Plumage coloration of the Lesser Snow Goose (*Anser caerulescens*) is a classic example of a melanin-based polymorphism. The extreme white and blue morph birds are shown here. A single identified locus (*MC1R*) has been identified as the major determinant of this variation. (B) Pale and dark morphs of the Arctic Skua or Parasitic Jaeger (*Stercorarius parasiticus*) vary in degree of melanization. As in the Lesser Snow Goose, intermediate phenotypes between the extreme morphs occur, and the *MC1R* locus is the major contributor to color variation. *Credits:* (A) Peter LaTourrette; (B) Harold Stiver.

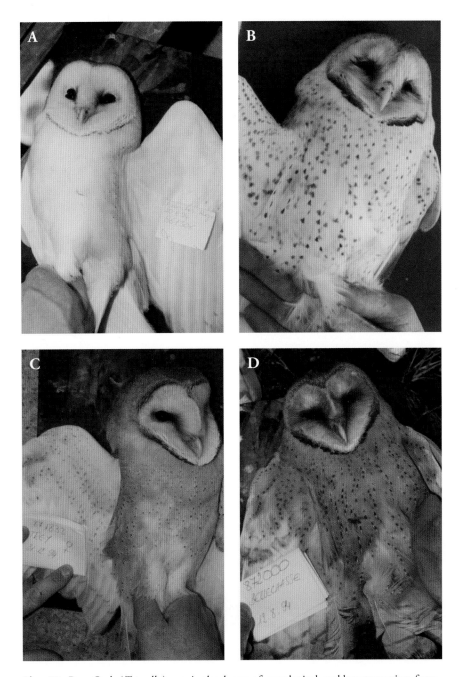

Plate 29. Barn Owls (*Tyto alba*) vary in the degree of eumelanin-based breast spotting, from (A) immaculate to (B) heavily marked with black spots. Individuals also vary in the darkness of breast feathers from (A, B) white to (C, D) dark reddish-brown. Both plumage traits are heritable, with neither environmental challenges nor body condition altering expression. Spottiness and breast darkness are genetically correlated, with darker reddish-brown birds displaying more and larger black spots. *Credit:* Alexandre Roulin.

Plate 30. Carotenoid-based coloration can be shaped by environmental conditions during molt. (A) Cedar Waxwings (*Bombycilla cedrorum*) grow orange rather than typical yellow tail feathers when they ingest rhodoxanthin during molt. (B) Egyptian Vultures (*Neophron percnopterus*) get the pigments for their face coloration by ingesting ungulate feces, and this face color varies with the carotenoid content of feces. (C) Expression of the yellow coloration of the bill and plumage of American Goldfinches (*Carduelis tristis*) has been shown to be affected by carotenoid access, nutrition, and parasites. *Credits:* (A, C) Robert Mulvihill/Powdermill Avian Research Center; (B) Manuel de la Riva.

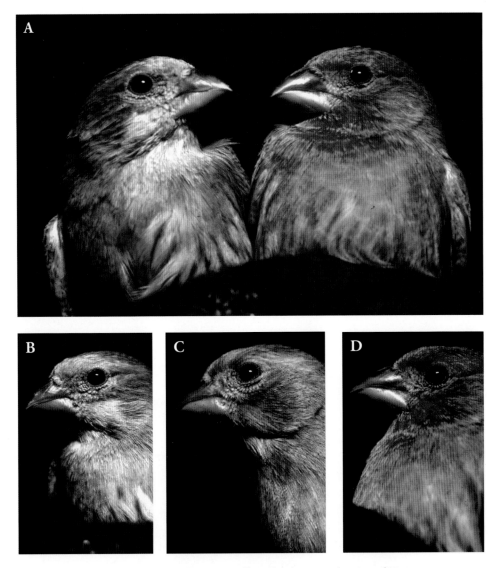

Plate 31. Dietary access to carotenoid pigments affects the plumage coloration of House Finches (*Carpodacus mexicanus*), which vary in nature (A) from pale yellow to bright red. (B) Males that were fed only millet and sunflower seeds during molt grew yellow plumage. (C) Males fed seeds with the orange carotenoid β-carotene added to their water grew orange plumage. (D) Those fed seeds with the red carotenoid canthaxanthin added to their water grew red plumage. *Credit:* David Bay and Geoffrey E. Hill.

Plate 32. Structural plumage coloration is affected by parasites and nutrition during molt. (A) The structural blue/ultraviolet coloration of Eastern Bluebirds (*Sialia sialis*) is affected by nutrition during molt. (B) The brilliant iridescent coloration of male Wild Turkeys (*Meleagris gallopavo*) is affected by gut parasites during molt. *Credits:* (A) Mark Liu; (B) Maslowski/National Wild Turkey Federation.

Combined Structural and Pigmentary Colors

In the ornithological literature, there is the frequent observation that structural blue coloration combined with yellow pigments produces the green colors of avian plumage. This general relationship of complex combinations of structural and pigmentary colors can been seen in mutants of captive birds (e.g., Budgerigars [*Melopsittacus undulatus*]; Plate 17, Volume 2). Surprisingly, however, this relationship has rarely been examined chemically and microscopically. The recognition that structural colors of feather barbs and skin are coherently, rather than incoherently, scattered changes the role of the structural color in these structural/pigmentary combinations.

Dyck (1971b) has gone furthest by examining blue and green feathers of Peach-faced Lovebirds (*Agapornis roseicollis*). Dyck found that the effects of the combined short-wavelength structural color and long-wavelength pigmentary color were to produce a saturated hue that was at the long-wavelength end of the reflectance spectrum of the structural color alone. The superficial absorbance of shorter wavelengths by the yellow pigment reduced the coherent scattering at the shorter wavelength end of the structural reflectance spectrum. Furthermore, Dyck (1971b) and Prum et al. (1999b) showed that the spongy medullary nanostructures of the green feathers were larger in periodicity, and thus tuned to produce a longer wavelength, green structural color. Essentially the structural color was much more influential than the pigmentary color in producing the final hue. Subsequently, Dyck (1978) described the spongy medullary nanostructures of numerous olive green birds, but did not identify or characterize the associated pigments.

Prum et al. (1994, 1999a) and Prum and Torres (2003b) identified purely green structural colors from the skin of a number of avian species, including the Velvet Asity and Bare-throated Bellbird, showing that there is no inherent limit to the production of longer-wavelength structural colors by coherent scattering. However, Prum and Torres (2003b) also identified a number of species with highly saturated longer-wavelength skin colors that are produced by a combination of structural and pigmentary mechanisms, including Cabot's Tragopan, Toco Toucan (Plate 11), and the Yellow-bellied Wattle-eye (*Dyaphorophyia concreta;* Platysteridae; Plate 9). In these instances, the pigments were not identified, but were hypothesized to be carotenoids in solution in lipid vacuoles of epidermal cells. The underlying collagen nanostructures were large enough in periodicity to produce a purely structural color of the observed hue. Apparently, the contribution of the superficial pigment to the hue was to provide additional absorbance of shorter visible wavelengths. Interestingly, coherent scattering of a saturated structural color prevented the production of a typical carotenoid hue with a broad peak of reflectance at longer wavelengths. The result is a uniquely saturated yellow or orange color that is quite distinct from those of carotenoid pigments alone.

Nothing is known in birds about the potential combinations of structural colors with other pigments, such as porphyrins and pterins. Such exotic combinations are particularly plausible in the avian iris (Oliphant 1981, 1987a,b; Oliphant and Hudon 1993).

In conclusion, structural and pigmentary mechanisms can combine to provide hues that are unavailable among purely pigmentary avian colors; for example, saturated long-wavelength hues (Prum and Torres 2003b). It is important that notions about the combined interactions of structural and pigmentary colors be consistent with the known mechanisms of structural color production. Interaction between structural and pigmentary colors should be a rich area of future research.

Development of Avian Structural Colors

Avian structural colors are produced by coherent scattering from nanostructures that require extraordinarily precise bioengineering to create an appropriate visual signal. Given the color discriminating abilities of birds (Vorobyev and Osorio 1998), an error in the average dimensions of the color-producing nanostructure of 20 nm (or roughly a millionth of an inch) will produce a visibly distinct color, result in an inferior signal, and perhaps result in lower mating success (Chapter 4, Volume 2) or social status (Chapter 3, Volume 2). Underlying elements to both the consistency and diversification of these structurally colored signals are the heritability and development of the coherent scattering nanostructures. Unfortunately, the development of avian color-producing nanostructures has not been well studied, but there are a few possible generalizations that can guide future research.

Many color-producing nanostructures are self-assembled. Self-assembly is the growth of larger scale structures determined by molecular interactions and kinetics. Many self-assembled nanostructures are composed of protein polymers, including both feather β-keratin and collagen. The development of extracellular collagen fibers and intracellular β-keratin matrices involves the production of protein filaments that must then bond to a polymer composed of numerous protein filaments. During self-assembly, the details of the molecular-scale interactions ultimately produce the nanometer-scale variations in structure. Self-assembly of nanostructures from filamentous proteins is not understood for any molecular system, but some tantalizing and compelling mechanistic insights can be made from the chemistry of block co-polymers. Experimental co-polymers are mixtures of multiple, self-assembled, polymer-forming hydrocarbons (or heteropolymers). The heterogeneous mixture of two types of polymer-forming molecules creates a series of attractive and repellent forces that cause disorder at cold or warm temperatures (i.e., kinetic extremes), but produces nanometer-scale order under intermediate kinetic conditions. The nanostructural patterns created by self-assembled block co-polymers are strikingly similar to the patterns seen in spongy medullary feather β-keratin. It is very likely that the identical kinetics—the sum of the attractive and repellent physical interactions among a mixture of various polymer-forming protein filaments—underlies the development of feather structural colors.

The structural colors produced by melanosome-keratin arrays in feather barbules require a particularly complex development and are not likely to be entirely self-assembled. The melanosomes are in melanocytes, or pigment cells of neural crest origin, that migrate into the dermal pulp of the developing feather germ and peripherally among the developing cells of the feather barb ridges (Strong 1902; Watterson 1942; Prum and Williamson 2002; Chapter 6). The melanocytes extend pseudopodia out among the various barb ridge cells, and the melanosomes are moved by protein filaments to the tips of the melanocyte pseudopodia. Apparently as a consequence of the interactions among the melanocytes and individual feather keratinocytes, completed melanosomes are transferred to or taken up by individual keratinocytes. Melanosome transfer takes place before the synthesis and polymerization of keratin in the feather keratinocytes. As synthesis of β-keratin takes place, the melanosomes are moved into place in the keratinocytes by some unknown mechanism. This process gives rise to all feather melanin patterns (Prum and Williamson 2002; Chapter 6). With multilayer barbule structural colors, the production of an appropriate color requires the development of melanosomes of the appropriate dimensions, transfer of those melanosomes to the appropriate keratinocytes, and then the deposition of the appropriate amount of β-keratin between the melanosomes to produce the correct nanoperiodicity and color. Essentially, nothing is known about the cellular mechanisms involved in melanosome size and shape determination, melanosome transfer mechanism, or melanosome position determination with the keratinocytes

(Chapter 6). Furthermore, the hollow melanosomes that have evolved in many families of birds with structurally colored barbules are apparently unique to Aves. Little is known about how these hollow melanosomes develop, except for an outstanding early descriptive study (Durrer and Villiger 1967). In summary, to evolve their vivid structural colors, birds have evolved extraordinary developmental coordination between melanocytes and keratinocytes. Single layer, thick-film nanostructures that create oily black plumages require only the specification of the thickness of the superficial keratin layer by the positioning of the underlying melanosomes. Thus these thick-film systems are inherently simpler to develop.

Condition Dependence of Structural Coloration

There has been a recent explosion of interest in the potential for condition dependence of and honest signaling by ornamental bird coloration, including structural colors. Although all known avian structural colors work by the common physical mechanism of coherent scattering, the development and physical composition of the structural colors of avian skin, feather barbules, feather barbs, and eyes are very different and yield different opportunities for condition-dependent expression. Even within feathers, the structural colors produced by spongy medullary β-keratin of barb rami, laminar or crystal-like arrays of melanosomes and β-keratin in barbules, or a single, superficial layer of β-keratin above a layer of melanosome in the barbules (i.e., iridescent, oily or glossy black plumages) are all anatomically very distinct and likely require entirely distinct developmental and physiological mechanisms to produce condition dependence. Results from one anatomical system are unlikely to generalize simply to any others. Thus new results should be interpreted conservatively.

Based on the physics and anatomy of structural color production, different aspects of color—hue, brilliance, saturation, and other aspects of reflectance—are not all equally likely to be subject to condition dependence. Empirical studies of the effects of nutrition and parasites on structural coloration in birds are reviewed in Chapter 12. Here I make some comments about the potential for, and constraints on, quality indication of structural colors. Because many color-producing nanostructures are developed through molecular self-assembly (see above), the critical dimensions of the arrays that determine peak hue are likely controlled by molecular interactions and regulation of protein expression in particular classes of cells. For example, examining the transition between structurally colored turquoise and blue patches created by spongy barb keratin on the primary feathers of Lilac-breasted Roller (*Coracias benghalensis;* Coraciidae), it is easy to see that developmental decisions about whether to produce blue or turquoise nanostructures are produced on an autonomous cell-by-cell basis (Raman 1935). In this instance, the differences in hue between different cells are not determined by continuous variation among cells in some simple parameter, but apparently by two alternative developmental programs leading to the two distinct nanostructures. In contrast, the iridescent black colors are a consequence of the thickness of that single superficial keratin layer. These systems would be relatively easy to determine and evolve. Furthermore, hue could change if feather wear reduces the thickness of this superficial layer, which is likely occurring in iridescent swifts and swallows.

At larger spatial scales, structural colors could vary because of the quantity of the nanostructural arrays and the presentation of those arrays to ambient light and observers. Variation in the overall thickness or number of nanostructures could affect the brilliance (total reflectance) or the chroma (the saturation or purity of hue) of a structural color. Variation in the shape of feather barbs or barbules may produce variations in the brilliance of the hue. Thus an appreciation of the biology of structural colors indicates that brilliance and chroma are more susceptible to developmental perturbation than is hue.

To date, the majority of papers on condition dependence of structural colors have focused on glossy black species, such as those of *Volatinia, Molothorus,* and *Ptilonorhynchus* (Doucet 2002; McGraw et al. 2002; Doucet and Montgomerie 2003a,b), which are all likely produced with single superficial keratin layer. These single-film systems may be easier to perturb developmentally than other anatomical systems because of their simplicity, but for the same reason, the results cannot be generalized to other types of structural colors.

Furthermore, all the structurally colored species examined for condition dependence to date are members of species or clades with substantial delayed plumage maturation: Blue-black Grassquit (*Volatinia jacarina*), Brown-headed Cowbird (*Molothorus ater;* Plate 7), Satin Bowerbird (*Ptilonorhynchus violaceous;* Plate 8), Eastern Bluebird (*Sialia sialis;* Plate 32), and Blue Grosbeak (*Passerina caerulea;* Plate 7) (Keyser and Hill 1999, 2000; Doucet 2002; Doucet and Montgomerie 2003a,b; Siefferman and Hill 2003). None of the field studies has explicitly examined the relationship between age and structural coloration to determine what components of delayed plumage maturation may confound the mostly indirect evidence of condition dependence. Most of the evidence is correlational, and has yet to exclude the role of age in signal variation.

Although the realization of the potential for condition dependence of carotenoid signals in birds is a rather new discovery in ornithology, this condition dependence was well understood in aviculture for more than a century. In contrast, however, there are no such anecdotes about structural colors from the centuries of captive rearing of a wide variety of structurally colored birds. The absence of observations of condition dependence in the anecdotal literature strongly implies that any condition-dependent effects on structural coloration are likely to be quite modest. This conclusion is further supported by the observation that no color-producing nanostructures are constructed from any limited nutrients. Thus they lack the direct production costs or constraints that might prevent the evolution or development of counterfeit signals. The self-assembly inherent in many color-producing nanostructures also limits developmental perturbability, rather than enhancing it, despite their small size.

To date, there is only a single experimental study of condition dependence in structural colors in Brown-headed Cowbirds (McGraw et al. 2002). No investigations have attempted to examine the physical basis of the individual variations in structural coloration that has been proposed to be condition related. Shawkey et al. (2003) were able to show some nanostructural correlates with color variation in the Eastern Bluebird. However, such studies are tremendously difficult because the margins of error in quantitative electron microscopy are on the same order of magnitude as the variations among individual birds. Also, the structural colors are produced by nanoscale arrays that may be produced by thousands of cells in a feather barb, many barbs in a feather, and many feathers in a plumage patch. So far, sampling has been strictly limited to a few cells, in a few barbs of a few feathers at most. To accurate describe these variations, a great deal of careful, quantitative electron microscopy is required.

Summary

Structural colors are produced by physical interactions of light waves with nanometer-scale structures in the organism. These colors provide numerous displays that are not available by pigmentary mechanisms. Understanding of the anatomy and physics of structural colors has expanded tremendously in the past 50 years, with the application of electron microscopy. Only recently, however, have published studies led to an understanding of the structural colors of avian feather barbs, skin, and eyes.

All known chromatic structural colors of birds are produced by the same physical mechanism—coherent scattering (i.e., constructive interference).

These colors differ only in the nanostructure, anatomical organization, and physical composition of the arrays. In contrast, white feathers are produced by incoherent scattering of all ambient light waves. Coherently scattering arrays have three basic types of nanostructure: laminar, crystal-like, or quasi-ordered. Prominent iridescence—change in hue with angle of observation or illumination—is created by those structural colors that have laminar or crystal-like nanostructure, such as the structural colors of feather barbules. Quasi-ordered arrays generally do not produce iridescence, because they are appropriately nanostructured to coherently back-scatter the same color from multiple directions. Therefore coherent scattering is not synonymous with iridescence. Structural colors of feathers vary in anatomy and optical properties in different parts of the feathers. Structural colors of feather barbules are created by arrays of melanin granules and β-keratin. The granules can vary tremendously in shape and sometimes include a hollow air vacuole. These colors are often brilliant, highly saturated, and iridescent (e.g., in hummingbirds, sunbirds, trogons, peacocks).

A distinct class of barbule structural colors includes the single-film colors that are produced by the superficial layer of β-keratin above a layer of melanosomes, which generally produce blackish plumages with an iridescent sheen that may have multiple reflectance peaks in the visible spectrum.

In contrast to barbules, the structural colors of feather barbs are produced by coherent scattering from the matrix of spongy β-keratin and air in the medullary cells of the barb rami. The spongy keratin occurs in two forms, in which the air bubbles are either spherical cavities or a complex network of channels. The larger the air cavities or channels, the longer wavelength the color. The structural colors of avian skin are produced by coherent scattering from arrays of parallel collagen fibers in the skin. These structures have evolved numerous times independently in birds. In some lineages (e.g., *Tragopan, Ramphastos, Dyaphorophyia*), a combination of structural yellow or orange and carotenoid pigments produce saturated, long-wavelength hues that are unavailable among pigments alone.

References

Andersson, S. 1999. Morphology of UV reflectance in a Whistling-thrush: Implications for the study of structural colour signalling in birds. J Avian Biol 30: 193–204.

Auber, L. 1957. The distribution of structural colors and unusual pigments in the Class Aves. Ibis 99: 463–476.

Auber, L. 1971/1972. Formation of "polyhedral" cell cavities in cloudy media of bird feathers. Proc R Soc Edinburgh 74: 27–41.

Bagnara, J. T. 1998. Comparative anatomy and physiology of pigment cells in non-mammalian tissues. In J. J. Nordlund, R. E. Boissy, V. J. Hearing, R. A. King, and J. P. Ortonne, ed., The Pigmentary System—Physiology and Pathophysiology, 9–40. Oxford: Oxford University Press.

Bagnara, J. T., and M. E. Hadley. 1973. Chromatophores and Color Change. New Jersey: Prentice-Hall.

Bellairs, R., M. L. Harkness, and R. D. Harkness. 1975. The structure of the tapetum of the eye of the sheep. Cell Tissue Res 157: 73–91.

Benedek, G. B. 1971. Theory of transparency of the eye. Appl Opt 10: 459–473.

Bohren, C. F., and D. R. Huffman. 1983. Absorption and Scattering of Light by Small Particles. New York: John Wiley and Sons.

Bragg, W. H., and W. L. Bragg. 1915. X-rays and Crystal Structure. London: G. Bell.

Briggs, W. L., and V. E. Henson. 1995. The DFT. Philadelphia: Society for Industrial and Applied Mathematics.

Brink, D. J., and N. G. van der Berg. 2004. Structural colours from the feathers of the bird *Bostrychia hagedash*. J Phys D Appl Phys 37: 813–818.

Camichel, C., and H. Mandoul. 1901. Des colorations bleues et vertes de la peau des Vertébrés. Compt Rend Séanc l'Acad Sci 133: 826–828.

Craig, A. J. F. K., and A. H. Hartley. 1985. The arrangement and structure of feather melanin granules as a taxonomic character in African starlings (Sturnidae). Auk 102: 629–632.

Cuthill, I. C., A. T. D. Bennett, J. C. Patridge, and E. J. Maier. 1999. Plumage reflectance and the objective assessment of avian sexual dichromatism. Am Nat 160: 183–200.

Dorst, J., G. Gastaldi, R. Hagege, and J. Jacquemart. 1974. Différents aspects des barbules de quelques Paradisaeidés sur coupes en microscopie électronique. Compt Rend Séanc l'Acad Sci 278: 285–290.

Doucet, S. M. 2002. Structural plumage coloration, male body size, and condition in Blue-black Grassquit. Condor 104: 30–38.

Doucet, S. M., and R. Montgomerie. 2003a. Multiple sexual ornaments in satin bowerbirds: Ultraviolet plumage and bowers signal different aspects of male quality. Behav Ecol 14: 503–509.

Doucet, S. M., and R. Montgomerie. 2003b. Structural plumage colour and parasites in Satin Bowerbirds *Ptilonorhynchus violaceus:* Implications for sexual selection. J Avian Biol 34: 237–242.

Doucet, S., M. D. Shawkey, M. K. Rathburn, H. L. Mays, and R. Montgomerie. 2004. Concordant evolution of plumage colour, feather microstructure and a melanocortin receptor gene between mainland and island populations of a fairy wren. Proc R Soc Lond B 271: 1663–1670.

Durrer, H. 1962. Schillerfarben beim Pfau (*Pavo cristatus* L.). Verhandl Naturf Ges Basel 73: 204–224.

Durrer, H. 1965. Bau und Bildung der Augfeder des Pfaus. Rev Suisse Zool 72: 264–411.

Durrer, H. 1977. Schillerfarben der Vogelfeder als Evolutionsproblem. Denkschr Schweiz Naturforsch Ges 14: 1–126.

Durrer, H. 1986. The skin of birds: Colouration. In J. Bereiter-Hahn, A. G. Matoltsky, and K. S. Richards, ed., Biology of the Integument 2: Vertebrates, 239–247. Berlin: Springer-Verlag.

Durrer, H., and W. Villiger. 1966. Schillerfarben der Trogoniden. J Ornithol 107: 1–26.

Durrer, H., and W. Villiger. 1967. Bildung der Schillerstruktur beim Glanzstar. Zeitschr Zellforsch 81: 445–456.

Durrer, H., and W. Villiger. 1968. Schillerfarben der Nectarvögel (Nectariniidae). Rev Suisse Zool 69: 801–814.

Durrer, H., and W. Villiger. 1970a. Schillerfarben der Stare (*Sturnidae*). J Ornithol 111: 133–153.

Durrer, H., and W. Villiger. 1970b. Schillerradien des Goldkuckucks (*Chrysococcyx cupreus* (Shaw)) im Elektronenmikroskop. Zeitschr Zellforsch 109: 407–413.

Durrer, H., and W. Villiger. 1975. Schillerstruktur des Kongopfaus (*Afropavo congensis* Chapin 1936) im Electronmikroskop. J Ornithol 116: 94–102.

Dyck, J. 1971a. Structure and colour-production of the blue barbs of *Agapornis roseicollis* and *Cotinga maynana*. Zeitschr Zellforsch 115: 17–29.

Dyck, J. 1971b. Structure and spectral reflectance of green and blue feathers of the Lovebird (*Agapornis roseicollis*). Biol Skrift 18: 1–67.

Dyck, J. 1973. Feather structure: The surface of the barbs and barbules. Zool Jb Anat 90: 550–566.

Dyck, J. 1976. Structural colours. Proc Int Ornithol Congr 16: 426–437.

Dyck, J. 1978. Olive green feathers: Reflection of light from the rami and their structure. Anser 3 (Suppl.): 57–75.

Dyck, J. 1979. Winter plumage of the Rock Ptarmigan: Structure of the air-filled barbules and function of the white colour. Dansk Orn Foren Tidsskr 73: 41–58.

Dyck, J. 1987. Structure and light reflection of green feathers of fruit doves (*Ptilinopus* spp.) and an Imperial Pigeon (*Ducula concinna*). Biol Skrift 30: 2–43.

Dyck, J. 1999. Feather morphology at the ultrastructural level. Acta Ornithol 34: 132–134.

Ferris, W., and J. T. Bagnara. 1972. Reflecting pigment cells in the dove iris. In V. Riley, ed., Pigmentation: Its Genesis and Biological Control, 181–192. New York: Appleton-Century-Crofts.

Finger, E. 1995. Visible and UV coloration in birds: Mie scattering as the basis of color in many bird feathers. Naturwissenschaften 82: 570–573.

Fox, D. L. 1976. Animal Biochromes and Structural Colors. Berkeley: University of California Press.

Frank, F. 1939. Die Färbung der Vogelfeder durch Pigment und Struktur. J Ornithol 87: 426–523.

Frank, F., and H. Ruska. 1939. Übermikroskopische Untersuchung der Blaustruktur der Vogelfeder. Naturwissenschaften 27: 229–230.

Ghiradella, H. 1991. Light and colour on the wing: Structural colours in butterflies and moths. Appl Opt 30: 3492–3500.

Gill, F. B. 1995. Ornithology, second edition. New York: W. H. Freeman and Co.

Greenewalt, C. H., W. Brandt, and D. Friel. 1960a. The iridescent colors of hummingbird feathers. J Opt Soc Am 50: 1005–1013.

Greenewalt, C. H., W. Brandt, and D. Friel. 1960b. The iridescent colors of hummingbird feathers. Proc Am Phil Soc 104: 249–253.

Häcker, V., and G. Meyer. 1902. Die blaue Farbe der Vogelfedern. Arch Mikrosk Anat Entw Mech 35: 68–87.

Hays, H., and H. Habermann. 1969. Note on bill color of the Ruddy Duck, *Oxyura jamaicensis rubida*. Auk 86: 765–766.

Hecht, E. 1987. Optics, second edition. Reading, MA: Addison-Wesley.

Herring, P. J. 1994. Reflective systems in aquatic animals. Comp Biochem Physiol A 109: 513–546.

Huxley, A. F. 1968. A theoretical treatment of the reflexion of light by multi-layer structures. J Exp Biol 48: 227–245.

Huxley, J. 1975. The basis of structural colour variation in two species of *Papilio*. J Entomol A 50: 9–22.

Johnsen, S. 2001. Hidden in plain sight: The ecology and physiology of organismal transparency. Biol Bull 201: 301–318.

Keyser, A. J., and G. E. Hill. 1999. Condition-dependent variation in the blue-ultraviolet coloration of a structurally based plumage ornament. Proc R Soc Lond B 266: 771–777.

Keyser, A. J., and G. E. Hill. 2000. Structurally based plumage coloration is an honest signal of quality in male Blue Grosbeaks. Behav Ecol 11: 202–209.

Kinoshita, S., S. Yoshioka, and K. Kawagoe. 2002. Mechanisms of structural colour in the *Morpho* butterfly: Cooperation of regularity and irregularity in an iridescent scale. Proc R Soc Lond B 269: 1417–1421.

LaBastille, A., D. G. Allen, and L. W. Durrel. 1972. Behavior and feather structure of the Quetzal. Auk 89: 339–348.

Land, M. F. 1972. The physics and biology of animal reflectors. Prog Biophys Molec Biol 24: 77–106.

Lee, D. W. 1997. Iridescent blue plants. Am Sci 85: 56–63.

Li, J., X. Yu, X. Hu, C. Xu, X. Wang, X. Liu, and R. Fu. 2003. Coloration strategies in peacock feathers. Proc Natl Acad Sci USA 100: 12576–12578.

Lucas, A. M., and P. R. Stettenheim. 1972. Avian Anatomy: Integument. Agriculture Handbook 362. Washington, DC: U.S. Department of Agriculture.

Lythgoe, J. N. 1974. The structure and phylogeny of iridescent corneas in fishes. In M. A. Ali, ed., Vision in Fishes: New Approaches in Research, 253–262. New York: Plenum Press.

Macleod, H. A. 1986. Thin-film Optical Filters, second edition. Bristol: Adam Hilger.

Mandoul, H. 1903. Recherches sur les colorations tégumentaires. Ann Sci Nat B Zool 8 Serie 18: 225–463.

Mason, C. W. 1923a. Structural colors of feathers. I. J Phys Chem 27: 201–251.

Mason, C. W. 1923b. Structural colors of feathers. II. J Phys Chem 27: 401–447.

McGraw, K. J. 2004. Multiple UV reflectance peaks in the iridescent neck feathers of pigeons. Naturwissenschaften 91: 125–129.

McGraw, K. J., E. A. Mackillop, J. Dale, and M. Hauber. 2002. Different colors reveal different information: How nutritional stress affects the expression of melanin- and structurally based ornamental plumage. J Exp Biol 205: 3747–3755.

Nassau, K. 1983. The Physics and Chemistry of Color. New York: John Wiley and Sons.

Oehme, H. 1969. Vergleichende Untersuchungen über die Färbung der Vogeliris. Biol Zentr 88: 3–35.

Oliphant, L. W. 1981. Crystalline pteridines in the stromal pigment cells of the iris of the Great Horned Owl. Cell Tissue Res 217: 387–395.

Oliphant, L. W. 1987a. Observations on the pigmentation of the pigeon iris. Pigment Cell Res 1: 202–208.

Oliphant, L. W. 1987b. Pteridines and purines as major pigments of the avian iris. Pigment Cell Res 1: 129–131.

Oliphant, L. W., and J. Hudon. 1993. Pteridines as reflecting pigments and components of reflecting organelles in vertebrates. Pigment Cell Res 6: 205–208.

Oliphant, L. W., J. Hudon, and J. T. Bagnara. 1992. Pigment cell refugia in homeotherms—The unique evolutionary position of the iris. Pigment Cell Res 5: 367–371.

Osorio, D., and A. D. Ham. 2002. Spectral reflectance and directional properties of structural coloration in bird plumage. J Exp Biol 205: 2017–2027.

Parker, A. R. 1999. Invertebrate structural colours. In E. Savazzi, ed., Functional Morphology of the Invertebrate Skeleton, 65–90. London: John Wiley and Sons.

Prum, R. O. 1988. Historical relationships among avian forest areas of endemism in the Neotropics. Proc IXX Int Ornithol Congr 2: 2562–2572.

Prum, R. O., and V. R. Razafindratsita. 1997. Lek behavior and natural history of the Velvet Asity *Philepitta castanea* (Eurylaimidae). Wilson Bull 109: 371–392.

Prum, R. O., and R. H. Torres. 2003a. A Fourier tool for the analysis of coherent light scattering by bio-optical nanostructures. Int Comp Biol 43: 591–602.

Prum, R. O., and R. H. Torres. 2003b. Structural colouration of avian skin: Convergent evolution of coherently scattering dermal collagen arrays. J Exp Biol 206: 2409–2429.

Prum, R. O., and R. H. Torres. 2004. Structural colouration of mammalian skin: Convergent evolution of coherently scattering dermal collagen arrays. J Exp Biol 207: 2157–2172.

Prum, R. O., and S. Williamson. 2002. Reaction-diffusion models of within feather pigmentation patterning. Proc R Soc Lond B 269: 781–792.

Prum, R. O., R. L. Morrison, and G. R. Ten Eyck. 1994. Structural color production by constructive reflection from ordered collagen arrays in a bird (*Philepitta castanea*: Eurylaimidae). J Morphol 222: 61–72.

Prum, R. O., R. H. Torres, S. Williamson, and J. Dyck. 1998. Coherent light scattering by blue feather barbs. Nature 396: 28–29.

Prum, R. O., R. H. Torres, C. Kovach, S. Williamson, and S. M. Goodman. 1999a. Coherent light scattering by nanostructured collagen arrays in the caruncles of the Malagasy asities (Eurylaimidae: Aves). J Exp Biol 202: 3507–3522.

Prum, R. O., R. H. Torres, S. Williamson, and J. Dyck. 1999b. Two-dimensional Fourier analysis of the spongy medullary keratin of structurally coloured feather barbs. Proc R Soc Lond B 266: 13–22.

Prum, R. O., S. Andersson, and R. H. Torres. 2003. Coherent scattering of ultraviolet light by avian feather barbs. Auk 120: 163–170.

Prum, R. O., J. Cole, and R. H. Torres. 2004. Blue integumentary structural colours of dragonflies (Odonata) are not produced by incoherent Tyndall scattering. J Exp Biol 207: 3999–4009.

Raman, C. V. 1935. The origin of colours in the plumage of birds. Proc Ind Nat Acad Sci A 1: 1–7.

Rawles, M. E. 1960. The integumentary system. In A. J. Marshall, ed., Biology and Comparative Physiology of Birds, 189–240. New York: Academic Press.

Rutschke, E. 1966a. Die submikroskopische Struktur schillernder Federn von Entenvögeln. Zeitschr Zellforsch 73: 432–443.

Rutschke, E. 1966b. Über den Bau und die Färbung der Vogelfeder. Monats Ornithol Vivarienk 13: 291–31.

Schmidt, W. J., and H. Ruska. 1961. Elektronenmikroskopische Untersuchung der Pigmentgranula in den schillernden Federstahlen der Taube *Columba trocaz* H. Zeitschr Zellforsch 55: 379–388.

Schmidt, W. J., and H. Ruska. 1962a. Über das schillernde Federmelanin bei *Heliangelus* und *Lophophorus*. Zeitschr Zellforsch 57: 1–36.

Schmidt, W. J., and H. Ruska. 1962b. Tyndallblau-struktur von federn im Elektronenmikroskop. Zeitschr Zellforsch 56: 693–708.

Schorger, A. W. 1966. The Wild Turkey; Its History and Domestication. Norman: University of Oklahoma Press.

Shawkey, M. D., A. M. Estes, L. M. Siefferman, and G. E. Hill. 2003. Nanostructure predicts intraspecific variation in ultraviolet-blue plumage colour. Proc R Soc Lond B 270: 1455–1460.

Siefferman, L. M., and G. E. Hill. 2003. Structural and melanin coloration indicate parental effort and reproductive success in male Eastern Bluebirds. Behav Ecol 14: 855–861.

Srinivasarao, M. 1999. Nano-optics in the biological world: Beetles, butterflies, birds, and moths. Chem Rev 99: 1935–1961.

Strong, R. M. 1902. The development of color in the definitive feather. Bull Mus Comp Zool 40: 147–185.

Tièche, M. 1906. Über benigne Melanome ("Chromatophore") der Haut—"blaue Naevi." Virch Archiv Pathol Anat Physiol 186: 216–229.

van de Hulst, H. C. 1981. Light Scattering by Small Particles. New York: Dover.

Vorobyev, M., and D. Osorio. 1998. Receptor noise as a determinant of colour thresholds. Proc R Soc Lond B 265: 351–358.

Watterson, R. L. 1942. The morphogenesis of down feathers with special reference to the developmental history of melanophores. Physiol Zool 15: 234–259.

Wetmore, A. 1917. Certain secondary sexual characters in the male Ruddy Duck, *Erismatura jamaicensis (Gmelin)*. Proc US Natl Mus 52: 479–482.

Young, A. T. 1982. Rayleigh scattering. Physics Today 35: 42–48.

8

Mechanics of Uncommon Colors: Pterins, Porphyrins, and Psittacofulvins

KEVIN J. MCGRAW

The main pigments found in the integument of birds are the carotenoids (Chapter 5) and the melanins (Chapter 6). However, these are not the pigments responsible for the red color in Eclectus Parrot (*Eclectus roratus*) feathers (Plate 15), the yellow in Great Horned Owl (*Bubo virginianus*) eyes (Plate 14), or the blue color of American Robin (*Turdus migratorius*) eggs (Plate 13). There are at least five other classes of endogenously synthesized colorants that have been identified in bird feathers, bare parts, and other colorful structures, such as egg yolk and shells:

1. Porphyrin pigments and their relatives, which include heme in red blood-filled tissues; porphyrins in the brown feathers of bustards, goatsuckers, and owls; turacin and turacoverdin in turacos; and biliverdin in blue eggshells;
2. Pterin (or pteridine) pigments and their relatives, including the yellow, orange, red, and white pigments in the irises of blackbirds, starlings, pigeons, and owls;
3. Psittacofulvin pigments that give red, orange, and yellow color to parrot feathers;
4. Flavin pigments found in the yellow egg yolks of birds; and
5. Undescribed pigments, such as the fluorescent biochromes that color the yellow plumage of penguins and the downy natal plumage of many birds.

Here I review the physical, chemical, and biological properties of these less common groups of colorants. They are often but not always restricted to certain orders or families of birds, but like carotenoids and melanins, can generate colors that are used as social or sexual signals (see Part I, Volume 2). I estimate that we have confirmed the biochemical or structural nature of colors in only 5% of bird species worldwide, so the above list of bird pigments should not be considered exhaustive, but rather a summary of the information accumulated to date. With the advent of such sophisticated chromatographic techniques as high-performance liquid chromatography (HPLC), this should be an active time for research on animal pigments. The more we search, the more likely we are to discover other unique types of integumentary colorants in birds.

Porphyrins and Their Relatives

Porphyrins color the forests and grasslands green and the blood of animals red (Marks 1969). Both chlorophyll and heme are representatives of this large group of pigments that is united by the presence of nitrogen-containing pyrrole rings (hence the name "pyrrolic zoochromes" also given to porphyrins by Needham 1974). Most porphyrins are tetrapyrrolic (having four aromatic N-heterocyclic rings), but differ in whether the pyrroles form super-rings or open chains, as well as in the presence of various metal ions found at the molecule's core (Figure 8.1). Classes of porphyrins that are used as colorants by birds include the natural porphyrins, metalloporphyrins, and bilins. The natural porphyrins consist mainly of uroporphyrin, coproporphyrin, and protoporphyrin. The three primary members of the colorful metalloporphyrins in birds are heme, turacin, and turacoverdin. The bilins include biliverdin.

Like most pigments, complete conjugated double-bonding gives many of these molecules their vibrant array of colors, from the red turacin in turacos (Plate 12) to the blue biliverdin in egg shells (Plate 13). However, certain large porphyin polymers can act much like melanins do to absorb most light wavelengths and give a rusty or brown appearance, as in brown egg shells (Plate 13) or owl plumage (Plate 14). In some instances, porphyrins are bound to proteins (e.g., heme in hemoglobin), and protein complexing not only influences the color of many porphyrins but also quenches their usual intense red fluorescence (much like the presence of copper and iron will; Fox and Vevers 1960; Needham 1974). As a mixed bag of biochemicals, there are few chemical tests that diagnose all porphyrins. They generally are hydrophobic and can be

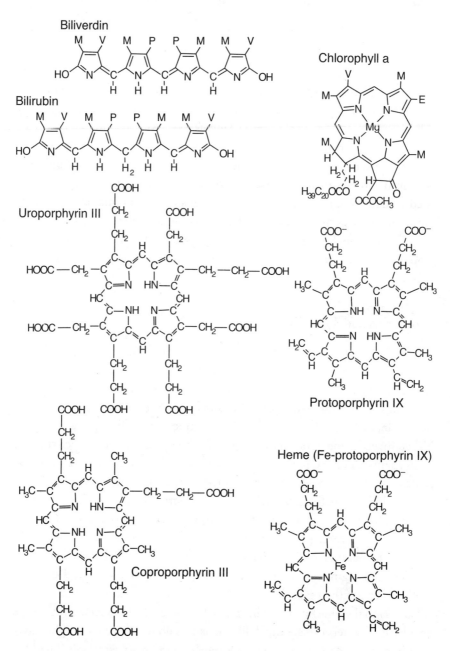

Figure 8.1. Chemical structures of the various porphyrins. Note the variation in how the four highly conjugated pyrrole rings form straight-chained (e.g., the bilins) versus super-ring structures. Magnesium is incorporated into the tetrapyrrole nucleus of chlorophyll, whereas heme contains iron. In biliverdin, bilirubin, and chlorophyll a, M = methyl group, V = vinyl, and E = ethyl. The lactim tautomer of bilirubin is depicted.

Mechanics of Uncommon Colors 357

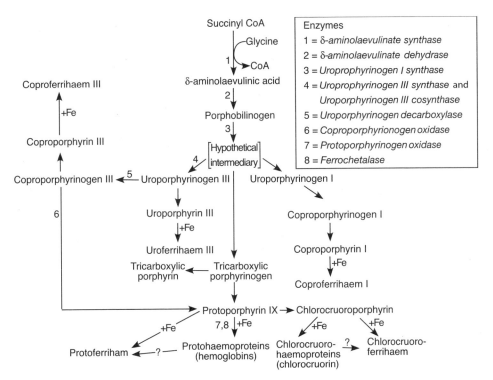

Figure 8.2. Synthesis pathways of the various porphyrins (adapted from Needham 1974), with more recently described enzymes that catalyze each step added. CoA = coenzyme A.

extracted from feathers using such solvents as acetic acid (for feathers; With 1978) or acidified methanol (for eggshells; With 1973).

Birds, like other animals, manufacture their own porphyrins. All porphyrins are ultimately derived from the same two precursors—succinyl Coenzyme A (CoA; an intermediate of the Krebs cycle) and glycine (a nonessential amino acid)—that are condensed to form δ-aminolaevulinic acid (Figure 8.2). This reaction is thought to be the rate-limiting step, and malonate is known to promote porphyrinogenesis by protecting succinyl CoA in the Krebs cycle (Lascalles 1964). The three main natural porphyrins listed above are formed first in the pathway (from porphyrinogens; Thiel 1968; Needham 1974). Hemes and turacins are synthesized directly from natural porphyrins via the simple enzyme-aided addition of iron (Ponka 1999) and copper (Nicholas and Rimington 1954; With 1957), respectively. Finally, bilins result from oxidative degradation of heme, with biliverdin being the initial open-chain catabolite and bilirubin the reduced end-product (Lemberg and Legge 1949; Otterbein et al. 2003).

Although heme is a comparatively widespread red colorant among vertebrates (from axolotl gills to facial blushing in humans; Leary et al. 1992), outside of birds, the only other porphyrins that have been described from the integument of animals are bilins from the scales and fins of certain fishes (e.g., wrasse, sculpin; Fontaine 1941; Yamaguchi et al. 1976) and natural porphyrins from European Hedgehog (*Erinsceus europaeus*) spines (Derrien and Turchini 1925). This rarity is not because other animals cannot synthesize porphyrins, however, as bile pigments and the natural porphyrins (as intermediates to heme and bilins) are common in the internal tissues and fluids of vertebrates (Fox and Vevers 1960). Interestingly, the placenta of dogs is also pigmented with a bilin (uteroverdin), suggesting a reproductive role for these compounds (Lemberg and Barcroft 1932).

Natural Porphyrins

Eggshells

Through the early 1900s, it was largely assumed that the "meat spots" of bird eggs were colored by blood (or hemoglobin; Schwartz et al. 1975). Natural porphyrins were first described from the cuticle layer of brown eggshells in chickens, ducks, geese, and owls over 80 years ago (initially termed "ooporphyrin"; Fischer and Kogl 1923), and since have been reported from the eggs of nearly 100 species spanning just over half of all avian orders, including quail (Poole 1965), gulls (Baird et al. 1975), swifts, and several songbirds (e.g., blackbird, sparrow, shrike; Miksik et al. 1994; reviewed in Kennedy and Vevers 1976; Lang and Wells 1987). Excluding porphyrins, no other brown (or blue; see below) pigments have been described from eggshells.

There is disagreement as to which types of porphyrins are found in eggshells. Three of the larger studies on eggshell porphyrins document protoporphyrin IX (the immediate, iron-lacking precursor of heme; see more below) as the lone brown pigment (Kennedy and Vevers 1976; Miksik et al. 1994, 1996), but others have found appreciable amounts of coproporphyrin, uroporphyrin, and such intermediates as pentacarboxylic porphyrin (Kennedy and Vevers 1973; With 1973; Baird et al. 1975). This variability may be due to different conditions under which birds are raised (e.g., captive versus wild; Miksik et al. 1994, 1996) or to different chemical extraction and analytical techniques.

Female birds synthesize natural porphyrins destined for eggshells in the oviduct (Poole 1966). Lucotte et al. (1975) was among the first to demonstrate that uterine extracts from Japanese Quail (*Coturnix japonica*) are capable of producing porphyrins in vitro when administered the precursor δ-aminolevulinic acid. Pigment granules have been documented from the surface cells of the shell gland (in hens and gulls; Baird et al. 1975) and the ciliated apical cells of the shell gland pouch (in quail; Tamura et al. 1965), but to my knowledge, the precise cellular location of eggshell porphyrin biosynthesis is still unknown. Shell pigmentation in chicken (Nys et al. 1991) and quail hens (Woodard and Mather 1964; Poole 1965) occurs approximately 3 hours before egg laying.

Several researchers have suggested that physiological factors (e.g., laying order, female condition) are likely to affect the extent to which females pigment their eggs with porphyrins (Miksik et al. 1994, 1996; Moreno and Osorno 2003). Sex-steroid hormones such as progesterone are known to affect pigment production in shell glands of Japanese Quail (Soh and Koga 1994). The only data available on the condition-dependence of porphyrin-based eggshell coloration showed that egg color in hens can be affected by behavioral disturbances during egg laying, but this was because stressed females retained eggs in the oviduct longer and deposited an additional layer of white calcium carbonate atop the porphyrins, making the eggs appear paler (Walker and Hughes 1998). Regarding direct mechanisms of porphyrin production and deposition, Schwartz et al. (1975) found that treating chicken hens with drugs that suppress coccidial infections markedly reduced the red color of egg meat spots, perhaps by altering porphyrin-biosynthetic enzymes. Spot patterning is also known to have a strong genetic basis in some species (Collias 1993; Gosler et al. 2000).

Much is known of the adaptive significance of variation in avian eggshell pigment patterns (reviewed in Cott 1940; Underwood and Sealy 2002). Many birds develop cryptically colored eggs to reduce predation risk or, as is the case for brood parasites, pigment their eggshells in a pattern that mimics the host(s) and reduces the likelihood of rejection. However, these adaptations do not explain why birds use these unusual pigments as egg colorants, as opposed to the ubiquitous melanins that generate similar visible colors. Bakken et al. (1978) studied the thermal consequences of porphyrin pigmentation in the eggs of 25 bird species from nine families and found that, unlike most pigments, porphyrins (including the blue bilins described below) do not absorb infrared light, which constitutes nearly half of all incident solar energy. The authors speculated that birds use these pigments in eggshells to prevent their eggs from overheating, which can be a threat to species that nest in hot climates or exposed areas and regularly leave their eggs unattended. Kennedy and Vevers (1976) similarly noted that pure white eggs are limited to species that nest in trees (e.g., parrots), under eaves (e.g., swifts) or foliage (e.g., pigeons), and by species that continually attend them (e.g., penguins). It would be exciting to rigorously test this hypothesis by examining the ecological (e.g., habitat, latitude), behavioral, and phylogenetic correlates of egg pigmentation.

Natural porphyrins can also act as antioxidants and prooxidants (Afonso et al. 1999; McGraw 2005a), which is true of any highly conjugated molecule (e.g., carotenoids) and depends on the concentration at which these pigments are found (e.g., causing oxidative damage only at pharmacological levels; Krinsky 2001). Moreno and Osorno (2003) offered the idea that females may advertise their condition and genetic quality to males by producing these compounds and coating their eggs with them. This hypothesis has intuitive appeal, because there may be maternal costs to sacrificing precursors ultimately destined for heme synthesis or to inducing oxidative stress, as well as maternal benefits if these pigments can serve as cytoprotectants (to mothers, not to embryos, as the pigments are located on the outer cuticle of the shell) at physiological levels. However, we still await studies that test whether natural porphyrin coloration of eggs function as an intraspecific signal of quality per se.

Feathers

Less is known about the natural porphyrins in bird plumage. These fluorescent reddish-brown pigments were first recovered from feathers around the same time as the eggshell porphyrins were identified (Derrien and Turchini 1925; Völker 1938, 1939). They are known to occur in at least 13 different orders of birds, but most notably from owls (Order Strigiformes; Plate 14), goatsuckers (Order Caprimulgiformes), and bustards (Order Gruiformes; Plate 12; With 1967, 1978). Unlike eggshells, for which protoporphyrin is the major if not the only porphyrin, With (1978) found no evidence of protoporphyrin in feathers and instead documented a predominance of coproporphyrin, along with smaller amounts of uroporphyrin and other intermediates, in the down of a Tawny Owl (*Strix aluco*) and a Great Bustard (*Otis tarda*). To date, there have been no investigations into how or why eggshell and feather tissues vary so widely in porphyrin composition.

Völker's collaborator, Thiel (1968), conducted the lone study on the nature of feather-porphyrin synthesis. His work indicated that these colorants in bustards are derived from porphyrin precursors in erythrocytes. Because these compounds are naturally synthesized by the liver and other metabolically active tissues in the body (Battersby 2000), it is conceivable that these pigments are delivered to and taken up by maturing feathers for pigmentation. Until cellular studies of feather follicles are conducted, however, we cannot rule out the possibility that these biochromes are made at the developing, highly vascularized feather germs themselves.

No biochemical investigations have been undertaken on these pigments for over 25 years, and even the most recent study used thin-layer chromatography on feathers from one owl and one bustard (With 1978). It seems reasonable to expect that we will learn much more about the chemical nature, fluorescent properties, taxonomic distribution, or visual-signaling functions of feather porphyrins from larger-scale analyses using such modern techniques as HPLC.

Moreover, no one has yet speculated on the function of natural porphyrins in feathers. No age or sexual dimorphisms have been reported. It is curious, however, that porphyrins are common in the downy feathers of several nocturnal species. These species, adapted to be active during cooler times of day, may be faced with thermoregulatory challenges that, like those facing eggshells, are best tackled with porphyrins.

Metalloporphyrins

Zinc-, magnesium-, manganese-, iron-, and copper-porphyrin chelates are abundant in living things (Maines 1997). Iron complexes, such as heme (iron-protoporphyrin IX), are fundamental components of cytochromes and globins, and magnesium-porphyrin conjugates, such as chlorophyll, function as photoreceptors for photosynthesis. Only occasionally do metalloporphyrins give coloration to animals like birds, however. Where birds circulate blood at exposed tissues (e.g., combs, wattles, mouthparts), the chromoprotein hemoglobin is often the source of red porphyrin pigmentation (Lucas and Stettenheim 1972). Even less common are the red copper-porphyrin complex and the green pigment that color the feathers of turacos (Plate 12; Moreau 1958).

Hemoglobin

The red combs, wattles, and caruncles of Domestic Chickens (*Gallus gallus*; Plate 26) and Wild Turkeys (*Meleagris gallopavo*; Plate 32; Laruelle et al. 1951) are classic examples of highly vascularized bare parts that are colored by hemoglobin-rich blood. Surprisingly, these are among the few red-colored fleshy tissues that have been examined biochemically and microstructurally for their pigmentary basis (see Prum and Torres 2003 for the structural basis of other avian flesh colors). In most instances, researchers have simply assumed, based on the rapid brightening of such red-colored areas as the mouthflush of begging nestling birds, that changes in blood flow regulate color (e.g., Kilner 1997).

Not all red bare parts in birds are colored strictly by hemoglobin, however. Red irises can harbor colorful pterin pigments (discussed in more detail below), and red beaks and legs commonly are enriched with carotenoids (Czeczuga 1980; Chapter 5). In fact, contrary to the aforementioned assumptions about hemoglobin, other researchers have presumed that reddish skin or bare parts are based, at least in part, on carotenoids (Götmark and Ahlström 1997; von Schantz et al. 1999; Saino et al. 2000, 2003). In certain instances, such as for the comb of the Capercaillie (*Tetrao urogallus*; Egeland et al. 1993) and the wattles of Ring-necked Pheasants (*Phasianus colchicus*; Brockmann and Völker 1934), there is biochemical evidence that red carotenoids, such as astaxanthin, are present in fleshy tissues. Yet in other tissues, the contributions of carotenoids versus hemoglobin to red color are less apparent, particularly for nestling mouthparts. These tissues require a closer look.

The first difficulty with identifying the chemical identity of mouthpart colors is that several areas of the nestling mouth are vibrantly colored, including the gape, flanges, tongue, and horny sheath (Plate 13; Swynnerton 1916; Ficken 1965) and there may be very different mechanisms for pigmentation in these regions. For example, carotenoids may bestow color on yellow hard parts, whereas hemoglobin may pigment the red soft parts (Wetherbee 1961; Hunt et al. 2003). The second problem with determining the pigmentary basis of mouth colors is that yellow dietary carotenoids (rarely red) are generally transported through blood (McGraw 2004), so it is difficult to discern the extent to which circulating yellow carotenoids contribute to the red color of vascularized regions. Saino et al. (2000) studied gape coloration in nestling Barn Swallows (*Hirundo rustica*) and showed that birds supplemented with a yellow carotenoid (lutein) maintained a redder gape when faced with an immune challenge than did unsupplemented chicks (Figure 8.3). In fact, carotenoids could have contributed directly to the color of blood in the mouth, boosted the health and vigor of birds, and increased blood flow to the gape. Determining the pigmentary (as well as structural; Prum and Torres 2003) basis of color in avian fleshy parts should be a priority for biologists interested in understanding the costs and information content of these variable signals.

Mechanics of Uncommon Colors 363

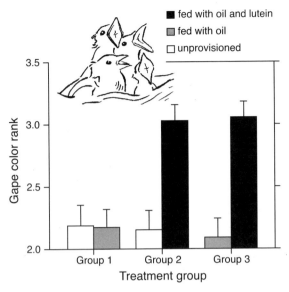

Figure 8.3. Effect (mean ± standard error) of dietary carotenoid supplementation on mouth-flush coloration in nestling Barn Swallows. Nestlings were given a daily lutein dose every other day between hatch and day 10. Lutein was dissolved in a peanut-oil vehicle; sham-control ("fed with oil") birds were given equal doses of oil, whereas "unprovisioned" birds received no supplement but were otherwise similarly handled by the researchers. All birds were subjected to a humoral immune challenge (with sheep red blood cells). Higher gape-color ranks indicate redder mouths. Siblings were randomly assigned to one of the three treatment groups, so data are presented to illustrate pairwise differences in gape color between nestmates. The gape color ranks were not significantly different for Group 1 ($z = 0.4$, $n = 40$, $P > 0.05$; Wilcoxon matched pairs, signed-ranks test), but both of the other comparisons yielded significant differences (Group 2: $z = 3.6$, $n = 43$, $P = 0.0004$; Group 3: $z = 4.0$, $n = 42$, $P = 0.0001$). Redrawn from Saino et al. (2000).

It was recently discovered in several passerines that red nestling mouths (Hunt et al. 2003) and skin (Jourdie et al. 2004) reflect strongly in the ultraviolet (UV). Based on the UV-visible absorbance properties of hemoglobin, however, which indicate strong absorbance below 500 nm (Takatani and Graham 1987; Suzuki 1998), these pigments should not be responsible for such coloration. Instead, I suspect a structural origin to this reflectance, as has been described for other UV colors in birds (Chapter 7).

There are a few instances in which hemoglobin has also been characterized from the eyes of birds (Plate 14). Bronzed Cowbirds (*Molothrus aeneus*), for example, have vivid red eyes that do not contain either red carotenoid or pterin

Table 8.1. Distribution of Two Copper Porphyrin Pigments, Turacin and Turacoverdin, from the Feathers of Turacos

Species	Turacin?	Turacoverdin?	Habitat	Other plumage colors
Ross's Turaco (*Musophaga rossae*)	Crest	None	Gallery forest	
Violet Turaco (*Musophaga violacea*)	Crest	None	Gallery forest	
Ruwenzori Turaco (*Musophaga johnstoni*)	Wings	Head, neck, breast	Gallery forest	
Purple-crested Turaco (*Musophaga porphyreolopha*)	Wings	Breast (reduced)	Gallery forest	
White-crested Turaco (*Tauraco leucolophus*)	Wings	Breast (reduced)	Gallery forest	
Livingstone's Turaco (*Tauraco livingstonii*)	Wings	Head, neck, breast	Wet forest	
Fischer's Turaco (*Tauraco fischeri*)	Crest	Head, neck, breast	Wet forest	
White-cheeked Turaco (*Tauraco leucotis*)	Wings	Head, neck, breast	Wet forest	
Red-crested turaco (*Tauraco erythrolophus*)	Crest	Head, neck, breast	Wet forest	
Bannerman's Turaco (*Tauraco bannermani*)	?	?	?	
Guinea Turaco (*Tauraco persa*)	Wings	Head, neck, breast	Wet forest	
Schalow's Turaco (*Tauraco schalowi*)	Wings	Head, neck, breast	Wet forest	
Knysna Turaco (*Tauraco corythaix*)	Wings	Head, neck, breast	Wet forest	
Black-billed Turaco (*Tauraco schuettii*)	Wings	Head, neck, breast	Wet forest	
Yellow-billed Turaco (*Tauraco macrorhynchus*)	Wings	Head, neck, breast	Wet forest	
Hartlaub's Turaco (*Tauraco hartlaubi*)	Wings	Head, neck, breast	Wet forest	
Prince Ruspoli's Turaco (*Tauraco ruspolii*)	Wings	Head, neck, breast	Wet forest	
Great Blue Turaco (*Corythaeola cristata*)	Vent	Breast stripe (reduced)	Wet forest	
Grey Go-away Bird (*Corythaixoides concolor*)	None	None	Acacia woodland	Pure gray body
White-bellied Go-away Bird (*Corythaixoides leucogaster*)	None	None	Acacia woodland	Gray dorsum, white belly
Bare-faced Go-away Bird (*Corythaixoides personata*)	None	None	Acacia woodland	Black, white, and gray
Western Gray Plantain-eater (*Crinifer piscator*)	None	None	Savannah	Slate gray
Eastern Gray Plantain-eater (*Crinifer zonurus*)	None	None	Savannah	Slate gray

Source: Adapted from Moreau (1958).

pigments (Hudon and Muir 1996). Instead, they have enlarged venous sinuses on the anterior surface of the iris that are blood-engorged (Oliphant 1988). Both male and female Bronzed Cowbirds have red irises (although sex difference in color has never been systematically quantified), but the iris in juveniles is dull brown. The commonality and use of hemoglobin-based eye color as a visual signal in birds remains to be determined.

Few researchers have considered the physiological costs and benefits of hemoglobin-based coloration in birds (see Part I, Volume 2, for behavioral advantages). Heme and hemoglobin, like the porphyrins mentioned above, are also considered to have potent antioxidant activity (Giulivi and Davies 1990; Otterbein et al. 2003), so animals may use them to signal their level of oxidative protection. Hemoglobin also plays the more traditional and critical role in gas transport. Oxygen availability and uptake are important determinants of hemoglobin synthesis in vertebrates (Richmond et al. 1951; Hammel and Bessman 1965)—birds that develop the reddest colors may be revealing their metabolic efficiency. Begging in nestlings (Jurisevic et al. 1999) and competitive behavior in birds (Hammond et al. 2000) cause a substantial increase in metabolic rate, so there may be very good reasons why birds use hemoglobin-based colors as begging and dominance signals (e.g., the comb of male Red Junglefowl [*Gallus gallus*]; Plate 26; Ligon et al. 1990).

Turacin and Turacoverdin

In only one order and family of birds—the turacos (Order Musophagiformes, Family Musophagidae) from Africa—have metalloporphyrins been unequivocally described from plumage (Plate 12). Oddly enough, these pigments were the first ever chemically characterized colorants of feathers (Church 1870). There is also recent evidence from spectrophotometric and solubility tests that green pigments in the feathers of Northern Jacanas (*Jacana spinosa*) and two partridge species (Blood Pheasant [*Ithaginis cruentus*] and Crested Wood-partridge [*Rollulus roulroul*]) are closely related, if not identical to, turacoverdin (Dyck 1992).

Of the two types of feather metalloporphyrins, the red pigment, turacin, found in the wing or crest feathers of all but five of the 23 turaco species (Table 8.1), is best characterized. We still only have basic physicochemical information about turacoverdin—the green pigment found in the ventral plumage (e.g., head, neck, breast) of these same taxa (with the exception of two species; Table 8.1). Those species that deposit turacin in feathers also deposit turacoverdin. In some species (e.g., Ross's Turaco [*Musophaga rossae*]), black melanins are co-deposited with these porphryins and mask their presence (Moreau 1958). All green and red feathers in this group are believed to be colored by these pigments. Green pigments are extremely rare in birds (see the section on undescribed pigments, below); in most instances, green color is generated either by a purely structural phenomenon (e.g., the iridescent head of male Mallards [*Anas platyrhynchos;* Plate 15, Volume 2]) or by a combined yellow pigmentary (e.g., carotenoids) and blue structural mechanism (Dyck 1966, 1971, 1978, 1987).

Turacin, formerly known as "turacoporphyrin" (Church 1892), is a copperporphyrin conjugate (With 1957). Despite early claims that it was composed of uroporphyrin I (Fischer and Hilger 1923, 1924), it contains almost pure uroporphyrin III (Rimington 1939; Nicholas and Rimington 1954), which is the only known natural occurrence of this compound. There was speculation that red-colored cuckoos (close relatives of the turacos, once grouped in the same order) might also pigment themselves with turacin, but this claim was refuted early on (Church 1913). Some accounts have suggested that pigments in mussel shells (genus *Pteria*) are identical to turacin, but there is no published evidence to this effect (Moreau 1958).

Some of turacin's other chemical properties are still unresolved. Needham (1974) asserted that this pigment is soluble in pure water, which is remarkable, if birds intend to preserve their feather coloration (and not lose it to rainwater, bathing, or preening) for sexual or camouflage purposes throughout the year. He used this line of thinking to suggest that feather turacin may serve as a direct form of aposematic coloration—a direct line of chemical defense that is distasteful to predators. However, Moreau (1958) argues that turacin can only be leached from feathers under alkaline conditions.

Kruckenberg (1882) and later Moreau (1958) and Dyck (1992) undertook some of the only investigations done to date into the chemical nature of turacoverdin. It was shown to be less soluble in basic solutions than is turacin (Moreau 1958) and was initially thought to contain little, if any, copper (Kruckenberg 1882). Later spectroscopic analyses, however, demonstrated high copper (and low iron) content in green feathers from Knysna (*Tauraco corythaix;* Moreau 1958) and Schalow's Turacos (*T. schalowi;* Dyck 1992). Actually, two turacoverdin pigments that differ slightly in polarity may together be responsible for green color in turacos (Moreau 1958). Extracted turacin has been noted to take on a green hue when exposed to light, oxygen, or strong bases (Church 1870), which was used by one author to suggest that turacoverdin is an oxidized metabolite of turacin (Fox 1953). Dyck (1992) went on to support this chemical relationship between these two pigments based on spectral properties, the co-occurrence of the two pigments (turacin never occurs in a species without turacoverdin, and the two pigments can be intermingled in some plumage patches), that both pigments contain copper, and their similar microscopic arrangements in feather cells. The absorbance curve of turacoverdin peaks at blue wavelengths and in the long-wave range above yellow, thus giving its green appearance; there appears to be little UV reflectance by this pigment (Dyck 1992).

To manufacture turacin and turacoverdin for plumage, turacos must accumulate and use large quantities of copper. As primarily arboreal species, they get this trace-mineral from a diet that is rich in fruits, flowers, buds, and other plant parts. Based on pigment concentrations in food and feathers, Church (1913) and Moreau (1958) estimated that 3 months' worth of fruit intake go into producing the pigment present in the freshly grown plumage of *T. corythaix*. This dietary link to plumage color then raises the questions of (1) whether turaco foods are especially enriched in copper compared to other birds; (2) whether turacos are particularly effective at extracting copper from foods; and (3) whether the turaco species that lack red and green pigmentation have a diet that is comparatively deficient in copper, absorb less from the diet, or lack the enzymes that facilitate copper incorporation into the uroporphyrin super-ring.

There have been no formal tests of the functional significance of turacin and turacoverdin coloration. Moreau (1958) made the observation that those turaco species inhabiting forests are more likely to be green in color (Table 8.1), potentially affording themselves concealment from predators. However, this claim requires a rigorous phylogenetic and biochemical follow-up study. There is no evidence for sexual dichromatism in this order, but again, this is based on human perception, and no spectrophotometric or biochemical studies have been conducted to objectively test for sex differences in coloration. Turacos may, in fact, use these colors as sexual or social advertisements, for we now know plenty of monochromatic species that use bright colors as quality indicators (e.g., auklets, motmots). There may even be physiological and biochemical benefits to forming these colorful chelates; copper, like porphyrins, can be physiologically damaging to birds when accumulated at high concentrations (Klasing 1998). As proposed for metals and melanins (McGraw 2003), turacos might detoxify the high copper levels ingested with porphyrins (Keilin and McCosker 1961) and advertise the protection they have afforded themselves by depositing turacin in feathers. Why all turacos seem to pigment the same regions of the wing feathers (Moreau 1958) with turacin is also mysterious.

Bilins

Bustards, ibises, gulls, pheasants, and thrushes are among the many groups of birds that lay blue or blue-green eggs. These eggshell colors attracted the attention of some of the earliest animal biochemists (Wicke 1858; Sorby 1875). Although Kruckenberg (1883) believed the shell pigment was identical to biliverdin, it was not until Lemberg's (1934) purification of crystalline biliverdin that its identity was confirmed. These pigments are named for their presence and pigmenting action in vertebrate bile fluid. Because of their affinity for transition metals, it is no surprise that these and other porphyrins are found closely associated with calcium in the shell cuticle (Kennedy and Vevers 1973).

Kennedy and Vevers (1973) first characterized biliverdin IXα from the blue eggs of a strain of Domestic Chicken (Araucano fowl) and later documented it from the blue, blue-green, and olive-colored eggshells of more than 30 avian species (Kennedy and Vevers 1976). The zinc chelate of this pigment is also present in half of the species with biliverdin-colored eggs. In most birds, blue color is accompanied by brown spotting, so these shells contain protoporphyrin as well. There is one rare instance in which a bilin, known as "mesobiliviolin," was detected along with biliverdin in the eggshell of a Black-bellied Bustard (*Lissotis melanogaster*). To date, the only avian eggshells found to contain pure biliverdin are those of American Robins (Plate 13; Kennedy and Vevers 1976).

The bilins are a large class of readily oxidized and reduced intermediates formed from heme degradation. The linear tetrapyrrole biliverdin (Figure 8.1) is specifically the result of heme oxidation by heme oxidase, which opens the heme super-ring (Figure 8.2; Bauer and Bauer 2000). The straight-chained, highly conjugated nature of bilins gives them their bright coloration. Unlike their natural porphyrin counterparts, they do not have strong fluorescent activity unless they are conjugated with metals (Needham 1974). Biliverdin present in bile is formed primarily in the liver (Maines 1988), but it seems that at least some female birds also synthesize this pigment peripherally. Poole (1965) detected biliverdin along with other natural porphyrins in the oviduct of laying Japanese Quail.

Several hypotheses, including both mimicry and camouflage, have been advanced over the past century to explain the adaptive significance of blue bird eggs (Underwood and Sealy 2002). Many songbirds that build open-cup nests in trees, for example, lay blue eggs that might appear to a predator from above as holes in the green vegetation (Lack 1958). In one of the only empirical tests of this idea, however, darkened Song Thrush (*Turdus philomelos*) eggs were not taken at higher rates by predators than were lightened eggs (Götmark 1992; Weidinger 2001). Some ground-nesting birds also have brilliant blue eggs (e.g., the Great Tinamou [*Tinamus major*]), and in this communally nesting species, it is possible that blue colors function to attract conspecifics to nests (P. Brennan, pers. comm.). These reasons would not necessarily account for the intraspecific variation in egg color that we see in several birds, however. Dramatic egg-color dimorphisms have even been reported, in which whole clutches can be composed of either all white eggs or all blue eggs in the Crow Tit (*Paradoxornis webbiana*; Kim et al. 1995). Females may uniquely color their eggs to improve discrimination of their own eggs (e.g., Jackson 1992; Lyon 2003) or to advertise their condition and genetic worth to males (reviewed by Moreno and Osorno 2003; see above). This second idea may be especially important in biliverdin pigmentation systems, because heme concentration in blood limits the rate at which bilins are produced (Elbirt and Bonkovsky 1999) and because biliverdin can act as a valuable antioxidant (Kaur et al. 2003). Thus there may be reliable molecular costs and benefits that allow these colors to be honest signals of quality. To this end, Moreno et al. (2004) showed in a recent study that male Pied Flycatchers (*Ficedula hypoleuca*; Plate 27) provisioned nestlings more when their mates laid bluer eggs (Figure 8.4). This behavior suggests that males use egg color to preferentially allocate resources to high-quality mates.

Biliverdin is photolabile and can be destroyed by light exposure (Needham 1974). Blue egg colors often fade considerably over time when stored in museums (Bakken et al. 1978): caution should be exercised when relying on stored material for analyses. Biliverdin absorbs maximally in the UV range of the spectrum (λ_{max} = 372–376 nm; Kennedy and Vevers 1973; Needham 1974), so there should be few, if any, UV color signals generated by this pigment on eggshells. This idea is supported by a recent comparative study showing that visible, but not UV, light reflectance from blue passerine eggs was significantly associated with parental investment and mating system (Soler et al. 2005; but see Cherry and Bennett 2001 for the importance of UV in parasitic brown eggs).

No other blue pigments have been documented in birds. Other blues found in feathers and bare parts are structural colors (Chapter 7).

Pterins and Their Relatives

The pterins (or pteridines) are a group of nitrogenous, heterocyclic compounds that are catabolic byproducts of purines—the nucleotides, such as adenine and

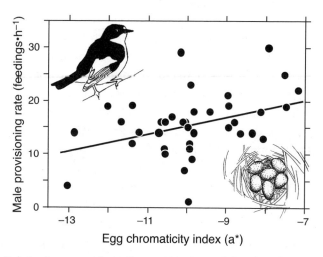

Figure 8.4. Relation between male nestling provisioning and the color of his social mate's eggs in paired Pied Flycatchers. The measure of color (a*), from CIELAB color-space dimensions, is green chroma; higher scores indicate deeper green colors. Redrawn from Moreno et al. (2004).

guanine, that are the fundamental building blocks of DNA and RNA (Hurst 1980). The two-pyrimidine–ring backbone (called the "pteridine skeleton") is consistent across pterins, and these pigments vary primarily in the degree of linear or cyclic substitution (Figure 8.5), which confers their characteristic hue. Pterins are usually brightly colored red, orange, and yellow pigments, but some can be colorless (Fox 1976). These compounds were first described as conspicuous integumentary colorants in the eyes and wings of various insects and butterflies (reviewed in Pfleiderer 1994); in fact, the name "pterin" was originally derived from their prevalence in lepidopterins. Among vertebrates, they occur in the skin and eyes of many fishes (e.g., Henze et al. 1977; Grether et al. 2001), amphibians (e.g., Obika and Bagnara 1963; Bagnara and Obika 1965; Frost and Bagnara 1977), and reptiles (e.g., in *Anolis* dewlaps; Ortiz et al. 1962; Macedonia et al. 2000). In birds, they have only been confirmed as colorants of the eye (Oliphant and Hudon 1993). Hawks, doves, owls, and woodpeckers are some of the common avian groups with colorful, pterin-based eyes (Oliphant 1987a; Plate 14). The purine precursors of pterins, especially guanine, also confer the shiny white and silver colors in the eyes of such birds as starlings (see the section below).

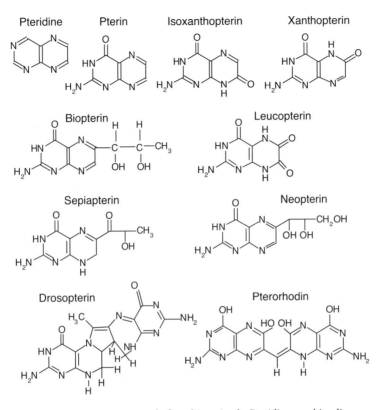

Figure 8.5. Pterin structures commonly found in animals. Pteridines are bicyclic compounds made up of a pyrimidine and a pyrazine ring. The pteridine backbone is common to pterins; their distinguishing feature from other pteridines and their derivatives is the 2-amino-4-oxo structure.

Pterins

Pterins are chemically and cellularly distinct from other red, orange, and yellow pigments in birds in that they are: (1) soluble in mild acids (e.g., acidified ethanol) and bases (e.g., NaOH, KOH), (2) fluorescent (imparting a yellow or blue color) under UV light (Needham 1974), (3) predominantly absorbant in the UV range (Figure 8.6), and (4) exist in crystalline, granular form in pigment-containing cells known as "chromatophores" (Oliphant et al. 1992). Poikilotherms also house pterins in chromatophores that are scattered across the dermis and allow the animals to rapidly change color (Bagnara and Hadley

1973). However, birds have largely lost these pigment-bearing cells in the integument (with the exception of dermal melanocytes; Chapter 6), and chromatophores are found in dense populations only in the avian iris—the specific tissue in the eye where pterins accumulate (Oliphant et al. 1992).

Colorful pterins come in only a dozen or so forms in vertebrates, and from bird irises, only one of these has been unequivocally identified—xanthopterin, the yellow pigment found in *Colias* butterfly wings—and only from a handful of species, including the Great Horned Owl (Plate 14; Oliphant 1981), the European Starling (*Sturnus vulgaris*; Plate 23; Hudon and Oliphant 1995), and a half dozen icterids (Hudon and Muir 1996). The colorless pterins hypoxanthine and leucopterin have also been detected in these birds. There is evidence that red eyes contain red pterins (perhaps pterorhodin in Red-eyed Vireos [*Vireo olivaceus*; Plate 28, Volume 2]; Hudon and Muir 1996), but more research is needed on characterizing the types of pterins found in avian irises (for HPLC methods, see Hudon and Muir 1996).

The chemical nature of avian eye colors confused biologists throughout much of the twentieth century. Early studies in chickens showed that their yellow eye coloration is due to the presence of xanthophyll carotenoids (Hollander and Owen 1939a). For the next several decades, it was then assumed that carotenoids were responsible for brightly colored eyes in birds (Oliphant 1987a). However, one of the first studies of eye pigmentation in birds showed that the granular yellow substances in the eyes of Rock Pigeons (*Columba livia*) were not of a carotenoid origin (Bond 1920; also see Hollander and Owen 1939b). It was not until Oehme's (1969) seminal studies of eye color in nearly 150 bird species from 18 different families that the presence of colorful pterins was confirmed from birds with red, orange, and yellow eyes. Far fewer birds were found to use carotenoids as eye colorants (e.g., diving ducks, the Great Blue Heron [*Ardea herodias*]) than pterins (more than 170 species now confirmed; Oliphant 1987a). Outside of pterins and their relatives, eye color can also be due to melanins, in the case of black or brown irises (e.g., Red-winged Blackbirds [*Agelaius phoeniceus*; Plate 11, Volume 2]), and in rarer cases, due to hemoglobin (Oliphant 1988; see above).

Most of the attention to the colorful pterins in birds has been at the ultrastructural level, for which authors have described the arrangement and composition of organelles and pigment crystals from the irises of different birds (reviewed in Oliphant and Hudon 1993). Crystalline pterin granules are stored in pigment organelles (known as "pterinosomes") that occur in the stromal cells of the anterior epithelium of the iris (Oliphant 1988). These pigments

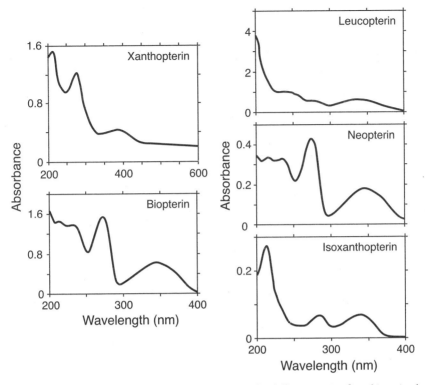

Figure 8.6. Characteristic UV light-absorbance spectra for different pterins found in animal integuments. Absorbance is expressed in arbitrary units that depend on the solvent used. Redrawn from Win (2000).

are arranged as bundles of rod-shaped angular crystals, generating a highly reflective layer akin to the iridophores of fish, amphibians, and reptiles (Ferris 1968; Tillotson and Oliphant 1990). These crystals have been termed "reflecting platelets" or "reflecting xanthophores" (Oliphant 1981, 1987b).

In contrast, much less work has been devoted to understanding the broader-scale adaptive functions, environmental or physiological mechanisms, and phylogenetic patterns of pterin coloration. Sex- and age-specific variability in pterin-based eye coloration has been noted in a number of species, which hints at the social- or sexual-signaling information potentially contained in these traits (e.g., Snyder and Snyder 1974; Scholten 1999; Bortolotti et al. 2003). Female Lesser Scaup (*Aythya affinis*), for example, acquire yellower eyes as they age (Trauger 1974). This age-related pattern is common for pteridine-colored animals as distantly related to birds as fruit flies (Tomic-Carruthers et al. 1996). Eye color often changes as birds molt into their breeding plumage, even for species that take more than a year to complete the transition (e.g., Bald Eagles [*Haliaeetus leucocephalus*], Herring Gulls [*Larus argentatus*]). European Starling sexes can be distinguished by a ring of yellow pigmentation in the eyes of females but not in males. In the only quantitative study of this sort, adult male Brewer's Blackbirds (*Euphagus cyanocephalus*) had more colorful yellow eyes than did females or juveniles of either sex, and concordantly had more and larger pigment organelles in their irides (Hudon and Muir 1996). An important point in all this work, however, is that we still have not used UV-VIS reflectance spectrophotometry to quantify pterin-based eye color in birds. Given the limitations of human vision and the interesting fluorescence characteristics of these molecules, it is quite possible that we are underestimating the types and amounts of color variation in avian irises. Spectral analyses should be incorporated into future behavioral tests of pterin signaling.

If iris colors are used as signals, what might control their production? As indicated above, pterins are endogenously synthesized from purines, as well as from other intermediate products of adenine or guanine metabolism (at least in amphibians; Stackhouse 1966; Figure 8.7). Perhaps enzyme-driven biosynthesis of pterins or their substrates is energy demanding. Might these precursors be costly to shunt away from other vital processes, including cell proliferation and neurotransmitter synthesis (Blakley 1969)? Pterins also act as antioxidants in humans (e.g., Gieseg et al. 1995) and amphibians (Zvetkova 1999) and serve immunostimulatory roles so vital that pterin levels in urine are used as biomarkers of patient health (reviewed in McGraw 2005a), so there is the opportunity for birds to signal their health state directly with their pterin-based iris colors. Because carotenoids are now known to serve as valuable photoprotectants in the avian eye (e.g., in quail; Thomson et al. 2002), pterins may yet serve this function too. Finally, there appear to be interesting phylogenetic distributions and clusters of birds that deposit pterins in their eyes. It would seem valuable to investigate the ecological and evolutionary correlates of pterin coloration systems across birds.

Purines

Many of the white- or silver-colored displays of the skin, scales, and eyes of lower vertebrates contain purines (Bagnara et al. 1988). This includes the "eye-shine" of fishes, generated by guanine-based retinal layers that constitute the

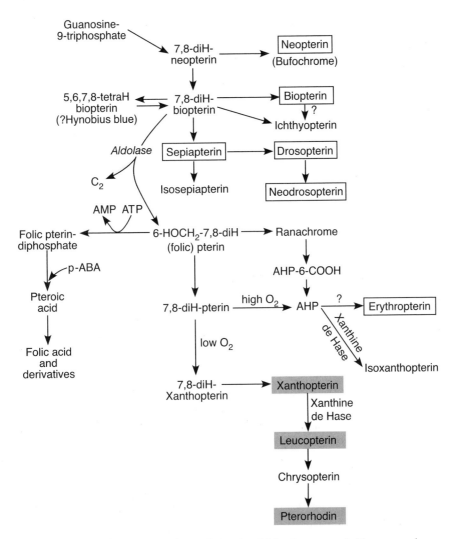

Figure 8.7. Proposed synthetic pathways for pterins. Molecules surrounded by an open box are those that have been reported as integumentary colorants in animals other than birds; grayed boxes denote those that have been documented from colorful avian tissues. Adapted from Needham (1974).

tapetum lucidum (Takei and Somiya 2002), but not those of mammals, which have similar cell ultrastructure but typically contain riboflavin and zinc (Elliot and Futterman 1963). Birds incorporate two main purine bases—guanine and hypoxanthine—into their irises to confer these same hues (Oliphant

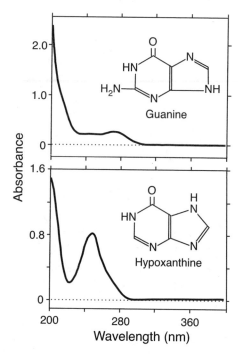

Figure 8.8. Chemical structures and absorbance spectra of two common purines reported from the avian eye—guanine and hypoxanthine. Purines are characterized by the fusion of a six- and a five-member nitrogen-containing ring, in contrast to the bicyclic six-membered pteridine rings. Unlike guanine, hypoxanthine is not a building block of DNA, but is instead an intermediate formed during adenosine metabolism. Absorbance is expressed in arbitrary units that depend on the solvent used.

1987a; Figure 8.8; Plate 14). Guanine is the more common and concentrated of the two (e.g., see data from icterids in Hudon and Muir 1996).

Purines, like pterins, are deposited into chromatophores in the avian iris, exist in crystalline structure, and serve as reflective platelets (Bagnara et al. 1988). Their cell and organelle ultrastructure are remarkably similar to the iridophores of lower vertebrates (Bagnara and Hadley 1973). They have, in fact, been termed "iridorphores" in certain species (e.g., Inca dove [*Columbina inca*]; Ferris and Bagnara 1972), because the rod-shaped crystals exhibit an iridescent sheen. Where iridescence is lacking, however, they are called "leucophores" (e.g., in the pigeon iris; Oliphant 1987b). Because of the extent to which light interference is responsible for the color that purines impart, these are often referred to as "white structural colors" (Needham 1974; Chapter 7).

They scatter a broad range of visible-light wavelengths and thus appear white. Apigmented white feathers in birds, also considered to be structural colors, seem to be generated by a different mechanism of incoherent scattering (Chapter 7), although no one has ever tested for the presence of guanine in white bird feathers. The white, basal portions of feathers in certain species (e.g., penguins) fluoresce strongly under UV light (pers. obs.), which suggests the presence of a white or colorless compound, such as a pterin or purine.

In some instances, purines co-exist with pterins in reflecting platelets, brightening their usual red, orange, or yellow appearance (Hudon and Muir 1996). In fact, guanine is often the major component of colorful red and yellow eyes, found in far greater concentrations than pterins in icterids (Hudon and Muir 1996). The diagnostic chemical feature that sets purines apart from pterins is that they absorb only very short wavelengths of light (for guanine, λ_{max} = 246 and 273 nm; for hypoxanthine, λ_{max} = 249 nm); pterins invariably also absorb light above 300 nm (Needham 1974).

There are few accounts or considerations of the mechanisms and functions of iris purines. Several species are candidates for purine-based signaling, however. Adult African Pied Starlings (*Spreo bicolor*), for example, have a white iris, whereas the eyes of juveniles are brown (Sweijd and Craig 1991). Sex and age differences in guanine concentrations have also been reported in Brewer's Blackbird, with males having higher levels than females, and adults having more than juveniles (Hudon and Muir 1996). These differences did not hold for all purines (e.g., hypoxanthine), however, suggesting a specific guanine-based pigmentation strategy in these birds.

Flavins

As nitrogen-based, fluorescent, heterocyclic compounds, flavins (also termed "isoalloxazines" or "benzopteridines") are sister chemicals to pterins. A flavin is composed of a pterin backbone that is condensed with a benzene ring (hence their designation as benzopteridines; Figure 8.9). Riboflavin and its derivatives (flavin mononucleotide, and flavin adenine dinucleotide) are the most widespread flavins in nature (Rivlin 1975). Riboflavin, a yellow molecule due to its strong absorption of short-wave light (discussed in more detail below), is the primary flavin that gives color to wild animals. Certain fishes (e.g., Atlantic Sturgeon [*Acipenser oxyrhinchus*]; Courts 1960), toads (e.g., Japanese Common Toad [*Bufo bufo japonicus*]; Obika and Negishi 1972), and salamanders (e.g., Mexican Axolotl [*Ambystoma mexicanum*]; Frost et al. 1984;

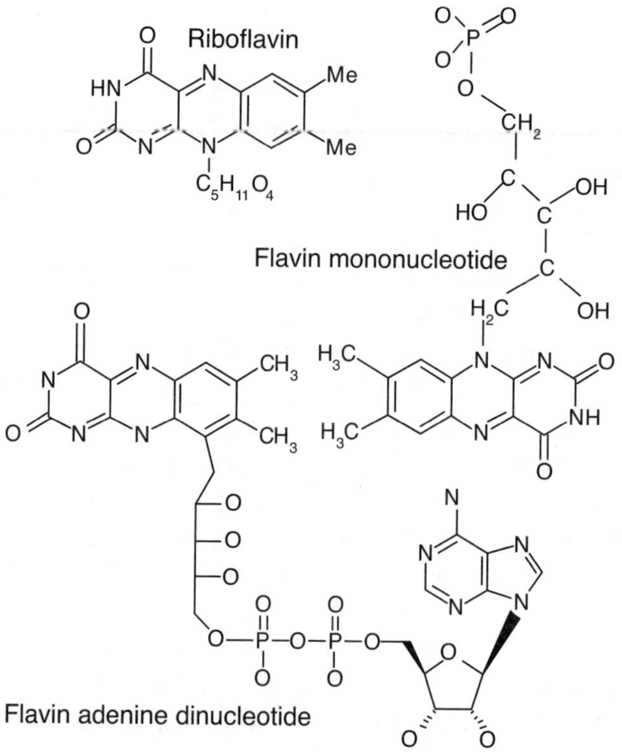

Figure 8.9. Chemical structure of flavins. Note the fully conjugated benzene rings—a unique characteristic for this class of biochromes.

Bukowski et al. 1990) use riboflavin to give patches of skin a yellow or olive (when co-deposited with melanin) tint. Green snake skin (e.g., in Green Mambas [*Dendroaspis viridis*]; Blair and Graham 1954; Blair 1957; Villela and Thein 1967) also contains high concentrations of riboflavin. In birds, however, this pigment is only known to occur along with carotenoids in egg yolks and albumen (Gliszczynska-Swiglo and Koziolowa 2000). To my knowledge, riboflavin has only been quantified from chicken eggs, although as a vitamin (discussed in more detail below), there is reason to believe that riboflavin is present in the yolks of all birds.

Pure riboflavin in aqueous solution has a complex absorbance spectrum, with four distinct peaks at 220, 266, 375, and 447 nm. Because of its fluorescent nature, riboflavin also emits light at longer wavelengths (~530 nm). It is highly light sensitive in solution and easily bleaches when analyzed under normal laboratory lighting conditions. It is quite soluble, although unstable, in alkaline solutions. Preferred extraction methods from yolk and other animal products involve an initial lipid extraction with methanol:dichloromethane, followed by riboflavin recovery using ammonium acetate, pH 6.0 (Gliszczynska and Koziolowa 1998; Gliszczynska-Swiglo and Koziolowa 2000). Reliable HPLC-based methods have been devised for analyzing flavins in food (Gliszczynska-Swiglo and Koziolowa 2000).

Riboflavin, also known as vitamin B_2, must be acquired by vertebrates from the diet. Microbes and plants synthesize riboflavin from purine precursors, such as guanine (Rivlin 1975), much like pterins. As a vitamin, riboflavin is critical for growth and survival. It functions as an integral co-enzyme in redox reactions that metabolize carbohydrates, fats, and proteins and provide energy to animals (Rivlin 1975; Klasing 1998). At low dietary riboflavin levels, hens lose weight and produce fewer, lighter, and less viable eggs (Squires and Naber 1993). Without it in egg yolk, eggs fail to hatch (White 1996).

In both the old (Romanoff and Romanoff 1949) and more recent (Burley and Vadehra 1989) literature on the nutritive value and determinants of coloration in avian eggs, there is remarkably little mention of the pigmenting role of riboflavin in egg yolk. Riboflavin is also present in moderate concentrations (~3 μg/g) in chicken albumen and is known to give egg white a greenish-yellow tint (Gliszczynska-Swiglo and Koziolowa 2000); in fact, the intensity of riboflavin pigmentation in egg albumen is directly tied to the concentration of riboflavin in the diet (Naber and Squires 1993). However, free riboflavin is even more concentrated in chicken egg yolk (~4 μg/g) than in the white (Gliszczynska-Swiglo and Koziolowa 2000), and yet throughout the literature, primacy is given to carotenoids as control agents of yolk coloration (e.g., Blount et al. 2000). Carotenoid levels virtually always exceed those of riboflavin in chicken yolks (range, 5–60 μg/g; Surai 2002). However, the extinction coefficient at λ_{max} of riboflavin (Koziol 1966) is three to five times higher than that of yolk carotenoids (Britton 1985), so there is reason to expect that riboflavin contributes a significant amount of color to yolks. In pilot studies, chicken hens fed carotenoid-free diets their entire lives still lay yolks that retain some yellow color, even though biochemical analyses yield no yolk carotenoids. Only when females are deprived of dietary riboflavin do they lay colorless yolks (K. Klasing, unpubl. data). Clearly we need additional studies, not just on poultry but in other birds as well, to understand how riboflavin influences yolk coloration.

Given the dearth of knowledge on this topic, it should serve as a prosperous avenue for future research—one that augments the current wave of interest in nongenetic sources of maternal investment in young, from carotenoids to steroid hormones to immunoglobulins (Mousseau and Fox 1998). With a strong functional role of yolk riboflavins in hand, we can now pursue the mechanisms underlying riboflavin deposition in eggs. Are there maternal costs to sacrificing riboflavin to eggs? Do certain eggs in the laying order or clutches during the year preferentially receive high riboflavin doses? How do such molecules as carotenoids and riboflavin co-vary in yolk to ultimately affect offspring health and development? There is some evidence for a heritable bases of riboflavin accumulation in eggs (White 1996), despite the strong environmental (dietary) contributions, so there should be ample opportunity for selection to favor those riboflavin investment strategies that maximize fitness.

Psittacofulvins

Also called "psittacins" or "parrodienes" (Morelli et al. 2003), this class of colorful pigments is limited to a single order of birds—actually, a single group of organisms on earth—the parrots (Order Psittaciformes). Among birds, parrots are perhaps the most colorful group, displaying the full spectrum of colors often in a single species (e.g., Eastern Rosellas [*Platycercus eximius;* Plate 28, Volume 2]; Forshaw 1989). Both sexes are usually brilliantly colored, but certain species, such as the Eclectus Parrot, are sexually dichromatic, with males displaying a green appearance but females nearly fully covered with red feathers (Plate 15). It is this group of red (along with orange and yellow) plumage colorants that has intrigued pigment biochemists and ornithologists for over a century.

Despite the belief still held by some parrot fanciers and veterinarians that carotenoids are responsible for the red, orange, and yellow colors of psittacine feathers, the unusual coloration system of parrots was discovered over a century ago. Kruckenberg (1882) was the first to isolate a series of unusual red and yellow pigments (the psittacofulvins) from the feathers of psittacines, including the Scarlet Macaw (*Ara macao;* Plate 15). Chemical tests demonstrated that, like carotenoid pigments, these were lipophilic compounds, but that their absorption characteristics were inconsistent with previously described carotenoids in bird feathers. Völker (1936, 1937, 1942) provided additional evidence that parrot colors are not derived from carotenoids. He found that some yellow parrot feathers fluoresce under UV light and that, unlike carotenoid-

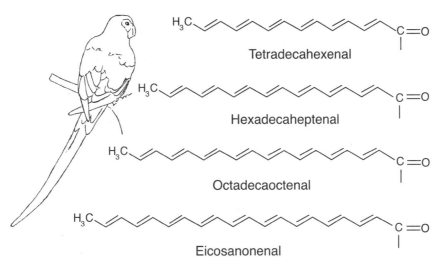

Figure 8.10. Proposed chemical structures of the four main red lipochromes in parrot feathers. Note that they differ only in the number of C-C double bonds in conjugation ($n = 6-9$). The presence of these components has now been characterized from the plumage of more than 45 parrot species (McGraw and Nogare 2004).

containing feathers, pigment incorporation completely lacks an intermediate lipid stage in the feather germ and is not dependent on the carotenoid content in the diet (for dietary origin of plumage carotenoids, see Brockmann and Völker 1934; Hill 2002; Chapter 12). For over 100 years, however, this was the only biochemical information available on these unique structures (Boles 1991).

Hudon and Brush (1992) recently resurrected this dormant line of research, showing with spectrophotometry that these pigments absorb light maximally at shorter wavelengths than do carotenoids and offering the first HPLC method for analyzing psittacofulvins. Veronelli et al. (1995) went on to employ sophisticated in situ Raman resonance spectroscopy to propose a molecular structure for parrot plumage pigments. They speculated that the color-producing mechanism in several psittacines was due to a linear polyene chain that contains seven to nine conjugated double bonds. Stradi et al. (2001) followed up this work and analyzed red pigment extracts from a Scarlet Macaw using HPLC coupled with UV-VIS and mass spectrometry. Consistent with earlier hypotheses, they isolated four particular polyenal components (Figure 8.10). They offered the idea that these polyenic chains may be synthesized via a polyketide pathway from acetyl CoA, which would be a novel biochemical pathway previously confirmed only in bacteria, fungi, and plants, or by enzymatic desaturation of fatty acids. There also may be an interesting bathochromatic shift in pigment color that is a function of the bond between psittacofulvins and feather keratin; the nature of this molecular interaction may be the primary means by which parrots generate their rainbow of plumage colors. Green colors in parrots, for example, are created by a yellow psittacofulvin and a blue structural component to the feather. These colors are different from the olive-greens found in, for example, European Greenfinches (*Carduelis chloris*) that are the result of melanin and carotenoid pigments co-deposited in plumage (Lucas and Stettenheim 1972).

Chemical analyses of red feather pigments have now been extended to nearly 50 species, representing the major lineages of parrots (McGraw and Nogare 2005). All of these parrots use the same suite of psittacofulvins to color their plumage red. Even the orange (e.g., from some Amazon parrots) and pink (e.g., from cockatoos) colors of parrot feathers are due to these same pigments at lower concentrations (McGraw and Nogare 2005). Some of these birds even show the ability to accumulate high levels of dietary carotenoid pigments in the blood, so their failure to use them as colorants—instead using psittacofulvins—is not because they do not have ample reservoirs of carotenoids, but because they are selectively excluding carotenoids at feather follicles (McGraw and Nogare 2004). In contrast, psittacofulvins are not present in the blood or liver and presumably are made directly at maturing feathers (McGraw and Nogare 2004).

Until very recently, red parrot plumage was omitted from the body of work on condition-dependence of avian coloration. Recent studies of Burrowing Parrots (*Cyanoliseus patagonus*), however, in which males display an ornamental patch of red abdominal feathers (Plate 16, Volume 2), show that these colors can reveal important information about an individual's quality and hence mate attractiveness or competitive ability. Masello and Quillfeldt (2003) showed that (1) pairs mated assortatively, based on the size of the abdominal patch (Figure 8.11); (2) males in better body condition grew larger red patches; and (3) the size of this red patch was larger in both sexes during a year when rainfall was more abundant. More studies are needed to both evaluate intrasexual variation in psittacofulvin-based plumage color in parrots and its potential role in visual communication.

What still has remained elusive in this work on parrot colors is the precise chemical structure of yellow psittacofulvins (Nemesio 2001; Plate 16). This

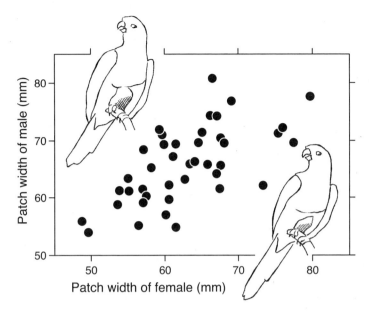

Figure 8.11. Relation between the size of the red, psittacofulvin-based abdominal breast patch of mated male and female Burrowing Parrots. Redrawn from Masello and Quillfeldt (2003).

topic should be a priority in future research on parrot colors, especially as there is good evidence now that parrots also use yellow plumage colors as social and sexual signals (Pearn et al. 2001; Arnold et al. 2002). Moreover, like virtually all of the other pigments outlined above, red psittacofulvins can exhibit antioxidant activity (Morelli et al. 2003). The role that this activity plays in individual health should provide further insights into how and why parrots are so colorful.

Still-Undescribed Pigments

Given that we have yet to chemically document the integumentary pigments in the overwhelming majority of all bird species, it is conceivable that there are several classes of colorants still waiting to be discovered. To date, we know of at least two specific groups of pigments with chemical properties that do not match those previously characterized from feathers and that we are just now beginning to understand.

Despite the general notion that penguins display a very basic black-and-white appearance, several species of penguins have yellow or yellow-orange decorative plumage features (Plate 17). King (*Aptenodytes patagonicus*) and Emperor Penguins (*Aptenodytes forsterii*), for example, develop brilliant patches of yellow-orange auricular and breast plumage. Rockhopper (*Eudyptes chrysocome*), Macaroni (*Eudyptes chrysolophus*), and Yellow-eyed Penguins (*Megadyptes antipodes*) develop post-orbital plumes that are yellow in color; male and female Yellow-eyed Penguins also have yellow irises. Although comparatively little is known of the mechanisms and functions of conspicuous colors in seabirds generally (O'Donald 1980; Jones et al. 2000), two recent studies suggest that penguins use these colors as sexual signals. Reducing the size of colored plumage patches in male King Penguins, for example, significantly delays the speed with which birds secure mates (Jouventin et al. 2005). Also, male and female Yellow-eyed Penguins with brighter colors are older, in better body condition, more likely to form pair bonds, and raise the most offspring in a season (Massaro et al. 2003). Massaro et al. (2003) made the assumption that carotenoids bestow color on the feathers and eyes of Yellow-eyed Penguins. However, it has more recently been found that yellow feathers in these and other penguins do not contain carotenoids (McGraw et al. 2004). Instead, yellow pigments in King, Emperor, Macaroni, and Yellow-eyed Penguin feathers are soluble in mild acids and bases, strongly absorb UV light, and fluoresce in the UV (McGraw 2005b). These characteristics are shared with those of pterin pigments described above in colorful avian irises, but we still await more detailed biochemical analyses, such as mass spectrometry or nuclear magnetic resonance, before we can confirm that these pigments are, in fact, pterins.

Penguins do not appear to be the only birds that harbor these pigments in feathers. Many birds, as young chicks, develop yellow colors in their downy plumage. These are most common in ducks (Order Anseriformes); shorebirds (Order Charadriiformes); gamebirds, such as quail and turkeys; and chickens (Order Galliformes). Some songbirds, such as Great Tits (*Parus major*) also have yellow in their plumage as nestlings, but these colors respond to dietary carotenoid manipulations and are likely to be carotenoid-based (Fitze et al. 2003). In contrast, preliminary biochemical tests on the pigments in the yellow and buff natal down of Domestic Chickens (Plate 17), Japanese Quail, and Wood Ducks (*Aix sponsa*) yield similar solubility, light-absorbance, and fluorescence properties to those found in yellow penguin feathers (McGraw et al. 2004; K. J. McGraw, unpubl. data). Uniquely fluorescent yellow plumage streaks have recently been identified under UV light in Wild Turkey poults as well (Sherwin and Devereux 1999). Interestingly, in this study, it was found that, under insufficient UV lighting, fluorescent plumage regions were the sites of the most damage received from pecking by other poults, suggesting that there is some potential signal value in these colors. Because these animals are still sexually immature, it will be exciting to see in what capacity these color signals function.

Although green pigments have been described from bird eggs and the feathers of turacos (see above), there appear to be some birds that use wholly different and still incompletely characterized green pigments in plumage. These plumage pigments occur specifically in two groups of ducks—the eiders (genus *Somateria* and *Arctonetta*) and the pygmy-geese (genus *Nettapus;* Auber 1957; Brush 1978). Dyck (1992) investigated the spectral reflectance patterns of nape feathers from the Common Eider (*Somateria mollissima*; Plate 17) and found that they did not match those of green turaco, partridge, or jacana feathers. Instead, the reflectance peak gradually increased from 400–500 nm and plateaued above 500 nm. This is the general shape of reflectance curves for yellow carotenoids, such as the xanthophylls (e.g., lutein, zeaxanthin; Brush 1978); in fact, Auber (1957) called these "green carotenoids." However, one author remarked that the spectrum and shape of the crystals isolated from this green plumage closely resembled a porphyrin (J. Hudon, pers. comm.).

Hall (1953) noted the presence of an unusual, noncarotenoid-based rusty-red plumage color in some populations of the White-shouldered Starling (*Sturnus sinensis*). Auber (1957) claimed this to be the only atypical pigment in all of the oscine birds, but this account has never been followed up.

Over the years, there have been pigments that were initially given new identities and names, but later found to be identical to one or a combination of known color-producing mechanisms. A good example is the violet plumage coloration in some species of cotingas (Order Passeriformes, Family Cotingidae), which was once thought to be generated by a type of pigment termed "cotingin" (Görnitz and Rensch 1924). This color has since been described as a combination of blue structural coloration and red carotenoid pigmentation (Brush 1969).

Conclusions

Several important themes emerge from the century or more of work on these less common types of endogenously produced pigments in birds. First, biochemical aspects of coloration in several groups of birds and for several types of pigment have either never been investigated or were last tackled several decades ago, prior to the modern advances in liquid chromatography. It would be fruitful to revisit several of these long-ago–characterized pigment classes, to better understand their synthetic origins and chromatic characteristics, to broaden our chemical investigations of color to those underrepresented avian classes, and to improve our understanding of how different colors have evolved and currently function. Second, it should now be apparent that one cannot judge the identity of a bird pigment simply from its color. If we consider yellow colors alone, for example, we see that they can be produced by at least six different mechanisms: melanin, carotenoid, pterin, psittacofulvins, iron-oxide staining (Chapter 9), and structural color (Chapter 7). Several recent studies in behavioral ecology have unfortunately used these misleading rules of thumb to assign the wrong pigment types to colors, and have gone on to draw conclusions about the evolution and function of their colorful trait under study without ever diagnosing its chemical nature. With the growing interest in how different types of pigment-based and structurally based colors are produced and function (e.g., McGraw et al. 2002), it is imperative that we appropriately match colors with their pigments, so that we can continue to accumulate a reliable knowledge base. Third, and perhaps most importantly, one of the major goals of this book is to integrate mechanistic and functional approaches and encourage multidisciplinary research on bird colors. This chapter in particular points out the series of knowledge gaps in both of these areas of avian color research for certain types of pigmentary ornaments. Future collaborations aimed at including pigment biochemistry, the biological action of these molecules, and the perception and use of colors as signals should be among the most profitable of all future lines of research on bird coloration.

Summary

Carotenoids and melanins are by no means the only pigments that birds use for coloration. Birds synthesize at least four other main classes of colorants: porphyrins, pterins, psittacofulvins, and flavins. Porphyrins are nitrogen-containing, pyrrolic molecules that include (1) natural porphyrins, such as protoporphyrin in brown eggshells and rusty-brown feathers of owls, bustards, and goatsuckers; (2) metalloporphyrins, such as heme in blood and turacins in the red and green feathers of turacos; and (3) bilins, such as biliverdin in blue eggshells. Natural porphyrins are manufactured from succinyl CoA and glycine, with metalloporphyrins derived directly from natural porphyrins via enzymatic additions of minerals, such as iron (heme) and copper (turacins), and bilins, the result of oxidative degradation of heme. These events largely occur in the liver, but have also been found in peripheral tissues, such as the oviduct (for eggshell pigments). There is preliminary evidence that heme and the egg colorants serve thermoregulatory or sexual-signaling functions, but more work is needed in this area on feather porphyrins. Pterin pigments are also nitrogen-rich, heterocyclic colorants, united by the presence of a two-pyrimidine-ring backbone (pteridine skeleton) and exhibiting diagnostic UV-fluorescence. Pterins, such as xanthopterin, color the irises of many birds (e.g., raptors, pigeons, blackbirds, starlings) red, orange, yellow, and even white. They are made from purine nitrogenous bases (e.g., guanine), which can also color eyes white, and are concentrated into unique "reflecting platelets" in the iris. They are also the suspected colorants of fluorescent yellow feathers in penguins and young gamebirds, shorebirds, and waterfowl. Although sexual dichromatism in pterin color is noted for several species, we await detailed studies on the signal content and function of these ornaments. Colorful flavins in birds are restricted to the presence of riboflavin (vitamin B_2)—a commonly overlooked yellow colorant of egg yolk and albumen. Flavins are chemical relatives of pterins, containing the base pterin backbone and a benzene ring. As a vitamin, riboflavin must be acquired from the diet and present in egg yolk for embryos to survive. It has only been documented from the eggs of chickens, but should generate considerable interest as a potentially limiting and valuable "maternal effect." Psittacofulvins are the red-to-yellow lipochromes found only in the colorful feathers of parrots. Red psittacolfulvins are linear polyenal aldehydes that are made by parrots directly at maturing feather follicles. Purported synthesis routes include a polyketide pathway from acetyl CoA or fatty-acid desaturation. The apparent invariance of red plumage within parrot species had led to few tests of its signaling function, but in the one well-studied species to date in this context (Burrowing Parrots), there is good reason to suspect a sexual- and quality-signaling function. Yellow psittacofulvin-colored plumage often fluoresces under UV light and is used in mate discrimination; the chemical structure of these molecules has not yet been determined.

In sum, there is an apparent need to understand more about how these relatively rare color features are produced, function, and evolve. By revealing more of their biochemical and physiological characteristics, we are likely to expose fundamental differences for how and why different birds use these different colorants in their integument.

References

Afonso, S., G. Vanore, and A. Batlle. 1999. Protoporphyrin IX and oxidative stress. Free Rad Res 31: 161–170.

Arnold, K. E., I. P. F. Owens, and A. J. Marshall. 2002. Fluorescent signaling in parrots. Science 295: 92.

Auber, L. 1957. The distribution of structural colours and unusual pigments in the class Aves. Ibis 99: 463–476.

Bagnara, J. T., and M. E. Hadley. 1973. Chromatophores and Color Change: The Comparative Physiology of Animal Pigmentation. Englewood Cliffs, NJ: Prentice-Hall.

Bagnara, J. T., and M. Obika. 1965. Comparative aspects of integumental pteridine distribution among amphibians. Comp Biochem Physiol 15: 33–49.

Bagnara, J. T., K. L. Kreutzfeld, P. J. Fernandez, and A. C. Cohen. 1988. Presence of pteridine pigments in isolated iridophores. Pigment Cell Res 1: 361–365.

Baird, T., S. E. Solomon, and D. R. Tedstone. 1975. Localisation and characterization of egg shell porphyrins in several avian species. Brit Poult Sci 16: 201–208.

Bakken, G. S., V. C. Vanderbilt, W. A. Buttemer, and W. R. Dawson. 1978. Avian eggs: Thermoregulatory value of very high near-infrared reflectance. Science 200: 321–323.

Battersby, A. R. 2000. Tetrapyrroles: The pigments of life. Nat Prod Rep 17: 507–526.

Bauer, M., and I. Bauer. 2000. Heme oxygenase-1: Redox regulation and role in the hepatic response to oxidative stress. Antiox Redox Sign 4: 749–758.

Blair, J. A. 1957. Pigments and pterins in the skin of the green mamba *Dendroaspis viridis*. Nature 180: 1371.

Blair, J. A., and J. Graham. 1954. The pigments of snake skins. 1. The isolation of riboflavin as a pigment of the skins of the green snakes *Philothamnus semivariegatus* and *Dispholidus typus*. Biochem J 56: 286–287.

Blakley, R. L. 1969. The Biochemistry of Folic Acid and Related Pteridines. North-Holland: Amsterdam.

Blount, J. D., D. C. Houston, and A. P. Møller. 2000. Why egg yolk is yellow. Trends Ecol Evol 15: 47–49.

Boles, W. E. 1991. Glowing parrots. Birds Int 3: 76–79.

Bond, C. J. 1920. On certain factors concerned in the production of eye color in birds. J Genet 9: 69–81.

Bortolotti, G. R., J. E. Smits, and D. M. Bird. 2003. Iris colour of American Kestrels varies with age, sex and exposure to PCBs. Physiol Biochem Zool 76: 99–104.

Britton, G. 1985. General carotenoid methods. Methods Enzymol 111: 113–149.

Brockmann, H., and O. Völker. 1934. Der gelbe Federfarbstoff des Kanarienvogels (*Serinus canaria canaria*) und das Vorkommen von Carotinoiden bei Vögeln. Hoppe-Seyl Z 224: 193–215.

Brush, A. H. 1969. On the nature of "cotingin." Condor 71: 431–433.

Brush, A. H. 1978. Avian pigmentation. In A. H. Brush, ed., Chemical Zoology, Volume X, 141–161. New York: Academic Press.

Bukowski, L., K. Erickson, and T. A. Lyerla. 1990. Characterization of the yellow pigment in the axanthic mutant of the Mexican axolotl, *Ambystoma mexicanum*. Pigment Cell Res 3: 123–125.

Burley, R. W., and D. V. Vadehra. 1989. The avian egg: Chemistry and biology. New York: John Wiley and Sons.

Cherry, M. I., and A. T. D. Bennett. 2001. Egg color matching in an African cuckoo, as revealed by ultraviolet-visible reflectance. Proc R Soc Lond B 268: 565–571.

Church, A. H. 1870. Researches on Turacin, an animal pigment containing copper. Proc R Soc Phil Trans 159: 627–636.

Church, A. H. 1892. Researches on Turacin, an animal pigment containing copper. Proc R Soc Phil Trans A 183: 511–530.

Church, A. H. 1913. Notes on turacin and turacin-bearers. Proc Zool Soc 1913: 639–643.

Collias, E. C. 1993. Inheritance of egg-color polymorphism in the village weaver (*Ploceus cucullatus*). Auk 110: 683–692.

Cott, H. B. 1940. Adaptive Coloration in Animals. London: Methuen.

Courts, A. 1960. Occurrence of riboflavin in sturgeon skin. Nature 185: 463–464.

Czeczuga, B. 1980. Carotenoids in the skin of certain species of birds. Comp Biochem Physiol B 62: 107–109.

Derrien, E., and J. Turchini. 1925. Nouvelles observations des fluorescences rouges chez les animeaux. Compt Rend Séanc Soc Biol Paris 92: 1030–1031.

Dyck, J. 1966. Determination of plumage colours, feather pigments, and structures by means of reflection spectrophotometry. Dan Ornithol Foren Tidsskr 60: 49–76.

Dyck, J. 1971. Structure and spectral reflectance of green and blue feathers of Rose-faced Lovebirds (*Agapornis roseicollis*). Biol Skr Dan Vid Selsk 18: 1–67.

Dyck, J. 1978. Olive green feathers: Reflection of light from the rami and their structure. Anser, Suppl. 3 (Proc 1st Nord Congr Ornithol): 57–75.

Dyck, J. 1987. Structure and light reflection of green feathers of fruit doves (*Ptilinopus* spp.) and an Imperial Pigeon (*Ducula concinna*). Biol Skr Dan Vid Selsk 30: 1–43.

Dyck, J. 1992. Reflectance spectra of plumage areas covered by green feather pigments. Auk 109: 293–301.

Egeland, E. S., H. Parker, and S. Liaaen-Jensen. 1993. Carotenoids in combs of Capercaillie (*Tetrao urogallus*) fed defined diets. Poult Sci 72: 747–751.

Elbirt, K. K., and H. L. Bonkovsky. 1999. Heme oxygenase: Recent advances in understanding its regulation and role. Proc Assoc Am Physicians 111: 438–447.

Elliot, J. H., and S. Futterman. 1963. Fluorescence in the tapetum of the cat's eye. Arch Ophthalmol 70: 531–534.

Ferris, W. 1968. Reflecting cells in the iris of the Inca Dove. J Cell Biol 39: 43a.

Ferris, W., and J. T. Bagnara. 1972. Reflecting pigment cells in the dove iris. In V. Riley ed., Pigmentation: Its Genesis and Biological Control, 181–192. New York: Appleton-Century-Crofts.

Ficken, M. S. 1965. Mouth color of nestling passerines and its use in taxonomy. Wilson Bull 77: 71–75.

Fischer, H., and J. Hilger. 1923. Zur Kenntnis der natürlichen Porphyrine (II). Über das Turacin. Hoppe-Seylers Z Physiol Chem 128: 167–174.

Fischer, H., and J. Hilger. 1924. Zur Kenntnis der natürlichen Porphyrine. 8. Mitt. Über das Vorkommen von Uroporphyrin (als Kupfersalz, Turacin) in das Turakusvögeln und den Nachweis von Koproporphyrin in der Hefe. Hoppe-Seylers Z Physiol Chem 138: 49–67.

Fischer, H., and F. Kogl. 1923. Zur Kenntnis der natürlichen Porphyrine (IV). Über das Ooporphyrin. Z Physiol Chem 131: 241–261.

Fitze, P. S., B. Tschirren, and H. Richner. 2003. Carotenoid-based colour expression is determined early in nestling life. Oecologia 137: 148–152.

Fontaine, M. 1941. Recherche sur quelques pigments seique et dermique de poisons marins (Labrides et Cyclopterides). Bull Inst Océanogr Monaco 38: 792.

Forshaw, J. M. 1989. Parrots of the World. Melbourne, Australia: Lansdowne Editions.

Fox, D. L. 1953. Animal biochromes. Cambridge, UK: Cambridge University Press.

Fox, D. L. 1976. Animal Biochromes and Structural Colours. Berkeley: University of California Press.

Fox, H. M., and G. Vevers. 1960. The Nature of Animal Colours. London: Sidgwick and Jackson.

Frost, S. K., and J. T. Bagnara. 1977. Developmental aspects of pteridine pigments in the Mexican leaf frog. Yale J Biol Med 50: 557.

Frost, S. K., L. G. Epp, and S. J. Robinson. 1984. The pigmentary system of developing axolotls. 1. A biochemical and structural analysis of chromatophores in wild-type axolotls. J Embryol Exp Morph 81: 105–125.

Gieseg, S. P., G. Reibnegger, H. Wachter, and H. Esterbauer. 1995. 7,8 dihydroneopterin inhibits low density lipoprotein oxidation in vitro. Evidence that this macrophage-secreted pteridine is an antioxidant. Free Rad Res 23: 123–136.

Giulivi, C., and K. J. Davies. 1990. A novel antioxidant role for hemoglobin. The comproportionation of ferrylhemoglobin with oxyhemoglobin. J Biol Chem 265: 19543–19560.

Gliszczynska, A., and A. Koziolowa. 1998. Chromatographic determination of flavin derivatives in baker's yeast. J Chromatogr A 822: 59–66.

Gliszczynska-Swiglo, A., and A. Koziolowa. 2000. Chromatographic determination of riboflavin and its derivatives in food. J Chromatogr A 881: 285–297.

Görnitz, K., and B. Rensch. 1924. Uber die violette Färbung der Vogelfedern. J Ornithol 72: 113–118.

Gosler, A. G., P. R. Barnett, and S. J. Reynolds. 2000. Inheritance and variation in eggshell patterning in the Great Tit *Parus major*. Proc R Soc Lond B 267: 2469–2473.

Götmark, F. 1992. Blue eggs do not reduce nest predation in the Song Thrush, *Turdus philomelos*. Behav Ecol Sociobiol 30: 245–252.

Götmark, F., and M. Ahlström. 1997. Parental preference for red mouth of chicks in a songbird. Proc R Soc Lond B 264: 959–962.

Grether, G. F., J. Hudon, and J. A. Endler. 2001. Carotenoid scarcity, synthetic pteridine pigments and the evolution of sexual coloration in Guppies (*Poecilia reticulata*). Proc R Soc Lond B 268: 1245–1253.

Hall, B. P. 1953. Colour varieties in *Sturnus sinensis* (Gmelin). Bull Brit Ornithol Club 73: 2–8.

Hammel, C. L., and S. P. Bessman. 1965. Control of hemoglobin synthesis by oxygen tension in a cell-free system. Arch Biochem Biophys 110: 622–627.

Hammond, K. A., M. A. Chappell, R. A. Cardullo, R. S. Lin, and T. S. Johnsen. 2000. The mechanistic basis of aerobic performance variation in red Junglefowl. J Exp Biol 203: 2053–2064.

Henze, M., G. Rempeters, and F. Anders. 1977. Pteridines in the skin of xiphophorine fish (Poeciliidae). Comp Biochem Physiol B 56: 35–46.

Hill, G. E. 2002. A Red Bird in a Brown Bag: The Function and Evolution of Colorful Plumage in the House Finch. New York: Oxford University Press.

Hollander, W. F., and R. D. Owen. 1939a. The carotenoid nature of yellow pigment in the chicken iris. Poult Sci 18: 385–387.

Hollander, W. F., and R. D. Owen. 1939b. Iris pigmentation in Domestic Pigeons. Genetica 21: 408–419.

Hudon, J., and A. H. Brush. 1992. Identification of carotenoid pigments in birds. Methods Enzymol 213: 312–321.

Hudon, J., and A. D. Muir. 1996. Characterization of the reflective materials and organelles in the bright irides of North American blackbirds (Icterinae). Pigment Cell Res 9: 96–104.

Hudon, J., and L. W. Oliphant. 1995. Reflective organelles in the anterior pigment epithelium of the iris of the European Starling *Sturnus vulgaris*. Cell Tissue Res 280: 383–389.

Hunt, S., R. M. Kilner, N. E. Langmore, and A. T. D. Bennett. 2003. Conspicuous, ultraviolet-rich mouth colours in begging chicks. Proc R Soc Lond B 270 (Suppl.): S25–S28.

Hurst, D. T. 1980. An Introduction to the Chemistry and Biochemistry of Pyrimidines, Purines and Pteridines. Chichester, UK: John Wiley and Sons.

Jackson, W. M. 1992. Estimating conspecific nest parasitism in the Northern Masked Weaver based on within-female variability in egg appearance. Auk 109: 435–443.

Jones, I. L., F. M. Hunter, and G. Fraser. 2000. Patterns of variation in ornaments of Crested Auklets *Aethia cristatella.* J Avian Biol 31: 119–127.

Jourdie, V., B. Moureau, A. T. D. Bennett, and P. Heeb. 2004. Ultraviolet reflectance by the skin of nestlings. Nature 431: 262.

Jouventin, P., P. M. Nolan, F. S. Dobson, and M. Nicolaus. 2005. Colored patches influence pairing in King Penguins. Ibis (in press).

Jurisevic, M. A., K. J. Sanderson, and R. V. Baudinette. 1999. Metabolic rates associated with distress and begging calls in birds. Physiol Biochem Zool 72: 38–43.

Kaur, H., M. N. Hughes, C. J. Green, P. Naughton, R. Foresti, and R. Motterlini. 2003. Interaction of bilirubin and biliverdin with reactive nitrogen species. FEBS Lett 543: 113–119.

Keilin, J., and P. J. McCosker. 1961. Reactions between uroporphyrin and copper and their biological significance. Biochim Biophys Acta 52: 424–435.

Kennedy, G. Y., and H. G. Vevers. 1973. Eggshell pigments of the Araucano Fowl. Comp Biochem Physiol B 44: 11–25.

Kennedy, G. Y., and H. G. Vevers. 1976. A survey of avian eggshell pigments. Comp Biochem Physiol B 55: 117–123.

Kilner, R. 1997. Mouth colour is a reliable signal of need in begging canary nestlings. Proc R Soc Lond B 264: 963–968.

Kim, C. H., S. Yamagishi, and P. O. Won. 1995. Egg-color dimorphism and breeding success in the Crow Tit (*Paradoxornis webbiana*). Auk 112: 831–839.

Klasing, K. C. 1998. Comparative avian nutrition. Wallingford, UK: CAB International.

Koziol, J. 1966. Studies on flavins in organic solvents. I. Spectral characteristics of riboflavin, riboflavin tetrabutyrate and lumichrome. Photochem Photobiol 5: 41–54.

Krinsky, N. I. 2001. Carotenoids as antioxidants. Nutrition 17: 815–817.

Kruckenberg, C. F. W. 1882. Der Federfarbstoffe der Psittaciden. Verg-Physiol Studien Reihe 2, Abtlg 2: 29–36.

Kruckenberg, C. F. W. 1883. Die Farbstoffe der Vögeleierschalen. Verg Physiol-Med Ges Würzb 17: 109–127.

Lack, D. 1958. The significance of the colour of turdine eggs. Ibis 100: 145–166.

Lang, M. R., and J. W. Wells. 1987. A review of eggshell pigmentation. World's Poultr Sci J 43: 238–246.

Laruelle, L., M. Beumont, and E. Legait. 1951. Recherches sur le mécanisme des changements de couleur des caroncules vasculaires du dindon (*Meleagris gallopavo* L.). Arch d'Anat Microsc Morphol Exp 40: 91–113.

Lascalles, J. 1964. Tetrypyrrole biosynthesis and its regulation. New York: Benjamin.

Leary, M. R., T. W. Britt, W. D. Cutlip, II, and J. L. Templeton. 1992. Social blushing. Psychol Bull 112: 446–460.

Lemberg, R. 1934. Bile pigments—VI. Biliverdin, uteroverdin and oocyan. Biochem J 28: 978–987.

Lemberg, R., and J. Barcroft. 1932. Uteroverdin, the green pigment of the dog's placenta. Proc R Soc Lond B 110: 362–372.

Lemberg, R., and J. W. Legge. 1949. Haematin compounds and bile pigments. New York: Interscience.

Ligon, J. D., R. Thornhill, M. Zuk, and K. Johnson. 1990. Male-male competition, ornamentation and the role of testosterone in sexual selection in red Jungle Fowl. Anim Behav 40: 367–373.

Lucas, A. M., and P. R. Stettenheim. 1972. Avian anatomy: Integument. Agriculture Handbook 362. Washington, DC: U.S. Government Printing Office.

Lucotte, G., M. Choussy, and M. Barbier. 1975. Polymorphism of egg-shell pigmentation in Japanese Quail (*Coturnix coturnix japonica*). 4. Porphyrin biosynthesis by diverse shell glands in dominant form. Compt Rend 169: 34–38.

Lyon, B. E. 2003. Egg recognition and counting reduce costs of avian conspecific brood parasitism. Nature 422: 495–499.

Macedonia, J. M., S. James, L. W. Wittle, and D. L. Clark. 2000. Skin pigments and coloration in the Jamaican radiation of *Anolis* lizards. J Herpetol 34: 99–109.

Maines, M. D. 1988. Heme oxygenase: Function, multiplicity, regulatory mechanisms, and clinical applications. FASEB J 2: 2557–2568.

Maines, M. D. 1997. The heme oxygenase system: A regulator of second messenger gases. Annu Rev Pharmacol Toxicol 37: 517–554.

Marks, G. S. 1969. Heme and chlorophyll: Chemical, biochemical, and medical aspects. London: D. Van Nostrand.

Masello, J. F., and P. Quillfeldt. 2003. Body size, body condition and ornamental feathers of burrowing parrots: Variation between years and sexes, assortative mating and influences on breeding success. Emu 103: 149–161.

Massaro, M., L. S. Davis, and J. T. Darby. 2003. Carotenoid-derived ornaments reflect parental quality in male and female Yellow-eyed Penguins (*Megadyptes antipodes*). Behav Ecol Sociobiol 55: 169–175

McGraw, K. J. 2003. Melanins, metals, and mate quality. Oikos 102: 402–406.

McGraw, K. J. 2004. Colorful songbirds metabolize carotenoids at the integument. J Avian Biol 35: 471–476.

McGraw, K. J. 2005a. Antioxidant function of many animal pigments: Consistent benefits of sexually selected colorants? Anim Behav 69: 757–764.

McGraw, K. J. 2005b. Not all red, orange, and yellow animal colors are carotenoid-based: The need to couple biochemical and behavioral studies of color signals. Proc Ind Nat Sci Acad (in press).

McGraw, K. J., and M. C. Nogare. 2004. Carotenoid pigments and the selectivity of psittacofulvin-based coloration systems in parrots. Comp Biochem Physiol B 138: 229–233.

McGraw, K. J., and M. C. Nogare. 2005. Distribution of unusual red feather pigments in parrots. Biol Lett 1: 38–43.

McGraw, K. J., E. A. Mackillop, J. Dale, and M. E. Hauber. 2002. Different colors reveal different information: How nutritional stress affects the expression of melanin- and structurally based ornamental coloration. J Exp Biol 205: 3747–3755.

McGraw, K. J., K. Wakamatsu, S. Ito, P. M. Nolan, P. Jouventin, et al. 2004. You can't judge a pigment by its color: Carotenoid and melanin content of yellow and brown feathers in swallows, bluebirds, penguins, and Domestic Chickens. Condor 106: 390–395.

Miksik, I., V. Holan, and Z. Deyl. 1994. Quantification and variability of eggshell pigment content. Comp Biochem Physiol A 109: 769–772.

Miksik, I., V. Holan, and Z. Deyl. 1996. Avian eggshell pigments and their variability. Comp Biochem Physiol B 113: 607–612.

Moreau, R. E. 1958. Some aspects of the Musophagidae. Part 3. Ibis 100: 238–270.

Morelli, R., R. Loscalzo, R. Stradi, A. Bertelli, and M. Falchi. 2003. Evaluation of the antioxidant activity of new carotenoid-like compounds by electron paramagnetic resonance. Drugs Exp Clin Res 29: 95–100.

Moreno, J., and J. L. Osorno. 2003. Avian egg colour and sexual selection: Does eggshell pigmentation reflect female condition and genetic quality? Ecol Lett 6: 803–806.

Moreno, J., J. L. Osorno, J. Morales, S. Merino, and G. Tomas. 2004. Egg colouration and male parental effort in the Pied Flycatcher *Ficedula hypoleuca*. J Avian Biol 35: 300–304.

Mousseau, T. A., and C. W. Fox. 1998. Maternal Effects as Adaptations. Oxford: Oxford University Press.

Naber, E. C., and M. W. Squires. 1993. Early detection of the absence of a vitamin premix in layer diets by egg albumen riboflavin analysis. Poult Sci 72: 1989–1993.

Needham, A. E. 1974. The Significance of Zoochromes. New York: Springer-Verlag.

Nemesio, A. 2001. Color production and evolution in parrots. Int J Ornithol 4: 75–102.

Nicholas, R. E. H., and C. Rimington. 1954. Isolation of unequivocal uroporphyrin III. A further study of turacin. Biochem J 49: 33–34.

Nys, Y., J. Zawadzki, J. Gautron, and A. D. Mills. 1991. Whitening of brown-shelled eggs: Mineral composition of uterine fluid and rate of protoporphyrin deposition. Poult Sci 70: 1236–1245.

Obika, M., and J. T. Bagnara. 1963. Pteridines as pigments in amphibians. Science 143: 485–487.

Obika, M., and S. Negishi. 1972. Melanosome of toad: A storehouse of riboflavin. Exp Cell Res 70: 293–300.

O'Donald, P. 1980. Sexual selection by female choice in a monogamous bird: Darwin's theory corroborated. Heredity 45: 201–217.

Oehme, H. 1969. Vergleichende Untersuchungen über die Färbung der Vogeliris. Biol Zentralbl 88: 3–35.

Oliphant, L. W. 1981. Crystalline pteridines in the stromal pigment cells of the iris of the Great Horned Owl. Cell Tissue Res 217: 387–395.

Oliphant, L. W. 1987a. Pteridines and purines as major pigments of the avian iris. Pigment Cell Res 1: 129–131.

Oliphant, L. W. 1987b. Observations of pigmentation of the pigeon iris. Pigment Cell Res 1: 202–208.

Oliphant, L. W. 1988. Cytology and pigments of non-melanophore chromatophores in the avian iris. In J. T. Bagnara ed., Advances in Pigment Cell Research, 65–82. New York: Liss.

Oliphant, L. W., and J. Hudon. 1993. Pteridines as reflecting pigments and components of reflecting organelles in vertebrates. Pigment Cell Res 6: 205–208.

Oliphant, L. W., J. Hudon, and J. T. Bagnara. 1992. Pigment cell refugia in homeotherms—The unique evolutionary position of the iris. Pigment Cell Res 5: 367–371.

Ortiz, E., L. H. Throckmorton, and H. G. Williams-Ashman. 1962. Drosopterins in the throat-fans of some Puerto Rican lizards. Nature 196: 595–596.

Otterbein, L. E., M. P. Soares, K. Yamashita, and F. H. Bach. 2003. Heme oxygenase-1: Unleashing the protective properties of heme. Trends Immunol 24: 449–455.

Pearn, S. M., A. T. D. Bennett, and I. C. Cuthill. 2001. Ultraviolet vision, fluorescence and mate choice in a parrot, the Budgerigar *Melopsittacus undulatus*. Proc R Soc Lond B 268: 2273–2279.

Pfleiderer, W. 1994. Nature pteridine pigments—Pigments found in butterflies wings and insect eyes. Chimia 48: 488–489.

Ponka, P. 1999. Cell biology of heme. Am J Med Sci 318: 241–256.

Poole, H. K. 1965. Spectrophotometric identification of eggshell pigments and timing of superficial pigment deposition in the Japanese Quail. Proc Soc Exp Biol Med 119: 547–551.

Poole, H. K. 1966. Relative ooporphyrin content and porphyrin forming capacity of wild-type and white-egg Japanese Quail uterine tissue. Proc Soc Exp Biol Med 122: 596–598.

Prum, R. O., and R. H. Torres. 2003. Structural colouration of avian skin: Convergent evolution of coherently scattering dermal collagen arrays. J Exp Biol 206: 2409–2429.

Richmond, J. E., K. I. Altman, and K. Salomon. 1951. The relation of oxygen uptake to hemoglobin synthesis. Science 113: 404–405.

Rimington, C. 1939. A re-investigation of turacin, the copper porphyrin pigment of certain birds belonging to the Musophagidae. Proc R Soc Lond B 127: 106–120.

Rivlin, R. S. 1975. Riboflavin. New York: Plenum Press.

Romanoff, A. L., and A. J. Romanoff. 1949. The Avian Egg. New York: John Wiley and Sons.

Saino, N., P. Ninni, S. Calza, R. Martinelli, F. De Bernardi, and A. P. Møller. 2000. Better red than dead: Carotenoid-based mouth coloration reveals infection in Barn Swallow nestlings. Proc R Soc Lond B 267: 57–61.

Saino, N., R. Ambrosini, R. Martinelli, P. Ninni, and A. P. Møller. 2003. Gape coloration reliably reflects immunocompetence of Barn Swallows. Behav Ecol 14: 16–22.

Scholten, C. J. 1999. Iris colour of Humboldt Penguins *Spheniscus humboldti*. Marine Ornithol 27: 187–194.

Schwartz, S., B. D. Stephenson, D. H. Sarkar, and M. R. Bracho. 1975. Red, white, and blue eggs as models of porphyrin and heme metabolism. Ann NY Acad Sci 244: 570–590.

Sherwin, C. M., and C. L. Devereux. 1999. Preliminary investigations of ultraviolet-induced markings on Domestic Turkey chicks and a possible role in injurious pecking. Br Poult Sci 40: 429–433.

Snyder, N. F. R., and H. A. Snyder. 1974. Function of eye coloration in North American accipiters. Condor 76: 219–222.

Soh, T., and O. Koga. 1994. The effects of sex steroid hormones on the pigment accumulation in the shell gland of Japanese Quail. Poult Sci 73: 179–185.

Soler, J. J., J. Moreno, J. M. Aviles, and A. P. Møller. 2005. Blue and green egg-color intensity is associated with parental effort and mating system in passerines: Support for the sexual selection hypothesis. Evolution 59: 636–644.

Sorby, H. C. 1875. On the colouring-matters of the shells of birds' eggs. Proc Zool Soc Lond 1875: 351–365.

Squires, M. W., and E. C. Naber. 1993. Vitamin profiles of eggs as indicators of nutritional status in the laying hen: Riboflavin study. Poult Sci 72: 483–494.

Stackhouse, H. L. 1966. Some aspects of pteridine biosynthesis in amphibians. Comp Biochem Physiol 17: 219–235.

Stradi, R., E. Pini, and G. Celentano. 2001. The chemical structure of the pigments in *Ara macao* plumage. Comp Biochem Physiol B 130: 57–63.

Surai, P. F. 2002. Natural antioxidants in avian nutrition and reproduction. Nottingham, UK: Nottingham University Press.

Suzuki, Y. 1998. Determination of human hemoglobin in blood based on its spectral change due to the solvent effect of ethanol. Analyt Sci 14: 1013–1016.

Sweijd, N., and A. J. F. K. Craig. 1991. Histological basis of age-related changes in iris color in the African Pied Starling (*Spreo bicolor*). Auk 108: 53–59.

Swynnerton, C. F. M. 1916. On the coloration of the mouths and eggs of birds. I. The mouths of birds. Ibis 4: 249–264.

Takatani, S., and M. D. Graham. 1987. Theoretical analysis of diffuse reflectance from a two-layer tissue model. IEEE Trans Biomed Eng 26: 656–664.

Takei, S., and H. Somiya. 2002. Guanine-type retinal tapctum and ganglion cell topography in the retina of a carangid fish, *Kaiwarinus equula*. Proc R Soc Lond B 269: 75–82.

Tamura, T., S. Fiyii, H. Kunisaki, and M. Yamane. 1965. Histological observations on the quail oviduct; with reference to pigment (porphyrin) in the uterus. J Fac Fish Anim Husb Hiroshima Univ 6: 37–57.

Thiel, H. 1968. Die Porphyrine in Vogelfedern. Untersuchungen über ihre Herkunft und Einlagerung. Zool Jb 95: 147–188.

Thomson, L. R., Y. Toyoda, A. Langner, F. C. Delori, K. M. Garnett, et al. 2002. Elevated retinal zeaxanthin and prevention of light-induced photoreceptor cell death in quail. Invest Ophthalmol Vis Sci 43: 3538–3549.

Tillotson, H. M., and L. W. Oliphant. 1990. Iris stromal pigment cells of the Ringed Turtle Dove. Pigment Cell Res 3: 319–323.

Tomic-Carruthers, N., D. C. Robacker, and R. L. Mangan. 1996. Identification and age-dependence of pteridines in the head of adult Mexican Fruit Fly, *Anastrepha ludens*. J Insect Physiol 42: 359–366.

Trauger, D. L. 1974. Eye color of female Lesser Scaup in relation to age. Auk 91: 243–254.

Underwood, T. J., and S. G. Sealy. 2002. Adaptive significance of egg coloration. In D. C. Deeming, ed., Avian Incubation, Behaviour, Environment, and Evolution, 280–298. Oxford: Oxford University Press.

Veronelli, M., G. Zerbi, and R. Stradi. 1995. In situ resonance Raman spectra of carotenoids in bird's feathers. J Raman Spectrosc 26: 683–692.

Villela, G. G., and M. Thein. 1967. Riboflavin in the blood serum, the skin and the venom of some snakes of Burma. Experientia 23: 722.

Völker, O. 1936. Über den gelben federfarbstoff des wellensittichs [*Melanopsittacus undulatus* (Shaw)]. J Ornithol 84: 618–630.

Völker, O. 1937. Über fluoreszierende, gelbe federpigmente bei papageien, eine neue klasse von federfarbstoffen. J Ornithol 85: 136–146.

Völker, O. 1938. Porphyrine in vogelfedern. J Orn Lpz 86: 436–456.

Völker, O. 1939. Porphyrins of feathers. Hoppe-Seyl Z 258: 1–5.

Völker, O. 1942. Die gelben und roten federfarbstoffe der papageien. Biol Zbl 62: 8–13.

von Schantz, T., S. Bensch, M. Grahn, D. Hasselquist, and H. Wittzell. 1999. Good genes, oxidative stress and condition-dependent sexual signals. Proc R Soc Lond B 266: 1–12.

Walker, A. W., and B. O. Hughes. 1998. Egg shell colour is affected by laying cage design. Br Poult Sci 39: 696–699.

Weidinger, K. 2001. Does egg colour affect predation rate on open passerine nests? Behav Ecol Sociobiol 49: 456–464.

Wetherbee, D. K. 1961. Observations on the developmental condition of neonatal birds. Am Midl Nat 65: 413–435.

White, III, H. B. 1996. Sudden death of chicken embryos with hereditary riboflavin deficiency. J Nutr 126: 1303S–1307S.

Wicke, W. 1858. Über des Pigment in den Eischalen der Vogel. Naumannia 8: 393–397.

Win, T. 2000. Isolation and structural identification of an alarm pheromone from the Giant Danio *Danio malabaricus* (Cyprinidae, Ostariophysi, Pisces). Ph.D. diss., University of Oldenburg, Oldenburg, Germany.

With, T. K. 1957. Pure unequivocal uroporphyrin III. Simplified method of preparation from turaco feathers. Scand J Clin Lab Invest 9: 398–401.

With, T. K. 1967. Darstellung von kristallizierten Koproporphyrin III als Kupfer-Komplexsalz aus den Daunen von *Otis tarda* (Grosstrappe). J Orn Lpz 108: 480–483.

With, T. K. 1973. Porphyrins in egg shells. Biochem J 137: 597–598.

With, T. K. 1978. On porphyrins in feathers of owls and bustards. Int J Biochem 9: 893–895.

Woodard, A. E., and F. B. Mather. 1964. The timing of ovulation, movement of the ovum through the oviduct, pigmentation and shell deposition in Japanese Quail (*Coturnix coturnix japonica*). Poult Sci 43: 1427–1432.

Yamaguchi, K., K. Hashimoto, and F. Matsuura. 1976. Identity of blue pigments obtained from different tissues of the Sculpin, *Pseudoblennius percoides* Günther. Comp Biochem Physiol B 55: 85–87.

Zvetkova, E. 1999. Ranopterins—Amphibia skin pteridines displaying hematopoietic, immunomodulatory, and macrophageal proliferative biological activities. Pteridines 10: 178–189.

9

Cosmetic and Adventitious Colors

ROBERT MONTGOMERIE

Almost all of the colors of bird feathers are acquired during a molt, via either pigment deposition or the development of keratin nanostructures (Chapters 5–8). In this chapter, however, I look at colors that are applied to, gathered by, or revealed by birds for use as sexual signals and camouflage (cosmetic colors), as well as colors that result from normal wear and tear (adventitious colors), such as soiling, staining, and abrasion. Although examples of such coloration in birds are rare and not fully understood, they provide us with nice insights into the adaptive significance of colors in general, and, like other exceptional phenomena in nature, help to probe some of the rules governing the significance of animal colors.

Cosmetic and adventitious colorants have been noticed on birds and other organisms for quite some time (Darwin 1871; Hingston 1933), but only very recently have these been considered as possible adaptations (Borgia 1986; Negro et al. 1999; Piersma et al. 1999; Montgomerie et al. 2001). Because only a few examples of these phenomena have really been studied in any depth, I have organized this chapter largely around some brief case studies. So far, there do not appear to be any general phylogenetic or ecological patterns to these examples that would lead to some useful predictions about their provenance. Instead, the cosmetic colors, in particular, are almost all quirky instances of birds taking advantage of a special opportunity to change their colors for camouflage or mate attraction, or to use colored objects that they find in their environments to communicate with conspecifics.

Cosmetic Colors

Birds can exogenously modify or produce colors in three ways: (1) by applying substances, such as soil or oils, to their plumage and bare parts; (2) by acquiring colored objects from their environment; and (3) by removing portions of feathers to reveal underlying colors. I review the mechanisms, functions, and prevalence of each of these color types below.

Applied Cosmetics

To date, there are only three reasonably well-studied examples of birds that apply substances to their feathers, in an apparently deliberate fashion, to change their color for an adaptive purpose. These include iron-oxide staining in Bearded Vultures (*Gypaetus barbatus*), uropygial-oil preening in Red Knots (*Calidris canutus*), and plumage soiling in Rock Ptarmigan (*Lagopus mutus*). There are also anecdotal accounts of staining from uropygial gland secretions in hornbills and gulls, discussed briefly here.

Bearded Vulture

In the wild, the plumage of adult male and female Bearded Vultures is a rich rusty-orange color on the neck, head, and underparts (Plate 18), and the coloration is highly variable among individuals (Brown and Bruton 1991; Negro et al. 1999). On the Mediterranean islands of Corsica and Crete, many Bearded Vultures lack this rusty coloration, and in captivity, it often eventually disappears (Negro et al. 1999). The rusty-orange coloration results from deliberate dyeing of the feathers with soils and mud rich in iron oxide (Fe_2O_3), and the intensity of the color increases with age (Frey and Roth-Callies 1994; Bertrán and Margalida 1999; Negro et al. 1999). The behavior of birds in captivity leaves no doubt about the staining being deliberate, as birds bathing in red mud rub themselves against the substrate and preen the mud into their plumage for up to an hour at a time (Frey and Roth-Callies 1994; Plate 18), whereas those bathing in water just plunge in and shake the water off their plumage (Negro et al. 1999). The application of red mud to plumage is apparently done in secrecy, as it has never been observed in the wild, despite more than 800 h of direct observation over a 3-year period (Brown and Bruton 1991).

Negro et al. (1999) argued that the deliberate staining by wild birds, the variation in intensity with age and sex, and the dominance relations among individuals all indicate that the rusty-orange color is used as a signal of social status and, because the staining is costly to obtain, may honestly advertise the quality of the bearer. For example, such a signal might reflect the ability of a Bearded Vulture to find, use, and even defend rare patches of iron oxide–stained soil (Negro et al. 1999).

It is also possible that the iron oxide deposited on the feathers serves some chemical function. Arlettaz et al. (2002), for example, suggested that these iron oxides might have an adaptive function if they are passively transferred from the parents' feathers to their eggs or chicks during incubation and brooding. They argued that such oxides could mitigate any negative effects of bacteria on egg or nestling survival, via their role as prooxidants, and might improve nestling health by mobilizing vitamin A. Negro et al. (2002), however, rejected this argument on the grounds that iron oxide is not very toxic to cultured cells, and there is as yet no evidence that it can even be absorbed through eggshells or bird skin. Moreover, juvenile Bearded Vultures also dye their feathers 5–6 years before reaching sexual maturity, at a time when iron oxide could serve no useful function for reproduction. Nonetheless, the prooxidant hypothesis provides a logical basis for the pursuit and defense of iron oxide-rich soils, and thus, by application to the plumage, a means of honestly advertising the ability of a bird to find such resources. Even such advertisement by juveniles could function to enhance their ability to attract or compete for high-quality mates if juveniles interact with potential mates, or rivals, years before overt courtship begins. It is also possible that the iron-oxide stains simply reduce the incidence of ectoparasites (Brown and Bruton 1991) or bacteria on the adults' feathers, which would diminish the detrimental effects of these agents on nestlings. All of these hypotheses are plausible, and not mutually exclusive, but there is not yet enough empirical data to allow any of them to be critically evaluated.

Red Knot

Like most birds, Red Knots actively apply uropygial gland secretions to their feathers and bare parts by preening. These secretions are mainly lipids (e.g., oils, waxes, vitamin D precursors, fatty acids) that make feathers more waterproof and flexible, delay feather wear, and keep bare parts supple and in good condition (Jacob and Ziswiler 1982). In Red Knots and probably most other waders (family Scolopacidae, Reneerkens et al. 2002; families Charadriidae and Haematopodidae, Reneerkens et al. 2005), and in female Mallards (*Anas platyrhynchos;* Kolattukudy et al. 1987), the waxes in the uropygial gland secretion are monoesters in the nonbreeding season, but change to diesters during a short period at the start of the breeding season. Monoester waxes have lower molecular weights than diesters, and the diesters in uropygial gland secretions are more viscous, tend to make the plumage more vivid or shinier, absorb less ultraviolet (UV) radiation, and are less volatile (Piersma et al. 1999; Reneerkens et al. 2002; Reneerkens and Korsten 2004). Piersma et al. (1999) argued that the seasonal timing of this switch from monoester to diester waxes in the Red Knot suggested a signaling function during courtship. Thus they reasoned that a plumage with these diester waxes preened into it—making the plumage look shinier—would serve as an honest signal of quality, presumably indicating the sorts of better health or foraging ability that would result in more of these waxes being produced by the uropygial glands. They also suggested that the Red Knot's cold, arctic breeding grounds would make the more viscous diester waxes harder to apply than monoesters, thus testing the abilities of birds to apply these waxes to their feathers and further demonstrating the bearer's quality.

Reneerkens et al. (2002, 2005), however, found that 24 other species of arctic- and temperate-zone-breeding sandpipers, plovers, and oystercatchers also had these diester waxes in their uropygial gland secretions during the breeding season. All of these waders begin to secrete diester waxes from the uropygial gland upon arrival on the breeding grounds and continue until the end of incubation, when the waxes change back to monoesters. Moreover, the only two species in their sample that had reduced male parental care showed much less (or none) of the diester waxes in males, suggesting that these waxes might serve some function during incubation, rather than during courtship. Reneerkens et al. (2002, 2005) suggested that the function of diester waxes is to reduce the odor of the incubating bird—diesters being less volatile than monoesters—thereby diminishing the predation risk from mammalian predators. This argument is bolstered by the occurrence of diester waxes in the uropygial gland secretions of female Mallards when they, too, are most vulnerable to predation—during the incubation period—but not at other times of the year (Kolattukudy et al. 1987).

Reneerkens and Korsten (2004) also showed that the diester waxes from Red Knots absorb some radiation in the UV range (particularly <350 nm). Thus the application of diester waxes via their uropygial secretions might also reduce UV damage to the feathers and bare parts of birds that breed (and court and incubate) in the arctic, where the days are long and the UV radiation intense. Moreover, reflectance spectra from the plumage of knots were not significantly different before and after the chemical removal of both diester and monoester waxes from their feathers (Reneerkens and Korsten 2004). This observation further suggests that the waxes do not have the signaling function originally proposed by Piersma et al. (1999) with respect to a noticeable change in color, although a change in the luster of the plumage might still be a useful signal (Reneerkens et al. 2005). The preen wax story is clearly more complicated than originally supposed (Piersma et al. 1999), and the various hypotheses to explain the change in wax composition now need to be tested experimentally in the field.

Rock Ptarmigan

In winter, the plumage of both sexes of the Rock Ptarmigan is white and immaculate, save for a few black tail feathers and a black eye line. To the human eye, at least, these birds are perfectly camouflaged against the arctic snow. In spring, however, as the snows melt and the brown-green tundra is exposed, females molt rapidly into their mottled-brown summer plumage, which renders them just as camouflaged as they were in the winter. However, males delay the spring molt until long after most of the snow has melted and are thus incredibly conspicuous for a few weeks in spring, during territory establishment and courtship (Huxley 1938; Plate 19).

When their mates are no longer fertilizable, male Rock Ptarmigan actively soil their plumage (Plate 19), resulting in a tenfold reduction in conspicuousness (Montgomerie et al. 2001; Figure 9.1). Montgomerie et al. (2001) argued that this deliberate application of dirt to plumage is an adaptation that, in a short period of time, reduces predation risk from aerial predators when the male's immaculate white plumage is no longer needed for territorial defense or mate attraction. Indeed, the predation rate on immaculate Rock Ptarmigan males in spring greatly exceeds that of females and most other birds (Gardarsson 1971, 1988), suggesting that this conspicuous plumage is very costly to males. Crypsis is certainly less costly to achieve by soiling plumage than by molting, and is quickly reversible if females again become sexually receptive (e.g., due to nest predation; Montgomerie et al. 2001).

Montgomerie et al. (2001) used these observations to gain some insights into the sexual selection function of the male's white plumage in spring. They argued that this highly conspicuous plumage is retained in spring because it honestly advertises a male's ability to avoid predation, an important aspect of male quality. Thus the cost of remaining immaculate is thought to be the risk of predation, particularly from Gyrfalcons (*Falco rusticolus*), whereas the cost of molting or getting dirty would be reduced mating success. Females who preferred to mate with conspicuous males would pass on whatever genes influenced

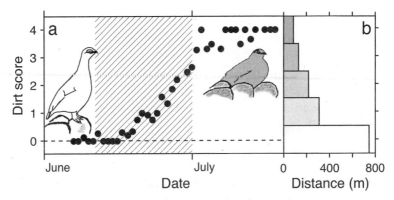

Figure 9.1. (a) Progress of plumage soiling by male Rock Ptarmigan observed during one field season in the Canadian high arctic. Dots show average dirt scores for all males observed on each day during the breeding season. Hatched area is the period during which females laid eggs. Insets show posture and appearance of males with immaculate (score = 0) and dirty (score = 4) plumage. (b) Maximum distance at which males with each dirt score were first detected by observers in the field. Shading indicates increasing dirtiness. Adapted from Montgomerie et al. (2001).

male quality to their offspring. Because males soiled their plumage when they had no more opportunities for reproduction, it seems clear that their conspicuous plumage is an important signal to either copulation partners (female choice) or territorial competitors (male-male competition), or both, and is thus maintained by sexual selection.

Hornbills

Male Great Hornbills (*Buceros bicornis*) use a vivid yellow secretion from their uropygial gland to stain their bill and casque, as well as the white feathers of their head, neck, rump, and wing-stripes (Hingston 1933; Plate 18). Because only males stain themselves, it seems likely that the color serves a function in sexual signaling, but there are no behavioral or morphological data available on any aspects of this unusual, cosmetic coloration.

Gulls

In winter, the belly and breast plumage in several species of gulls and terns (family Laridae) becomes suffused with a pink coloration (Plate 18; see Sibley 2000) that was long thought to have been applied to the surface of the feathers from the birds' uropygial gland secretions (Grant 1986). The color probably comes from carotenoids in the tissues of copepods and euphausiids that these birds eat during the winter while foraging on the ocean (Grant 1986), and in most species (except Ross's Gull [*Rhodostethia rosea*]; Plate 18), the color fades within 2–4 weeks of arrival on the breeding grounds (e.g., Burger and Gochfeld 1994). The fact that the color fades led many to believe that it must be applied to the feather surface. Two recent studies, however, have shown that, at least in Elegant Terns (*Sterna elegans;* Hudon and Brush 1990) and both Ring-billed (*Larus delawarensis*) and Franklin's Gulls (*L. pipixcan;* McGraw and Hardy 2005), this coloration is due to the carotenoid astaxanthin deposited inside the feathers during the molt. This temporary coloring is clearly an interesting phenomenon that needs further investigation, but it seems clear from this recent evidence that it is not a cosmetic color.

Display of Colored Objects

Given the incredible diversity of plumage colors and patterns in birds, it may be a little surprising to find that males of a few species also use vividly colored objects gathered from their environment in their courtship or territorial displays. So far this phenomenon has been reported in only three of the more than 200 bird families—in the fairy-wrens (family Maluridae) and bowerbirds (family Ptilonorhynchidae) of Australia and New Zealand, and in the motmots (family Momotidae) of tropical Central and South America. Although the collection of colored objects has been studied in only a few species so far, it is clear that it influences female choice in both bowerbirds and fairy-wrens, and that these objects have some properties that make them a different kind of ornament from plumage and bare part colors.

Bowerbirds

Males in 17 of the 20 bowerbird species clear display courts on which they arrange up to 12,000 colored objects (Frith and Frith 2004) and on which most species also build bowers, structures of varying size and complexity made from sticks and grass stems (Borgia 1986). Although it has long been obvious that bowers are used in courtship (Darwin 1871), their relevance to the advertisement of male quality has only recently been studied in a few species (e.g., Borgia 1986; Madden 2003a), and much remains to be learned. Across bowerbird species, there appears to be a general negative correlation between the colorfulness of male plumage and the complexity of bowers and decorations,

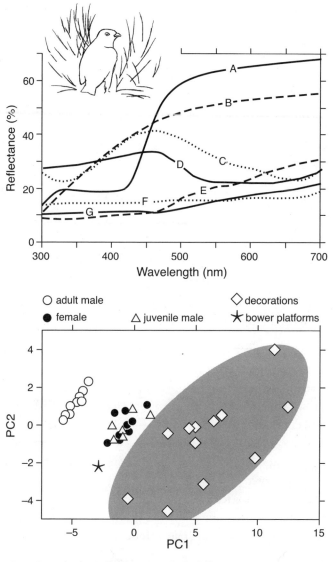

Figure 9.2. Satin Bowerbirds: Reflectance spectra (top) from seven common natural bower decorations in the bowers of males on Mt. Baldy, Queensland, Australia. A = yellow Sulfur-crested Cockatoo (*Cacatua galerita*) feather; B = white Sulfur-crested Cockatoo feather; C = blue *Solanum* berry; D = purplish Crimson Rosella (*Platycercus elegans*) feather; E = yellowish-brown orchid stem; F = brownish cicada exoskeleton; G = brownish snail shell. Bottom graph shows comparison of spectral characteristics of adult males, juvenile males, and females, as well as mean decoration scores for each bower, and the mean score of all bower platforms (n = 12). Principal Component scores (PC1 and PC2) were derived from spectra shown in the top graph; shaded region is the 90% density ellipse for the mean decoration scores. Unpublished data from S. Doucet and R. Montgomerie.

suggesting that plumage ornamentation may have declined as bowers evolved (Gilliard 1969; Kusmierski et al. 1997).

The most extensive and intensive research on bowers and bowerbirds has been conducted on Satin Bowerbirds (*Ptilonorhynchus violaceus*) in eastern Australia (Borgia 1986, 1995; Patricelli et al. 2003; Coleman et al. 2004). Males of this species decorate their bower platform with a large number of natural and manmade objects (Borgia 1986; Plate 20), the colors and numbers of which are correlated with various aspects of male quality (Doucet and Montgomerie 2003; Patricelli et al. 2003) and influence female choice (Coleman et al. 2004). Blue, yellow, and brownish objects often predominate, with flowers, leaves, stems, feathers, insect parts, and snails common in bowers in remote forested areas, but drinking straws, clothespins, string, flagging tape, and small toys increasing in frequency the nearer the bower is to human activities (R. Montgomerie, pers. obs.). Analysis of the colors of more than 250 decorations of over 20 different types from 12 Satin Bowerbird bowers in the rainforest on Mt. Baldy, Queensland, shows that the color of most decorations contrasts with that of both the male's and the female's plumage, as well as with the color of the bower platform itself (S. Doucet and R. Montgomerie, unpubl. data; Figure 9.2; Plate 20). These observations suggest that objects are chosen by males to augment their plumage display and to be easily distinguishable from the platform on which they are arranged.

When given a choice, male Satin Bowerbirds prefer to decorate their bowers with blue and yellow objects gathered from their environment or stolen from their neighbors' bowers (S. Doucet and R. Montgomerie, unpubl. data). For example, given a choice of otherwise identical blue, green, yellow, and red feathers, all 10 males that were tested chose blue feathers first and added them to their own bower decorations; five males also took some yellow feathers, four took some green feathers, only one male took orange feathers, and none took red. Thus the number and color of decorations on a male's bower are the result of the choices that he makes from the colored objects available in his environment, as well as his ability to protect his choices from being stolen by neighbors (S. Doucet and R. Montgomerie, unpubl. data).

The number and color of decorations on a male's bower is correlated with both his body size and his ectoparasite load (Doucet and Montgomerie 2003), suggesting that females could use bower decorations as a signal of these aspects of male quality. Because females prefer to visit, court, and copulate with males that have had the number of blue objects on their bowers experimentally augmented (Coleman et al. 2004), the colors of a male's bower decorations appear to be an important component of his "extended phenotype," strongly influenced by sexual selection, and potentially advertising some aspects of male quality that females can pass on to their offspring.

Males of many other bowerbird species also use characteristic types and colors of decorations on their bowers (Frith and Frith 2004). Spotted Bowerbirds (*Chlamydera maculata;* Plate 20) in central Queensland, for example, prefer green *Solanum* berries (Madden 2003a), and the number of these berries on a male's bower accurately predicts his mating success (Madden 2003b). In southern Queensland, however, Spotted Bowerbird males that exhibit red, pink, or purple glass on their bowers enjoy higher mating success (Borgia and Meuller 1992).

In New Guinea, male Vogelkop Bowerbirds (*Amblyornis inornatus*) decorate their immense bowers with piles of colored objects, often grouped by color. Males of this species on Mount Wandammen use black fungi, red and orange fruits and flowers, blue butterfly wings, and beetles as decorations, whereas males in the Kumawa Mountains, more than 200 km away, use large green pandanus leaves and either grey, brown, black, or white stones and shells (Diamond 1984). Because both populations appear to have the same ornaments available in their environments, Diamond (1984) argued that the actual colors used by male Vogelkop Bowerbirds as decorations may have a cultural component within populations, which would make a general interpretation of their color choices difficult. Alternatively, ambient light and background colors may differ between these localities (e.g., Endler and Théry 1996), influencing the colors of decorations chosen such that they maximally contrast with the bower platform (Uy and Endler 2004; S. Doucet and R. Montgomerie, unpubl. data).

We are clearly just beginning to understand how bowerbirds use the colors of decorations in their courtship displays. Like any of the other cosmetic colors described in this chapter, they are easily changed in the short term and may thus be a good indicator of current condition and health (Doucet and Montgomerie 2003). Unlike all other colors displayed by birds, however, they can be, and often are, stolen by neighbors for the latters' own use (Borgia and Meuller 1995; S. Doucet and R. Montgomerie, unpubl. data), so the number and quality of the decorations in a male's bower are also affected by his current dominance status. Although the types and colors displayed by males might well be influenced by learning from other males and transmitted culturally (Diamond 1984), it remains to be seen whether there are universal properties of the colors that males within a species use as decorations and whether those properties have a consistent adaptive function.

Cosmetic and Adventitious Colors

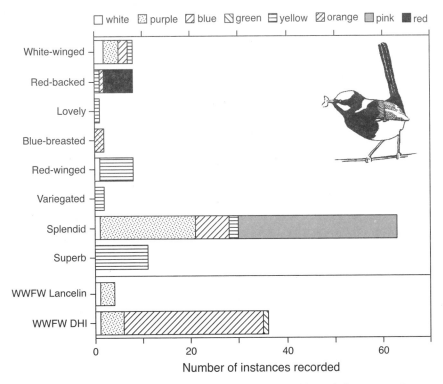

Figure 9.3. Frequency distribution of flower-petal colors carried by male fairy-wrens. Data are the number of instances that a petal of each color has been observed being carried by each fairy-wren species. Data for the upper eight species listed are from Rowley and Russell (1997: Table 5.4). Data for White-winged Fairy-wrens on Dirk Hartog Island (WWFW DHI) and the mainland (WWFW Lancelin) of Western Australia are from Rathburn and Montgomerie (2003).

Fairy-wrens

Despite being among the most brilliantly colored birds, male fairy-wrens often carry a small colorful flower or fruit in their bill during courtship (Figure 9.3; Plate 21), especially when visiting a female on a nearby territory, even outside the breeding season (Rowley and Russell 1997). Such behavior has been recorded in eight of the nine Australian fairy-wren species, with either yellow or orange petals carried by every species, but blue, pink, and purple petals predominating in the best-studied species (Figure 9.3).

In White-winged Fairy-wrens (*Malurus leucopterus*) in Western Australia, black nuptial males of the subspecies *leucopterus* on Dirk Hartog Island (Plate 21) most often carried blue petals (Rathburn and Montgomerie 2003). In contrast, cobalt blue males of the subspecies *leuconotus* (Plate 21) on the nearby mainland rarely carried petals at all, but when they did, purple or pink were the most common colors observed (Figure 9.3). This difference in color choice was obtained even though there was no obvious difference in the petal colors available to the birds (Rathburn and Montgomerie 2003). As Rowley and Russell (1997) suggested, the preferred colors of these courtship props might well contrast with male plumage color, thereby enhancing his display. Unfortunately this petal-carrying behavior is very difficult to observe and thus remains enigmatic.

Motmots

In at least three species of motmot (family Momotidae), both sexes have been recorded performing a "leaf-display," during which a large yellow (or, less often, green) leaf is grasped at its midpoint by the tip of the bill, and held there for up to 20 minutes (Skutch 1964; T. Murphy pers. comm.). Because this display is seen only in members of a mated pair, and only during the breeding season, it probably serves some reproductive function, although it is clearly not a courtship display that only precedes copulation (T. Murphy, pers. comm.). In Turquoise-browed Motmots (*Eumomota superciliosa*), the display is most often seen during, or immediately after, territorial disputes with neighbors, suggesting that it may somehow be related to dominance interactions. A similar display has also been recorded in Blue-crowned (*Momotus momota*; Skutch 1964) and Russet-crowned (*M. mexicanus*) Motmots (T. Murphy, pers. comm.), and may well occur in all nine species in the family. Because the birds do not use leaves in their burrow nests, or in any other fashion, the origins of this display are enigmatic, but the specific preference for yellow leaves suggests that leaf color may be important in the display.

Revealed Plumage Colors

In at least a few species, incidental or deliberate abrasion of the plumage removes some feather tips to reveal an underlying plumage color. This phenomenon has been noted only in starlings (family Sturnidae) and a number of finches, sparrows, and buntings (families Fringillidae and Passeridae), in which dark or buffy feather tips wear away during the winter nonbreeding season to reveal a more colorful plumage beneath, just before the breeding season begins (Newton 1972; Cabe 1993; Willoughby et al. 2002). The taxonomic distribution of this method of exposing breeding plumage colors in songbirds has not yet been documented and may be more widespread than the few anecdotal reports would indicate.

The European Starling (*Sturnus vulgaris*) provides what is probably the most striking example of a change in plumage color due to feather abrasion but this phenomenon seems never to have been studied. Starlings undergo a single complete molt in the summer, with juveniles beginning to molt 4–6 weeks after leaving the nest and adults beginning in late June or early July (Cabe 1993). Following this molt, all starlings have large cream-to-cinnamon colored spots at the tips of some body feathers, giving the bird a very spotted appearance (Plate 23). During the next 6 months, these spotted feather tips gradually wear away so that, by the start of the next breeding season, starlings have a mainly black, glossy-looking plumage (Plate 23), often retaining a few small buffy spots (Cabe 1993). The mechanisms, progression, and reasons for this change in color by abrasion are completely unknown.

The abrasion of feather tips to reveal an underlying color has been documented quantitatively only in male House Sparrows (*Passer domesticus;* Møller and Erritzoe 1992), and in both sexes of Lawrence's Goldfinch (*Carduelis nivalis;* Willoughby et al. 2002). House Sparrows have a single annual molt in the autumn, after which the feathers on and around the bottom and side margins of their black throat patch (bib) have 1- to 6-mm-long whitish tips (Plate 23). These light feather tips gradually wear down, apparently as a result of daily preening with the bill (Møller and Erritzoe 1992). When the feather tips break off, more of the black feathers beneath become exposed, progressively enlarging the visible area of the black bib (Bogliani and Brangi 1990; Plate 23)—a badge that is associated with male dominance (Møller 1990). Møller and Erritzoe (1992) showed that a male with the largest badge (1,000 mm^2) would have it fully visible by January, whereas a male with the smallest badge (400 mm^2) would not have it fully exposed until March. Thus they argued that badge size might be an honest indicator of male quality, and that this method of exposing the badge gradually by preening allows males flexibility in the acquisition of their full breeding plumage. Such flexibility would allow males to gauge their dominance status and adjust badge size accordingly, thereby avoiding costly aggression that might ensue if their badge size advertised that they were more dominant (aggressive) than they really were. These ideas need to be tested experimentally.

Lawrence's Goldfinches have only one molt per year (Willoughby et al. 2002), in the autumn. Following the molt, grayish and olive-brown feather tips on the backs and breasts of both males and females gradually wear away

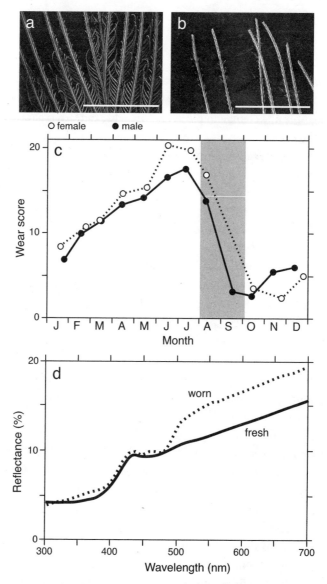

Figure 9.4. Lawrence's Goldfinch: (a,b) Photomicrographs of worn breast feathers of a male captured on 16 May; (a) shows proximal parts of terminal barbs used in calculating the wear scores shown in (c), and (b) shows the distal ends of those same barbs (scale bars = 500 μm); (c) seasonal variation in wear scores from selected male and female feathers (shaded area shows period of molt); (d) mean reflectance spectra from both freshly molted and worn (abraded) back plumage of males. Worn back feathers reflect significantly more in the green-red portion of the spectrum (500–700 nm), making the back plumage look more yellowish. Photos and graphs adapted from Willoughby et al. (2002).

(Figure 9.4), revealing feather parts with more saturated yellowish colors underneath—a change in color that had previously attributed to a late winter molt before the breeding season (Davis 1999). Willoughby et al. (2002) nicely documented the seasonal progress of this abrasion, the changes in colors, and the details of barbule breakage (Figure 9.4a,b), but whether this abrasion is passive or actively promoted by the birds is unknown. Immediately following molt, the yellow breast feathers of males, for example, are brownish in color at the tips of the barbules. As these brownish barbules gradually break off, the rachillae become increasingly denuded at the tips of the barbs, thus enhancing the yellow color of the back (Figure 9.4c,d) and enlarging the size of the yellow breast as the summer breeding season approaches. The rachillae of these yellow breast feathers are unusually thick on males, presumably resisting wear and further loss of yellow color (Willoughby et al. 2002).

Male Purple Finches (*Carduelis purpureus*) also have red carotenoid-rich feathers on their crown that are tipped with whitish barbules that break off as the breeding season approaches, making their crown more intensely red (Dwight 1900). Although it is clear that the structure of these crown feathers is modified in such a way (reduced melanin, flattened barbs) to increase the exposure of the feather pigments (Brush and Seifried 1968), the mechanism of feather abrasion in this species has never been studied.

Colors are also revealed by abrasion in two closely related high arctic songbirds, the Snow Bunting (*Plectrophenax nivalis;* Lyon and Montgomerie 1995) and the Lapland Longspur (*Calcarius lapponicus;* Hussell and Montgomerie 2002), both in the family Emberizidae). In both species, the fall (prealternate; August–September) molt brings in white- or gray-tipped feathers on the head and neck, such that the winter plumage has little exposed black (Plate 22). As spring approaches, the whitish feather tips (up to 7 mm long) wear or break off, exposing the underlying black part of the feather and dramatically changing the appearance of the males (Plate 22). On the breeding grounds, male Snow Buntings have often been seen deliberately rubbing their heads on crusty snow, apparently trying to hasten the removal of the feather tips (Lyon and Montgomerie 1995).

My survey of the species accounts in Poole (2004) shows that a few other North American species have been recorded to change their plumage by abrasion, but either to a lesser extent than, or not as well studied as, those described above. Male Blue Grosbeaks (*Passerina caerulea;* Plate 7; Blake 1969) and Mallards (Plate 15, Volume 2; Drilling et al. 2002), for example, gradually wear buffy feather tips on their blue body and chestnut breast feathers, respectively, before the breeding season begins, making these regions more uniformly colored. In the Eastern (*Sturnella magna*) and Western Meadowlark (*S. neglecta*), subapical spots on the back feathers and light-colored edgings on all of the body feathers wear off during the winter, making the plumage colors much more vivid by the start of the breeding season (Lanyon 1994, 1995). In Sanderlings (*Calidris alba*), white fringes on the feathers of the crown, mantle, and upper scapulars wear off during the breeding season, so that by July, when the eggs have already hatched, these plumage regions are mainly black (MacWhirter et al. 2002). This particular change in plumage coloration thus has no obvious sexual function, but may result in better thermoregulation after the incubation period, during which time the birds may need the feather fringes to enhance crypsis.

The reason that some birds expose plumage colors by abrasion rather than molt is unknown, but it does seem to provide a means by which males can change their plumage display before the breeding season without the energetic expense of a molt. In those species, the completeness of feather-tip removal may also provide an honest signal of a male's recent health and dominance status (as in House Sparrows) and could potentially be a useful way for females or rival males to assess a male's quality.

Adventitious Colors

In the course of a bird's normal daily activities, many of its feathers undoubtedly undergo some physical changes from exposure to sun, rain, wind, abrasion, dirt, and the action of microorganisms. These agents are all likely to have a measurable effect on feather color with time, but there have been few quantitative data published on this topic to date. In this section, I draw mostly on anecdotal observations and a few very recent quantitative studies to summarize what is known about this subject.

Dirt and Stains

Some colors seem to be acquired passively and often appear to have no obvious adaptive function. For example, the head and neck plumage of several species of waterfowl (mostly geese and swans; family Anatidae), the bellies of loons (*Gavia* spp.) and phalaropes (*Phalaropus* spp.), and the backs of Sandhill Cranes (*Grus canadensis*) often acquire a rusty, red-orange hue (Plate 24), mainly through the fall and winter months (Kennard 1918). Thus these birds often arrive on their spring breeding grounds looking very different than they did after their last molt, at the end of the previous breeding season. As in Bearded Vultures (Plate 18), the red-orange color is caused by iron oxide that stains the tips of feathers (Grinnell 1910; Kennard 1918; Höhn 1955) and is most visible on white plumage on heads (e.g., Snow Geese [*Chen caerulescens*], Emperor Geese [*C. canagica*]; Plate 24) and bellies (e.g., Red-throated Loon [*Gavia stellata*), Red-necked Phalarope [*Phalaropus lobatus*]; Kennard 1918).

Höhn (1955: 414) reported that the brownish-red stains on the backs of Sandhill Cranes (Plate 24) might be caused by "the birds' placing water weeds on the back," suggesting that the staining could be deliberate and thus have an adaptive, cosmetic function in that species. Indeed, Lynch (1996) watched cranes dab iron-rich mud and rotting vegetation on their feathers, and noted that the staining made an incubating bird much harder to see. In most other species, however, the stains are thought to be simply an incidental byproduct of the birds' activities (Grinnell 1910; Kennard 1918). For example, staining is most common on feathers that regularly come into contact with water—on the bellies of birds that float on the surface and on the heads of birds that forage with their heads submerged (Plate 24), particularly those that forage in soft mud (Kennard 1918). The degree of staining also varies considerably among individuals within and between populations (Plate 24), presumably depending upon the degree of contact with iron-rich waters and mud (Kennard 1918). Even if these stains are acquired passively, they might still have some signal value, particularly for individual recognition (Chapter 2, Volume 2) given the considerable within-population variation that has been reported (Kennard 1918). Staining might also provide some chemical benefits (as has been proposed for Bearded Vultures) or have some detrimental effects on feather quality, but these possibilities have never been examined.

Among North American birds, at least 28 of the more than 700 species have been recorded with stains on their plumage (information extracted from Poole 2004). In addition to the iron-oxide stains described above, nonpasserines are also sometimes stained by their nesting, roosting, and foraging substrates. The white plumage of Downy Woodpeckers (*Picoides pubescens*), for example, has been reported to be stained by red soils, plants like Black Walnut (*Juglans nigra*) and mulberry (*Morus* spp.), tannins in waterlogged roosting cavities, and artificial stains on the wooden sidings of buildings (Jackson and Ouellet 2002). Fruits and berries are the most common source of plumage staining in pigeons, doves, and passerines. In one study, approximately 20% of Tennessee Warblers (*Vermivora peregrina*) feeding on *Combretum* flowers had their throat and face stained pink to orange (Ficken and Ficken 1962). Even a predominantly carnivorous bird, such as the Eskimo Curlew (*Numenius borealis*), was reported to sometimes have its vent, bill, legs, throat, and breast stained with deep purple berry juice (Coues 1861).

The plumage of many other birds gets very dirty with time (Box 9.1), but, as with staining, this dirtiness seems to serve no positive adaptive function and might indeed be detrimental to thermoregulation, feather structure, and/or an individual's attractiveness to potential mates. The considerable effort expended by birds preening and bathing (Cotgreave and Clayton 1994) suggests that keeping the feathers clean is important, and the accumulation of dirt might well be a sign of poor health or condition.

UV Damage, Abrasion, and Bacterial Degradation

Feather colors might also change after molt as a result of (1) UV damage, (2) abrasion of their fine structure, or (3) keratin-degrading enzymes from bacteria on their surfaces (Shawkey and Hill 2004). For example, Örnborg et al. (2002) showed that the structural color (blue) on the crown of adult male and female Blue Tits (*Parus caeruleus;* Plate 18; Volume 2) in Sweden changed from October (immediately after molt) through June (when nestlings were being fed). In particular, they found that hue of crown feathers shifted toward longer wavelengths as the season progressed, being in the ultraviolet part of the spectrum immediately after molt and moving steadily into the blue region by the chick-feeding period (Figure 9.5). Brightness and chroma also changed during this period, although there were some differences between years and study sites where the measurements were taken. Detailed analysis of the structure of feathers plucked from the birds' crowns throughout the study period showed significant effects of abrasion on barb length and the number of broken barbs on a feather (Figure 9.5), but no effect on any of the color variables measured (J. Örnborg and S. Andersson, unpubl. data). Örnborg et al. (2002) suggested that the color changes that they documented might be related to an increase in the amount of dirt and oils on feathers as the season progresses, rather than resulting from abrasion, but this idea still needs to be tested. Alternatively, exposure to the sun might progressively damage some of the keratin nanostructuring that produces UV reflectance (Prum et al. 1999), causing at least a shift in hue.

The carotenoid-based plumage color patches on male House Finches (*Carpodacus mexicanus;* Plate 31; McGraw and Hill 2004) and Great Tits (*Parus*

Box 9.1. Dirty Birds

Most birds look immaculate nearly all of the time, the result of a daily regimen of preening and bathing that maintains their feathers in the best possible condition. This meticulous grooming is not so surprising, given the importance of plumage for insulation, shedding water, flight, and, of course, signaling. Indeed, most birds that have been studied in any detail devote at least a small proportion of their time every day to feather maintenance (Cotgreave and Clayton 1994).

Some species, however, get very dirty during the breeding season, in a manner that may not confer any adaptive advantage but will have a large impact on perceived and measured colors. This phenomenon has not yet been seriously studied in birds, so I provide three examples here from my own experience preparing bird specimens for a museum where we routinely washed skins in a mild detergent. House Sparrows, Pine Grosbeaks (*Pinicola enucleator*), and Evening Grosbeaks (*Coccothrauses vespertinus*) all get strikingly dirty during their breeding seasons (at least). In general, dust-bathing helps clean feathers (Vestergaard et al. 1999), so it is unlikely that any of this dirtiness is acquired directly by this means. Because urban House Sparrows generally have dirtier plumage than those living in rural areas, dirty living conditions are probably responsible for their dirty plumage in cities. Similarly, grosbeaks presumably become soiled incidentally from resins in the leaves of evergreen trees in which they forage and nest and not from any deliberate soiling of their plumage.

To look at the effects of such dirt on plumage color in more detail, I measured reflectance spectra from the breasts of washed and unwashed males of each species at the Royal Ontario Museum (Toronto, Canada), matching birds within species for year and month of collection to minimize the effects of specimen storage and the season in which they were collected. In House Sparrows, the effect of dirt on their white breast/belly feathers is both to reduce the total reflectance (brightness) to about half the value for clean feathers ($P < 0.05$, repeated measures ANOVA) and to increase the UV chroma (relative plumage reflectance in the UV range; Figure B9.1). Thus very dirty House Sparrows appear to reflect relatively more UV from their white plumage than do clean birds because the non-UV portion of the spectrum is masked by dirt more than is the UV region below 350 nm.

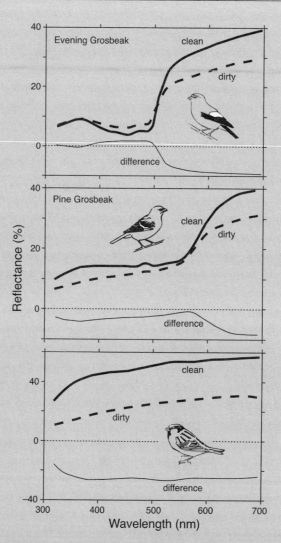

Figure B9.1. Reflectance curves from the breast feathers of washed and unwashed museum specimens of male Evening Grosbeaks (Plate 25), Pine Grosbeaks (Plate 25), and House Sparrows (Plate 23). Curves shown are average reflectance spectra from three washed (clean) and three unwashed (dirty) males of each species, where the curve from each individual is calculated as the average from five spectral readings from haphazardly chosen locations on each bird's breast. The difference between the clean and dirty curves is also shown. See Plate 25 for photos of patches of washed and unwashed plumage from each species.

Ed: Replacement figure used. Pls. advise.

> The effect of resins on grosbeak plumage colors is somewhat complex. In Pine Grosbeaks, which have red breast plumage, the dirtiness associated with increased resin on the feathers results in a shift in hue toward yellow-orange and away from red, and a significant reduction in reflectance in both the blue (<500 nm) and red (>620 nm) regions of the spectrum (Figure B9.1). Thus dirtier male Pine Grosbeaks appear to have less red in their plumage. In Evening Grosbeaks, which have yellow breast plumage, there is almost no change in hue resulting from a dirty plumage (R. Montgomerie, unpubl. data), but reflectance is reduced in the red portion of the spectrum and increased slightly in the blue region, due to the resins on the feathers (Figure B9.1).

major; Plate 11, Volume 2; Senar 2004) also change hue with time. In individual male House Finches, a change in hue from reddish to more yellowish, in the period from January to July, was correlated with the time between measurements (Figure 9.6), but was not related to the male's age or his original hue (McGraw and Hill 2004). The color of these patches also became significantly more saturated during this period, but did not change in brightness. McGraw and Hill (2004) suggested that this change in color may be due either to abrasion or soiling of the feather surface, or to the degradation of carotenoid pigments by continued exposure to UV irradiation or bacterial enzymes. All of these mechanisms are plausible, not mutually exclusive, and very much in need of further study.

The feathers of all birds probably get progressively more worn, faded, and dirty between molts, and this circumstance should be taken into account, especially when comparing individuals at different periods following a molt. In species with brown plumage, the changes in colors due to fading and abrasion can be considerable (e.g., Fox Sparrow [*Passerella iliaca*]; Weckstein et al. 2002], making characterization of their plumage colors difficult. Thus when analyzing plumage colors (Chapter 3), it is wise to be aware of that some of the measured variation might not represent intrinsic variation in feather color due to pigments (Chapters 5, 6, and 8) or keratin nanostructures (Chapter 7). Feathers can be readily plucked and washed to check for such effects (Box 9.1), but to date, this potential source of variation in plumage color signals has not

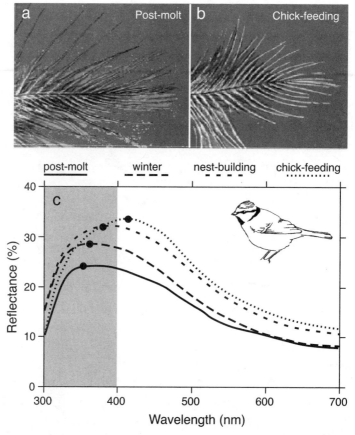

Figure 9.5. Photographs of crown feathers of a Blue Tit taken during (a) post-molt and (b) chick-feeding periods; note the lack of barbules on the rachillae of feathers during the chick-feeding period. Adapted from J. Örnborg and S. Andersson (unpubl. data). (c) Reflectance spectra of the crown measured at four stages following molt; solid dots show reflectance maxima (R_{max}) and indicate that hue (wavelength at R_{max}) increases through the season, possibly due to UV damage or dirt accumulation; shaded area shows the UV region. Note that the proportion of the area under the curve in the UV region (i.e., UV chroma) decreases as the season progresses. Adapted from Örnborg et al. (2002).

been studied in detail for any bird species. There is often a slight trend with brightness decreasing progressively after the molt (e.g., Örnborg et al. 2002; R. Montgomerie, pers. obs.), but rarely is it statistically significant or of large magnitude (usually <1% of total reflectance per month). Whether such trends are the result of increasing dirtiness, however, is not yet known.

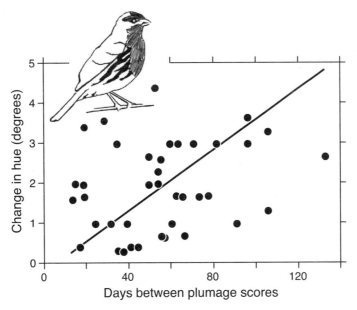

Figure 9.6. Carotenoid-based red plumage of male House Finches fades (increased hue score) progressively during the period from late January to mid-July. Data from individual males ($n = 41$) that were measured twice up to 135 days apart and also showed some degree of plumage fading between measurements. Model II regression line is shown. Adapted from McGraw and Hill (2004).

Conclusions

Although there are relatively few examples of colors that are applied, assembled, and exposed by birds to alter or enhance their appearance, there are some common threads among these types of color displays. First, all these forms of plumage coloration involve changes that take place after the molt, sometimes even months later. Unlike colors acquired via molt, the ability of an individual to change its color may be influenced by its current condition or health. Thus, when cosmetic and adventitious colors are used in sexual signals, they are always a more recent snapshot of the bearer's condition than are colors acquired by molt. For example, cosmetic and adventitious colors may be a more useful indicator of current parenting ability, as this behavior can change seasonally, depending on health and condition. Bare part colors, in contrast, seem to be rarely modified exogenously, possibly because these colors can be changed in the short term and already are good indicators of current condition. Second, many cosmetic and adventitious colors can change relatively rapidly, potentially indicating a change in condition (e.g., Red Knot, bowerbirds, Bearded Vulture) or mating status (e.g., Rock Ptarmigan). Finally, almost all cosmetic and adventitious colors are influenced by selection on behavior, so variation among individuals must therefore reflect differences in behavior, whether it be dominance (Bearded Vulture), searching and spatial memory (Bearded Vulture, bowerbirds, fairy-wrens), or social interactions (Rock Ptarmigan, House Sparrow).

Clearly, cosmetic and adventitious colors need more study. How, for example, does intraspecific variation in these colors influence reproductive success via mate acquisition (e.g., bowerbirds, Great Hornbill, Bearded Vulture, House Sparrow) or survival (e.g., Rock Ptarmigan, Red Knot)? What effects do abrasion, UV irradiation, and bacterial degradation have on seasonal change in colors (e.g., Blue Tit, House Finch)? And are such changes somehow minimized by the birds, and if so, why? Are cosmetic and adventitious colors more widespread in birds than has so far been revealed?

Summary

In this chapter, I review three unusual ways that birds actively acquire and use cosmetic and revealed coloration for display or camouflage: (1) by applying dirt, stains, and waxes to their plumage, (2) by gathering colored objects from their surroundings, and (3) by removing feather tips through abrasion and preening to reveal the hidden colors beneath. I also summarize what little is known about adventitious changes in colors with time, due to soiling, staining, abrasion, UV irradiation, and bacterial damage. There are few examples of each of these phenomena, but research on such color changes has given us some intriguing insights into color displays in general, and sexual selection in particular. For example, cosmetic colors can usually be acquired as needed, often long after the molt. For that reason, they might be more accurate signals of current health and condition than either feather pigments or structural colors that are acquired by molting, usually long before they are used in displays.

Great Hornbills, Bearded Vultures, and Rock Ptarmigan all apply substances to their plumage that change the color in a way that is probably influenced by sexual selection. Colored objects gathered during courtship by most species of bowerbirds and fairy-wrens have clearly been influenced by sexual selection, as evidenced by the timing of these displays, their correlation with male quality, and a few studies that show how they influence female choice.

Colors that are revealed by active abrasion of feather tips that differ in color from the base of the feather have been reported in European Starlings, several cardueline finches, House Sparrows, Lapland Longspurs, and Snow Buntings. This phenomenon has been studied very little and may be much more widespread, as it provides a potentially inexpensive method of converting a dull, cryptic winter plumage to a vivid breeding ornament without the need for a costly molt.

Adventitious agents that can have cumulative effects over time, such as dirt, iron oxide, UV irradiation, bacterial enzymes, and abrasion, also have a measurable effect on colors and may need to be taken into account when comparing the reflectance spectra of individuals and species. The feathers of many waterfowl, for example, acquire orange-red stains from iron oxide when they forage extensively in iron-rich water and mud. Although such adventitious color changes are probably not actively promoted by the bird, they may well indicate aspects of quality and health that provide a useful signal to conspecifics.

References

Arlettaz, R., P. Christe, P. F. Surai, and A. Møller. 2002. Deliberate rusty staining of the plumage in the Bearded Vulture: Does function precede art? Anim Behav 64: F1–F3.

Bertrán, J., and A. Margalida. 1999. Copulatory behavior of the Bearded Vulture. Condor 101: 164–168.

Blake, C. H. 1969. Notes on the Indigo Bunting. Bird Banding 40: 133–139.

Bogliani, G., and A. Brangi. 1990. Abrasion of the status badge in the male Italian Sparrow *Passer italiae*. Bird Study 37: 195–198.

Borgia, G. 1986. Sexual selection in bowerbirds. Sci Am 254: 92–101.

Borgia, G. 1995. Why do bowerbirds build bowers? Am Sci 83: 542–547.

Borgia, G., and U. Meuller. 1992. Bower destruction, decoration stealing, and female choice in the Spotted Bowerbird (*Chlamydera maculata*). Emu 92: 11–18.

Brown, C. J., and A. G. Bruton. 1991. Plumage colour and feather structure of the Bearded Vulture (*Gyptaeus barbatus*). J Zool (Lond) 223: 627–640.

Brush, A. H., and H. Seifried. 1968. Pigmentation and feather structure in genetic variants of the Gouldian Finch, *Poephila gouldiae*. Auk 85: 416–430.

Burger, J., and M. Gochfeld. 1994. Franklin's Gull (*Larus pipixcan*). In A. Poole, ed., The Birds of North America Online, retrieved 25 November 2004, from http://bna.birds.cornell.edu/. Ithaca, NY: Cornell Laboratory of Ornithology.

Cabe, P. R. 1993. European Starling (*Sturnus vulgaris*). In A. Poole, ed., The Birds of North America Online, retrieved 25 November 2004, from http://bna.birds.cornell.edu/. Ithaca, NY: Cornell Laboratory of Ornithology.

Coleman, S., G. L. Patricelli, and G. Borgia. 2004. Variable female preferences drive complex male displays. Nature 428: 742–745.

Cotgreave, P., and D. H. Clayton. 1994. Comparative analysis of time spent grooming by birds in relation to parasite load. Behaviour 131: 171–187.

Coues, E. 1861. Notes on the ornithology of Labrador. Proc Acad Nat Sci Phila 13: 215–257.

Darwin, C. 1871. The Descent of Man, and Selection in Relation to Sex. London: John Murray.

Davis, J. N. 1999. Lawrence's Goldfinch (*Carduelis lawrencei*). In A. Poole, ed., The Birds of North America Online, retrieved 25 November 2004, from http://bna.birds.cornell.edu/. Ithaca, NY: Cornell Laboratory of Ornithology.

Diamond, J. 1984. The bower builders. Discover 5: 52–58.

Doucet, S. M., and R. Montgomerie. 2003. Multiple sexual ornaments in Satin Bowerbirds: Ultraviolet plumage and bowers signal different aspects of male quality. Behav Ecol 14: 503–509.

Drilling, N., R. Titman, and F. McKinney. 2002. Mallard (*Anas platyrhynchos*). In A. Poole, ed., The Birds of North America Online, retrieved 25 November 2004, from http://bna.birds.cornell.edu/. Ithaca, NY: Cornell Laboratory of Ornithology.

Dwight, J. 1900. The sequence of plumages and moults of the passerine birds of New York. Ann NY Acad Sci 13: 73–360.

Endler, J. A., and M. Théry. 1996. Interacting effects of lek placement, display behavior, ambient light, and color patterns in three Neotropical forest-dwelling birds. Am Nat 148: 421–452.

Ficken, M. S., and R. W. Ficken. 1962. The comparative ethology of the wood warblers: A review. Living Bird 1: 103–122.

Frey, H., and N. Roth-Callies. 1994. Zur Genese der Haftfarbe (Rostfärbung durch Eisenoxid) beim Bartgeier, *Gyptaeus barbatus*. Egretta 37: 1–22.

Frith, C. B., and D. W. Frith. 2004. The Bowerbirds *Ptilorynchidae*. Oxford: Oxford University Press.

Gardarsson, A. 1971. Food Ecology and Spacing Behavior of Rock Ptarmigan *Lagopus mutus* in Iceland. Ph.D. diss., University of California, Berkeley, Berkeley.

Gardarsson, A. 1988. Cyclic population changes and some related events in Rock Ptarmigan in Iceland. In A. T. Bergerud, and M. W. Gratson, ed., Animal Populations in Relation to Their Food Resources, 465–519. Minneapolis: University of Minnesota Press.

Gilliard, E. T. 1969. Birds of Paradise and Bowerbirds. London: Weidenfeld & Nicolson.

Grant, P. J. 1986. Gulls: A Guide to Identification, second edition. Calton, UK: T. and A. D. Poyser.

Grinnell, J. 1910. Birds of the 1908 Alexander Alaska Expedition with a note on the avifaunal relationships of the Prince William Sound District. Univ Calif Publ Zool 5: 361–428.

Hingston, R. W. G. 1933. The Meaning of Animal Colour and Adornment. London: Edward Arnold.

Höhn, E. O. 1955. Evidence of iron staining as the cause of rusty discoloration of normally white feathers in Anserine birds. Auk 72: 414.

Hudon, J., and A. H. Brush. 1990. Carotenoids produce flush in the Elegant Tern plumage. Condor 92: 798–801.

Hussell, D. J. T., and R. Montgomerie. 2002. Lapland Longspur (*Calcarius lapponicus*). In A. Poole, ed., The Birds of North America Online, retrieved 25 November 2004, from http://bna.birds.cornell.edu/. Ithaca, NY: Cornell Laboratory of Ornithology.

Huxley, J. S. 1938. Darwin's theory of sexual selection and the data subsumed by it, in the light of recent research. Am Nat 72: 416–433.

Jackson, J. A., and H. R. Ouellet. 2002. Downy Woodpecker (*Picoides pubescens*). In A. Poole, ed., The Birds of North America Online, retrieved 25 November 2004, from http://bna.birds.cornell.edu/. Ithaca, NY: Cornell Laboratory of Ornithology.

Jacob, J., and V. Ziswiler. 1982. The uropygial gland. In D. S. Farner, J. R. King, and K. C. Parkes, ed., Avian Biology, Volume 4, 199–324. New York: Academic Press.

Kennard, F. H. 1918. Ferruginous stains on waterfowl. Auk 35: 123–132.

Kolattukudy, P. E., S. Bohnet, and L. Rogers. 1987. Diesters of 3-hydroxy fatty acids produced by the uropygial glands of female Mallards uniquely during the mating season. J Lipid Res 28: 582–588.

Kusmierski, R., G. Borgia, A. Uy, and R. H. Crozier. 1997. Labile evolution of display traits in bowerbirds indicates reduced effects of phylogenetic constraints. Proc R Soc Lond B 264: 307–313.

Lanyon, W. E. 1994. Western Meadowlark (*Sturnella neglecta*). In A. Poole, ed., The Birds of North America Online, retrieved 25 November 2004, from http://bna.birds.cornell.edu/. Ithaca, NY: Cornell Laboratory of Ornithology.

Lanyon, W. E. 1995. Eastern Meadowlark (*Sturnella magna*). In A. Poole, ed., The Birds of North America Online, retrieved 25 November 2004, from http://bna.birds.cornell.edu/. Ithaca, NY: Cornell Laboratory of Ornithology.

Lynch, W. 1996. A is for Arctic. Toronto: Firefly Books.

Lyon, B. E., and R. Montgomerie. 1995. Snow Bunting (*Plectrophenax nivalis*) and McKay's Bunting (*Plectrophenax hyperboreus*). In A. Poole, ed., The Birds of North America Online, retrieved 25 November 2004, from http://bna.birds.cornell.edu/. Ithaca, NY: Cornell Laboratory of Ornithology.

MacWhirter, B., P. Austin-Smith, Jr., and D. Kroodsma. 2002. Sanderling (*Calidris alba*). In A. Poole, ed., The Birds of North America Online, retrieved 25 November

2004, from http://bna.birds.cornell.edu/. Ithaca, NY: Cornell Laboratory of Ornithology.

Madden, J. 2003a. Male Spotted Bowerbirds preferentially choose, arrange and proffer objects that are good predictors of mating success. Behav Ecol Sociobiol 53: 263–268.

Madden, J. 2003b. Bower decorations are good predictors of mating success in the Spotted Bowerbird. Behav Ecol Sociobiol 53: 269–277.

McGraw, K. J., and L. S. Hardy. 2005. Astaxanthin is responsible for the pink plumage flush in Franklin's and Ring-billed Gulls. J Avian Biol (in press).

McGraw, K. J., and G. E. Hill. 2004. Plumage color as a dynamic trait: Carotenoid pigmentation of male House Finches (*Carpodacus mexicanus*) fades during the breeding season. Can J Zool 82: 734–738.

Møller, A. P. 1990. Sexual behavior is related to badge size in the House Sparrow *Passer domesticus*. Behav Ecol Sociobiol 27: 23–29.

Møller, A. P., and J. Erritzoe. 1992. Acquisition of breeding coloration depends on badge size in male House Sparrows *Passer domesticus*. Behav Ecol Sociobiol 31: 271–277.

Montgomerie, R., B. Lyon, and K. Holder. 2001. Dirty ptarmigan: Behavioral modification of conspicuous male plumage. Behav Ecol 12: 429–438.

Negro, J. J., A. Margalida, F. Hiraldo, and R. Heredia. 1999. The function of the cosmetic coloration of Bearded Vultures: When art imitates life. Anim Behav 58: F14–F17.

Negro, J. J., A. Margalida, M. J. Torres, J. M. Grande, F. Hiraldo, and R. Heredia. 2002. Iron oxides in the plumage of Bearded Vultures: Medicine or cosmetics? Anim Behav 64: F5–F7.

Newton, I. 1972. Finches. New York: Taplinger.

Örnborg, J., S. Andersson, S. C. Griffith, and B. C. Sheldon. 2002. Seasonal changes in a ultraviolet structural colour signal in Blue Tits, *Parus caeruleus*. Biol J Linn Soc 76: 237–245.

Patricelli, G. L., J. A. C. Uy, and G. Borgia. 2003. Multiple traits interact: Attractive bower decorations facilitate attractive behavioral displays in Satin Bowerbirds. Proc R Soc Lond B 270: 2389–2395.

Piersma, T., M. Dekker, and J. S. S. Damsté. 1999. An avian equivalent of make-up? Ecol Letters 2: 201–203.

Poole, A., ed. 2004. The Birds of North America Online, retrieved 25 November 2004, from http://bna.birds.cornell.edu/. Ithaca, NY: Cornell Laboratory of Ornithology.

Prum, R. O., R. Torres, C. Kovach, S. Williamson, and S. M. Goodman. 1999. Coherent light scattering by nanostructured collagen arrays in the caruncles of the Malagasy asities (Eurylaimidae: Aves). J Exp Biol 202: 3507–3522.

Rathburn, M., and R. Montgomerie. 2003. Breeding biology and social structure of White-winged Fairy-wrens (*Malurus leucopterus*): Contrasts between island and mainland subspecies having different plumage phenotypes. Emu 103: 295–306.

Reneerkens, J., and P. Korsten. 2004. Plumage reflectance is not affected by preen wax composition in Red Knots *Calidris canutus*. J Avian Biol 35: 405–409.

Reneerkens, J., T. Piersma, and J. S. S. Damsté. 2002. Sandpipers (Scolopacidae) switch from mono- to diester preen waxes during courtship and incubation, but why? Proc R Soc Lond B 269: 2135–2139.

Reneerkens, J., T. Piersma, and J. S. S. Damsté. 2005. Discerning adaptive value of seasonal variation in preen waxes: Comparative and experimental approaches. Proc 23rd Int Ornithol Congr (in press).

Rowley, I., and E. Russell. 1997. Fairy-wrens and Grasswrens *Maluridae*. Oxford: Oxford University Press.

Senar, J. C. 2004. Mucho Más Que Plumas. Monografies del Museu de Ciéncias Naturales 2. Barcelona: Museu de Ciéncias Naturales.

Shawkey, M. D., and G. E. Hill. 2004. Feathers at a fine scale. Auk 121: 652–655.

Sibley, D. A. 2000. The Sibley Guide to Birds. New York: Alfred A. Knopf.

Skutch, A. F. 1964. Life history of the Blue-crowned Motmot (*Momotus momota*). Ibis 106: 321–332.

Uy, J. A. C., and J. A. Endler. 2004. Modification of the visual background increases the conspicuousness of Golden-collared Manakin displays. Behav Ecol 15: 1003–1010.

Vestergaard, K. S., B. I. Damm, U. K. Abbott, and M. Bildsøe. 1999. Regulation of dustbathing in feathered and featherless domestic chicks: The Lorenzian model revisited. Anim Behav 58: 1017–1025.

Weckstein, J. D., D. E. Kroodsma, and R. C. Faucett. 2002. Fox Sparrow (*Passerella iliaca*). In A. Poole, ed., The Birds of North America Online, retrieved 25 November 2004, from http://bna.birds.cornell.edu/. Ithaca, NY: Cornell Laboratory of Ornithology.

Willoughby, E. J., M. Murphy, and R. L. Gorton. 2002. Molt, plumage abrasion, and color change in Lawrence's Goldfinch. Wilson Bull 114: 380–392.

III

Controls and Regulation of Expression

10

Hormonal Control of Coloration

REBECCA T. KIMBALL

Numerous studies have examined the evolution of avian color patterns, particularly the evolution of sexual dichromatism (Part II, Volume 2). In contrast, fewer studies have examined the physiological mechanisms that determine sexual, seasonal, or age-related dichromatism. Those studies that have examined the mechanisms controlling dichromatism have been restricted to a limited number of taxa (for reviews, see Domm 1939; Witschi 1961; Vevers 1962; Kimball and Ligon 1999). However, understanding the mechanisms that regulate patterns of dichromatism and intrasexual variation in coloration has implications for understanding the function of these traits. Coloration is an important visual signal in birds and can provide such information as an individual's age, sex, or sexual status, as well as information about the individual's potential quality as a mate or rival (Part I, Volume 2). It is these types of information that have helped stimulate much interest in avian coloration. For example, good-genes models of sexual selection predict that females should select males using traits that accurately reflect male condition (e.g., Andersson 1994; Ligon 1999); however, this prediction requires understanding whether certain traits, such as plumage or bare-part coloration, accurately reflect male quality (Zuk 1991). In addition, an understanding of the proximate control of trait expression may result in the establishment of novel hypotheses to explain the evolution of some coloration patterns (e.g., whether bright male plumage is ancestral; Kimball and Ligon 1999).

Because males and females are largely influenced by the same physiological, ecological, and environmental conditions, sexual dichromatism must be controlled by other factors. Hormones, which can vary sexually, seasonally, and ontogenetically, have been targeted as the primary mechanism controlling dichromatism of plumage and bare parts in birds (Ralph 1969). In addition to affecting dichromatism, hormones may also affect variation in coloration in a particular sex or age-class, although this area has received less attention. Here I review the hormonal factors that are known to affect coloration of plumage and nonplumage traits and discuss the functional and evolutionary implications of these patterns.

Hormones and Coloration

Studies examining the mechanisms regulating coloration have primarily included (1) seasonally or permanently sexually dichromatic species, (2) monochromatic species in which both sexes alter plumages seasonally, and (3) monochromatic species in which adults and juveniles exhibit different plumages. These studies have largely focused on adults (but see Strasser and Schwabl 2004), and thus only provide information about the role of hormones and coloration after development has been completed. Whether hormones also have organizational effects on coloration remains to be examined. Experimental methods used to elucidate the mechanisms regulating coloration patterns include castration of males or females (gonadectomy) and supplementation with one or more hormones. For many species, both gonadectomies and hormone supplementation have been used, either in the same study or in different studies on the same species. A smaller number of studies have examined the control of dichromatism by exchanging skin grafts or gonadal tissue between males and females, although these have largely been done in species in which gonadectomy and/or hormone supplementation have also been examined.

In addition to experimental studies, many anecdotal observations exist of individuals that exhibit coloration inconsistent with their sex and age, or exhibit coloration characteristic of both sexes (e.g., bilateral gynandromorphs). Although the specific hormones involved cannot be determined by these kinds of observations, the resulting color patterns frequently support conclusions reached via experimental studies, thus providing additional information regarding color control. There are also studies that have correlated levels of hormones (particularly testosterone) with color expression. However, these studies do not always indicate that there is a direct relationship between the hormone and plumage coloration, particularly as such studies are often done at times of the year when molt is not occurring. In addition, both testosterone levels and coloration may be dependent on a third variable, such as nutrition, rather than on one another.

Several hormones have been examined with respect to avian coloration, including the steroidal androgens (e.g., testosterone) and estrogens, as well as the peptide hormone, luteinizing hormone (LH). These hormones are involved in reproduction and are part of the same hormonal cascade (Johnson and Everitt 2000; Nussey and Whitehead 2001). The hypothalamus releases gonadotropin-releasing hormone, which stimulates the release of LH and follicle-stimulating hormone from the anterior pituitary. In both sexes, LH stimulates secretion of androgens from the gonads. In females, LH also stimulates the action of aromatase, which converts androgens (particularly androstenedione and testosterone) into estrogens. Negative-feedback loops exist between the secretion of LH from the pituitary and the steroid hormones from the gonads. In addition to secretion of steroid hormones from the gonads, birds are also known to secrete steroid hormones from the brain (Tsutsui and Schlinger 2001) and androgens from the adrenal gland (Boswell et al. 1995; Schlinger et al. 2004). Due to the intertwined and complex nature of these hormonal pathways, the specific hormones that control color may not be readily identifiable in some cases, and published results may need to be interpreted cautiously (Box 10.1). In addition, for some species, the coloration of plumage and nonplumage traits does not appear to be affected by hormones (at least in adults).

Although hormones may frequently affect the presence or absence of color patterns and the intensity of coloration, the specific type of color produced depends on other factors. Black, brown, and gray colors are due to the deposition of melanin pigments (Chapter 6). Carotenoid pigments give rise to red, orange, and yellow colors (Chapter 5). Blue, green, violet, ultraviolet, and white feathers derive their color from the microstructure of the feather (Chapter 7). Many species have color patches that involve a mixture of color types (e.g., both melanin and carotenoid patches). Regardless, environmental factors, such as food access and parasites, can affect all of these color displays (Chapter 12). When hormones affect coloration, they may do so by affecting pigment synthesis, pigment use (deposition or withdrawal), or the formation of specific microstructures.

Box 10.1. Experimental Limitations

Studies examining the hormonal control of avian coloration are necessarily limited in scope, meaning that no one study provides a complete picture of hormonal control of coloration in a species. For example, some studies have examined only a single sex (usually males), used only a single approach (e.g., gonadectomy but not hormone supplementation), or examined only a subset of the possibilities (e.g., supplemental testosterone was given but other hormones were not supplemented or examined). Gonadectomy alone may suggest that steroid hormones (androgens and/or estrogens) are involved in the development of colorful plumage. However, the absence of a change in coloration following gonadectomy does not rule out a role for steroid hormones due to extragonadal steroidogenesis (e.g., Boswell et al. 1995; Tsutsui and Schlinger 2001; Schlinger et al. 2004). Removal of all gonadal tissue, particularly ovarian tissue, is difficult. Although most studies including gonadectomies have performed autopsies to look for residual gonad, this practice is not universal. Many of these studies were done before it was possible to obtain pure and consistently quantified hormones. For example, the hormones administered to birds were often extracted from urine (human or horse), but specific hormones were not further purified, so study subjects may have received a mixture of all hormones present in the original extract. Conclusions from such studies may be affected by impurities, or levels of hormones that are outside normal ranges.

Interactions and interconversions among hormones can also be problematic. Many of these hormones interact in negative-feedback loops, so the release of one hormone, such as androgens or estrogens, can lead to inhibition of other hormones, such as LH (Ball and Bentley 2000). In vertebrates, the enzyme aromatase converts androgens into estrogen, which may lead to a conclusion that testosterone determines coloration, when the agent may actually be estrogen. Careful study can tease apart such complications (e.g., if supplemental estrogens give a similar response as supplemental testosterone, it is likely that the testosterone is being aromatized).

Hormones are also involved in other aspects of molt. For example, depending on the species, androgens can inhibit or delay molt (e.g., Duttmann et al. 1999; Stoehr and Hill 2001) and stimulate or accelerate molt (e.g.,

Peters et al. 2000). These interactions, and differing roles for hormones, make it difficult to determine whether a specific hormone is directly affecting coloration or is instead affecting other processes that are regulating color patterns.

Skin grafts have generally yielded mixed results, and authors using this technique have often concluded that dichromatism is affected by a combination of hormonal and nonhormonal factors. However, when both skin grafts and other methods have been employed in the same species, hormones appear to be the primary factor regulating overall color patterns. Why these differences have been obtained is unclear, but it may be due to immune responses or other side effects of transplantation. In some studies, molt was induced by plucking (e.g., Witschi 1961). This practice was implemented when the individual would not have normally molted at that time (e.g., for some species in which females molt once a year whereas males molt twice, investigators have examined what plumage females would molt into if a second molt was induced) or because the manipulations inhibited molt. There is no evidence that plucking influences the induced plumage, but such effects could exist.

The complications discussed here mean that some of the conclusions reached from some of these studies may, at times, be incorrect. However, for many species, multiple methods were used, and when all studies are taken together, the results strongly support a single mechanism controlling coloration. For other species, limited studies can be augmented by more careful research on related species that resulted in similar conclusions, providing some degree of confidence that the correct mechanism was identified.

Hormones and Plumage Coloration

Dichromatism

Experimental data on the hormonal control of plumage coloration exist for only five avian orders. The different mechanisms affecting plumage dichromatism are not randomly distributed among these orders (Table 10.1). Four of the orders appear to have a single mechanism that controls plumage dichromatism; only in order Passeriformes has more than one mechanism

Table 10.1. Experimental Studies Examining Hormonal Effects on Sexual, Seasonal, and Age-Related Plumage Dichromatism

Species	Type of data[a]	Conclusion[b]	Reference
Struthioniformes			
Ostrich (*Struthio camelus*)	Castrate M, F	Estrogen dependence	Duerden (1919)
Galliformes			
Gambel's Quail (*Callipepla gambelii*)	Castrate M	Not androgens	Hagelin and Kimball (1997); Hagelin (2001)
Scaled Quail (*Callipepla squamata*[c])	Castrate M	Not androgens	Hagelin (2001)
Lady Amherst Pheasant (*Chrysolophus amherstiae*)	Hormone supplement; skin grafts	Estrogen dependence	Vevers (1954)
Common Quail (*Coturnix coturnix*)	Castrate M, F; hormone supplement	Estrogen dependence	Kannankeril and Domm (1968); Warner (1970)
Domestic Chicken (*Gallus gallus*)	Castrate M, F; hormone supplement; gonadal implants; skin grafts	Estrogen dependence	Domm (1939); Witschi (1961); George et al. (1981)
Willow Ptarmigan (*Lagopus lagopus*)	Castrate M; hormone supplement	Unclear	Stokkan (1979a,b)
White-tailed Ptarmigan (*Lagopus leucurus*)	Hormone supplement	Unclear	Höhn and Braun (1980)
Wild Turkey (*Meleagris gallopavo*)	Castrate M, F	Estrogen dependence	Scott and Payne (1934); van Oordt (1936)
Ring-necked Pheasant (*Phasianus colchicus*)	Castrate M, F; hormone supplement; skin grafts[c]	Estrogen dependence?	Danforth (1937a,c); Morejohn and Genelly (1961)
Reeves's Pheasant (*Syrmaticus reevesi*)	Castrate M, F; hormone supplement; skin grafts[c]	Estrogen dependence	Danforth (1937b)
Anseriformes			
Blue-winged Teal (*Anas discors*)	Castrate M, F; hormone supplement	Estrogen dependence	Greij (1973)
Mallard (*Anas platyrhynchos*)	Castrate M, F; hormone supplement; gonadal implants; skin grafts	Estrogen dependence	Goodale (1910, 1918); Walton (1937); Caridroit (1938) Mueller (1970); Endler et al. (1988); Haase and Schmedemann (1992); Haase (1993); Haase et al. (1995)

Charadriiformes

Species	Method	Result	Reference
Herring Gull (*Larus argentatus*)	Castrate M; hormone supplement	Androgen dependence	Boss (1943)
Laughing Gull (*Larus atricilla*)	Castrate M, F; hormone supplement	Androgen dependence	Noble and Wurm (1940)
Black-headed Gull (*Larus ridibundus*)	Castrate M	Androgen dependence	van Oordt and Junge (1933); Groothuis and Meeuwissen (1992)
Red-necked Phalarope (*Phalaropus lobatus*)	Hormone supplement	Androgen dependence	Johns (1964)
Wilson's Phalarope (*Phalaropus tricolor*)	Hormone supplement	Androgen dependence	Johns (1964)
Ruff (*Philomachus pugnax*)	Castrate M; hormone supplement	Androgen dependence	van Oordt and Junge (1934); Lank et al. (1999)

Passeriformes

Species	Method	Result	Reference
Gouldian Finch (*Chloebia gouldiae*)	Hormone supplement	Not estrogens or androgens	Crew and Munro (1938)
Red Avadavat (*Amandava amandava*)	Castrate M, F; hormone supplement	LH dependence	Thapliyal and Tewary (1961, 1963)
Brewer's Blackbird (*Euphagus cyanocephalus*)	Hormone supplement	Not estrogens	Danforth and Price (1935)
Yellow-crowned Bishop (*Euplectes afer*)	Hormone supplement	LH dependence	Ralph et al. (1967b); Ortman (1967)
Orange Bishop (*Euplectes franciscanus*)	Castrate M, F; hormone supplement	LH dependence	Witschi (1936, 1961); Ortman (1967)
Superb Fairy-wren (*Malurus cyaneus*)	Hormone supplement	Androgen dependence?	Peters et al. (2000)
Chestnut Munia (*Lonchura atricapilla*)	Castrate M, F	Non-hormonal?	Saxena and Thapliyal (1961)
House Sparrow (*Passer domesticus*)	Castrate M, F; hormone supplement; skin grafts	Non-hormonal	Keck (1934); Mueller (1977)
Indigo Bunting (*Passerina cyanea*)	Castrate M, F; hormone supplement	LH dependence	Witschi (1935); Witschi (1961)
Baya Weaver (*Ploceus philippinus*)	Castrate M, F	LH dependence	Thapliyal and Saxena (1961)
Red-billed Quelea (*Quelea quelea*)	Castrate M, F; hormone supplement	LH dependence	Witschi (1961)
Paradise Whydah (*Vidua paradisea*)	Castrate M, F; hormone supplement	LH dependence	Witschi (1961); Ortman (1967); Ralph et al. (1967b)

a. F, female, M, male.

b. Androgen dependence, bright plumage develops in the presence of androgens; dull plumage develops in the absence of androgens; Estrogen dependence, dull plumage develops in the presence of estrogen, bright plumage develops in the absence of estrogen; LH, luteinizing hormone; LH dependence, bright plumage develops in the presence of LH, dull plumage develops in the absence of LH; Nonhormonal, bright and dull plumages (in adults) develop in the absence of hormones; Not androgens, bright and dull plumages develop in the absence of androgens; Not estrogens, bright and dull plumages develop in the absence of estrogens; Not estrogens or androgens, bright and dull plumages develop in the absence of estrogens or androgens.

c. *Callipepla squamata castanogastris* exhibits slight plumage dichromatism and was included in Hagelin (2001).

been observed. In the following section, I describe in more detail the different mechanisms that control plumage dichromatism and the orders in which they occur.

Estrogen-Dependent Dichromatism:
Struthioniformes, Galliformes, and Anseriformes

The presence or absence of estrogens determines plumage color and pattern in dichromatic species from three avian orders: Struthioniformes, Galliformes, and Anseriformes. Presence of estrogens leads to production of a dull, femalelike plumage, whereas its absence results in a bright, malelike plumage. In both sexes, removal of the gonads results in assumption of the bright, cock plumage. Estrogen supplements to either males or females during the molt results in femalelike plumage, regardless of whether the gonads are present. In contrast, treatment with androgens does not affect plumage development in either sex. In these species, therefore, the bright coloration of male feathers does not reflect male hormonal status. Instead, the brightly colored malelike plumage is the "default" condition that results in the absence of ovarian hormone input (Ligon et al. 1990; Ligon and Zwartjes 1995; Owens and Short 1995). In females, the presence of the dull plumage is informative and indicates that the ovaries were functioning normally at the time of molt.

Among paleognathous birds, Ostriches (*Struthio camelus*) have the most sexually dichromatic plumage. The plumage presumably contains melanins and varies from brown and white in females to black and white in males. It is also the only species of paleognath for which information on the control of plumage dichromatism exists. Although this species has not been studied extensively, both experimental (Table 10.1) and observational data (Fitzsimons 1912) suggest that dichromatism in this species is dependent on the presence or absence of estrogen.

Many studies examining hormonal control of plumage dichromatism have been conducted on galliform birds (Table 10.1). In this order, adults of most species undergo a single annual molt following reproduction (but see below for an exception). Males retain their colorful plumage all year. Plumage coloration in galliforms is due to melanins and structural colors, and not carotenoids (Chapters 5–7). Extensive studies on the Domestic Chicken (*Gallus gallus*), as well as studies on pheasants, partridges, turkeys, and both Old and New World quail, have generally demonstrated that female plumage develops in the presence of estrogens, whereas the specialized and colorful plumage of males develops in the absence of ovarian hormones (Plate 26). The extensive work with this group, with multiple independent studies on several species, demonstrates that estrogen dependence (and androgen independence) is characteristic of many species in this order. Studies in Ring-necked Pheasants (*Phasianus colchicus*) are less clear, as fully feminized plumage does not always develop when giving skin grafts to females or with the supplementation of estrogen (Danforth 1937a,c; Morejohn and Genelly 1961), although estrogen clearly has some affect on plumage coloration in Ring-necked Pheasants. Observational data, such as females in malelike or partially malelike plumage (but not males in femalelike plumage), also support the idea that estrogen is the primary hormone involved in plumage dichromatism in galliforms (Harrison 1932; Hagelin and Kimball 1997). Observations also suggest that plumage is estrogen dependent in other galliforms, including the Indian Peafowl (*Pavo cristatus;* Plate 8), Northern Bobwhite (*Colinus virginianus*), and California Quail (*Callipepla californica;* Hunter 1780; Brodkorb and Stevenson 1934; Buchanan and Parkes 1948; Crawford et al. 1987), but experimental studies are still needed for these taxa.

The best evidence for the role of estrogens in determining plumage type comes from extensive experimental studies of hen-feathered roosters in Sebright and Campine chickens. In these breeds, a retrotransposon insertion into the promoter of the autosomal aromatase gene results in constitutive expression of aromatase. The increased aromatase activity converts much of the circulating androgens to estrogens in extragonadal tissues, such as the skin (George et al. 1981; Wilson et al. 1987; Matsumine et al. 1990; Matsumine et al. 1991). Therefore healthy males producing normal levels of androgens will have high levels of estrogens at the feather follicles, which causes males to molt into a femalelike, rather than malelike, plumage (Plate 26).

Two galliform species that do not fit this pattern are the Willow Ptarmigan (*Lagopus lagopus*) and White-tailed Ptarmigan (*L. leucurus*). Among galliforms, male ptarmigan are unique in that they undergo up to four molts per year (some authors only recognize three molts), including into a pure white plumage in winter and into several pigmented plumages during the spring, summer, and fall (Plate 19). Castration and hormone-supplementation experiments have been performed on Willow Ptarmigan (Stokkan 1979a,b; Höhn and Braun 1980; Figure 10.1), whereas only hormone supplements have been examined in White-tailed Ptarmigan (Höhn and Braun 1980). The existing studies, involving gonadectomy and hormone supplementation (with androgens, LH, thyroxin, follicle-stimulating hormone, and α-melanocyte-stimulating hormone), suggest that androgens and LH may be involved in the deposition

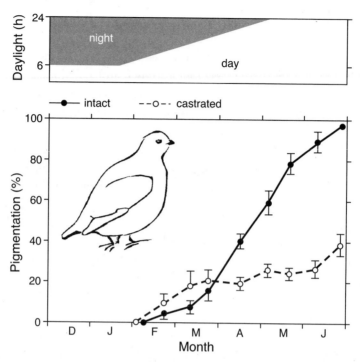

Figure 10.1. Percentage (mean ± standard error) of the head and back appearing pigmented in intact ($n = 9$) and castrated ($n = 5$) male Willow Ptarmigan experiencing a springlike increase in day length (illustrated in top graph, where night is shaded). Redrawn from Stokkan (1979a).

of pigment in the male breeding plumage (Stokkan 1979a,b; Höhn and Braun 1980), although not all results were consistent. Other pigmented plumages, however, can develop in castrated individuals, suggesting that androgens are not necessary for pigmentation per se (Stokkan 1979a,b; Höhn and Braun 1980). Females, which lack the breeding plumage but otherwise are similar to males, are also hypothesized to require a combination of hormones for the pigmented plumages, including follicle-stimulating hormone, which does not appear to be involved in male plumage coloration (Höhn and Braun 1980). The white plumage assumed by both sexes during the winter is thought to develop in the absence of hormonal stimulation. The possible role of estrogen in the plumage development of either sex has not yet been examined in this species and thus cannot be ruled out (Owens and Short 1995). More work remains to be done in ptarmigan to determine what hormones are directly involved in determining the color patterns for each plumage and how these hormones interact with genetic and other factors. Due to these complex hormonal patterns, such studies may be difficult and will require the administration of individual hormones and the use of hormone blockers, as well as combinations of these compounds to determine what role hormones have on the development of different plumages in both males and females.

Among the anseriforms, only two species of ducks, both in the genus *Anas*, have been examined experimentally (Table 10.1). In the genus *Anas*, males of species from the northern temperate zones annually undergo two molts, alternating between a bright (alternate) and an eclipse (basic) plumage. As with galliforms, both structural and melanin colors are present. In these species, as in the galliforms, estrogens appear to regulate plumage coloration (Table 10.1). Gonadectomy of both male and female ducks results in maintenance of the bright, alternate plumage all year, whereas supplementation with estrogens during the molt results in assumption of the eclipse plumage. The most detailed studies involve work on the Mallard (*Anas platyrhynchos;* Plate 15, Volume 2). Injection of androgens causes castrated male Mallards to molt into eclipse plumage, although it is suggested that this molt is a result of aromatization of androgens into estrogens (Haase and Schmedemann 1992; Haase 1993). Haase and Schmedemann (1992) examined other possible mechanisms for the control of plumage dichromatism in the Mallard. For example, LH seems unlikely to control color, as males that replace plucked feathers when natural LH levels are high molt into the eclipse plumage, whereas males normally molt into the bright, alternate plumage when natural LH levels are low. Supplemental estrogens (which cannot be converted back into androgens) result in the basic, eclipse plumage in both males and females. Therefore Haase and Schmedemann (1992) concluded that estrogen-dependent plumage dichromatism is the only hypothesis supported by all of the data. The aromatization of androgens into estrogens in male ducks is supported by the high levels of circulating estrogens during the late spring and early summer, when males molt into eclipse plumage (Höhn and Cheng 1967; Humphreys 1973; Donham 1979). Observational data suggest estrogen-dependent plumage in one additional species, the Northern Pintail (*Anas acuta*), in which four females were observed with partially masculinized plumage (Chiba et al. 2004). These females had low levels of estrogen (relative to control females) and showed ovarian degeneration, suggesting that the masculinized plumage was likely due to low levels of estrogen during molt.

Androgen-Dependent Dichromatism: Charadriiformes and Passeriformes

Androgens are important in determining plumage dichromatism in the order Charadriiformes (Table 10.1). In contrast to taxa in which estrogens affect plumage coloration, among the charadriiforms, dull plumage develops in the absence of the gonads, whereas the more brightly colored plumage requires the presence of androgens. For example, in males of the Ruff (*Philomachus pugnax;* Plate 7, Volume 2), androgens are required for development of the ornamental feathers at the neck (van Oordt and Junge 1934), and testosterone supplementation induces females to develop malelike feathers (Lank et al. 1999). Among the sex-role-reversed phalaropes, males are dull and females exhibit bright plumage coloration. In the two species that have been studied, the Red-necked Phalarope (*Lobipes lobatus*) and Wilson's Phalarope (*Phalaropus tricolor*), androgens are necessary for females to attain their bright, alternate (breeding) plumage (Johns 1964). This response appears to be mediated in the skin, as the skin of female Wilson's Phalaropes has a greater capacity to convert testosterone to an active metabolite (5α-dihydrotestosterone) than does the skin of males (Schlinger et al. 1989), at least for one dichromatic and melanin-containing region. In addition, there is little expression of aromatase in the skin of either sex, suggesting that the plumage of females is not due to the conversion of androgens to estrogens by aromatase activity (Schlinger et al. 1989), but rather to the action of androgens themselves.

Gulls are not sexually dichromatic. However, there is seasonal dichromatism, as well as differences in the alternate plumage between juveniles and adults. Experimental studies in three species of gulls, Laughing Gull (*Larus atricilla*), Herring Gull (*L. argentatus*), and the Black-headed Gull (*L. ridibundus*) have demonstrated that androgens are necessary for the development of the adult alternate plumage (Table 10.1). In two of these studies, both males and females were experimentally manipulated, and it appears that androgens are necessary for attainment of the alternate plumage in both sexes (Noble and Wurm 1940; Boss 1943; Groothuis and Meeuwissen 1992). A role for androgens has also been suggested for the development of alternate plumage in males and females in the Western Gull (*L. occidentalis*), as high levels of testosterone have been observed in females of this species during the breeding season (Wingfield et al. 1980).

Although androgen-dependent plumage is common in the charadriiforms (six species from four genera), recent data also suggest that androgens may be involved in plumage dichromatism in a passerine as well (Peters et al. 2000).

Hormonal Control of Coloration

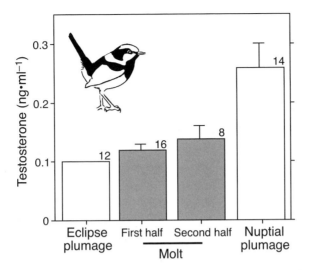

Figure 10.2. Comparison of testosterone levels (mean ± standard error) in male Superb Fairy-wrens in eclipse plumage, undergoing the prenuptial molt and in completed nuptial plumage. Males were captured in late winter and early spring; numbers above bars are sample sizes. Redrawn from Peters et al. (2000).

In the Superb Fairy-wren (*Malurus cyaneus*), males have both a dull, basic plumage as well as a bright, alternate plumage, and molt into the bright plumage is correlated with elevations in testosterone. In addition, testosterone supplementation to males induces molt into the bright plumage, whereas removal of the testosterone supplement causes cessation of molt (Peters et al. 2000; Figure 10.2). Although the data currently suggest that development of the bright plumage is most likely due to testosterone, more work is needed to definitively conclude that androgens affect plumage coloration in this species. An alternative explanation for the Superb Fairy-wren is that testosterone stimulates molt, but other factors may have determined the coloration of the plumage. In this alternative model, the bright plumage might reflect the conversion of androgens to estrogens due to aromatase activity (however, the development of bright plumage in response to estrogens would be novel) or due to LH or nonhormonal factors.

Luteinizing Hormone–Dependent Dichromatism: Passeriformes

The effects of the pituitary gonadotropin LH on plumage dichromatism have been well studied in several species of passerine birds (e.g., the weavers) in which males undergo two molts per year (Table 10.1). In these species, males wear a bright (alternate) plumage during the breeding season and a dull (basic) plumage during the nonbreeding season. Females, in contrast, molt only once per year and remain in their dull, basic plumage all year (e.g., Witschi 1961). Gonadectomy of either sex can result in both the brightly colored, alternate plumage, assumed during the prealternate molt (this molt must be induced in females by plucking), and the basic plumage. Because both plumages can be produced in the absence of the gonads, and no effects are seen with supplementation of estrogens or androgens, it is suggested that neither of these hormones directly affects plumage type.

For these passerines, it is the presence of LH that results in assumption of alternate plumage, and the absence of this hormone results in basic plumage (Witschi 1961). Males undergo prealternate molt in the spring, when levels of LH (and androgens) are high in both sexes; females, however, do not molt at this time. Following the breeding season, both sexes undergo the prebasic molt, when LH (and androgen) levels in both sexes are low, and the dull plumage develops. Injection of LH, but not androgens, generally results in the assumption of the bright plumage. The levels of LH necessary to stimulate production of the malelike plumage differ among studies, even in the same species (Ortman 1967). Exactly how LH acts to affect plumage coloration is not known. It appears that local injections of LH result in a systemic response (Hall et al. 1965), although what is involved in that systemic response has been difficult to elucidate (Ralph et al. 1967a).

There is also some observational data on LH control of dichromatism in passerines. A Blackpoll Warbler (*Dendroica striata*) was observed in femalelike plumage, although the bird was behaviorally a male and had been banded in a previous year in male plumage (Rimmer and Tietz 2001). This phenomenon could have occurred if LH levels had been abnormally low during molt, which might result from pituitary problems or inhibition of LH due to excessive levels of estrogens or androgens. While in female plumage, the male sang and tended a nest in which it was the putative father, suggesting that androgen levels were normal at that time (although androgen levels may not have been normal at the time of molt). Female passerines have also been observed in malelike alternate plumage, including a Rufous-sided Towhee (*Pipilo erythrophthalmus;* Bergtold 1916) and a Bay-breasted Warbler (*Dendroica castanea;* Stoddard 1921). Females could grow malelike plumage if they molt at a time when they have naturally high levels of LH (e.g., during the spring in northern temperate areas), or are secreting abnormally high levels of LH. Tumors that secrete LH have been observed in mammals (Snyder and Sterling 1976), and similar tumors in birds could lead to unusually high LH levels. In species with estrogen-dependent plumage, females in aberrant, malelike plumage are unlikely to breed, because this condition arises when the ovary is producing little or no estrogen. In contrast, there may not be barriers to female reproduction in species in which nongonadal hormones are involved. Supporting this idea, the female passerines that have been observed in male plumage have been found with well-developed ovaries and partially developed ova (Bergtold 1916; Stoddard 1921). Although these observations do not provide unequivocal evidence for the role of LH in plumage dichromatism, they do suggest that, at least for these taxa, steroidal hormones are not involved in plumage coloration.

Gynandromorphs, Mosaics, and Control of Dichromatism

Gynandromorphs—individuals in which half of the body exhibits the plumage of one sex while the other half exhibits the plumage of the opposite sex—have been observed in a variety of avian taxa, including galliforms, anseriforms, and passerines (e.g., see references in Crew and Munro 1938; Patten 1993; Agate et al. 2003; Plate 27). In addition, mosaics, in which there are both malelike and femalelike feather tracts, have been observed in galliforms (Crawford et al. 1987; Hagelin and Kimball 1997), falconiforms (Parrish et al. 1987; Tella et al. 1996), and passerines (Summers and Kostecke 2004). Because hormones circulate throughout the body, the presence of gynandromorphs and mosaics has been used to suggest genetic or other nonhormonal control of plumage dichromatism (e.g., Cock 1960; Witschi 1961; Agate et al. 2003). However, gynandromorphs and mosaics have been observed in species for which hormonal control of plumage coloration has been well established (e.g., galliforms and anseriforms). To explain this phenomenon, Lillie (1931) proposed a hypothesis consistent with estrogen-dependent plumage dichromatism. In many gynandromorphs, one side of the body (usually that with the malelike plumage) is larger than the other (hemihypertrophy). The sensitivity of feathers to the presence of estrogen varies in Domestic Chickens and Mallards (Juhn et al. 1931; Endler et al. 1988), and may be related to feather growth rate (Juhn et al. 1931), such that faster-growing feathers are less sensitive to estrogen. Thus the more rapidly growing feathers on the male side of the body might be less sensitive to estrogen and so develop as though estrogen were not present, whereas the slower-growing feathers on the female side of the body are sensitive to the estrogen and develop into female plumage (Lillie

1931). A similar explanation could be applied to mosaics. However, there are several other possible explanations. For example, there may be differential expression of aromatase or hormone receptors on the different regions of the body that could lead to the observed patterns. Unfortunately, none of these hypotheses has been tested, making it difficult to determine the causes of gynandromorphy or to use this information to better understand the development of plumage dichromatism.

Coloration Type and Hormones Affecting Coloration

From the studies mentioned above, it is not clear whether these hormones differentially affect the type of plumage coloration that results (e.g., melanin versus carotenoid). Of interest is that carotenoids are not known to occur in any of the species for which estrogen-dependent plumage coloration has been observed (but see below for an example of estrogen control of carotenoid-based bare-part coloration), nor are carotenoids present in any of the charadriiforms known to have androgen-dependent plumage coloration. Carotenoids are present in passerines, and thus may occur in some species with androgen-dependent plumage coloration (although the Superb Fairy-wren lacks red or yellow feathers). However, given the many orders for which the hormonal mechanisms are unknown, it is possible that estrogen- or androgen-dependent coloration may occur in species exhibiting carotenoid-based plumages. The passerine species that have been examined include all types of feathers, with most species exhibiting melanins, as well as carotenoids and/or structural colors. It is known that LH and estrogens can upregulate tyrosinase, a critical enzyme in melanin synthesis (Okazaki and Hall 1965; Hall 1966, 1969). However, LH must at least affect deposition of carotenoids as well. Most likely, hormones turn on specific pathways that lead to a suite of plumage characteristics, but they may not be directly related to the type of coloration that develops.

Intrasexual Variation

In addition to their role in shaping gross plumage dichromatism, hormones can also have more subtle (to the human eye) effects on plumage coloration. Although not as easy to detect as those situations in which hormones determine male- or femalelike patterns, hormone profiles may influence the color variation that is seen in a sex or age class in natural populations (e.g., Gonza-

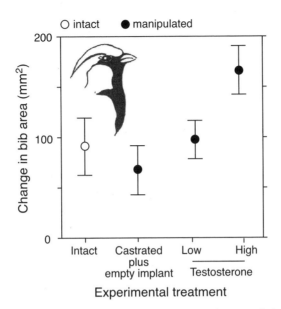

Figure 10.3. Effect of manipulation on the change in bib area during molt (mean ± standard error) in captive male House Sparrows. Redrawn from Evans et al. (2000).

lez et al. 2001; Stoehr and Hill 2001). Thus understanding how hormones affect variation in color brightness, intensity, and extent is very important in understanding the role of plumage coloration in intra- and intersexual signaling.

Although the presence or absence of male-specific plumage in House Sparrows (*Passer domesticus*; Plate 23) is not affected by hormone supplementation (Table 10.1), hormones may control levels of color expression among males. Several studies have found that supplementation of testosterone may result in larger black badges, whereas supplementation with antiandrogens reduces badge size in adults (Evans et al. 2000; Buchanan et al. 2001; Gonzalez et al. 2001; Figure 10.3). Injections of testosterone into eggs also affects badge size once juveniles attain their adult plumage, although such injections did not affect whether an individual molted into malelike or femalelike plumage (Strasser and Schwabl 2004). A recent study found that males experiencing more aggressive interactions molted into larger badges (McGraw et al. 2003). This study and those indicating testosterone can affect badge size (even months after testosterone supplementation; Strasser and Schwabl 2004) suggest that the effects of androgens on plumage may be complex, possibly involving a combination of direct and indirect (i.e., social) factors.

Hormones may also affect the degree of carotenoid-based red coloration in House Finches (*Carpodacus mexicanus;* Plate 31), although the mechanisms that determine dichromatism are not well understood in this species. Castration of males results in femalelike plumage, as does supplementation with estrogen (Tewary and Farner 1973). In addition, during the breeding season, redder males have higher testosterone levels than do duller males (Duckworth et al. 2004), although it is not clear whether testosterone during the breeding season correlates with testosterone levels during molt. Even though these studies suggest a possible role for androgens in the development of the red male plumage, an experimental study found that testosterone supplementation results in males that are actually drabber (displaying less red pigmentation) than controls, even among captive males whose diets contained supplemental carotenoids (Stoehr and Hill 2001). At this time, it is not clear why the results of these studies disagree, but it may be that additional hormones are involved, or that the effect of androgens differs seasonally or in response to other hormones.

Other hormones may also influence intrasexual variation in colors. The role of thyroxin has primarily been examined with regard to its effect on molt, but thyroxin may also affect deposition of melanin pigmentation in some species. For example, Miller (1935) found that supplemental thyroxin altered some plumage colors in House Sparrows, although the results varied in different parts of the body. When males were administered thyroxin, feathers that were normally brown or chestnut became gray, the pale belly feathers darkened, and the black badge was replaced with gray feathers. Conflicting results in response to thyroxin (sometimes administered as thyroid rather than purified thyroxin) have been seen across studies, even within a species. For example, in Domestic Chickens, supplementation of thyroid has led to melanism in some cases and albinism in others (reviewed in Miller 1935), although how much this difference is affected by the purity of the extracts or the exact dosing is not known. The thyroid affects gonadal activity, but the activity differs in different species and can depend on other aspects of the individual (Dawson and Thapliyal 2001). These differing responses make it difficult to establish which effects of thyroxin on plumage coloration are consistent across taxa.

Unfortunately, at this time, there is not enough information available to understand the range of modifying effects, either direct or indirect, of hormones on individual variation in plumage coloration.

Hormones and Coloration of Nonplumage Traits

There are many types of nonplumage traits in birds that are dichromatic between the sexes, seasons, or among age classes, or vary in color intensity in these groups. These include the color of the eye, eye rings, bills, legs, or featherless regions of the face and neck (including specialized structures in these regions, e.g., combs, wattles). As with plumage, the effect of hormones on the color of nonplumage traits has only been examined in relatively few species from a small number of orders (Table 10.2).

Dichromatism

In contrast to plumage, in which hormonal pathways may be involved in different orders, the general pattern for nonplumage traits is that coloration (particularly dichromatism) is dependent on androgens (Table 10.2).

However, there are exceptions to the ubiquity of androgen-dependent control of nonplumage traits. For example, seasonal and sexual expression of bill color in the Red-billed Quelea (*Quelea quelea;* Plate 6, Volume 2) is dependent on estrogen, in contrast to other studied species, in which bill color is exclusively dependent on androgens (Table 10.2). Although plumage dichromatism in queleas is typical for passerines and appears to be dependent on LH (Table 10.2), queleas are unusual in that aggression is also thought to be mediated by LH (Lazarus and Crook 1973) rather than by androgens (as for other passerines; e.g., Wingfield et al. 2000). Thus queleas have several endocrinological differences from the majority of passerines that have been studied. LH may also affect bill coloration in the Paradise Whydah (*Vidua paradisea*). Castration and LH supplementation indicate that the black bill of the male is dependent on LH (Witschi 1961). LH has also been suggested to affect bill coloration in the Red Avadavat (also called Lal Munia; *Amandava amandava*), although this idea has not been examined thoroughly (Thapliyal and Gupta 1984).

There is some evidence that hormonal control of bill color may also occur in the avian order Anseriformes, but the results in this group are less clear. In wild and domesticated Mallards, castration of females can cause bill color to change, although the bills never become completely masculine (Goodale 1918; Domm 1939), and this response is not universal (Goodale 1910). Castration of males does not affect bill color (Goodale 1916). These studies might suggest that estrogen has some affect on bill color (leading to femalelike bill

Table 10.2. Experimental Studies Examining Hormonal Mechanisms Affecting Coloration of Nonplumage Traits

Species	Trait	Type of data	Reference
Struthioniformes			
Ostrich (*Struthio camelus*)	Leg color	Castrate M, F	Duerden (1919)
Galliformes			
Domestic Chicken (*Gallus gallus*)	Comb and wattle color	Castrate M, F; hormone supplement; gonadal implants	Domm (1939); Witschi (1961); Zuk et al. (1995)
Wild Turkey (*Meleagris gallopavo*)	Head coloration	Castrate M, F	Scott and Payne (1934); van Oordt (1936)
Ring-necked Pheasant (*Phasianus colchicus*)	Wattle color	T supplement	Morejohn and Genelly (1961)
Gruiformes			
Common Moorhen (*Gallinula chloropus*)	Shield color	T supplement	Eens et al. (2000)
Charadriiformes			
Herring Gull (*Larus argentatus*)	Bill color	Castrate M; hormone supplement	Boss (1943)
Laughing Gull (*Larus atricilla*)	Bill and leg color	Castrate M, F; hormone supplement	Noble and Wurm (1940)
Black-headed Gull (*Larus ridibundus*)	Bill and leg color	Castrate M	van Oordt and Junge (1933)
Ruff (*Philomachus pugnax*)	Eye tubercules	Castrate M; hormone supplement	van Oordt and Junge (1934); Lank et al. (1999)
Psittaciformes			
Budgerigar (*Melopsittacus undulatus*)	Cere color	T supplement	Nespor et al. (1996)
Passeriformes			
American Goldfinch (*Carduelis tristis*)	Bill color	Castrate M; hormone supplement	Mundinger (1972)
Bobolink (*Dolichonyx oryzivorus*)	Bill color	T supplement	Engels (1959)
Orange Bishop (*Euplectes franciscanus*)	Bill color	Castrate M, F	Witschi (1935, 1936, 1961)
Chestnut-shouldered Petronia (*Petronia xanthocollis*)	Bill color	Castrate M	Tewary et al. (1985)
House Sparrow (*Passer domesticus*)	Bill color	Castrate M, F; hormone supplement	Keck (1934); Haase (1975)
Indigo Bunting (*Passerina cyanea*)	Bill color	Castrate M, F; hormone supplement	Witschi (1935); Witschi (1961)
European Starling (*Sturnus vulgaris*)	Bill color	Castrate M, F; hormone supplement	Witschi and Miller (1938); De Ridder et al. (2002)
Zebra Finch (*Taeniopygia guttata*)	Bill color	Castrate M; hormone supplement	Cynx and Nottebohm (1992); McGraw (2003)

Note: Androgens are thought to regulate coloration in all instances.

colors); however, implantation of ovarian tissue in castrated males produced no changes in bill color, even though plumage was affected (Goodale 1918). In Blue-winged Teal (*Anas discors*), the number of spots on the bill was reduced in females that were castrated or supplemented with either estrogen or testosterone (Greij 1973). Thus, at this time, there is insufficient data to be certain of the relationship between bill color and hormones in anseriforms.

Coloration Type and Hormones Affecting Coloration

Similar to plumage coloration, coloration of nonplumage traits, particularly bill color, is often due to melanins and carotenoids. For example, the black coloration in the bills of House Sparrows is due to a deposition of melanins in response to androgens (e.g., Keck 1934; Haase 1975). Other taxa show a different pattern. For example, in European Starlings (*Sturnus vulgaris*; Plate 23) and American Goldfinches (*Carduelis tristis*; Plate 30), androgens cause a withdrawal of melanins (Witschi 1961; Mundinger 1972). The removal of melanins, combined with deposition of carotenoids (the hormonal mechanism regulating this is not well known; Mundinger 1972), results in bills that are yellow/orange during the breeding season. Still other factors can affect coloration in some nonplumage traits. For example, the color of the comb and wattles of the Red Junglefowl (*Gallus gallus*; Plate 26) are due to increased blood flow in these tissues, which occurs in response to androgens (Hardesty 1931; Lucas and Stettenheim 1972). So even though androgens are frequently involved in the coloration of nonplumage traits, it is clear that androgens turn on different biochemical pathways and colors in different tissues and different species.

Intrasexual Variation

Unfortunately, most studies that have examined the relationship between hormones and coloration of nonplumage regions have focused on large-scale color changes (e.g., between a black versus a yellow bill). Much less attention has been paid to the role of hormones in affecting intrasexual variation in the color of those traits, even though such variation may be important in such areas as mate choice (e.g., Johnson et al. 1993). It is known that androgens have an important role in regulating the intensity of red coloration in the combs of junglefowl (e.g., Zuk et al. 1995). In addition, supplementation of testosterone to female Ring-necked Pheasants resulted in redder wattles (Morejohn and

Genelly 1961). Given the potential importance of intrasexual variation in the coloration of these types of traits, further investigation into the role hormones might play in coloration is likely to be fruitful.

Importance of Understanding Hormonal Control of Coloration

A consideration of the hormonal basis of avian coloration can lead to a variety of insights about avian evolution (Box 10.2), as well as to improved understanding of the information content of color signals. One area that has been a focus of attention is the role of coloration in sexual selection. For example, good-genes models predict that females should select males using traits that reflect male condition (Andersson 1994; Ligon 1999). One hypothesis for how hormone-dependent traits might serve as a signal of male quality is the immunocompetence-handicap hypothesis (Folstad and Karter 1992). This hypothesis is based on the idea that androgens can suppress immune function, so that males with high levels of androgens might be more vulnerable to pathogens. Thus only high-quality males with either well-developed immune systems or the energetic resources to tolerate pathogen stress can afford to produce high levels of androgens. Tests of the immunocompetence hypothesis have yielded mixed results (Roberts et al. 2004), although the studies supporting the hypothesis suggest that, at least in some cases, androgen levels may provide information about male immunocompetence.

Consistent with the hypothesis that androgen-dependent traits provide information about individual quality, androgen-dependent nonplumage traits, such as bills or specialized structures like wattles and combs, appear to be important in sexual selection in a number of bird species (Chapter 4, Volume 2). Among those species in which the hormonal control of coloration is known, sexual selection studies have demonstrated a role for androgen-dependent nonplumage traits in a range of different species (Ligon et al. 1990; Zuk et al. 1990, 1995; Johnson et al. 1993; Buchholz 1995). For example, Red Junglefowl females base their mating decisions at least in part on male comb size and color (Zuk et al. 1990, 1995). Coloration of the comb correlates with testosterone levels (Zuk et al. 1995), indicating that combs provide an accurate assessment of male hormonal condition. Female American Goldfinches prefer males with brighter yellow bills (Johnson et al. 1993), the coloration of which is dependent on androgens (Mundinger 1972). In particular, androgen-dependent nonplumage traits can indicate current health status, as androgen levels and fleshy colors can become depressed in individuals that are ill (e.g., Verhulst et al. 1999).

Androgen-dependent plumage may also be important in sexual selection. In Superb Fairy-wrens, in which androgens appear to stimulate molt and may affect coloration (Peters et al. 2000), females prefer to mate with males that molt earlier (Mulder and Magrath 1994). Female preference for plumage coloration has also been reported in another species of fairy-wren, the Red-backed Fairy-wren (*Malurus melanocephalus;* Karubian 2002), and bright male plumage in this species seems likely to respond similarly to androgens. However, the coloration on the red back is probably due to carotenoids, making it difficult to determine whether females might be selecting on the possible androgen-dependent aspects of the plumage or the carotenoid coloration (Chapter 4, Volume 2). Regardless of whether fairy-wrens turn out to be a good test of the role of androgen-dependent plumage in sexual selection, establishing correlations between the mechanisms affecting plumage coloration and sexual selection has the potential to be a fruitful area of research.

Whether other hormones might provide similar levels of information about individual quality or condition is not yet known. However, traits that are not dependent on hormones have the potential to convey less, or at least different, information about individual quality or condition than do androgen-dependent traits (Morgan 1919; Ligon et al. 1990; Ligon and Zwartjes 1995; Owens and Short 1995). Consistent with this idea, females of some galliform and anseriform species appear to pay little attention to male plumage that develops in the absence of hormones (e.g., Buchholz 1995; Ligon and Zwartjes 1995; Omland 1996a,b; Ligon et al. 1998; Hagelin and Ligon 2001).

Understanding the hormone dependence of avian coloration can have benefits beyond those conceptually linked to sexual selection. For example, environmental endocrine disrupters, such as estrogen-mimics, could lead to changes in coloration for those taxa with estrogen-dependent traits (Ottinger et al. 2002). Endocrine disruptors are a common form of chemical pollution, and exposure to such disruptors could have negative consequences on fitness by altering some sex-specific signals. Knowledge of the links between hormones and color displays and how endocrine disruptors interfere with such systems also provides an opportunity to assess the degree of environmental contamination. Surveying contamination by assessing coloration of traits could be done easily in species in which individuals can be sexed readily from a distance using other means (e.g., song or other nonplumage traits). In a study on Tree Swallows (*Tachycineta bicolor;* Plate 27, Volume 2), subadult females had more

Box 10.2. Evolutionary Pathways of Hormonal Control of Coloration

Although the data are limited, it appears that the proximate mechanisms that control plumage dichromatism are largely conserved in broad taxonomic groups, making it possible to examine the evolutionary history of these mechanisms. Most morphological and molecular analyses suggest that paleognathous birds are the most basal lineage of living birds (e.g., Sibley and Ahlquist 1990; Braun and Kimball 2002). Among neognathous birds, the position of the orders Galliformes and Anseriformes are basal and generally thought to form a clade, whereas the orders Charadriiformes and Passeriformes are more derived (e.g., Sibley and Ahlquist 1990; Braun and Kimball 2002). Relationships in the passerines are not fully resolved, although most of the passerines examined in Table 10.1 fall into the Passeroidea; the exception is the Superb Fairy-wren, which belongs to the Corvida (Sibley and Ahlquist 1990; Barker et al. 2001). Using this phylogeny, it appears that estrogen-dependent plumage dichromatism is present in the deepest-branching avian lineages (Figure B10.1). Control of plumage dichromatism by estrogens may have been lost in later lineages, with other mechanisms having subsequently evolved (Figure B10.1).

An examination of the proximate mechanisms of plumage dichromatism among living species can provide clues regarding the evolution of dichromatism, assuming that monochromatism is ancestral to dichromatism among the earliest birds. To evolve estrogen-dependent plumage dichromatism, the most parsimonious pathway begins with bright coloration in both sexes, followed by selection for duller color in one sex (Figure B10.2a, top of figure). This scenario is in contrast to arguments based on sexual selection that assume that dichromatism is a result of selection for brighter coloration in one sex, presumably from a dull, monochromatic ancestor, rather than selection for dull coloration from a brightly colored ancestor, as these data suggest (see also Chapter 10, Volume 2 for further discussion of ancestral color displays in birds).

Alternatively, to evolve estrogen-dependent plumage dichromatism from an ancestral condition in which both sexes were dull would require two steps —the evolution of a bright plumage that develops in the absence of hormones, and the evolution of estrogen-dependence for the existing duller plumage. To achieve these conditions, either of two pathways could be fol-

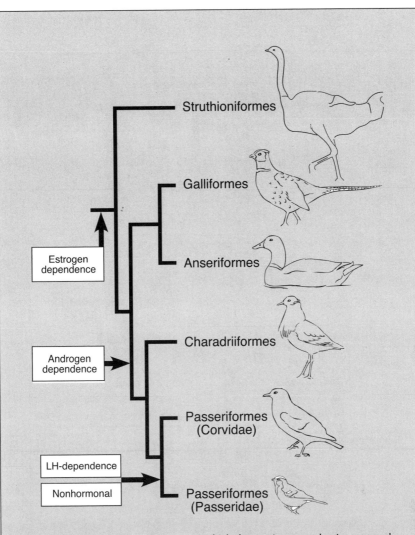

Figure B10.1. Phylogeny of the taxa in which the proximate mechanisms controlling plumage coloration are known. Transitions between different mechanisms are indicated.

lowed (Figure B10.2a, Paths A and B). Both of these scenarios require an additional evolutionary step over that proposed earlier.

To continue this line of reasoning, the evolution of androgen-dependent and LH-dependent plumage dichromatism, in which brighter plumage is dependent on the presence of a hormone, is most likely to have evolved from

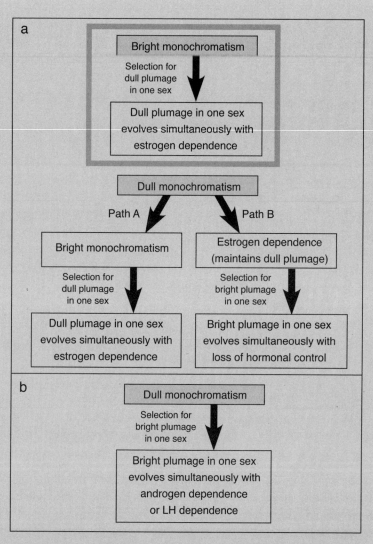

Figure B10.2. Schematics of the evolution of dichromatism. (a) To evolve estrogen-dependent plumage dichromatism, the most parsimonious pathway (top) begins with bright, monochromatic plumage. Assuming the initial state was dull monochromatism (bottom), estrogen-dependent plumage dichromatism can evolve through two mechanisms, path A or path B. (b) The most parsimonious pathway to evolve either androgen-dependent or LH-dependent plumage dichromatism.

a condition of duller monochromatism (Figure B10.2b). In this way, parallel to the evolution of dull coloration in species with estrogen-dependence, the novel (brighter) plumage evolves in concert with the novel mechanism (hormone dependence), resulting in the observed situation.

These scenarios explain transitions to dichromatism, although there have also been transitions to monochromatism. For species exhibiting estrogen-dependent plumage dichromatism, transitions to bright monochromatism could occur through a mutation that prevented expression of the estrogen receptors at the feather follicle. However, to transition to bright monochromatism in species with either androgen- or LH-dependent plumage coloration, both sexes must molt at times when the appropriate hormone is present. An alternative mechanism would be any mutation that mimics the effect of the hormone, such as changes that constitutively turn on the appropriate biochemical pathway (analogous to the change in the aromatase promoter that results in hen-feathered roosters [Plate 26]).

There are several pathways that could lead to dull monochromatic plumage from a dichromatic state. In species with estrogen-dependent dichromatism, any mutation that increased estrogen at the feather follicle in both sexes would result in a dull, femalelike plumage in both sexes. However, estrogens are highly pleiotropic, and such mutations may be deleterious in natural populations. Another pathway to dull monochromatism could be through a shift in the timing of molt, such that it occurs when estrogen levels are high (as in wild-type *Anas;* Humphreys 1973; Donham 1979). For species with androgen- or LH-dependent plumage dichromatism, dull monochromatism could occur if molt occurs at a time when hormone levels are low. For these species (unlike those with estrogen-dependent plumage dichromatism), a mutation causing reduced expression of the appropriate receptor at the feather follicle would also result in dull coloration in both sexes.

The predominance of androgen-dependent nonplumage traits among species and across orders suggests one of two possible scenarios. First, androgen-dependent secondary sexual traits may be primitive among birds, but other mechanisms (e.g., estrogen- or LH-dependence) may have subsequently evolved in the passerines. An alternative is that there has been evolution of multiple hormonal mechanisms in many avian lineages, but selection has favored maintenance of androgen-dependent traits.

adultlike plumage in areas with high polychlorinated biphenyl (PCB) contamination than in PCB-free areas (McCarty and Secord 2000). Unfortunately, the mechanisms affecting plumage coloration in Tree Swallows are not known, so it is not clear why this correlation may exist.

Conclusions

Although understanding the hormonal and nonhormonal regulators of coloration is important for addressing many other evolutionary and ecological issues, there is still much that is not understood about endocrine effects on avian coloration. Unfortunately, the experimental data currently available have been gathered from a very limited sample of avian biodiversity. Even in some of the better-studied groups, such as the passerines, more rigorous sampling is clearly needed. Mechanisms in addition to those discussed in this chapter may occur in taxa not yet examined, and the frequency of shifts among different mechanisms of plumage color may have evolved more frequently than current sampling suggests. Careful experimental studies that distinguish among several alternative hypotheses need to be conducted and placed in an explicit phylogenetic framework (e.g., Badyaev and Hill 2003). Use of carefully quantified and purified hormones, as well as new chemicals that act as hormone antagonists or inhibit specific pathways, will facilitate such studies. For example, there are now aromatase inhibitors (e.g., Vaillant et al. 2003; Moore et al. 2004), antiandrogens (e.g., Gonzalez et al. 2001; Moore et al. 2004), and estrogen-blockers (e.g., Lupu 2000) that have been demonstrated to be effective in birds. Molecular markers, such as mRNAs for aromatase, 5α-reductase, and hormone receptors, are another set of tools that have the potential to provide information on the hormonal basis of avian coloration. These approaches should stimulate new research on the control of avian coloration, and have the potential to provide a greater understanding of the mechanisms regulating coloration.

Summary

Coloration of plumage and bare parts in birds develops in response to both hormonal and nonhormonal factors. Several different hormones are known to affect coloration, including estrogens, androgens, and LH. To develop sexual plumage dichromatism, the presence of estrogen during molt results in a dull, femalelike plumage among ostriches, waterfowl, and most galliform birds.

Androgens and LH, when present during molt, result in a bright, malelike plumage in charadriiforms (androgens only) and passerines (both androgens and LH). Hormones may also affect the expression of intrasexual variation in plumage coloration (particularly androgens), although this phenomenon has not been well studied. Coloration of nonplumage traits (bills, ceres, leg color, and specialized structures [e.g., combs]) is dependent on androgens in most studied species, although LH and estrogen are known to affect bill color in a few species. As with plumage, little is known about the effects of hormones on intrasexual variation in the color of bare parts. Although understanding the effects of hormones on coloration can provide insights into the information contained in such signals, much work still remains to be done to fully understand the relationship between hormones and coloration of plumage and nonplumage traits.

References

Agate, R., W. Grisham, J. Wade, S. Mann, J. Wingfield, et al. 2003. Neural, not gonadal, origin of brain sex differences in a gynandromorphic finch. Proc Natl Acad Sci USA 100: 4873–4878.

Andersson, M. 1994. Sexual Selection. Princeton, NJ: Princeton University Press.

Badyaev, A. V., and G. E. Hill. 2003. Avian sexual dichromatism in relation to phylogeny and ecology. Annu Rev Ecol Syst 34: 27–49.

Ball, G., and G. Bentley. 2000. Neuroendocrine mechanisms mediating the photoperiodic and social regulation of seasonal reproduction in birds. In K. Wallen and J. Schneider, ed., Reproduction in Context: Social and Environmental Influences on Reproduction, 129–158. Cambridge, MA: MIT Press.

Barker, F. K., G. F. Barrowclough, and J. G. Groth. 2001. A phylogenetic hypothesis for passerine birds: Taxonomic and biogeographic implications of an analysis of nuclear DNA sequence data. Proc R Soc Lond B 269: 295–308.

Bergtold, W. H. 1916. Pseudo-masculinity in birds. Auk 33: 439.

Boss, W. R. 1943. Hormonal determination of adult characters and sex behavior in Herring Gulls (*Larus argentatus*). J Exp Zool 94: 181–209.

Boswell, T., M. Hall, and A. Goldsmith. 1995. Testosterone is secreted extragonadally by European quail maintained on short days. Physiol Zool 68: 967–984.

Braun, E. L., and R. T. Kimball. 2002. Examining basal avian divergences with mitochondrial sequences: Model complexity, taxon sampling, and sequence length. Syst Biol 51: 614–625.

Brodkorb, P., and J. Stevenson. 1934. Additional northeastern Illinois notes. Auk 51: 100–101.

Buchanan, F. W., and K. C. Parkes. 1948. A female Bob-white in male plumage. Wilson Bull 60: 119–120.

Buchanan, K., M. Evans, A. Goldsmith, D. Bryant, and L. Rowe. 2001. Testosterone influences basal metabolic rate in male House Sparrows: A new cost of dominance signaling? Proc R Soc Lond B 268: 1337–1344.

Buchholz, R. 1995. Female choice, parasite load and male ornamentation in Wild Turkeys. Anim Behav 50: 929–943.

Caridroit, F. 1938. Recherches experimentales sur les rapports entre testicules, plumage d'éclipse et mues chez le canard sauvage. Trav Stn Zool Wimereaux 13: 47–67.

Chiba, A., H. Sakai, M. Sato, R. Honma, K. Murata, and F. Sugimori. 2004. Pituitary-gonadal axis and secondary sex characters in the spontaneously masculinized Pintail, *Anas acuta* (Anatidae, Aves), with special regard to the gonadotrophs. Gen Comp Endocrinol 137: 50–61.

Cock, A. G. 1960. Four new half-and-half mosaic fowls. Genet Res 1: 275–287.

Crawford, J. A., P. J. Cole, and K. M. Kilbride. 1987. Atypical plumage of a female California Quail. Calif Fish Game 73: 245–247.

Crew, F., and S. Munro. 1938. Gynandromorphism and laternal asymmetry in birds. Proc R Soc Edinburgh 58: 114–135.

Cynx, J., and F. Nottebohm. 1992. Testosterone facilitates some conspecific song discriminations in castrated Zebra Finches (*Taeniopygia guttata*). Proc Natl Acad Sci USA 89: 1376–1378.

Danforth, C. H. 1937a. Artificial gynandromorphism and plumage in *Phasianus*. J Genet 34: 497–506.

Danforth, C. H. 1937b. An experimental study of plumage in Reeves Pheasants. J Exp Zool 77: 1–11.

Danforth, C. H. 1937c. Responses of feathers of male and female pheasants to theelin. Proc Soc Exp Biol Med 36: 322–324.

Danforth, C. H., and J. B. Price. 1935. Failure of theelin and thyroxin to affect plumage and eye-color of the Blackbird. Proc Soc Expl Biol Med 32: 675–678.

Dawson, A., and J. Thapliyal. 2001. The thyroid and photoperiodism. In A. Dawson and C. Chaturvedi, ed., Avian Endocrinology, 141–151. Pangbourne, UK: Alpha Science International.

De Ridder, E., R. Pinxten, V. Mees, and M. Eens. 2002. Short- and long-term effects of male-like concentrations of testosterone on female European Starlings (*Sturnus vulgaris*). Auk 119: 487–497.

Domm, L. V. 1939. Modifications in sex and secondary sexual characters in birds. In E. Allen, ed., Sex and Internal Secretions: A Survey of Recent Research, 227–327. Baltimore: Williams and Wilkins.

Donham, R. S. 1979. Annual cycle of plasma luteinizing hormone and sex hormones in male and female Mallards (*Anas platyrhynchos*). Biol Reprod 21: 1273–1285.

Duckworth, R. A., M. T. Mendonca, and G. E. Hill. 2004. Condition-dependent sexual traits and social dominance in the House Finch. Behav Ecol 15: 779–784.

Duerden, J. E. 1919. Crossing the North African and South African Ostrich. J Genet 8: 155–198.

Duttmann, H., S. Dieleman, and T. G. G. Groothuis. 1999. Timing of moult in male and female Shelducks *Tadorna tadorna:* Effects of androgens and mates. Ardea 87: 33–39.

Eens, M., E. Van Duyse, and L. Berghman. 2000. Shield characteristics are testosterone-dependent in both male and female Moorhens. Horm Behav 37: 126–134.

Endler, B., B. Rahmig, and E. Rutschke. 1988. On the hormonal regulation of the sexual and seasonal dimorphism of the Mallard (*Anas platyrhynchos*). In H. Ouellet, ed., Acta XIX Congress Internationalis Ornithologici, 2234–2239. Ottawa: University of Ottawa Press.

Engels, W. L. 1959. The influence of different day lengths on the testes of a transequitorial migrant, the Bobolink (*Dolichonyx oryzivorus*). In R. Withrow, ed., Photoperiodism and Related Phenomena in Plants and Animals, 759–766. Washington, DC: American Association for the Advancement of Science.

Evans, M. R., A. R. Goldsmith, and S. R. A. Norris. 2000. The effect of testosterone on antibody production and plumage coloration in male House Sparrows (*Passer domesticus*). Behav Ecol Sociobiol 47: 156–163.

Fitzsimons, F. W. 1912. A hen Ostrich with plumage of a cock. Agr J Univ South Africa 4: 380.

Folstad, I., and A. J. Karter. 1992. Parasites, bright males, and the immunocompetence handicap. Am Nat 139: 603–622.

George, F. W., J. F. Noble, and J. D. Wilson. 1981. Female feathering in sebright cocks is due to conversion of testosterone to estradiol in skin. Science 213: 557–559.

Gonzalez, G., G. Sorci, L. Smith, and F. de Lope. 2001. Testosterone and sexual signaling in male House Sparrows (*Passer domesticus*). Behav Ecol Sociobiol 50: 557–562.

Goodale, H. D. 1910. Some results of castration in ducks. Biol Bull 20: 35–56.

Goodale, H. D. 1916. Gonadectomy in Relation to the Secondary Sexual Characters of some Domestic Birds. Washington, DC: Carnegie Institution of Washington.

Goodale, H. D. 1918. Feminized male birds. Genetics 3: 276–295.

Greij, E. D. 1973. Effects of sex hormones on plumages of the Blue-winged Teal. Auk 90: 533–551.

Groothuis, T., and G. Meeuwissen. 1992. The influence of testosterone on the development and fixation of the form of displays in two age classes of young Black-headed Gulls. Anim Behav 43: 189–208.

Haase, E. 1975. The effects of testosterone propionate on secondary sexual characters and testes of House Sparrows, *Passer domesticus*. Gen Comp Endocrinol 26: 248–252.

Haase, E. 1993. Zur Wirkung von Androgenen auf die Induktion des Schlichtkleids bei kastrierten Stockerpeln (*Anas platyrhynchos*). J Ornithol 134: 191–195.

Haase, E., and R. Schmedemann. 1992. Dose-dependent effect of testosterone on the induction of eclipse coloration in castrated wild Mallard drakes (*Anas platyrhynchos* L.). Can J Zool 70: 428–431.

Haase, E., S. Ito, and K. Wakamatsu. 1995. Influences of sex, castration, and androgens on the eumelanin and pheomelanin contents of different feathers in wild Mallards. Pigment Cell Res 8: 164–170.

Hagelin, J. C. 2001. Castration in Gambel's and Scaled Quail: Ornate plumage and dominance persist, but courtship and threat behaviors do not. Horm Behav 39: 1–10.

Hagelin, J. C., and R. T. Kimball. 1997. A female Gambel's Quail in partial male plumage. Wilson Bull 109: 544–546.

Hagelin, J. C., and J. D. Ligon. 2001. Female quail prefer testosterone-mediated traits, rather than the ornate plumage of males. Anim Behav 61: 465–476.

Hall, P. F. 1966. Tyrosinase activity in relation to plumage color in weaver birds (*Steganura paradisaea*). Comp Biochem Physiol A 18: 91–100.

Hall, P. F. 1969. Hormonal control of melanin synthesis in birds. Gen Comp Endocrinol 2 (Suppl.): 451–458.

Hall, P. F., C. L. Ralph, and D. L. Grinwich. 1965. On the locus of action of interstitial cell-stimulating hormones (ICSH or LH) on feather pigmentation of African weaver birds. Gen Comp Endocrinol 5: 552–557.

Hardesty, M. 1931. The structural basis for the response of the comb of the brown leghorn fowl to the sex hormones. Am J Anat 47: 277–323.

Harrison, J. M. 1932. A series of nineteen pheasants (*Phasianus colchicus* L.) presenting anomalous secondary sexual characters in association with changes in the ovaries. Proc Zool Soc: 193–203.

Höhn, E. O., and C. E. Braun. 1980. Hormonal induction of feather pigmentation in ptarmigan. Auk 97: 601–607.

Höhn, E. O., and S. C. Cheng. 1967. Gonadal hormones in Wilson's Phalarope (*Steganopus tricolor*) and other birds in relation to plumage and sex behavior. Gen Comp Endocrinol 8: 1–11.

Humphreys, P. N. 1973. Behavioural and morphological changes in the Mallard and their relationship to plasma oestrogen concentrations. J Endocrinol 58: 353–354.

Hunter, J. 1780. Account of an extraordinary pheasant. Phil Trans R Soc London 70: 527–535.

Johns, J. E. 1964. Testosterone-induced nuptial feathers in phalaropes. Condor 66: 449–455.

Johnson, K., R. Dalton, and N. Burley. 1993. Preferences of female American Goldfinches (*Carduelis tristis*) for natural and artificial male traits. Behav Ecol 4: 138–143.

Johnson, M. H., and B. J. Everitt. 2000. Essential Reproduction, Fifth edition. Oxford: Blackwell Scientific.

Juhn, M., G. H. Faulkner, and R. G. Gustavson. 1931. The correlation of rates of growth and hormone threshold in the feathers of fowls. J Exp Zool 58: 69–107.

Kannankeril, J. V., and L. V. Domm. 1968. The influence of gonadectomy on sexual characters in the Japanese Quail. J Morph 126: 395–412.

Karubian, J. 2002. Costs and benefits of variable breeding plumage in the Red-backed Fairy-wren. Evolution 56: 1673–1682.

Keck, W. N. 1934. The control of the secondary sex characters in the English Sparrow *Passer domesticus* (Linnaeus). J Exp Zool 67: 315–341.

Kimball, R. T., and J. D. Ligon. 1999. Evolution of avian plumage dichromatism from a proximate perspective. Am Nat 154: 182–193.

Lank, D. B., M. Coupe, and K. E. Wynne-Edwards. 1999. Testosterone-induced male traits in female Ruffs (*Philomachus pugnax*): Autosomal inheritance and gender differentiation. Proc R Soc Lond B 266: 2323–2330.

Lazarus, J., and J. H. Crook. 1973. The effects of luteinizing hormone, oestrogen and ovariectomy on the agonistic behaviour of female *Quelea quelea*. Anim Behav 21: 49–60.

Ligon, J. D. 1999. The Evolution of Avian Breeding Systems. Oxford: Oxford University Press.

Ligon, J. D., and P. W. Zwartjes. 1995. Ornate plumage of male Red Junglefowl does not influence mate choice by females. Anim Behav 49: 117–125.

Ligon, J. D., R. Thornhill, M. Zuk, and K. Johnson. 1990. Male-male competition, ornamentation and the role of testosterone in sexual selection in Red Junglefowl. Anim Behav 40: 367–373.

Ligon, J. D., R. Kimball, and M. Merola-Zwartjes. 1998. Mate choice in Red Junglefowl: The issues of multiple ornaments and fluctuating asymmetry. Anim Behav 55: 41–50.

Lillie, F. R. 1931. Bilateral gynandromorphism and lateral hemihypertrophy in birds. Science 74: 387–390.

Lucas, A. M., and P. R. Stettenheim. 1972. Avian Anatomy: Integument. Washington, DC: U.S. Government Printing Office.

Lupu, C. 2000. Evaluation of side effects of tamoxifen in Budgerigars (*Melopsittacus undulatus*). J Avian Med Surgery 14: 237–242.

Matsumine, H., J. D. Wilson, and M. J. McPhaul. 1990. Sebright and campine chickens express aromatase p-450 messenger RNA inappropriately in extraglandular tissues and in skin fibroblasts. Mol Endocrinol 4: 905–911.

Matsumine, H., M. A. Herbst, S.-H. I. Ou, J. D. Wilson, and M. J. McPhaul. 1991. Aromatase mRNA in the extragonadal tissues of chickens with the henny-feathering trait is derived from a distinctive promoter structure that contains a segment of a retroviral long terminal repeat. J Biol Chem 266: 19900–19907.

McCarty, J., and A. Secord. 2000. Possible effects of PCB contamination on female plumage color and reproductive success in Hudson River Tree Swallows. Auk 117: 987–995.

McGraw, K. J. 2003. The Physiological Costs of Being Colorful in the Zebra Finch. Ph.D. diss., Cornell University, Ithaca, NY.

McGraw, K. J., J. Dale, and E. Mackillop. 2003. Social environment during molt and the expression of melanin-based plumage pigmentation in male House Sparrows (*Passer domesticus*). Behav Ecol Sociobiol 53: 116–122.

Miller, D. S. 1935. Effects of thyroxin on plumage of the English Sparrow, *Passer domesticus* (Linnaeus). J Exp Biol 71: 293–309.

Moore, I., B. Walker, and J. Wingfield. 2004. The effects of combined aromatase inhibitor and anti-androgen on male territorial aggression in a tropical population of Rufous-collared Sparrows, *Zonotrichia capensis*. Gen Comp Endocrinol 135: 223–229.

Morejohn, G. V., and R. E. Genelly. 1961. Plumage differentiation of normal and sex-anomalous Ring-necked Pheasants in response to synthetic hormone implants. Condor 63: 101–110.

Morgan, T. H. 1919. The Genetic and the Operative Evidence Relating to Secondary Sexual Characters. Washington, DC: Carnegie Institute of Washington.

Mueller, N. S. 1970. An experimental study of sexual dichromatism in the duck *Anas platyrhynchos*. J Exp Zool 173: 263–268.

Mueller, N. S. 1977. Control of sex differences in the plumage of the House Sparrow, *Passer domesticus*. J Exp Zool 202: 45–48.

Mulder, R. A., and M. J. L. Magrath. 1994. Timing of prenuptial molt as a sexually selected indicator of male quality in Superb Fairy-wrens (*Malurus cyaneus*). Behav Ecol 5: 393–400.

Mundinger, P. C. 1972. Annual testicular cycle and bill color change in the Eastern American Goldfinch. Auk 89: 403–419.

Nespor, A. A., M. J. Lukazewicz, R. J. Dooling, and G. F. Ball. 1996. Testosterone induction of male-like vocalizations in female Budgerigars (*Melopsittacus undulatus*). Horm Behav 30: 162–169.

Noble, G. K., and M. Wurm. 1940. The effect of hormones on the breeding of the Laughing Gull. Anat Rec Suppl. 78: 50–51.

Nussey, S. S., and S. A. Whitehead. 2001. Endocrinology: An Integrated Approach. Oxford: BIOS Scientific Publishers.

Okazaki, K., and P. F. Hall. 1965. The action of interstitial cell-stimulating hormone upon tyrosinase activity in the weaver bird (*Steganura paradisaea*). Biochem Biophys Res Comm 20: 667–673.

Omland, K. E. 1996a. Female Mallard mating preferences for multiple male ornaments. I. Natural variation. Behav Ecol Sociobiol 39: 353–360.

Omland, K. E. 1996b. Female Mallard mating preferences for multiple male ornaments. II. Experimental variation. Behav Ecol Sociobiol 39: 361–366.

Ortman, R. 1967. The performance of the Napoleon Weaver, the Orange Weaver, and the Paradise Whydah in the weaver finch test for luteinizing hormones. Gen Comp Endocrinol 9: 368–373.

Ottinger, M., M. Abdelnabi, M. Quinn, N. Golden, J. Wu, and N. Thompson. 2002. Reproductive consequences of EDCs in birds—What do laboratory effects mean in field species? Neurotox Teratol 24: 17–28.

Owens, I. P. F., and R. V. Short. 1995. Hormonal basis of sexual dimorphism in birds: Implications for new theories of sexual selection. Trends Ecol Evol 10: 44–47.

Parrish, J., J. Stoddard, and C. White. 1987. Sexually mosaic plumage in a female American Kestrel. Condor 89: 911–913.

Patten, M. A. 1993. A probably bilateral gynandromorphic Black-throated Blue Warbler. Wilson Bull 105: 695–698.

Peters, A., L. Astheimer, C. Boland, and A. Cockburn. 2000. Testosterone is involved in acquisition and maintenance of sexually selected male plumage in Superb Fairy-wrens, *Malurus cyaneus*. Behav Ecol Sociobiol 47: 438–445.

Ralph, C. L. 1969. The control of color in birds. Am Zool 9: 521–530.

Ralph, C. L., D. L. Grinwich, and P. F. Hall. 1967a. Hormonal regulation of feather pigmentation in African weaver birds: The exclusion of certain possible mechanisms. J Exp Zool 166: 289–294.

Ralph, C. L., D. L. Grinwich, and P. F. Hall. 1967b. Studies of the melanogenic response of regenerating feathers in the weaver bird: Comparison of two species in response to gonadotrophins. J Exp Zool 166: 283–287.

Rimmer, C., and J. Tietz. 2001. An adult male Blackpoll Warbler in female-like plumage. J Field Ornith 72: 365–368.

Roberts, M. L., K. L. Buchanan, and M. R. Evans. 2004. Testing the immunocompetence handicap hypothesis: A review of the evidence. Anim Behav 68: 227–239.

Saxena, R., and J. Thapliyal. 1961. Plumage control in Black-headed Munia (*Munia atricapilla*). Naturwissenschaften 48: 652.

Schlinger, B. A., A. J. Fivizzani, and G. V. Callard. 1989. Aromatase, 5α and 5β-reductase in brain, pituitary and skin of the sex-role reversed Wilson's Phalarope. J Endocrinol 122: 573–581.

Schlinger, B. A., N. I. Lane, W. Grisham, and L. Thompson. 2004. Androgen synthesis in a songbird: A study of Cyp17 (17α-hydroxylase/C17,20-lyase) activity in the Zebra Finch. Gen Comp Endocrinol 113: 46–58.

Scott, H. M., and L. F. Payne. 1934. The effect of gonadectomy on the secondary sexual characters of the Bronze Turkey (*M. gallopavo*). J Exp Zool 69: 123–133.

Sibley, C. G., and J. E. Ahlquist. 1990. Phylogeny and classification of birds: A study in molecular evolution. New Haven, CT: Yale University Press.

Snyder, P. J., and F. H. Sterling. 1976. Hypersecretion of LH and FSH by a pituitary adenoma. J Clin Endocrinol Metab 42: 544–550.

Stoddard, H. L. 1921. Female Bay-breasted Warbler in male plumage. Auk 38: 117.

Stoehr, A., and G. Hill. 2001. The effects of elevated testosterone on plumage hue in male House Finches. J Avian Biol 32: 153–158.

Stokkan, K. A. 1979a. The effect of permanent short days and castration on plumage and comb growth in male Willow Ptarmigan (*Lagopus lagopus*). Auk 96: 682–687.

Stokkan, K. A. 1979b. Testosterone and day length–dependent development of comb size and breeding plumage of male Willow Ptarmigan (*Lagopus lagopus lagopus*). Auk 96: 106–115.

Strasser, R., and H. Schwabl. 2004. Yolk testostersone organizes behavior and male plumage coloration in House Sparrows (*Passer domesticus*). Behav Ecol Sociobiol 56: 491–497.

Summers, S. G., and R. M. Kostecke. 2004. Female Brown-headed Cowbird with partial male plumage. Wilson Bull 116: 293–294.

Tella, J., J. Donazar, and F. Hiraldo. 1996. Variable expression of sexually mosaic plumage in female Lesser Kestrels. Condor 98: 643–644.

Tewary, P. D., and D. S. Farner. 1973. Effect of castration and estrogen administration on the plumage pigment of the male House Finch (*Carpodacus mexicanus*). Am Zool 13: 1278.

Tewary, P. D., P. M. Tripathi, and B. K. Tripathi. 1985. Effects of exogenous gonadal steroids and castration on photoperiodic responses of the Yellow-throated Sparrow *Gymnorhis xanthocollis* (Burton). Indian J Exp Biol 23: 426–428.

Thapliyal, J. P., and B. B. P. Gupta. 1984. Thyroid and annual gonad development, body weight, plumage pigmentation, and bill color cycles of Lal Munia, *Estrilda amandava*. Gen Comp Endocrinol 55: 20–28.

Thapliyal, J. P., and R. N. Saxena. 1961. Plumage control in Indian Weaver Bird (*Ploceus philippinus*). Naturwissenschaften 48: 741–742.

Thapliyal, J. P., and P. D. Tewary. 1961. Plumage in Lal Munia (*Amandava amandava*). Science 134: 738–739.

Thapliyal, J. P., and P. D. Tewary. 1963. Effect of estrogen and gonadotropic hormones on the plumage pigmentation in Lal Munia (*Estrilda amandava*). Naturwissenschaften 50: 529.

Tsutsui, K., and B. A. Schlinger. 2001. Steroidogenesis in the avian brain. In S. Dawson and C. M. Chaturvedi, ed., Avian Endocrinology, 59–77. Pangbourne, UK: Alpha Science International.

Vaillant, S., D. Guemene, M. Dorizzi, C. Pieau, N. Richard-Mercier, and J. Brillard. 2003. Degree of sex reversal as related to plasma steroid levels in genetic female chickens (*Gallus domesticus*) treated with fadrozole. Mol Repro Devel 64: 420–428.

van Oordt, G. J. 1936. The effect of gonadectomy on the secondary sexual characters of the turkey. Arch Port Sci Biol 5: 205–211.

van Oordt, G. J., and G. C. A. Junge. 1933. The influence of the testis hormone on the development of ambosexual characters in the Blackheaded Gull (*Larus ridibundus*). Acta Brev Neerl Physiol 3: 15–17.

van Oordt, G. J., and G. C. A. Junge. 1934. The relation between the gonads and the secondary sexual characters in the Ruff (*Philomachus pugnax*). Bull Soc Biol Lettonie 4: 141–146.

Verhulst, S., S. J. Dieleman, and H. K. Parmentier. 1999. A tradeoff between immunocompetence and sexual ornamentation in domestic fowl. Proc Natl Acad Sci USA 96: 4478–4481.

Vevers, H. G. 1954. The experimental analysis of feather pattern in the Amherst Pheasant, *Chrysolophus amherstiae* (Leadbeater). Trans Zool Soc Lond 28: 305–349.

Vevers, H. G. 1962. The influence of the ovaries on secondary sexual characters. In S. Zuckerman, ed., The Ovary, 263–289. New York: Academic Press.

Walton, A. 1937. On the eclipse plumage of the Mallard (*Anas platyrhynca platyrhyncha*). J Exp Biol 14: 440–447.

Warner, R. L. 1970. Endocrine control of sexually dimorphic plumage in Japanese Quail (*Coturnix coturnix japonica*). Ph.D. dissertation, University of California, Davis.

Wilson, J. D., M. Leshin, and F. W. George. 1987. The Sebright bantam chicken and the genetic control of extraglandular aromatase. Endocr Rev 8: 363–376.

Wingfield, J. C., A. Newman, G. L. J. Hunt, and D. S. Farner. 1980. Androgen in high concentrations in the blood of female Western Gulls, *Larus occidentalis*. Naturwissenschaften 67: 514–515.

Wingfield, J. C., J. D. Jacobs, A. D. Tramontin, N. Perfito, S. Meddle, D. L. Maney, and K. Soma. 2000. Toward an ecological basis of hormone-behavior in reproduction of birds. In K. Wallen, and J. E. Schneider, ed., Reproduction in Context: Social and Environmental Influences on Reproduction, 85–128. Cambridge, MA: MIT Press.

Witschi, E. 1935. Seasonal sex characters in birds and their hormonal control. Wilson Bull 47: 177–188.

Witschi, E. 1936. Effect of gonadotropic and oestrogenic hormones on regnerating feathers of Weaver Finches (*Pyromelana franciscana*). Proc Soc Expl Biol Med 35: 484–489.

Witschi, E. 1961. Sex and secondary sexual characteristics. In A. J. Marshall, ed., Biology and Comparative Physiology of Birds, 115–168. New York: Academic Press.

Witschi, E., and R. A. Miller. 1938. Ambisexuality in the female starling. J Exp Zool 79: 475–487.

Zuk, M. 1991. Sexual ornaments as animal signals. Trends Ecol Evol 6: 228–231.

Zuk, M., R. Thornhill, J. D. Ligon, K. Johnson, S. Austad, et al. 1990. The role of male ornaments and courtship behavior in female mate choice of Red Jungle Fowl. Am Nat 136: 459–473.

Zuk, M., T. S. Johnsen, and T. Maclarty. 1995. Endocrine-immune interactions, ornaments and mate choice in Red Jungle Fowl. Proc R Soc Lond B 260: 205–210.

11

Genetic Basis of Color Variation in Wild Birds

NICHOLAS I. MUNDY

Knowledge of the genetic basis of coloration is fundamental to many branches of avian ecology and evolution, but in spite of its importance, our understanding of the genetic control of the coloration of wild birds is currently limited. Genetic mechanisms underlying variation in sexually selected traits are crucial to testing alternative models of sexual selection (Andersson 1994). Adaptive evolution of coloration depends on genetic changes, and the genetic architecture of coloration may bias or constrain the direction of evolution in different lineages (Chapter 8, Volume 2). Predictions of evolutionary trajectories depend on knowledge of the heritabilities of colorful traits and genetic correlations among them, and how these evolve under varying selective regimes. In broader terms, the genetics of avian color variation provides an interesting contrast with the genetics of coloration in other vertebrates, where most work to date has been performed in mammals (Jackson 1997). It is also an excellent potential model of the evolution of development; in particular, the extraordinary diversity of feather coloration is attracting attention as a model of the mechanisms of pattern generation (Prum and Williamson 2002).

Human interest in the inheritance of coloration in birds extends back over many centuries and perhaps even millennia due to its association with the captive breeding of birds. Breeders would go to great lengths to obtain a desired color—the use of the Red Siskin (*Carduelis cucullata*) to produce red-factor Common Canaries (*Serinus canaria*) is a clear example of applied plumage genetics (Birkhead 2003). Inheritance patterns were regularly recorded by the nineteenth century, and Darwin's keen interest in the inheritance of plumage color among pigeon breeds played a prominent role in his theory of natural selection (Darwin 1859). It was not until the rediscovery of Mendel's work that a framework for interpreting these observations became available.

Attention to the genetics of color variation among wild birds per se is much more recent. Following substantial work on polymorphisms from the 1950s through to the late-1980s (Buckley 1987), interest in the genetics of the coloration of wild birds declined in the late twentieth century, and it is only recently that it has regained momentum. In contrast, work on the genetics of quantitative color traits only began in the late 1980s (Møller 1989) and still remains poorly studied.

Studies of single-locus and polygenic effects have separate methodologies and have been applied to different questions, so I begin this review by considering them separately. This summary is followed by a section on the genetic basis of population and species differences in coloration. The main discussion is largely restricted to plumage color variation, as there is little information on other colored body parts (e.g., bills, bare skin, eyes).

Single-Locus/Oligogenic Control of Color Variation

Many color traits show inheritance patterns that follow simple Mendelian principles involving one locus or a few loci. The review by Buckley (1987) is still an excellent introduction to this topic. Here I provide a brief overview of earlier work before presenting newer results, particularly from molecular genetics of melanin pigmentation, in more detail.

A substantial body of work exists on the genetic basis of coloration in domesticated species of birds, which has had an important impact on the study of genetic mechanisms of color variation in the wild. The inheritance patterns of large numbers of Mendelian coloration loci have been elucidated in such species as the Domestic Chicken (*Gallus gallus;* Plate 26), Japanese Quail (*Coturnix japonica*), Wild Turkey (*Meleagris gallopavo;* Plate 32), Mallard (*Anas platyrhynchos;* Plate 15, Volume 2), Rock Pigeon (*Columba livia*), Budgerigars (*Melopsittacus undulatus*), and a variety of cage songbirds, including canaries and Zebra Finches (*Taeniopygia guttata;* Plate 27; Lancaster 1963; Johnston and Janiga 1995; Landry 1997). These studies provide valuable information on basic genetic mechanisms for both pigmentary and structural colors. A simple example is that the presence of yellow and blue coloration in budgerigars (Plate 17, Volume 2), which are single-locus mutants from the wild-type green, shows that there is independent genetic control of yellow psittacofulvin (Chapter 8) and structural coloration (Chapter 7) in this species. Mendelian loci in domesticated species are also helpful for suggesting candidate loci that may be operating in the wild, particularly when color variation in the wild is mimicked by a color mutation in captivity, which is the usual starting point for the isolation of genes involved in pigmentation at the molecular level. A review of such work is beyond the scope of this chapter, but several examples relevant to wild birds are mentioned.

In wild populations of birds, the majority of evidence for the role of single-locus control of plumage color comes from plumage polymorphisms. The segregation of alleles for different color morphs in the same population provides a natural laboratory for studying the genetic basis of coloration. Other evidence comes from the occurrence of unusually colored individuals (e.g., leucistic or melanistic individuals), and studies of the genetic transmission of color variation among subspecies or among species in naturally occurring hybrids or in captive breeding (discussed in more detail below).

Color Polymorphisms

Color polymorphisms are generally defined as the presence in an interbreeding population of individuals of the same sex and age that have distinct coloration (Buckley 1987). The discrete categories of plumage coloration are commonly referred to as "phases" or "morphs" (Chapter 2, Volume 2). The distinction between color polymorphism and quantitative variation in coloration is not always straightforward, and indeed, the two can grade into one another. For example, in the Common Buzzard (*Buteo buteo*) the three common morphs (pale, intermediate, and dark) represent peaks of frequency along a continuous scale of variation from the palest to the darkest phenotype (O. Krüger, pers. comm.). However, in the context of this chapter, what is important is whether control of the plumage variation is oligogenic (i.e., involving one or a few genes) or polygenic.

In a survey of avian polymorphism, Galeotti et al. (2003) documented 334 species, or approximately 3.5% of all birds, with polymorphisms. The presence of polymorphisms is unevenly distributed across avian orders, with Strigiformes, Ciconiiformes (including the traditional Falconiformes), Cuculiformes, and Galliformes strongly overrepresented. The majority of polymorphisms involves the adult plumage of both sexes, but polymorphisms of immature birds, and of a single sex in adult birds, also occur (e.g., male Madagascar Paradise Flycatchers [*Terpsiphone mutata*]; Mulder et al. 2002).

Polymorphisms involving changes in melanin distribution alone are by far the most frequent. When two morphs are present, the most common combinations are rufous and gray, brown and buff, rufous and brown, and black and white, whereas in species with three morphs, rufous-gray-brown is most prevalent (Galeotti et al. 2003). Although the pigments involved have rarely been quantified, from the color variation it can tentatively be assumed that many examples involve changes in both eumelanin and phaeomelanin distribution (Chapter 6); some involve primarily eumelanin, whereas changes involving mostly phaeomelanin are rare. Work on quantifying which melanins vary in polymorphisms would be helpful in elucidating the genetic mechanisms involved, as different genetic changes have differential effects on the two pigments (see below).

A few polymorphisms are known that affect coloration that is not melanin-based. A sex-linked locus controls yellow/red carotenoid-based mask color in Gouldian Finches (*Poephila gouldiae;* Plate 18, Volume 2; Southern 1946). The polymorphism in Orange-fronted Parakeets (*Cyanoramphus malherbi*) presumably involves variation in psittacofulvins (McGraw and Nogare 2005). I am unaware of any good candidates for polymorphism in structural coloration. Many examples of changes in melanin distribution involve changes in the size of white patches (e.g., Lesser Snow Goose [*Anser c. caerulescens*]; Plate 28), sometimes affecting the whole bird, but in the absence of evidence to the contrary, it is simplest to assume that the differences are solely due to melanin deposition without any change in ultrastructure.

Polymorphisms that simultaneously involve changes in melanin and carotenoid distribution are rare. The only clear example appears to be the Gouldian Finch, in which one locus controls variation between the black- and red-masked morphs (Southern 1946). Other apparent cases involve shifts between yellow and black coloration in Bananaquits (*Coereba flaveola;* Wunderle 1981) and Evening Grosbeaks (*Coccothraustes vespertinus;* Plate 25; Hudon 1997). However, in these examples, the variation is probably due to melanin deposition alone, highlighting the importance of carefully defining the physicochemical basis of color changes.

Genetic Basis of Plumage Polymorphisms

Although the presence of discrete color morphs strongly suggests genetic control, it is only by observing transmission of phenotypes across generations that such a genetic basis can be directly demonstrated. Moreover, studies of genetic transmission can indicate the specific mode of inheritance, such as the number of loci involved, whether the loci are autosomal or sex-linked, and the dominance relationships among alleles. If more than one locus is involved, then the linkage relationships among loci, and the presence and nature of epistatic interactions among alleles at different loci, can also be determined. This information is critical for understanding the evolutionary dynamics of a trait and is useful for identifying putative homologous genetic mechanisms for color variation among different species.

Table 11.1 summarizes transmission studies of plumage polymorphisms in wild and feral populations of birds. The table provides data for just 7% of all polymorphic species and constitutes a biased taxonomic sample (e.g., no Strigiformes, Cuculiformes, or Galliformes are included). The species are grouped according to the likely type of pigment involved, whether the polymorphism occurs in juveniles or adults, and the dominance relationships of the alleles. Most of these polymorphisms can be attributed to a single locus, or at least a single locus accounts for a large part of the phenotypic variation seen. The majority of these studies was carried out before the era of DNA fingerprinting, which means that some error in paternity—and in some cases also maternity—assignment may have occurred.

Melanism (i.e., increased deposition of, predominantly, eumelanin) is the most common polymorphism represented. Considering adult and juvenile cases together, it usually exhibits dominant inheritance (9 of 17), with partial dominance (4 of 17) and recessive inheritance (4 of 17) being rarer. Most of the loci involved are autosomal.

Among wild birds, epistatic interactions between the effects of separate loci on plumage traits have been identified in feral pigeons and Gouldian Finches (Table 11.1). In pigeons, the melanic allele at the Spread locus is epistatic to all alleles at the C-series locus, to which the Spread locus is also physically linked (Johnston and Janiga 1995). In the Gouldian Finch, there is a complex series of epistatic interactions between the two alleles at each of the two loci, one of which is autosomal and the other sex-linked (Buckley 1987). Epistasis among pigmentation loci is a common feature in domestic species, in which it is relatively easy to detect.

A dramatic case of plumage polymorphism occurs in the Ruff (*Philomachus pugnax*). Males have extremely variable breeding plumage (Plate 7, Volume 2), but there is a basic dichotomy between territorial males with colorful ruffs and head tufts, which establish display areas on the lek, and satellite males with white ruffs and head tufts, which adopt an alternative mating strategy.

Table 11.1. Inheritance of Avian Polymorphisms in the Wild

Species	Age	Eumelanin	Phaeomelanin	Carotenoid	Dominance	A/SL	Reference
Common Guillemot (*Uria aalge*)	Adult	+			M>L	A	Jefferies and Parslow (1976)
Little Shag (*Phalacrocorax melanoleucos brevirostris*)	Adult	+			M>L	A	Dowding and Taylor (1987)
Indian Reef Heron (*Egretta gularis*)	Adult	+			M>L	A	Naik and Parasharya (1983)
Eleonora's Falcon (*Falco eleonorae*)	Adult	+	+		M>L	A	Wink et al. (1978)
Bananaquit (*Coereba flaveola*)	Adult	+			M>L	A	Wunderle (1981)
New Zealand Fantail (*Rhipidura fuliginosa*)	Adult	+			M>L	A	Craig (1972)
Lesser Snow Goose (*Anser caerulescens caerulescens*)	Adult	+			M=L	A	Cooke and Cooch (1968); Rattray and Cooke (1984)
Arctic Skua (*Stercorarius parasiticus*)	Adult	+			M=L	A	O'Donald and Davis (1959); O'Donald (1983)
Common Buzzard (*Buteo buteo*)	Adult	+	+		M=L	A	Krüger et al. (2001)
Variable Oystercatcher (*Haematopus unicolor*)	Adult	+			M=L	A	Baker (1973)
Giant Petrel (*Macronectes giganteus*)	Adult	+			L>M	A	Shaughnessy (1970)
Great White/Blue Heron (*Ardea herodias*)	Adult	+			L>M	A	Meyerriecks (1957)
Blackcap (*Sylvia atricapilla*)	Adult	+			L>M	A	Berthold et al. (1996)
Ruff (*Philomachus pugnax*)	Adult (male)	+	+		L>M	A	Lank et al. (1995)
Mute Swan (*Cygnus olor*)	Juvenile	+			M>L	A	Munro et al. (1968)
Ross' Goose (*Anser rossi*)	Juvenile	+			M>L	A	Cooke and Ryder (1971)
Eastern Screech Owl (*Otus asio*)	Adult	+	+		M=P	A	Hrubant (1955)
Ferruginous Hawk (*Buteo regalis*)	Adult	+	+		M>P	A	Schmutz and Schmutz (1981)
White-crowned Sparrow (*Zonotrichia albicollis*)	Adult		+		L>P	K	Thorneycroft (1975)
Rock Pigeon (*Columba livia*)	Adult	+ (C-series)	+		M>L	A	Johnston and Janiga (1995)
	Adult	+ (Spread)	+		M>L	A	
	Adult				P>M>Br	SL	
Gouldian Finch (*Poephila gouldiae*)	Adult	+		+	R>M	SL	Southern (1946)
	Adult			+	R>Y	A	Murray (1963)
Orange-fronted Parakeet (*Cyanoramphus malherbi*)	Adult	Psittacofulvin?			Y>O	A	Taylor et al. (1986)

Source: Most of the information in this table is from Roulin (2004: Table 1).

Notes: A, autosomal; Br, brown; dominance, dominance relationships; K, karyotype; L, light; M, (eu)melanic; O, orange; P, phaeomelanic; R, red; SL, sex-linked; Y, yellow; >, dominant-recessive alleles; =, partially dominant (additive) alleles.

Interestingly, inheritance of the white head ruff and satellite-specific behavior always co-vary, indicating tight linkage disequilibrium between the loci determining color and behavior (Lank et al. 1995).

Maintenance of Plumage Polymorphisms

The occurrence of genetically based phenotypic polymorphisms poses a classical question in ecological genetics of how the polymorphism is maintained in the face of inevitable allelic loss by genetic drift or stabilizing or directional selection. Potential mechanisms involved include gene flow, assortative or disassortative mating, selection on different morphs in different environments, overdominant selection, and frequency-dependent selection (Hartl and Clark 1997). Long-term demographic studies are required to tease out the various influences, so it is not surprising that, except for a few well-studied examples, the factors maintaining polymorphisms are still poorly understood. What is clear, however, is that different mechanisms may be involved in different species (Lank 2002). Work on adaptive advantages of polymorphisms is the subject of a recent, thorough review (Roulin 2004). Here I highlight two contrasting case studies and then consider recent comparative studies.

Perhaps the best-understood case of the evolutionary dynamics of a polymorphism is that of the Lesser Snow Goose, studied over many years by Cooke and colleagues (Cooke et al. 1995). Here the occurrence of polymorphic breeding colonies containing blue- (melanic) and white-morph birds (Plate 28) is almost certainly the result of recent population expansion, with blue and white populations formerly being allopatric (Cooke et al. 1988). Despite extensive studies, no fitness differences have been identified among blue- and white-phase birds at the La Pérouse Bay colony in Hudson Bay (Cooke et al. 1985; Rockwell et al. 1985). Mate choice is strongly influenced by exposure to parental color as goslings, which leads to a high frequency of positive assortative matings while populations are mixing. The overall picture, therefore, is one of gradual introgression of melanic alleles into white populations, which is slowed down by assortative mating. Many questions still remain, however. For example, was there an advantage for melanism when it first arose, and do white and blue morphs have different fitnesses in more extreme northerly environments?

A strong contrast is provided by recent work on a population of Common Buzzards in Germany (Krüger et al. 2001). In this population, intermediate-morph birds occupy higher-quality territories, have higher survival rates, and higher reproductive success than the extreme morph (dark and light) birds.

Hence heterozygote advantage (heterosis or overdominance) of intermediate-morph birds appears to be operating, although the proximate factors leading to the advantages are unknown. A surprising feature is that positive assortative mating by morph color occurs, which is maladaptive for extreme morphs, because it reduces the frequency of intermediate-morph offspring produced by extreme-morph birds (Krüger et al. 2001). This system is probably evolutionarily unstable, and genetic modifiers that lead to the inheritance of the intermediate morph in all birds would be under strong selection.

Comparative studies have been underutilized in the study of plumage polymorphisms. Recently the first comparative studies that use methods to explicitly control for shared phylogenetic history have been performed (Harvey and Pagel 1990). A common theme of these studies has been to consider two main hypotheses: disruptive selection and apostatic selection. Disruptive selection on coloration requires different adaptive benefits for different morphs in different environments. Apostatic selection involves the use of color polymorphism to evade the formation of "avoidance images" by prey (or "search images" by predators), and has been a prominent but controversial hypothesis for explaining polymorphisms in such predatory birds as raptors, gulls, and skuas (Paulson 1973; Rohwer and Paulson 1987). In the comparative studies, birds and mammals, but not reptiles and insects, are considered to be prey that might use an avoidance image to escape capture.

Three studies are broadly in agreement in finding relationships consistent with disruptive selection. In an analysis across all polymorphic species of birds, significant relationships were found between the presence of polymorphism and high habitat diversity, a day/night activity pattern, and a partially migratory or nomadic behavior; all of which were interpreted as evidence for disruptive selection (Galeotti et al. 2003; Figure 11.1). These results were confirmed and extended in a follow-up study of 91 pairs of closely related species of raptors, owls, and nightjars in which one species was monomorphic and one polymorphic (Galeotti and Rubolini 2004). There was no evidence for apostasis, but polymorphic species had significantly larger distributions, higher habitat diversity, higher climate variability, more diverse activity patterns, and were more migratory than monomorphic species. In a study on a broad range of raptor and owl species, increased polymorphism was related to increased sexual plumage dimorphism, larger population size, and larger species-range size, and there was no evidence for apostatic selection (Fowlie and Krüger 2003).

Conflicting results have been obtained in Roulin and Wink's (2004) study concentrating on the two genera of raptors (*Accipiter* and *Buteo*) with the

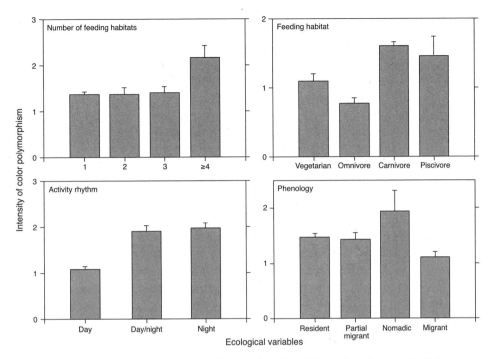

Figure 11.1. Relation between color and different ecological variables, using phylogenetically independent contrasts. Two indices of environmental variability (number and type of feeding habitats), in addition to activity rhythm, are significantly related to degree of polymorphism (Galeotti et al. 2003). Data shown are mean + standard error for each ecological variable.

highest levels of polymorphism. These authors found that polymorphic species in these genera more often prey on mammals, but not birds, insects, or reptiles, than do monomorphic species, and that polymorphic species have longer wings than monomorphic species. These results were interpreted as being consistent with apostatic selection. The reasons for the differing results of these studies are unclear, but they considered different ranges of species and scored polymorphism in different ways. Further work using more precise ecological variables is desirable.

Chromosomal Basis of Plumage Color Polymorphisms

Variation in gross structure of chromosomes is much rarer than variation in single genes and is not expected to be a common contributor to color variation.

However, there is a unique and well-known example of association between changes in the karyotype and plumage in the White-throated Sparrow (*Zonotrichia albicollis;* Thorneycroft 1966, 1975), which was the first case in which a plumage polymorphism was associated with an identified change in genetic material. Individuals with one or two copies of a variant form of chromosome 2 with a pericentric inversion (2^m or II^m) have a white headstripe in breeding plumage, whereas individuals with two normal copies of chromosome 2 have a tan headstripe in breeding plumage. Surprisingly, the variant chromosome morphologies are not identifiable when birds are in winter plumage (Vardy 1971). The polymorphism is maintained by disassortative mating, and a recent study found good evidence that tan- and white-striped males adopt different reproductive strategies (Tuttle 2003). No further correlations between karyotype and color polymorphism have been discovered, and for the moment, this example represents a genetic curiosity.

Molecular Genetic Basis of Plumage Color Polymorphisms

What are the genetic changes at specific pigmentation loci that cause differences in coloration among individuals in wild populations? Recent advances have enabled this question to be directly addressed for the first time for melanin-based coloration. In contrast, there is currently no information in either wild or captive birds about the molecular genetic control of variation in carotenoids, other classes of pigment, and structural coloration.

The first pigmentation gene to be isolated in birds was the gene for the enzyme tyrosinase in chickens, which catalyzes the first step in the synthesis of both eumelanin and phaeomelanin (Ochii et al. 1992). Loss-of-function mutations in tyrosinase cause albinism, with a complete loss of melanin in skin, feathers, iris, and retina. Such a dramatic change in phenotype would obviously be strongly selected against in nature, as daylight vision is impaired. It remains to be demonstrated whether variation in tyrosinase causing milder changes in plumage phenotype has been involved in plumage evolution.

Isolation of a gene associated with plumage melanism came in the late 1990s. In two important papers, the melanocortin-1 receptor gene (*MC1R*) was isolated in chickens, and a point substitution in the gene was found to be associated with melanism (Takeuchi et al. 1996, 1998). The MC1R is a 7-transmembrane G-protein coupled receptor expressed in melanocytes that had previously been studied in mice and several other domestic mammals (Robbins et al. 1993; Klungland et al. 1995; Marklund et al. 1996). The

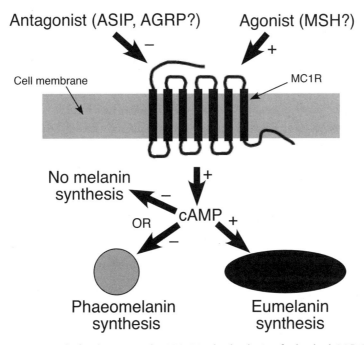

Figure 11.2. Control of melanogenesis by *MC1R* in the developing feather bud. MC1R is depicted in the cell membrane of a melanocyte, and the extracellular surface is up. Only the key decision point determined by MC1R activity is shown. Both MC1R agonist and antagonists are poorly known in birds, and many other genes are involved in eumelanogenesis and phaeomelanogenesis (Chapter 6). Abbreviations: AGRP, agouti-related protein; ASIP, agouti signaling protein; MSH, melanocyte stimulating hormone.

MC1R plays a critical role in determining what type of melanin is synthesized during feather and hair development (Figure 11.2). When the MC1R is stimulated by a melanocortin agonist, such as α-MSH, eumelanin synthesis in eumelanosomes occurs via a complex cascade of events that is initiated by a rise in intracellular cyclic adenosine monophosphate (cAMP; Chapter 6). In the absence of agonist or presence of antagonist (agouti signaling protein [ASIP] in mammals), the low intracellular cAMP typically causes phaeomelanin to be synthesized, as occurs in chickens and domestic mammals. The physiological agonist and antagonist (if present) of MC1R in birds are poorly known. In all domestic species, including chickens, *MC1R* alleles causing increased eumelanin synthesis (melanism) are inherited in a dominant or partially dominant fashion, whereas *MC1R* alleles causing increased phaeomelanin synthesis (phaeomelanism) are inherited as recessives, which can provide an important clue to the possible involvement of *MC1R* in melanism in other species.

The phenotypic changes caused by different *MC1R* alleles in domestic chickens mirror some of those seen in wild birds, and so a key question is whether variation at the *MC1R* locus has a role in plumage color evolution. Recent work in our group has shown that this is indeed the case. In three unrelated species with melanic polymorphisms, the Bananaquit, Lesser Snow Goose, and Arctic Skua (Parasitic Jaeger in North America, *Stercorarius parasiticus*), melanism is associated with point substitutions in *MC1R* (Theron et al. 2001; Mundy et al. 2004; Figure 11.3; Plate 28). Significantly, comparative data from neutral markers (mitochondrial DNA and nuclear microsatellite loci) show that these associations are specific to the *MC1R* locus.

There are some striking similarities and revealing differences in the relationship between *MC1R* and coloration in the three species (Figure 11.3). In all three species, the phenotypic change is presumably a change from the absence of melanin in certain patches of feathers to an increase in primarily eumelanin deposition in those feathers. This differs from the situation mentioned above in chickens and domestic mammals, in which the variation is from predominantly eumelanin to predominantly phaeomelanin. Different nonsynonymous substitutions in *MC1R* occur in the three species. Although it has not been demonstrated that these substitutions are causal, a plausible case can be made for each of them affecting MC1R action. Most notably, the amino acid change (Glu92Lys) in the Bananaquit has already been described in chickens and mice (Robbins et al. 1993; Takeuchi et al. 1998), where it causes melanism by making the MC1R constitutively active (i.e., continuously active even in the absence of agonist; King et al. 2003).

In Bananaquits, there are two morphs with no intermediates: the melanic morph is completely black and the melanic *MC1R* allele is inherited as a dominant. In contrast, although pale and dark morphs can be defined in Lesser Snow Geese (white and blue phases) and Arctic Skuas (pale and intermediate/dark phases), there is substantial quantitative variation in the degree of melanism among blue geese and intermediate/dark skuas. Interestingly, there is a correlation between copy number of variant *MC1R* alleles and the degree of melanism in both geese and skuas. Homozygotes for the variant allele have more melanic phenotypes, whereas all of the palest melanic-morph individuals are heterozygous (Figure 11.3). However, there is still phenotypic variation in a *MC1R* genotypic class, demonstrating that factors (genetic and/or environmental) other than *MC1R* variation affect the plumage phenotype, as

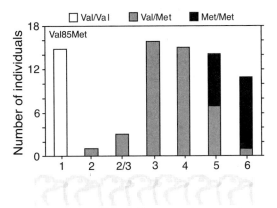
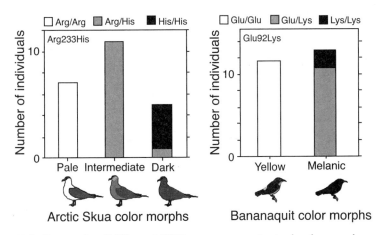

Figure 11.3. Frequencies of different *MC1R* genotypes occuring in the plumage phenotypes in single populations of three species, the Lesser Snow Goose, Arctic Skua, and Bananaquit. Redrawn from Mundy (2005).

previously suggested in genetic studies of wild populations of these species (O'Donald 1983; Rattray and Cooke 1984).

A notable finding is that MC1R has different effects on patterning in the three species (See the thumbnail sketches in Figure 11.3). In Bananaquits, there is a gross change in melanin distribution over the whole body, whereas variation in geese and skuas is more gradual, but in qualitatively different ways. In Arctic Skuas, there is essentially a gradual increase in melanization throughout a particular portion of the plumage (neck, breast, and belly). In contrast, in Snow Geese, there is a gradual increase in the amount of the body melanized, with relatively little change in the amount of melanin per feather (although the amount does vary at the boundary of melanized and nonmelanized areas). These results suggest that patterning mechanisms are lineage-specific and that results from one species may not be directly applicable to an unrelated species.

Reconstructing Plumage Evolution from Variation at Pigmentation Loci

Knowledge of the actual genetic changes responsible for plumage variation provides a novel method for reconstructing plumage evolution, which can be contrasted with traditional methods that map existing plumage variation onto a phylogeny according to some model of plumage evolution (Christidis et al. 1988; Price and Pavelka 1996; Omland and Lanyon 2000; Driskell et al. 2002). In Bananaquits, Snow Geese, and Arctic Skuas, the evolutionary relationships of all *MC1R* alleles were reconstructed using all variable nucleotide sites, enabling the position of the presumed melanism-causing mutation to be examined in relationship to other alleles. In all three cases, the mutation leading to the variant *MC1R* allele associated with melanism occurred in a derived part of the tree (Theron et al. 2001; Mundy et al. 2004). This observation provides strong evidence that melanism itself is a derived trait in these species and that the ancestral form was nonmelanic.

How Broad is the Role of *MC1R* in Plumage Evolution?

We have seen that *MC1R* controls qualitatively different melanic plumage evolution in three distantly related species. This phenomenon is in many ways surprising: the MC1R occupies a key position in the regulation of melanogenesis, the causal substitutions are apparently in the coding region rather than in the regulatory regions, and in mice, there are now over 120 loci known to be involved in coat-color determination (Bennett and Lamoreux 2003). Many pigmentation loci in mice have negative pleiotropic effects in other systems, so the most likely reason for the repeated involvement of *MC1R* is that pleiotropy is less marked at this locus (Mundy 2005). Although *MC1R* is expressed in some cells of the mammalian immune system, including neutrophils and macrophages (Star et al. 1995; Catania et al. 1996), the only good evidence for a function of *MC1R* in vertebrates unrelated to pigmentation is a poorly defined role in analgesia in mice and humans (Mogil et al. 2003).

An important issue for future research is to determine the scope of *MC1R* involvement in plumage variation. So far a role of *MC1R* has been demonstrated in intraspecific polymorphisms in eumelanin distribution over broad body regions in adults of three sexually monomorphic species. What remains to be determined is whether *MC1R* is important in creating color differences among subspecies or species, changes in phaeomelanin as well as eumelanin distribution, in sex-specific differences, and in juvenile plumages. In addition, the role of *MC1R* in regulating color across different body regions is poorly understood.

Two recent studies have begun to address these issues. Males of mainland subspecies of the White-winged Fairy-wren (*Malurus leucopterus*) in western Australia have blue nuptial plumage, whereas males on Dirk Hartog Island have black nuptial plumage (Plate 21). The feather ultrastructure responsible for the blue coloration in the mainland species is also present in the island subspecies, but in the island subspecies, it is obscured by extra melanin pigment, strongly suggesting that the ancestral coloration of the island population was blue (Doucet et al. 2004). The *MC1R* is differentiated among these subspecies, implying that it may be responsible for this case of sexually dimorphic variation in melanin deposition (Doucet et al. 2004). The detailed pattern of *MC1R* variation, however, is unexpected for simple involvement of *MC1R* in melanism: four out of five of the variable amino acid sites in *MC1R* have evolved along the lineage to the mainland subspecies, and the single amino acid variant that is derived in the island subspecies (Val166Ile) shows a pattern across the subspecies that is inconsistent with dominance of a melanic allele at *MC1R*. Neutral divergence between the populations remains a possible alternative explanation for these results, and it will be important to assess variation at unlinked loci in future work.

Differences in melanin-based coloration among species have been investigated in Old World leaf warblers (*Phylloscopus;* Plate 29, Volume 2) and provide the first example in which variation in the *MC1R* coding region is not associated with changes in melanization (MacDougall-Shackleton et al. 2003). The differences in fine-scale melanization among *Phylloscopus* species include the presence or absence of crown stripes, wing bars, and rump patches; these do not co-vary with variation in *MC1R*. It would be surprising from a developmental standpoint if variation in the coding sequence of a major gene had a discrete small-scale effect on patterning; such phenotypic variation is more likely due to substitutions affecting gene regulation.

As mentioned previously, the dominance relationships of melanic alleles provide an important clue to the possible role of *MC1R*. Returning to Table 11.1, species in which melanism is inherited as a dominant or partially dominant trait are good candidates for *MC1R* involvement, but *MC1R* is unlikely to be involved in cases for which melanism is recessive (e.g., Giant Petrels [*Macronectes giganteus*], and Great Blue/White Herons, [*Ardea herodias*]).

Future Work

The study of single-locus control of plumage-color variation at the molecular level is still in its infancy. The extensive overall homology between pigmentation genes controlling melanin deposition in birds and mammals, and even fish, together with the newly available chicken genome sequence, suggest that progress in isolating avian pigmentation loci will be rapid. However, there are likely also to be avian-specific pathways. For example, the two-dimensional patterns of feathers are far more complex than the essentially one-dimensional patterning found along hairs, which is surely reflected in increased complexity of genetic control of feather patterning that may well involve loci and genetic interactions unique to birds (Prum and Williamson 2002).

Color polymorphisms in domesticated species, particularly nongalliforms, represent an underutilized resource for the identification of novel pigmentation loci. Perhaps the most glaring example of our current ignorance is in the loci that control feather carotenoid deposition. Progress has been hampered by poor understanding of the some of the basic biochemical and physiological pathways of carotenoids that are destined for feather deposition, and the fact that mammals do not provide a model, as carotenoids are absent from hairs. Given the importance of carotenoids in sexually selected plumage traits, the isolation of such loci (e.g., those coding for enzymes that convert dietary carotenoids to other carotenoids) must be considered a high priority.

Polygenic Control of Plumage Color Variation

As most of the color traits of interest to behavioral and evolutionary ecologists are quantitative in nature, quantitative genetics of plumage coloration is clearly important. Quantitative genetics provides a theoretical framework for estimating the contribution of genes to any quantitative trait, where it is generally assumed that many loci of small effect are involved, as well as a variety of environmental factors. In addition to estimating the presence or absence of genetic effects, and the nature of the genetic contribution to a trait, the methodology is also useful in partitioning different kinds of environmental variance. Indeed, quantitative genetic methodology provides a powerful means of dissecting out different environmental effects even when additive genetic effects are small or absent. The role of environmental effects is currently an active area of research, but here I emphasize genetic effects. Falconer and Mackay (1996) provide a good introduction to quantitative genetic theory and methodology, and Merilä and Sheldon (2000) offer an excellent general review of quantitative genetic studies in birds, including studies of coloration.

The key parameter of interest in quantitative genetics studies is the narrow-sense heritability (h^2; hereafter shortened to "heritability"), which gives the proportional contribution of additive genetic variance to the total phenotypic variance. The major importance of heritability lies in its use in predicting the evolutionary response to selection on a trait—the higher the heritability, the higher the response and the faster the potential evolution of the trait. However, the finding of high heritabilities for certain traits in the face of strong selection presents a paradox that has yet to be resolved.

Heritability estimates in the wild have most commonly been obtained from parent-offspring regressions. The slope of the midparent-offspring regression, or twice the slope of the father-son regression (particularly relevant for sexually selected male traits) gives direct estimates of the heritability, if random mating is assumed. Because relatives can resemble one another both through shared genes and a shared environment (e.g., correlation between parent and offspring environment) it is desirable to control for this possibility in the experimental design. The usual way of achieving this is through a cross-fostering protocol in which nestlings are swapped soon after hatching to generate nests containing mixtures of genetic chicks and foster chicks. This design controls for shared rearing environment but not for maternal effects before egg laying. Failure to control for pre-laying maternal effects may be of particular importance in studying the quantitative genetics of secondary sexual characters, given evidence that pre-laying maternal effects may be related to male secondary sexual characters (e.g., Gil et al. 1999).

In practice, unless based on huge sample sizes, heritability estimates have large standard errors, and so a major question is whether heritability is significantly greater than zero, thus demonstrating an additive genetic contribution to the trait. It is important to remember that a particular heritability estimate is specific for the trait, sex, age, population, and season at which it is measured. For example, sustained directional selection on a trait is generally expected to

Table 11.2. Heritability Estimates for Plumage Coloration in Wild Birds

Species	Trait	Pigment	Method	h^2	SE	Reference
House Sparrow (*Passer domesticus*)	Throat badge size	Eumelanin	Father-son	0.60	±0.23*	Møller (1989)
			Father-son CF	-0.01	±0.15	Griffith et al. (1999)
Pied Flycatcher (*Ficedula hypoleuca*)	Back color	Eumelanin	Father-son	0.54	±0.42	Potti and Montalvo (1991)
			Father-son	0.50	±0.09*	Lundberg and Alatalo (1992)
			Father-son	-0.64	±0.36	Slagsvold and Lifjeld (1992)
Collared Flycatcher (*Ficedula albicollis*)	Forehead patch area	Absence of melanin	Father-son CF	0.44	±0.18*	Reported in Sheldon et al. (1997)
			Father-son	0.39	±0.09*	Qvarnström (1999)
			Full-sibling	0.72	±0.19*	Qvarnström (1999)
			Pt grandfather-grandson	0.40	±0.53	Qvarnström (1999)
			Mt grandfather-grandson	0.53	±0.45	Qvarnström (1999)
			Pt half-sibling	0.56	±0.54	Qvarnström (1999)
			Father-son	0.58	±0.15*	Hegyi et al. (2002)
			Pedigree AM, Y	0.40	±0.06*	Garant et al. (2004)
			Pedigree AM, O	0.47	±0.07*	Garant et al. (2004)
	Forehead patch width		Father-son	0.41	±0.07*	Merilä and Sheldon (2000)
	Wing patch area	Absence of melanin	Father-son Y	1.09	±0.28*	Török et al. (2003)
			Father-son O	0.53	±0.36	Török et al. (2003)
			Pedigree AM, Y	0.35	±0.05*	Garant et al. (2004)
			Pedigree AM, O	0.64	±0.04*	Garant et al. (2004)
Great Tit (*Parus major*)	Ventral badge size	Eumelanin	Father-son CF	1.44	±0.62	Norris (1993)
			Father-son CF	0.17	±0.12	Lemel and Wallin (1993)
Large Ground Finch (*Geospiza fortis*)	Color	Eumelanin	Father-son	0.62	±0.23	Grant (1990)
House Finch (*Carpodacus mexicanus*)	Color	Carotenoid	Father-son	0.84	±0.12	Hill (1991)
Barn Owl (*Tyto alba*)	Spottiness	Eumelanin	Midparent-offspring CF	0.81	±0.09*	Roulin and Dijkstra (2003)
	Color	Phaeomelanin	Midparent-offspring CF	0.65	±0.05*	Roulin and Dijkstra (2003)

Notes: AM, animal model; CF, cross-foster design; h^2, narrow-sense heritability; Mt, maternal; O, old; Pt, paternal; SE, standard error; Y, young.
*$P < 0.05$ that $h^2 > 0$.

lead a loss of additive genetic variance for the trait and a decline in heritability over time, so population differences in selection can lead to differences in heritability.

In addition to measuring heritability, quantitative genetic methodology can be used to estimate genetic correlations among different traits; that is, the degree to which the phenotypic correlation between two traits is due to shared additive genetic effects. Genetic correlations are important for predicting correlated responses to selection on traits and consequently their potential evolutionary trajectories (Lande 1979; Price et al. 1993; Lynch and Walsh 1998). They are also important for investigating evolutionary trade-offs. Unfortunately, they require even larger sample sizes than heritability estimates. In the context of coloration, there might be genetic correlation among different color traits or between a color trait and another trait. Genetic correlations may arise from pleiotropy (i.e., the multiple actions of a single locus) or through linkage disequilibrium among separate loci. The latter may occur as a selected evolutionary response to association between the two traits, but can also result from inbreeding or population structure.

Heritability for Plumage and Beak Coloration

Heritability estimates for natural variation in plumage coloration in wild birds are summarized in Table 11.2. Most studies concern sexually selected traits in males, and most involve melanin-based traits or unmelanized white patches. The estimates are clustered in the range of 0.40–0.80, and around half of them are significantly greater than zero. These studies have investigated only seven species, six of which are passerines.

It is notable that only a few of these studies controlled for parent-offspring environmental correlations (e.g., with cross-fostering experiments). In the absence of such controls, or a priori reasons for expecting such correlations to be minimal, the heritability estimates are more cautiously interpreted as upper bounds on the true heritability. An instructive example concerns House Sparrows (*Passer domesticus;* Plate 23). In an early study conducted without cross-fostering, a moderate, significant positive estimate of heritability for size of the male's black throat badge was obtained from father-son regression (Møller 1989). In contrast, a more recent study with cross-fostering surprisingly found no evidence for any additive genetic component to the size of the male badge (Griffith et al. 1999; Figure 11.4). There was a strong relationship between foster father and offspring, however, demonstrating a substantial

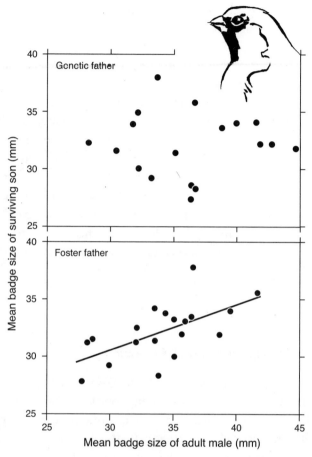

Figure 11.4. Father-son regressions for black throat patch ("badge") in House Sparrows. Badge size of sons is significantly correlated with badge size in foster father, but not genetic father, providing no evidence of a genetic component to this trait (Griffith et al., 1999).

environmental influence on the trait (Figure 11.4). These studies were performed at different times and in different populations, but the newer study certainly casts doubt that the heritability measured in the earlier study is solely attributable to additive genetic variance.

A study on factors contributing to the carotenoid-based yellow breast color and structural-based ultraviolet (UV)/blue tail color of nestling Blue Tits (*Parus caeruleus*; Plate 18, Volume 2) is important because it is the only study to report an additive genetic contribution to a structurally colored trait (Johnsen et al. 2003). Variance in the hue of the tail, but not breast color, was partly attributed to a significant paternal effect in maternal half-siblings, suggesting a genetic contribution to this trait. The authors argue that this effect is unlikely to be due to differential allocation by either the mother or father. Unfortunately, small sample sizes precluded an accurate estimate of heritability.

A rare study on heritability of bare-part (bill) color has been performed in captive Zebra Finches (Price and Burley 1993; Plate 27). This trait is of interest, as it is subject to opposing selection in the two sexes. Heritabilities based on parent-offspring regressions were estimated to be in the range of 0.34 and 0.75 and were mostly significant. Interestingly, there was a large genetic correlation between males and females, which is predicted to greatly retard the evolution to sex-specific optima. This study provides some of the only evidence for the genetic basis of bill color in birds.

Studies in *Ficedula* Flycatchers: Role of the Environment

The most extensive studies of quantitative genetics of coloration in wild populations were performed on male-specific traits in Collared Flycatchers (*Ficedula albicollis;* Plate 18, Volume 2) and their sister species, the Pied Flycatcher (*F. hypoleuca;* Plate 27). The white forehead patch is a sexually selected, condition-dependent indicator of viability in a Swedish population of Collared Flycatchers (Gustafsson et al. 1995). Its heritability has been quite consistently estimated at approximately 0.40 in several studies (Table 11.2). An estimate of the same heritability is somewhat higher (0.58) in a population in Hungary in which the trait is not condition-dependent (Török et al. 2003). In contrast, in a Norwegian population of Pied Flycatchers, in which the forehead patch is small and not under sexual selection, there was no significant heritability (Dale et al. 1999).

Several factors have been found to affect heritability of the forehead patch in the Swedish population. In a study on the effects of the natal environment, no significant heritability was found during unfavorable conditions (poor weather or large brood size; Qvarnström 1999), which suggested the presence of genotype-by-environment interactions. Using a large-scale climatic index, the North Atlantic Oscillation index (NAO), Garant et al. (2004) recently showed that forehead patch heritability was significantly higher in better environments over a 22-year period. In addition, they showed that the population mean phenotype has changed in response to the NAO, and that this change could be attributed to phenotypic plasticity in individual males.

Significant heritability has also been demonstrated for a second trait in male Collared Flycatchers, a depigmented patch in the primaries (Török et al. 2003; Garant et al. 2004), which shows an alternative pattern of condition-dependence in the Hungarian but not the Swedish population. In Sweden, the heritability of the wing patch was significantly higher than that for the forehead patch, and, unlike the forehead patch, its size and heritability were not affected by large-scale climatic conditions.

Together these results reveal many intricacies in the heritability of sexually selected traits. Heritabilities for the same trait can vary among different species, among different populations, and in different seasons in the same population. Heritabilities for different traits can not only differ but can respond in different ways to changes in the environment. The environment can have several different influences and, in particular, a low-quality environment can lead to lower heritability, which has been found in some other systems (Hoffmann and Merilä 1999). The importance of these environmental effects is that they predict that evolution will proceed faster in a better environment.

Genetic Correlations of Plumage Color

Only two studies appear to have demonstrated genetic correlations (r_G) between two plumage color traits. In the study on climatic effects on forehead and wing patches in Collared Flycatchers mentioned above, a small genetic correlation for the two traits was found for two age classes of males, and the correlation was significant for birds older than 1 year ($r_G = 0.12 \pm 0.02$; Garant et al. 2004).

In Barn Owls (*Tyto alba*), there is variation in both sexes for two traits on the ventral surface, the area covered by predominantly eumelanin spots ("spottiness"), and the degree of saturation of the ground coloration that is presumably due to phaeomelanin ("color"; see Plate 29, Volume 2). Both of these traits had high heritabilities in a cross-fostering design (Roulin and Dijkstra 2003; Table 11.2). Phenotypic correlations for spottiness and color were significant in males, and significantly higher in males than in females. Genetic correlations were not formally estimated, but two lines of evidence suggest that such a correlation is present at least in males: there was a significant relationship between color of male offspring and spottiness of their genetic father (Figure 11.5), and a significant association between plumage color and spottiness in siblings raised in different nests. These results strongly imply that there is no

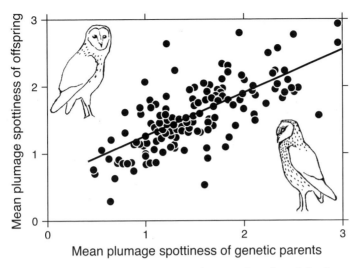

Figure 11.5. Mid-parent–offspring regression revealing significant heritability for ventral plumage spottiness ($h^2 = 0.65 \pm 0.05$) in Barn Owls. Spottiness is measured here as the square root of the mean percentage of plumage covered by black spots measured on the breast, belly, flanks, and the underside of the wings (Roulin and Dijkstra 2003).

trade-off in investment in color and spottiness (Roulin and Dijkstra 2003), suggesting that the two traits do not represent alternative evolutionary strategies.

In contrast to these demonstrations of genetic correlations is an absence of genetic correlation in two male-specific traits in Red-billed Queleas (*Quelea quelea;* Plate 6, Volume 2). The breast color of male queleas varies from yellow to red and is at least partly carotenoid-based, and the color of the melanized mask varies from white to black. A father-son regression suggests substantial heritability for the breast color, and the results also indicated heritability for mask color, but there is no phenotypic correlation between the two traits (Dale 2000).

Conclusions and Future Work

There is enormous scope for future work on the quantitative genetics of avian coloration. The few studies performed to date are strongly biased toward passerines, almost solely concerned with sexually selected traits in males, and have concentrated on melanin-based plumage traits.

As the studies in Table 11.2 show, most estimates of quantitative genetic parameters from wild populations are heritabilities with large associated standard errors. This shortcoming limits our ability to make definitive statements or to conduct quantitative comparisons among studies. To make comparisons among traits, populations, or environments, future studies should target sample sizes that are around an order of magnitude larger than most of those used in the studies in Table 11.2.

Although there is no real substitute for large samples when doing quantitative genetics, recent developments in the application of mixed-model statistical methodology to natural populations has increased the efficiency with which data can be used. The "animal model" approach uses a multivariate model in which all known relatives contribute to the estimation of genetic variances and co-variances, based on their position in a relatedness matrix (Kruuk 2004). Two additional advances that this method offers over traditional methodology are that (1) estimates are less biased by common environment effects and (2) it is possible to simultaneously control for a number of fixed effects, such as specific environmental influences. This technique has been used for just one published study of quantitative genetics of coloration so far (Garant et al. 2004), but an indication of its applicability is given in the many examples cited by Kruuk (2004).

There is currently great interest in identifying the chromosomal segments—the quantitative trait loci (QTL)—underlying additive genetic variance of quantitative traits in wild populations (Cheverud and Routman 1993; Erickson et al. 2004). The methodology uses trait-marker linkage analysis to estimate the number of QTLs involved and the size of their effect on the trait (i.e., the proportion of variance in the trait attributable to a particular QTL). This information would clearly provide important insights into the genetic architecture and mechanisms of plumage color evolution. QTL analysis is well established in domestic fowl, in which the ability to conduct controlled crosses provides many benefits, and there is now a substantial literature on QTLs involved in a range of traits related to production, such as body weight, growth rate, egg production, and behavioral traits (e.g., Kerje et al. 2003; Buitenhuis et al. 2004; Carlborg et al. 2004). An ultimate but extremely ambitious goal of the approach is to move from QTLs, which may represent large chromosomal segments, to phenotypically relevant variation at the single nucleotide level in individual genes (quantitative trait nucleotides [QTN]).

It would be technically feasible to perform a QTL analysis on a color trait in the wild. Given the investment in resources necessary for such a study, an important consideration would be confidence that a significant additive genetic variance for the trait is present in the study population. In addition to the standard approach using neutral molecular markers spaced across the genome, another strategy would be to examine the role of targeted candidate loci in contributing to variation in the trait.

Genetic Basis of Population, Subspecies and Species Differences in Plumage Color

Population divergence in coloration has been a pervasive feature of avian evolution that has generated the astonishing diversity that we see among species today. What are the genetic mechanisms responsible for this diversity? This simple question leads to many fundamental issues in evolutionary genetics, such as: Are macroevolutionary mechanisms among species a simple reflection of the microevolutionary changes occurring within species? Are similar genetic mechanisms used to generate similar phenotypic changes in different lineages? To what extent are lineage-specific differences in coloration due to the underlying genetic architecture? Are there genetic constraints on plumage color evolution? Unfortunately, we are a long way from being able to give informed answers to these questions. After briefly reviewing some of the direct evidence for genetic mechanisms underlying population divergence, I consider a few important issues in the genetic basis of plumage color evolution.

Evidence for the genetic basis of population divergence in coloration comes from a variety of sources. The most basic issue is whether observed differences are genetic in origin. For most phenotypic differences, especially those involving melanin, it is usually reasonable to assume that genes are involved. Differences in carotenoid-based pigmentation can be due to environmental differences, however (e.g., see Hill 1993), and as melanin-based colors can be condition dependent, it is worth at least considering that nongenetic causes may be involved in these traits as well. A common garden experiment, in which individuals from different populations are raised in the same environment, is a simple way to test for environmental influences on a plumage trait (e.g., Yeh 2004).

Hybrids, either those occurring naturally or generated in captivity, can be very informative about basic genetic mechanisms among populations or species. F_1 hybrids among the parental forms allow the presence of dominance to be determined. For example, naturally occurring hybrids between species of shrike (*Lanius*) that have different distributions of rufous and gray coloration on the back, rump, and tail—Isabelline (*L. isabellinus*) × Red-backed (*L. collurio*), and Woodchat (*L. senator*) × Red-backed—are gray throughout the dorsal surface (Lefranc and Worfolk 1997), showing that gray is dominant to rufous. F_1 hybrids are not informative, however, about the number of loci involved. This number can be estimated from the phenotypic complexity of backcrosses and F_1 intercrosses, which may occur naturally in hybrid zones or be generated in captivity.

The conclusions from these studies have recently been reviewed (Price 2002). In general, several to many loci appear to be involved in generating interspecific plumage differences in many cases, even among sister taxa, such as species of pheasant (Danforth 1950). An instructive example involves the distribution of color variation across three hybrid zones of Hermit (*Dendroica occidentalis*) and Townsend's Warblers (*D. townsendii;* Rohwer and Wood 1998). Seven out of eight color traits, such as flank, back and crown color, showed continuous variation across the hybrid zones, indicating polygenic inheritance. The remaining trait is face color, which is yellow in Hermit Warblers and black in Townsend's Warblers. The abrupt change in face color along the hybrid zones together with the absence of intermediates strongly suggest that the yellow face is inherited as a single-locus dominant, providing a rare example of a single locus that affects both carotenoid and melanin distribution. In addition to providing information on dominance and number of loci involved, hybrids offer considerable potential for dissecting the molecular genetic basis of interspecific differences in coloration.

Microevolution versus Macroevolution

Indirect evidence for a mechanistic link between micro- and macroevolution comes from several striking examples in which coloration in intraspecific polymorphisms mimics variation among congeneric species, for example, bushshrikes of the genus *Malaconotus* (Hall et al. 1966) and chats (*Oenanthe;* Mayr and Stresemann 1950). Another example is the oystercatchers (*Haematopus*). There are four extant species of black-colored oystercatchers, five species of pied oystercatchers (black dorsum with or without white patches, and white ventrum), and one polymorphic species, the Variable Oystercatcher (*H. unicolor*), which varies from pied to nearly all black (Baker 1973).

Comparative analysis of genes involved in intraspecific plumage evolution provides an important method to address this issue. However, the identification of phenotypically relevant changes in interspecific analyses is more challenging than, for example, in the case of a plumage polymorphism. This is because genetically isolated populations inevitably accumulate nucleotide substitutions across their genome that have no effects on the phenotype, leading to spurious genotype-phenotype associations. For this reason, it is desirable to compare data from putative pigmentation loci with those from neutrally evolving loci.

Cross-species comparison of the *MC1R* locus among skuas revealed that *MC1R* of the Great Skua (*Catharacta skua*) has accumulated many changes since its common ancestry with Arctic and Long-tailed Skuas (*Stercorarius longicaudus*) (Mundy et al. 2004). In addition, the presence of an amino acid–changing substitution at the same site implicated in melanism in Arctic Skuas suggests that parallel evolution in *MC1R* may contribute to the relatively dark phenotype of this species (Plate 28). These results are still preliminary, however, as comparative studies at neutral loci have yet to be performed. In contrast, as mentioned above, interspecific varation in *MC1R* among Old World leaf warblers does not correspond to fine-scale plumage changes (Mac-Dougall-Shackleton et al. 2003). Many more interspecific studies of this nature are expected to be performed in the near future.

Sexual Dichromatism

Sexual dimorphism in coloration is such an important feature of birds that it should be briefly considered in its genetic context—for a full discussion of the evolution of sexual dichromatism, see Badyaev and Hill (2003). There are two levels of genetic control that are important to keep distinct. First, dichromatism in a species is under genetic control by sex chromosomes. Second, the presence or absence of dichromatism in a species is itself genetically controlled. Although the genetic mechanisms controlling the presence of dichromatism are poorly understood, it seems likely that they are distinct in avian lineages with different hormonal mechanisms underlying dichromatism (Kimball and Ligon 1999; Chapter 10). Crosses between closely related species that differ in degree of dichromatism (e.g., among species in the mallard complex) provide a potential route for investigating the genetic basis of dichromatism, but such an approach does not appear to have been attempted.

Genetic Propensities and Constraints in the Evolution of Coloration

Do different avian lineages vary in their genetic potential for evolution of coloration? There are many ways in which lineages may differ in the evolvability of coloration, and these could include the ability to evolve a particular color trait de novo, the ability to evolve sexual dichromatism, and rates of evolution of particular color traits. The most striking cases are those in which novel pigments have evolved in a lineage (e.g., psittacofulvins in parrots, turacoverdins in turacos). Apart from lineage-specific pigments, the distribution of other types of coloration is highly nonrandom across avian taxa, which is suggestive of differences in genetic potential, although the alternative explanation that these differences are solely due to selection is difficult to rule out. For example, a survey of blue and iridescent structural colors revealed a relatively high incidence in passerines (Auber 1957).

Other types of genetic propensities could involve the evolution of sexual dichromatism, the rate of trait evolution, and the ability to repeatedly gain and lose a trait. For example, comparative phylogenetic studies have revealed repeated loss of sexual dichromatism in the mallard complex (*Anas;* Omland 1997) and remarkable lability and strong convergent evolution of plumage traits in the New World orioles (*Icterus;* Omland and Lanyon 2000; Chapter 10, Volume 2).

Are there genetic constraints on plumage evolution? Although comparative studies like those mentioned above can reveal what is evolutionarily possible, they are less than conclusive about what is not possible, as the absence of a trait may reflect the action of natural selection to eliminate the trait rather than a lack of genetic variation for it. Ideally, therefore, one requires a way of separating genetic propensity to produce a trait from the action of natural selection on the trait. An approximation comes from the results of domestication. Coloration is among the first traits to be selected for in domestication, and novel coloration is generally highly valued and hence maintained by artificial selection. The diversity of color phenotypes in captivity can therefore be viewed as a surrogate of the "phenotypic neighborhood" of coloration accessible to evolution that is one or a few mutational/recombinational steps removed from the standing genetic variation in the wild (Price 2002; Dichtel-Danjoy and Félix 2004).

Two examples illustrate this point. The absence of melanism in some common cage birds (e.g., Common Canaries, Zebra Finches, Budgerigars) strongly suggests a genetic constraint (Buckley 1987). Melanism occurs as a polymorphism or as a fixed trait in species in many avian orders, including passerines. Moreover, as discussed above, melanism in species from three orders, including the Passeriformes, has a common genetic basis that involves the same protein (MC1R), and melanic mutants at the *MC1R* locus have dominant or partially dominant inheritance and would be immediately noticed in captivity. In this case, it is highly unlikely that the constraint is due to absence of the *MC1R* gene, as it occupies a key place in the regulation of melanin synthesis, and melanin is present in these species. The constraint on melanism due to *MC1R* evolution in these species is more likely to be concerned with the regulation of *MC1R* expression, or deleterious actions of an overactive *MC1R*.

Similarly, the lack of red ketocarotenoid coloration in Common Canaries appears to be a genetic constraint. In this case, it is almost certainly the absence of one or more key factors (presumably including a carotenoid ketolase enzyme) that is involved, because the ability to produce the red coloration has been bred into Common Canaries via crosses with Red Siskins (Birkhead 2003).

Independence of Different Color Mechanisms

An important theme running through the studies presented here is that the different pigment systems and structural coloration are to a large extent under independent genetic control. This independence has the important consequence that coloration due to these various mechanisms is free to evolve relatively unconstrained at the genetic level, although of course the color of a particular area may involve interactions between more than one mechanism. Only two counterexamples are documented above—the mask polymorphism in Gouldian Finches and face color in Hermit and Townsend's Warblers. Interestingly, these examples both involve coordinated regulation of melanins and carotenoids on the head, suggesting that regulation of coloration in this region may be special, which might have arisen from the importance of the head for signaling.

Conclusions

The amount of attention given to the genetics of color variation is quite small compared with the enormous body of work on plumage coloration reviewed elsewhere in this volume. The new molecular tools should now enable rapid progress on single-locus mechanisms of melanin distribution that will provide important insights into many aspects of plumage evolution. Even for single-locus mechanisms, however, there are important gaps, particularly for carotenoid and structural pigmentation and bare-part coloration. Studies on quantitative genetics of coloration lag much further behind. An increased investment in such studies would be rewarding, because they are crucial for understanding the evolution of the majority of plumage traits.

Summary

In this chapter I consider the genetics of coloration in wild birds from the perspectives of single-locus and polygenic mechanisms within species, and the mechanisms underlying population and species divergence in color. The majority of information on single-locus Mendelian mechanisms has been gained from studies of intraspecific polymorphisms. Such polymorphisms have been estimated to occur in 3.5% of all avian species, and mostly involve changes in melanin distribution in plumage. The maintenance of polymorphisms has recently been studied in a comparative framework with control for phylogenetic effects for the first time. Conflicting results alternatively supporting disruptive selection or apostasis have been obtained in studies on raptors and owls.

The classical genetic basis of color polymorphisms in wild populations by examining transmission has only been ascertained in about 25 species. Single loci account for the majority of these examples. Melanism, which is the commonest phenotypic change among morphs in these studies, is more frequently inherited as a dominant rather than a partially dominant or recessive trait. Studies in domestic birds are beginning to isolate genes affecting coloration that are candidates for involvement in the evolution of coloration in wild populations. Recent work has focused on one such gene, *MC1R*, which encodes the melanocortin-1 receptor, which plays a critical role in determining relative levels of eumelanin and phaeomelanin synthesis in developing feather buds. Different single amino acid changes in *MC1R* are perfectly associated with melanism in three distantly related species with melanic polymorphisms, the Bananaquit, Lesser Snow Goose, and Arctic Skua. *MC1R* variation has different patterning effects in these species and shows that melanism is derived. *MC1R* may also be involved in determining melanism in an island subspecies of the White-winged Fairy-wren but does not have a role in interspecific variation in small demelanized patches in Old World leaf warblers. In contrast to these results on the genetics of melanin distribution, no loci affecting carotenoid or structural coloration or the coloration of bare parts have been isolated.

Despite the obvious importance of quantitative variation in coloration in plumage evolution, there have been relatively few studies on the quantitative genetics of coloration in birds, and these are concentrated on sexually selected male ornaments in a few species of passerine. Most studies have found moderate heritability for color traits. The most detailed studies have been on demelanized patches in male Collared Flycatchers. These studies have revealed the importance of the environment in affecting heritability and have suggested some interesting interpopulation differences in condition dependence and heritability of forehead versus wing patches.

Knowledge of the genetics of interspecific differences in color is poor but crucial for models of the evolution of coloration. An interesting area for future work is the degree to which different lineages vary in their genetic potential to evolve different colors. The absence of color traits in some species of cage bird that readily evolve in other lineages (e.g., melanism in Budgerigars, Common Canaries, and Zebra Finches) strongly suggests the existence of genetic constraints on plumage evolution.

Progress in defining the genetic changes underlying Mendelian traits is likely to be rapid. A much greater effort to elucidate quantitative genetic mechanisms of coloration is desirable in the future.

References

Andersson, M. 1994. Sexual Selection. Princeton, NJ: Princeton University Press.

Auber, L. 1957. The distribution of structural colours and unusual pigments in the class Aves. Ibis 99: 463–476.

Badyaev, A. V., and G. E. Hill. 2003. Avian sexual dichromatism in relation to phylogeny and ecology. Annu Rev Ecol Syst 34: 27–49.

Baker, A. J. 1973. Genetics of plumage variability in the Variable Oystercatcher (*Haematopus unicolor*). Notornis 20: 330–345.

Bennett, D. C., and M. L. Lamoreux. 2003. The color loci of mice—A genetic century. Pigment Cell Res 16: 333–344.

Berthold, P., G. Mohr, and U. Querner. 1996. The legendary "Veiled Blackcap" (Aves): A melanistic mutant with single-locus autosomal recessive inheritance. Naturwissenschaften 83: 568–570.

Birkhead, T. 2003. The Red Canary. The Story of the First Genetically Engineered Animal. London: Weidenfeld & Nicolson.

Buckley, P. A. 1987. Mendelian genes. In F. Cooke and P. A. Buckley, ed., Avian Genetics, 1–44. London: Academic Press.

Buitenhuis, A., T. Rodenburg, M. Siwek, S. Cornelissen, M. Nieuwland, et al. 2004. Identification of QTLs involved in open-field behavior in young and adult laying hens. Behav Genet 34: 325–333.

Carlborg, R., P. Hocking, D. Burt, and C. Haley. 2004. Simultaneous mapping of epistatic QTL in chickens reveals clusters of QTL pairs with similar genetic effects on growth. Genet Res 83: 197–209.

Catania, A., N. Rajora, F. Capsoni, F. Minonzio, R. Star, and J. Lipton. 1996. The neuropeptide α-MSH has specific receptors on neutrophils and reduces chemotaxis in vitro. Peptides 17: 675–679.

Cheverud, J., and E. Routman. 1993. Quantitative trait loci—Individual gene effects on quantitative characters. J Evol Biol 6: 463–480.

Christidis, L., R. Schodde, and P. Baverstock. 1988. Genetic and morphological differentiation and phylogeny in the Australo-Papuan scrubwrens (*Sericornis*, Acanthizidae). Auk 105: 616–629.

Cooke, F., and F. G. Cooch. 1968. The genetics of polymorphism in the goose *Anser caerulescens*. Evolution 22: 289–300.

Cooke, F., and J. Ryder. 1971. The genetics of polymorphism in the Ross' Goose (*Anser rossii*). Evolution 25: 483–496.

Cooke, F., C. S. Findlay, R. F. Rockwell, and J. A. Smith. 1985. Life history studies of the Lesser Snow Goose (*Anser caerulescens caerulescens*). III. The selective value of plumage polymorphism: Net fecundity. Evolution 39: 165–177.

Cooke, F., D. T. Parkin, and R. F. Rockwell. 1988. Evidence of former allopatry of the two color phases of Lesser Snow Geese (*Chen caerulescens caerulescens*). Auk 105: 467–479.

Cooke, F., R. F. Rockwell, and D. B. Lank. 1995. The Snow Geese of La Pérouse Bay. Oxford: Oxford University Press.

Craig, J. L. 1972. Investigation of the mechanism maintaining polymorphism in the New Zealand Fantail, *Rhipidura fuliginosa* (Sparrman). Notornis 19: 42–55.

Dale, J. 2000. Ornamental plumage does not signal male quality in Red-billed Queleas. Proc R Soc Lond B 267: 2143–2149.

Dale, S., T. Slagsvold, H. Lampe, and G.-P. Sætre. 1999. Population divergence in sexual ornaments: The white forehead patch of Norwegian Pied Flycatchers is small and unsexy. Evolution 53: 1235–1246.

Danforth, C. H. 1950. Evolution of plumage traits in pheasant hybrids, *Phasianus* × *Chrysolophus*. Evolution 4: 301–315.

Darwin, C. 1859. On the Origin of Species by Means of Natural Selection. London: John Murray.

Dichtel-Danjoy, M.-L., and M.-A. Félix. 2004. Phenotypic neighbourhood and micro-evolvability. Trends Genet 20: 268–276.

Doucet, S., M. Shawkey, M. Rathburn, H. M. Mays, Jr., and R. Montgomerie. 2004. Concordant evolution of plumage colour, feather microstructure, and a melanocortin receptor gene between mainland and island populations of a fairy-wren. Proc R Soc Lond B 271: 1663–1670.

Dowding, J., and M. Taylor. 1987. Genetics of polymorphism in the Little Shag. Notornis 34: 51–57.

Driskell, A. C., S. Pruett-Jones, K. A. Tarvin, and S. Hagevik. 2002. Evolutionary relationships among blue- and black-plumaged populations of the White-winged Fairy-wren (*Malurus leucopterus*). Aus J Zool 50: 581–595.

Erickson, D., C. Fenster, H. Stenøien, and D. Price. 2004. Quantitative trait locus analyses and the study of evolutionary process. Mol Ecol 13: 2505–2522.

Falconer, D. S., and T. F. C. Mackay. 1996. Introduction to Quantitative Genetics. Essex: Longman Group.

Fowlie, M. K., and O. Krüger. 2003. The evolution of plumage polymorphism in birds of prey and owls: The apostatic selection hypothesis revisited. J Evol Biol 16: 577–583.

Galeotti, P., and D. Rubolini. 2004. The niche variation hypothesis and the evolution of colour polymorphism in birds: A comparative study of owls, nightjars and raptors. Biol J Linn Soc 82: 237–248.

Galeotti, P., D. Rubolini, P. O. Dunn, and M. Fasola. 2003. Colour polymorphism in birds: Causes and functions. J Evol Biol 16: 635–646.

Garant, D., B. Sheldon, and L. Gustafsson. 2004. Climatic and temporal effects on the expression of secondary sexual characters: Genetic and environmental components. Evolution 58: 634–644.

Gil, D., J. Graves, N. Hazon, and A. Wells. 1999. Male attractiveness and differential testosterone investment in Zebra Finch eggs. Science 286: 126–128.

Grant, B. 1990. The significance of subadult plumage in Darwin's Finches. Behav Ecol 1: 161–170.

Griffith, S., I. Owens, and T. Burke. 1999. Environmental determination of a sexually selected trait. Nature 400: 358–360.

Gustafsson, L., A. Qvarnström, and B. Sheldon. 1995. Trade-offs between life-history traits and a secondary sexual character in male Collared Flycatchers. Nature 375: 311–313.

Hall, B. P., R. E. Moreau, and I. C. J. Galbraith. 1966. Polymorphism and parallelism in the African Bush-shrikes of the genus *Malaconotus* (including *Chlorophoneus*). Ibis 108: 161–181.

Hartl, D., and A. Clark. 1997. Principles of Population Genetics. Sunderland, MA: Sinauer.

Harvey, P., and M. Pagel. 1990. The Comparative Method in Evolutionary Biology. Oxford: Oxford University Press.

Hegyi, G., J. Török, and J. Tóth. 2002. Qualitative population divergence in proximate determination of a sexually selected trait in the Collared Flycatcher. J Evol Biol 15: 710–719.

Hill, G. E. 1991. Plumage coloration is a sexually selected indicator of male quality. Nature 350: 337–339.

Hill, G. E. 1993. Geographic variation in the carotenoid plumage pigmentation of male House Finches (*Carpodacus mexicanus*). Biol J Linn Soc 49: 63–86.

Hoffmann, A., and J. Merilä. 1999. Heritable variation and evolution under favourable and unfavourable conditions. Trends Ecol Evol 14: 96–101.

Hrubant, H. 1955. An analysis of the color phases of the Eastern Screech Owl, *Otus asio*, by the gene frequency method. Am Nat 89: 223–230.

Hudon, J. 1997. A non-melanic schizochroism in Alberta Evening Grosbeaks *Coccothraustes vespertinus*. Can Field Nat 111: 652–654.

Jackson, I. 1997. Homologous pigmentation mutations in human, mouse and other model organisms. Hum Mol Genet 6: 1613–1624.

Jefferies, D., and J. Parslow. 1976. The genetics of bridling in guillemots from a study of hand-reared birds. J Zool 179: 411–420.

Johnsen, A., K. Delhey, S. Andersson, and B. Kempenaers. 2003. Plumage colour in nestling Blue Tits: Sexual dichromatism, condition dependence and genetic effects. Proc R Soc Lond B 270: 1263–1270.

Johnston, R., and M. Janiga. 1995. Feral Pigeons. New York: Oxford University Press.

Kerje, S., O. Carlborg, L. Jacobsson, K. Schutz, C. Hartmann, et al. 2003. The twofold difference in adult size between the Red Junglefowl and white leghorn chickens is largely explained by a limited number of QTLs. Anim Genet 34: 264–274.

Kimball, R., and J. Ligon. 1999. Evolution of avian plumage dichromatism from a proximate perspective. Am Nat 154: 182–193.

King, M. K., M. C. Lagerström, R. Fredriksson, R. Okimoto, N. I. Mundy, et al. 2003. Association of feather colour with constitutively active melanocortin 1 receptors in chicken. Eur J Biochem 270: 1441–1449.

Klungland, H., D. I. Våge, L. Gomezraya, S. Adalsteinsson, and S. Lien. 1995. The role of melanocyte-stimulating hormone (MSH) receptor in bovine coat color determination. Mamm Genome 6: 636–639.

Krüger, O., J. Lindström, and W. Amos. 2001. Maladaptive mate choice maintained by heterozygote advantage. Evolution 55: 1207–1214.

Kruuk, L. 2004. Estimating genetic parameters in natural populations using the "animal model." Phil Trans R Soc B 359: 873–890.

Lancaster, F. 1963. The inheritance of plumage colour in the common duck (*Anas platyrhynchos* Linné). Bibliograph Genet 19: 317–404.

Lande, R. 1979. Quantitative genetic analysis of multivariate evolution, applied to brain: body size allometry. Evolution 32: 79–92.

Landry, G. 1997. The Varieties and Genetics of the Zebra Finch. Louisiana: Franklin.

Lank, D. B. 2002. Diverse processes maintain plumage polymorphisms in birds. J Avian Biol 33: 327–330.

Lank, D. B., C. Smith, O. Hanotte, T. Burke, and F. Cooke. 1995. Genetic polymorphism for alternative mating behaviour in lekking male ruff *Philomachus pugnax*. Nature 378: 59–62.

Lefranc, N., and T. Worfolk. 1997. Shrikes. A Guide to the Shrikes of the World. Sussex: Pica Press.
Lemel, J., and K. Wallin. 1993. Environmental and genetic influence on morphological characters. I. Parent-offspring analyses from cross-fostered Great Tits, *Parus major*. In J. Lemel, ed., Evolutionary and Ecological Perspectives of Status Signaling in the Great Tit (*Parus major* L.), 1–38. Göteborg, Sweden: University of Göteborg.
Lundberg, A., and R. Alatalo. 1992. The Pied Flycatcher. London: T & A Poyser.
Lynch, M., and B. Walsh. 1998. Genetics and Analysis of Quantitative Traits. Sunderland, MA: Sinauer.
MacDougall-Shackleton, E., L. Blanchard, and H. Gibbs. 2003. Unmelanized plumage patterns in Old World leaf warblers do not correspond to sequence variation at the melanocortin-1 receptor locus (*MC1R*). Mol Biol Evol 20: 1675–1681.
Marklund, L., M. J. Moller, K. Sandberg, and L. Andersson. 1996. A missense mutation in the gene for melanocyte-stimulating hormone receptor (*MC1R*) is associated with the chestnut coat color in horses. Mamm Genome 7: 895–899.
Mayr, E., and E. Stresemann. 1950. Polymorphism in the chat genus *Oenanthe* (Aves). Evolution 4: 291–300.
McGraw, K. J., and M. C. Nogare. 2005. Distribution of unique red feather pigments in parrots. Biol Lett 1: 38–43.
Merilä, J., and B. Sheldon. 2000. Heritability of lifetime reproductive success in the wild. Am Nat 155: 301–310.
Meyerriecks, A. 1957. Field observations pertaining to the systematic status of the Great White Heron in the Florida Keys. Auk 74: 469–478.
Mogil, J. S., S. G. Wilson, E. J. Chesler, A. L. Rankin, K. V. S. Nemmani, et al. 2003. The melanocortin-1 receptor gene mediates female-specific mechanisms of analgesia in mice and humans. Proc Nat Acad Sci USA 100: 4867–4872.
Møller, A. 1989. Natural and sexual selection on a plumage signal of status and on morphology in House Sparrows, *Passer domesticus*. J Evol Biol 2: 125–140.
Mulder, R. A., R. Ramiarison, and R. E. Emahalala. 2002. Ontogeny of male plumage dichromatism in Madagascar Paradise Flycatchers *Terpsiphone mutata*. J Avian Biol 33: 342–348.
Mundy, N. I. 2005. A window on the genetics of evolution: MC1R and plumage coloration in birds. Proc R Soc Lond B 272: 1633–1640.
Mundy, N. I., N. S. Badcock, T. Hart, K. Scribner, K. Janssen, and N. J. Nadeau. 2004. Conserved genetic basis of a quantitative plumage trait involved in mate choice. Science 303: 1870–1873.
Munro, R., L. Smith, and J. Kupa. 1968. The genetic basis of color differences observed in the Mute Swan (*Cygnus olor*). Auk 85: 504–505.

Murray, R. 1963. The genetics of the Yellow-masked Gouldian Finch. Avicult Mag 69: 108–113.

Naik, R., and B. Parasharya. 1983. Sequence of plumage changes and polymorphism in the Indian Reef Heron. Sandgrouse 5: 75–81.

Norris, K. 1993. Heritable variation in a plumage indicator of viability in male Great Tits *Parus major*. Nature 362: 537–539.

Ochii, M., A. Iio, H. Yamamoto, T. Takeuchi, and G. Eguchi. 1992. Isolation and characterization of a chicken tyrosinase cDNA. Pigment Cell Res 5: 162–167.

O'Donald, P. 1983. The Arctic Skua. A Study of the Ecology and Evolution of a Seabird. Cambridge: Cambridge University Press.

O'Donald, P., and P. Davis. 1959. The genetics of the colour phases of the Arctic Skua. Heredity 13: 481–486.

Omland, K. 1997. Examining two standard assumptions of ancestral reconstructions: Repeated loss of dichromatism in dabbling ducks (Anatini). Evolution 5: 1636–1646.

Omland, K., and S. Lanyon. 2000. Reconstructing plumage evolution in orioles (*Icterus*): Repeated convergence and reversal in patterns. Evolution 54: 2119–2133.

Paulson, D. 1973. Predator polymorphism and apostatic selection. Evolution 27: 269–277.

Potti, J., and S. Montalvo. 1991. Male color variation in Spanish pied flycatchers *Ficedula hypoleuca*. Ibis 133: 293–299.

Price, D., and N. Burley. 1993. Constraints on the evolution of attractive traits: Genetic (co)variance of Zebra Finch bill colour. Heredity 71: 405–412.

Price, T., and M. Pavelka. 1996. Evolution of a colour pattern: History, development and selection. J Evol Biol 9: 451–470.

Price, T., M. Turelli, and M. Slatkin. 1993. Peak shifts produced by correlated response to selection. Evolution 47: 280–290.

Price, T. D. 2002. Domesticated birds as a model for the genetics of speciation by sexual selection. Genetica 116: 311–327.

Prum, R., and S. Williamson. 2002. Reaction-diffusion models of within-feather pigmentation patterning. Proc R Soc Lond B 269: 781–792.

Qvarnström, A. 1999. Genotype-by-environment interactions in the determination of the size of a secondary sexual character in the Collared Flycatcher (*Ficedula albicollis*). Evolution 53: 1546–1572.

Rattray, B., and F. Cooke. 1984. Genetic modeling: An analysis of a colour polymorphism in the Snow Goose (*Anser caerulescens*). Zool J Linn Soc 80: 437–445.

Robbins, L. S., J. H. Nadeau, K. R. Johnson, M. A. Kelly, L. Roselli-Rehfuss, et al. 1993. Pigmentation phenotypes of variant extension locus alleles result from point mutations that alter MSH receptor function. Cell 72: 827–834.

Rockwell, R. F., C. S. Findlay, F. Cooke, and J. A. Smith. 1985. Life history studies of the Lesser Snow Goose (*Anser caerulescens caerulescens*). IV. The selective value of plumage polymorphism: Net viability, the timing of maturation, and breeding propensity. Evolution 39: 178–189.

Rohwer, S., and D. Paulson. 1987. The avoidance-image hypothesis and color polymorphism in *Buteo* hawks. Ornis Scand 18: 285–290.

Rohwer, S., and C. Wood. 1998. Three hybrid zones between Hermit and Townsend's Warblers in Washington and Oregon. Auk 115: 284–310.

Roulin, A. 2004. The evolution, maintenance and adaptive function of genetic colour polymorphism in birds. Biol Rev 79: 815–848.

Roulin, A., and C. Dijkstra. 2003. Genetic and environmental components of variation in eumelanin and phaeomelanin sex-traits in the Barn Owl. Heredity 90: 359–364.

Roulin, A., and M. Wink. 2004. Predator-prey relationships and the evolution of colour polymorphism: A comparative analysis in diurnal raptors. Biol J Linn Soc 81: 565–578.

Schmutz, S. M., and J. K. Schmutz. 1981. Inheritance of color phases of ferruginous hawks. Condor 83: 187–189.

Shaughnessy, P. D. 1970. The genetics of plumage phase dimorphism of the southern Giant Petrel *Macronectes giganteus*. Heredity 25: 501–506.

Sheldon, B., J. Merilä, A. Qvarnström, L. Gustafsson, and H. Ellegren. 1997. Paternal genetic contribution to offspring condition predicted by size of male secondary sexual character. Proc R Soc Lond B 264: 297–302.

Slagsvold, T., and J. Lifjeld. 1992. Plumage color is a condition-dependent sexual trait in male Pied Flycatchers. Evolution 46: 825–828.

Southern, H. 1946. Polymorphism in *Poephila gouldiae* Gould. J Genet 47: 51–57.

Star, R., N. Rajora, J. Huang, R. Stock, R. Catania, and J. Lipton. 1995. Evidence of autocrine modulation of macophage nitric-oxide synthase by alpha-melanocyte-stimulating hormone. Proc Natl Acad Sci USA 92: 8016–8020.

Takeuchi, S., H. Suzuki, S. Hirose, M. Yabuuchi, C. Sato, et al. 1996. Molecular cloning and sequence analysis of the chick melanocortin 1-receptor gene. Biochim Biophys Acta 1306: 122–126.

Takeuchi, S., H. Suzuki, M. Yabuuchi, and S. Takahashi. 1998. A possible involvement of melanocortin 1-receptor in regulating feather color pigmentation in the chicken. Biochim Biophys Acta 1308: 164–168.

Taylor, R., E. Heatherbell, and E. Heatherbell. 1986. The Orange-fronted Parakeet (*Cyanoramphus malherbi*) is a colour morph of the Yellow-crowned Parakeet (*C. auriceps*). Notornis 33: 17–22.

Theron, E., K. Hawkins, E. Bermingham, R. Ricklefs, and N. I. Mundy. 2001. The molecular basis of an avian plumage polymorphism in the wild: A point mutation in the melanocortin-1 receptor is perfectly associated with melanism in the Bananaquit (*Coereba flaveola*). Curr Biol 11: 550–557.

Thorneycroft, H. 1966. Chromosomal polymorphism in the White-throated Sparrow, *Zonotrichia albicollis* (Gmelin). Science 154: 1571–1572.

Thorneycroft, H. 1975. A cytogenetic study of the White-throated Sparrow *Zonotrichia albicollis*. Evolution 29: 611–621.

Török, J., G. Hegyi, and L. Garamszegi. 2003. Depigmented wing patch size is a condition-dependent indicator of viability in male Collared Flycatchers. Behav Ecol 14: 382–388.

Tuttle, E. 2003. Alternative reproductive strategies in the White-throated Sparrow: Behavioral and genetic evidence. Behav Ecol 14: 425–432.

Vardy, L. 1971. Color variation in the crown of the White-throated Sparrow *Zonotrichia albicollis* (Gmelin). Condor 73: 401–404.

Wink, M., C. Wink, and D. Ristow. 1978. Biologie des Eleonorenfalken (*Falco eleonorae*): 2. Zur Vererbung der Gefiederphasen (hell-dunkel). J Ornithol 119: 421–428.

Wunderle, J. M. 1981. An analysis of a morph ratio cline in the Bananaquit (*Coereba flaveola*) on Grenada, West Indies. Evolution 35: 333–344.

Yeh, P. J. 2004. Rapid evolution of a sexually selected trait following population establishment in a novel habitat. Evolution 58: 166–174.

12

Environmental Regulation of Ornamental Coloration

GEOFFREY E. HILL

To understand the signal content of avian color displays and the selective pressures that led to their evolution, we must investigate how specific environmental challenges affect different types of avian coloration. Compared to morphological features that are not subject to sexual selection, expression of the ornamental coloration of both plumage and bare parts tends to be highly variable and phenotypically plastic (Andersson 1994; Cuervo and Møller 2001; Figure 12.1). Such plasticity suggests that environmental conditions can have large disruptive or enhancing effects on the expression of color displays in birds.

If specific environmental variables have predictable effects on color displays, then by assessing ornament expression, conspecifics can gain valuable information about how individuals have dealt with environmental challenges and hence about their potential quality as mates or competitors (Zahavi 1977; Andersson 1994). This observation is the foundation of honest-signaling theory, and in the past decade or so, many research programs have focused on testing the effects of environmental variables on the expression of various types of color displays. Four broad classes of environmental factors have been shown to affect the expression of ornamental color displays in birds: (1) pigment access, (2) nutritional condition, (3) parasites, and (4) social environment. In this chapter, I review how each of these variables affects the production or maintenance of different types of color ornaments in birds and draw conclusions

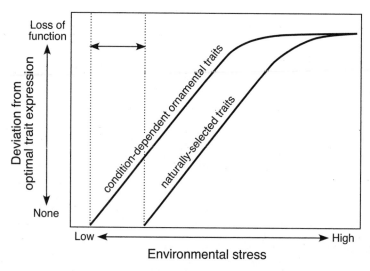

Figure 12.1. Hypothetical effects of environmental stress on sexually selected versus naturally selected traits. Sexually selected traits tend to be more phenotypically plastic, so that there is a range of environmental stress over which naturally selected traits are unaffected but sexually selected traits are negatively impacted (indicated by the arrow spanning the vertical dotted lines). This plasticity makes sexually selected traits sensitive indicators of individual condition. Redrawn from Hill (1995).

about the likely signal content of carotenoid, melanin, and structural coloration in birds. I also discuss the few studies that have been conducted on the other color types (porphyrins, pterins, and psittacofulvins).

Dietary Pigment Access

The effect of pigment intake on the expression of integumentary coloration in birds is a topic limited to carotenoid-based color displays. Melanin, porphyrin, pterin, and psittacofulvin pigments are manufactured by vertebrates from basic biological precursor molecules (Chapters 6 and 8). Only carotenoid pigments must be obtained from food as intact pigment molecules (Chapter 5), and consequently, of all avian coloration, only the expression of carotenoid-based coloration can be affected directly by the availability of dietary pigments. The question that has perplexed biologists for over a century is whether limited access to dietary carotenoids might affect expression of carotenoid pigmentation in wild birds (Hill 1994a; Hudon 1994).

Color Loss in Captivity as a Function of Pigment Access

It has long been known that a wide variety of vertebrates, from flamingos to trout, lose their bright yellow/orange/red coloration when they are kept in captivity (Comben 1976; Brush 1981). Even before carotenoid pigments were characterized biochemically or known to be the basis for yellow, orange, and red feather coloration, Keeler (1893) and Grinnell (1911) commented that House Finches (*Carpodacus mexicanus;* Plate 31) lose their bright red coloration and grow pale yellow plumage when brought into captivity. Völker (1938: 425) observed that canaries "become wholly white after having moulted" on a carotenoid-free diet. Later descriptions of ploceid finches (Kritzler 1943), flamingos (Fox 1962), spoonbills (Nieboer 1965), ibises (Nieboer 1965), woodpeckers (Test 1969), quetzals (Bruning 1971), barbets (Bruning 1971), cotingas (Bruning 1971), magpies (Bruning 1971), tanagers (Bruning 1971), waxwings (Witmer 1996), cardueline finches (McGraw and Hill 2001), and emberizid buntings (McGraw and Hill 2001) indicated that a wide range of avian species, when held in captivity on carotenoid-deficient diets, molted plumage coloration that was less red (more yellow) or with reduced chroma compared to the plumages displayed by wild birds. Reduced expression in captivity seems to be a property of carotenoid coloration in many animals, not just birds—fish that are yellow, orange, or red often shift to yellower hues with reduced chroma when maintained on carotenoid-deficient captive diets (Steven 1947; Nilsson and Andersson 1967; Pearson 1968; Kodric-Brown 1989; Wedekind et al. 1998; Grether 2000; Craig and Foote 2001).

Not all birds, however, show loss of carotenoid-based coloration in captivity on diets that seem carotenoid deficient (Figure 12.2). One reason that Zebra Finches (*Taeniopygia guttata;* Plate 27) are a favorite cage bird is that they maintain bright, carotenoid-based bill and leg coloration on a simple diet of seeds (Birkhead et al. 1999; McGraw et al. 2003b)—the same diet that causes reduced expression of plumage coloration in House Finches, American Goldfinches (*Carduelis tristis;* Plate 30), and Northern Cardinals (*Cardinalis cardinalis;* Plate 25, Volume 2; McGraw and Hill 2001). Similarly, wild-type Common Canaries (*Serinus canarius*) retain their yellow-green plumage (Völker 1938) and Red-billed Queleas (*Quelea quelea;* Plate 6, Volume 2) their bright red face and throat coloration (Dale 2000) on seed diets in captivity. These observations show that some birds can utilize the small amounts of dietary carotenoid pigments in seeds to fully express ornamental coloration, but even these species will lose their coloration if access to carotenoids drops below a

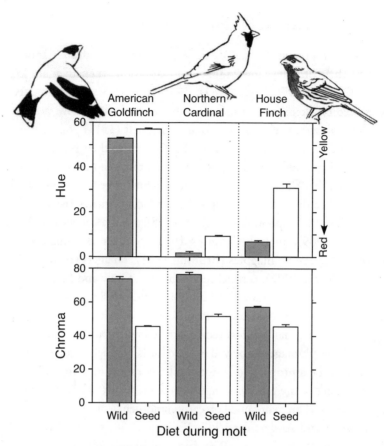

Figure 12.2. Comparison of the mean (+ standard error) hue and chroma of ornamental plumage coloration of male American Goldfinches, Northern Cardinals, and House Finches. "Wild" males underwent molt of ornamental plumage under free-living conditions. Males in the "seed" group underwent molt on a low-carotenoid diet of millet and sunflower seeds. On a seed diet, House Finches change hue more than American Goldfinches or Northern Cardinals do, but all three species show a significant loss of chroma. Drawn from data in Hill (1992) and McGraw et al. (2001).

---

certain level (Völker 1938). Such differences among species can be a function of whether they can use the pigments in seed for coloration (see Chapter 5), as well as the total amount of carotenoid pigment needed for display, but it might also reflect fundamental differences in the efficiency of carotenoid use among species. These differences among species in response to carotenoid deprivation should be the focus of future research.

Bird species also differ in the degree of plumage-color variability that they show following captive molt with standardized carotenoid intake. When groups of male House Finches were fed the same amounts of specific carotenoid pigments during molt, they converged on a similar appearance, with significantly less color variation than is observed among wild House Finches (Hill 1992, 1993). Such loss of plumage variation following captive molt, however, was not observed for captive American Goldfinches or Northern Cardinals (McGraw and Hill 2001). When males of those species were maintained on plain-seed diets in captivity, they grew plumage with reduced mean chroma compared to wild males of their species, but the chroma of these drabber males was just as variable as that of wild males (McGraw and Hill 2001; Figure 12.2). In addition, restricting males to a plain-seed diet caused a change in hue from red to yellow in House Finches and from red to orangish-red in Northern Cardinals, but it had no significant effect on the hue of the plumage of male American Goldfinches (Figure 12.2). In captive American Kestrels (*Falco sparverius*) maintained on a standardized diet of young cockerels, both males and females showed the same intensity and range of orange cere, lores, and leg coloration as wild kestrels (Bortolotti et al. 1996). Red-billed Queleas fed millet in captivity had the same distribution of color morphs, with the same coloration within morphs, as did wild birds (Dale 2000). And, in captive Red-legged Partridge (*Alectoris rufa*) maintained on a common diet with standardized carotenoid intake, males had brighter bills and legs than did females, and both males and females varied substantially in coloration (Villafuerte and Negro 1998).

Interestingly, in a subsequent feeding experiment with House Finches, when males were fed a precursor (β-cryptoxanthin) to the dominant red pigment (3-hydroxy-echinenone) in the plumage of wild males, they were more variable in plumage hue than were male House Finches in any previous feeding experiments, equaling wild males in variability (Hill 2000). These observations suggest that there is typically greater variation among individuals in expression of carotenoid coloration when they are fed precursors to integumentary pigments than when are fed pigments that can be used directly to color the integument. This result presumably reflects a cost of metabolic conversions of pigments and perhaps individual variability in the mechanisms for pigment metabolism (Hill 2002). Such a difference in whether dietary carotenoids are metabolized or deposited directly in the integument—as well as the types of metabolic conversions involved—could explain many of the differences in responses to captive diets between goldfinches, cardinals, queleas, kestrels, and partridge described above. Utilization of ingested carotenoids remains a poorly understood aspect of carotenoid physiology in birds (Chapter 5).

Biochemical Basis for Color Loss in Captivity

As described in Chapter 5, plumage coloration typically results from a mix of different types of carotenoid pigments deposited in integumentary structures. Studies show that depriving captive birds of dietary carotenoid pigments can cause the reduction of carotenoid-based coloration. But what are the biochemical bases for these observed changes in color? Are captive birds depositing the same mix of carotenoid pigments as their wild counterparts but at lower concentrations, or are they depositing different types of pigments? Is the change in plumage coloration in captive birds really the result of insufficient access to carotenoids, or is it a consequence of the disruption of the absorption, transport, metabolism, or deposition of carotenoids? A few studies have tried to resolve the mechanism for loss of carotenoid pigmentation in captive birds, but most of these questions remain unanswered.

How captivity causes color change in birds would seem to be a question that could be easily addressed with a few simple experiments. The problem, unfortunately, is that when birds are brought from the wild into captivity, they are exposed to a new and stressful environment that restricts their movement, alters their exposure to sunlight, reduces or intensifies interactions with conspecifics, and modifies their exposure to parasites. At the same time, the diets of captive birds are usually changed dramatically compared to what they consume in the wild. Thus the basic problem of deducing what causes color change in captive birds becomes a problem of simplifying the number of variables that may be affecting the expression of plumage coloration.

Völker (1934, 1938) was the first to feed captive birds specific carotenoid pigments and then record both the coloration and pigment composition of the feathers that the birds grew. He held male Common Canaries on carotenoid-free diets until the birds grew feathers lacking yellow carotenoid coloration. He then supplemented individual birds with either lutein, zeaxanthin, violaxanthin, β-carotene, or lycopene. Only lutein and zeaxanthin caused the males to grow yellow feathers. These experiments showed for the first time that specific carotenoid pigments were required for birds to grow feathers with carotenoid coloration, and they provided definitive evidence that inadequate access to appropriate carotenoid pigments could cause loss of coloration in captive birds. Moreover, these studies showed that the loss of plumage coloration was due to lower concentrations of carotenoid pigments being deposited in feathers and not to a change in the types of carotenoids being deposited.

In other cases, however, it appeared that even when captive birds had access to what seemed to be enough of the appropriate carotenoid pigments, they still grew plumage that was drabber than that of wild conspecifics. Weber (1961) proposed that loss of color in captive Common Redpolls (*Carduelis flammea*) was a consequence of physiological stress that prevented the birds from metabolically converting dietary carotenoid pigments into plumage pigments. He proposed that, in this case, physiological stress was a result of the inability of the redpolls to make display flights in small cages. To test this idea, Weber held redpolls in cages of three sizes, only the largest of which allowed the birds to make display flights. He reported that males in the largest cages grew brighter plumage than did those in the smaller cages (Weber 1961; Figure 12.3). Unfortunately, the cages differed not only in space available for flying, but also in vegetation and hence food resources. It is also possible that the crowded conditions in the smaller cages led to higher levels of parasitism or lower levels of food intake. For instance, coccidiosis, which is a ubiquitous disease of cardueline finches (see the next section below), can increase in prevalence and severity in crowded cage conditions (McGraw and Hill 2000). Thus although Weber's study suggested that cage-size per se affected the expression of plumage coloration, definitive experiments to support this hypothesis have yet to be conducted.

The loss of red coloration in captive House Finches, which was first noted in the literature more than a century ago (Keeler 1893), is apparently a consequence of the lack of suitable carotenoid precursors for the production of red plumage pigments. The primary red pigment in the feathers of male House Finches is 3-hydroxy-echinenone (Inouye et al. 2001), which is hypothesized to be the metabolic derivative of β-cryptoxanthin (Stradi 1998). The seeds that are typically used to maintain House Finches in captivity contain primarily zeaxanthin and lutein, with small amounts of β-cryptoxanthin and β-carotene (McGraw et al. 2001). Male House Finches apparently are not able to metabolically convert zeaxanthin or lutein to red plumage pigments. In support of the idea that lack of dietary β-cryptoxanthin constrains captive male House Finches from displaying colorful plumage, when captive males were supplemented with small amounts of β-cryptoxanthin during molt, they grew orange and red feathers (Hill 2000). Other red cardueline finches also use 3-hydroxy-echinenone as their primary red pigment (Stradi 1998; Stradi et al. 2001), and presumably these species would also lose their red coloration on a seed diet lacking a source of β-cryptoxanthin.

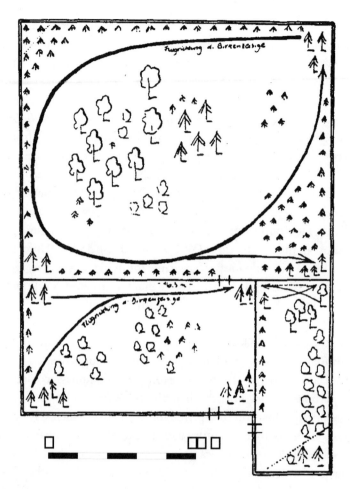

Figure 12.3. Experimental design used by Weber (1961) to test the effect of aviary size on the expression of carotenoid-based plumage coloration in Common Redpolls. Note the difference in the amount of vegetation (illustrated as trees and shrubs) in the three cages. Arrows show the flight paths of birds. Birds in the largest of the three aviaries grew more colorful plumage than did birds held in the other two cages. From Weber (1961).

Interestingly, male Northern Cardinals grow pink feathers when they are maintained on the same seed diets that cause male House Finches to turn yellow (McGraw and Hill 2001; Figure 12.2). The difference in response to this captive diet between House Finches and Northern Cardinals appears to be that cardinals are able to metabolically convert lutein to the red pigment α-doradexanthin and zeaxanthin to the red pigment astaxanthin (McGraw et al.

2001), both of which appear in their plumage. Similarly, American Goldfinches maintained on this same seed diet in captivity retain their species-typical yellow plumage coloration by converting lutein and zeaxanthin to canary xanthophyll A and B (McGraw et al. 2001; Figure 12.2). Thus captive House Finches grow drab yellow plumage in captivity because they are deprived of the appropriate precursors for red plumage pigments. However, Northern Cardinals and American Goldfinches grow drab plumage, but with a species-typical hue, because the precursor molecules provided in their captive diet, although of suitable kind, are not sufficient in quantity for the full expression of ornamental plumage.

In summary, the loss of bright plumage coloration by birds on captive diets is the result of a reduction in the overall quantities of carotenoids deposited in the integument or in the deposition of different carotenoids than are found in the integument of wild individuals. The former effect tends to result in reduction of chroma, whereas the latter can lead to a shift in hue as well as a reduction in chroma.

Carotenoid Enhancement

Change in color due to pigment access is not limited to the reduction of ornamental color display. Individuals are sometimes observed with plumage pigmentation that is more orange or red than had previously been observed in the species. Such carotenoid enhancement has long been noted in captive birds fed a variety of red foods (e.g., Beebe 1906; McGraw and Hill 2001; McGraw et al. 2001). The first mention in the literature of enhanced coloration in a wild bird concerned a Yellow Wagtail (*Motacilla flava*) that returned to Europe from wintering and molting quarters in Africa with orange rather than typical yellow plumage (Harrison 1963). It was suggested that this bird had attained its orange coloration by consuming red palm oil from a palm oil mill. Such oil, it was suggested, could have provided the bird with red carotenoid pigments that would not normally have been in its insectivorous diet (Harrison 1963; Serle 1964). Another case of enhanced carotenoid coloration in wild birds concerned several European Greenfinches (*Carduelis chloris*) in a small region of England. These birds had feathers with a reddish hue, unlike the yellow and yellow-green hues of typical greenfinches (Washington and Harrison 1969). It was suggested that this color change may have resulted from red carotenoid pigments derived from the berries of introduced yew trees in the region. Other cases of the sudden appearance of red or orange plumage in a species that is typically yellow have been reported for Bananaquits (*Coereba flaveola;* Hudon et al. 1996), Yellow-breasted Chats (*Icteria virens;* Plate 5; Mulvihill et al. 1992), Kentucky Warblers (*Oporornis formosus;* Mulvihill et al. 1992), White-throated Sparrows (*Zonotrichia albicollis;* Craves 1999), and Ring-billed Gulls (*Larus delawarensis;* Hardy 2003). A well-documented case of color enhancement concerns nestling White Storks (*Ciconia ciconia;* Plate 5). Adult storks have bright red bills and legs, but the bills and legs of nestlings are typically black with small patches of pale orange. After the introduction of a red crayfish from North America, however, nestlings that ingested the crayfish had bright red bills and legs, much brighter than had ever been observed before in wild birds (Negro et al. 2000). Biochemical analyses showed that the enhanced red coloration was due to the deposition of astaxanthin, the dominant pigment of the introduced crayfish, which had not previously been found in the skin of White Storks (Negro and Garrido-Fernandez 2000). Not all examples of color enhancement concern birds. Sea bass (*Cynoscion* sp.) with abnormally bright carotenoid pigmentation were netted in Tokyo Bay in 1979. This color enhancement was attributed to the sea bass eating more carotenoid-rich crabs (*Tritodynii horvathi*), which had appeared in the bay that year in unprecedented numbers (Goodwin 1984).

By far the best described and documented case of the enhancement of carotenoid color resulting from a change in diet concerns the colored band on the tip of the tail of the Cedar Waxwing (*Bombycilla cedrorum*). This band is typically yellow. Beginning in the early 1960s, individuals with orange bands on their tails were observed in the northeastern United States (Hudon and Brush 1989; Mulvihill et al. 1992; Witmer 1996; Plate 30). The change from yellow to orange was apparently the result of consumption of a red carotenoid, rhodoxanthin, found in the berries of Morrow's Honeysuckle (*Lonicera morrowii;* Brush 1990). The introduction of this honeysuckle to the northeastern United States coincided with the appearance of orange tails in waxwings (Mulvihill et al. 1992). Feeding experiments confirmed that, on a diet of Morrow's Honeysuckle, waxwings grew orange tail tips (Witmer 1996).

Both loss and enhancement of carotenoid coloration in birds are reversible. Bruning (1971) describes the use of carotenoid supplements to restore the bright, wild-type plumage coloration to a variety of birds that had become drab in captivity. Captive House Finches that had grown pale yellow feathers on a plain seed diet grew bright red feathers the next year when their diet was supplemented with the red pigment canthaxanthin during molt (Hill 2002). Conversely, bright red House Finches grew pale yellow feathers when fed a low-carotenoid seed diet the next year (Hill 2002). When a reddish European Greenfinch was captured and held in captivity through molt, it grew feathers with typical greenfinch coloration and no hint of red (Washington and Harrison 1969).

One might question the relevance of color enhancement to studies of the signal function of color displays—are not these simply a few oddities that have appeared due to special circumstances? Perhaps they are, but such cases might provide unique insight into how ornamental color displays can evolve. Take the case of orange tails in Cedar Waxwings. If some females show a preference for orange-tailed versus yellow-tailed males, then there could be selection on males to develop behavioral or biochemical mechanisms to grow tails with orange tips. Observing that a subtle shift in environment—in this case a human-assisted range expansion of a food plant—can cause a sudden and substantial change in color display helps us understand what is possible when we construct models for trait evolution (Part II, Volume 2). Moreover, the sudden access to abundant carotenoids could undercut the honesty of the color signal, with potential evolutionary consequences both for patterns of female mate preference and male ornament display (Hill 1994b).

Evidence for Carotenoid Limitation in Wild Birds

Endler (1980, 1983) first proposed that, because carotenoid pigments must be ingested, expression of carotenoid-based color displays would reflect, at least in part, the success of an individual in securing limited carotenoid resources. Endler focused specifically on Trinidadian Guppies (*Poecilia reticulata*) and carotenoids derived from algae, but the argument was easily applied to carotenoid coloration in birds (Hill 1990). Since it was first proposed, this idea has been contentious (Hill 1994a, 2002; Hudon 1994; Thompson et al. 1997).

In the lab, there is no doubt that depriving birds of dietary access to carotenoid pigments can cause the loss of carotenoid-based plumage coloration. Until recently, however, no convincing data were available to answer the question of whether carotenoid access affected intraspecific variation in the expression of integumentary coloration in wild birds (Hill 1994a; Hudon 1994). A growing number of field studies has now been published that support the idea that access to carotenoid pigments does affect carotenoid display in wild populations. To date, no study has been published showing that carotenoid pigments are so abundant in the diet of a wild vertebrate with an ornamental

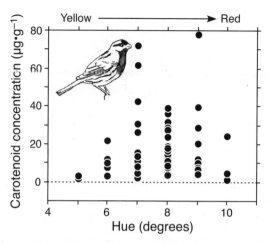

Figure 12.4. Relation between the carotenoid concentration in the gut of molting male House Finches and the hue of growing feathers. Redder males tended to have a higher concentration of carotenoids in their gut contents, supporting the hypothesis that dietary access to carotenoid pigments affects the expression of ornamental plumage color in wild House Finches. All males were collected in a 4-day period in San Jose, California. Redrawn from Hill et al. (2002).

carotenoid diplay that all individuals have access to more carotenoid pigments than are needed for ornament display.

The first direct test of the carotenoid limitation hypothesis as it relates to sexually selected plumage coloration was conducted by Hill et al. (2002), who compared the gut contents of male House Finches collected during molt to the coloration of growing feathers. They found a significant positive relationship between the hue of growing feathers and the carotenoid concentration in the gut contents of males; redder males tended to ingest more carotenoid pigments (Figure 12.4). Grether et al. (1999) showed a similar relationship between availability of dietary carotenoids and male integumentary color across populations of guppies.

McGraw et al. (2003b) systematically quantified food intake in Zebra Finches and related ingested carotenoids to circulating carotenoids and integumentary coloration. They first observed seed consumption in male and female Zebra Finches and found no sex differences in carotenoid intake, even though males have more brightly pigmented orange bills than do females (Plate 27). Despite similar intake of dietary carotenoids, females had much lower levels of carotenoid pigments circulating through their blood than did males. They concluded that, in the Zebra Finch, physiological processes, such as absorption or transport, are more important in determining the expression of bill coloration than is dietary access to carotenoid pigments. It remains uncertain why Zebra Finches were relatively insensitive to variation in access to dietary carotenoids, whereas other species seem very sensitive. The Zebra Finch study, however, concerned bill coloration; studies of other birds have focused on plumage coloration. It may be that dietary carotenoids have a greater effect on feather coloration than on bill coloration, but this idea needs to be rigorously tested.

Other evidence for carotenoid limitation in the wild is more indirect. Linville and Breitwisch (1997) observed a population-wide decrease in carotenoid-based red plumage coloration among male Northern Cardinals following a cold year in which fruit availability during molt was diminished. They speculated that fruit was a primary source of dietary carotenoids for male cardinals, and that the decline in fruit was the reason for the decline in plumage coloration. Hill (1993) invoked access to carotenoid resources to explain local and geographic variation in expression of carotenoid-based plumage coloration in male House Finches. Carotenoid resources in the various regions were not quantified, however, and other explanations, such as local parasite exposure, could not be ruled out. Similar geographic variation in the skin color of fish has also been attributed to differences in access to dietary carotenoids (Krogius 1981; Reimchen 1989; Craig and Foote 2001).

Negro et al. (2002) reported an unusual source of dietary carotenoids for Egyptian Vultures (*Neophron percnopterus*), the only Old World vulture with colorful facial skin (Plate 30). They observed that Egyptian Vultures search for and consume quantities of ungulate feces as a source of carotenoid pigments. Ungulate feces is a poor source of most nutrients, but a rich source of carotenoid pigments, and feces from ungulates feeding on green grass is richer in carotenoids than that from ungulates living in brown, dry pastures. Negro et al. (2002) found that vultures feeding in areas with green pastures had higher levels of circulating carotenoid pigments and brighter carotenoid-based facial coloration than did vultures feeding in areas lacking green pastures. Apparently, access to carotenoid-rich feces determines ornament expression in these vultures.

To summarize, a large body of evidence supports the idea that access to carotenoid pigments can have an effect on the expression of color displays in vertebrates. Despite skepticism by some authors (Hudon 1994; Thompson et al. 1997; Zahn and Rothstein 1999), all published studies are consistent with the idea that wild vertebrates are limited in their access to the carotenoid pigments needed to produce ornamental coloration. To really understand the

Box 12.1. Getting to the Bottom of the Control of Carotenoid-Based Color Expression

As presented in this chapter, aviary experiments have shown that nutritional condition, parasites, and access to dietary carotenoid pigments can all affect the expression of carotenoid-based plumage coloration. But for no population of birds do we know the relative importance of these variables to the display of carotenoid-based coloration. Throw in an unknown genetic effect and unknown maternal effects, and we really know little about the signal content of carotenoid-based ornamental plumage, which is the best understood of all plumage color ornaments. Hurdling the technical difficulties of simultaneously assessing multiple environmental effects on plumage coloration will not be easy, but it can be done.

What is needed is a detailed study of a population of birds with carotenoid-based plumage coloration through the molt period. These birds must be readily observable during foraging so that their food intake can be quantified. From known food intake, the carotenoid content, calories, and the nutritional value of their diet could be determined. Finally, the researcher must be able to repeatedly capture the study birds through the molt period to assess key parasites, most notably coccidia. Carotenoid access, general nutrition, and number of parasites will undoubtedly be related. What a bird ingests determines both its nutritional intake and its carotenoid access. Carotenoid intake and nutritional condition, in turn, will likely affect parasite resistance, and parasites such as coccidia will probably affect uptake of carotenoid pigments and other nutrients as well. Hence a path analysis would be most useful for tracing both the direct and indirect effects of each variable on coloration (Figure B12.1).

Such observations would be invaluable for working out the real effects of environmental variables on the expression of plumage coloration. This study, however, would necessarily ignore genetic and maternal effects (Figure B12.1). Building a model that also includes genetic and maternal effects will be much more difficult, requiring a breeding population of birds in which offspring do not all disperse. Paternity would have to be confirmed with genetic analysis. Testing for maternal effects would require careful monitoring of the nutritional and carotenoid content of eggs and levels of provisioning by parents.

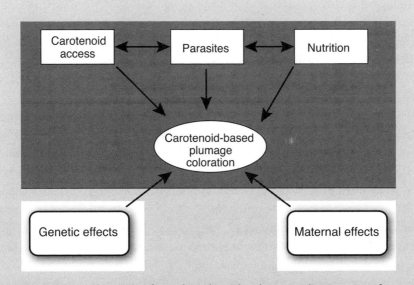

Figure B12.1. The primary factors hypothesized to determine the expression of carotenoid-based plumage coloration. Arrows show likely interactions. Boxes in the shaded area are the focus of this chapter. A primary research goal of biologists interested in the signal content of plumage coloration should be the gathering of data that will allow a path analysis of these factors.

The study outlined in the figure is specific to carotenoid-based coloration, but if the carotenoid-access box is deleted, the same approach could be used for melanin and structural coloration.

role that access to dietary carotenoids plays in the expression of plumage coloration among wild birds, we need to carefully quantify the dietary sources of and physiological demands for carotenoids. In this way we could compare the nutritional requirement for integumentary pigmentation to the levels of carotenoids consumed by individuals (Box 12.1).

Parentally Derived Carotenoids

The yellow/red downy and juvenile plumages of the nestlings of some (but not all, see Chapter 8) birds are pigmented with carotenoids (Partali et al. 1987).

Young birds also can have bills and legs pigmented by carotenoids. The pigments for such plumage and skin coloration must be acquired before a young bird can forage for itself, and thus these color displays are dependent on the carotenoids provided to it by parents.

One important source of carotenoids for a young bird is the yolk of its egg (Blount et al. 2000). The yolks of virtually all bird species are yellow or orange in part due to carotenoid content (Chapters 5 and 8). How much carotenoid a female is able to or chooses to deposit in an egg can affect the integumentary coloration of offspring. Such an effect of maternal carotenoids was shown in Domestic Chickens (*Gallus gallus;* Plate 17); chicks that hatched from carotenoid-depleted eggs had significantly paler legs than did chicks hatched from carotenoid-supplemented eggs (Koutsos et al. 2003). In a study of Zebra Finches, McGraw et al. (in press) showed that females on high-carotenoid diets during egg laying produced offspring with redder bills than did females on low-carotenoid diets, and this effect of maternal carotenoid access persisted until male offspring reached sexual maturity. More studies on the potential effects of maternal carotenoids on the carotenoid displays of offspring are clearly needed.

The carotenoid content of food provided to young birds by parents can also affect carotenoid displays. In the same study of chickens described above in which the carotenoid content of the egg significantly affected the leg coloration of chicks, Koutsos et al. (2003) observed that access to carotenoids during early growth outside the egg also had a significant effect on leg color. Several studies have shown an effect of carotenoid access during nestling growth on expression of juvenile plumage coloration. Slagsvold and Lifjeld (1985) showed that Great Tit (*Parus major;* Plate 11, Volume 2) nestlings that developed in deciduous forests grew juvenile plumage with higher chroma than did nestling that developed in spruce forests. The feathers of the former group of nestling had higher levels of lutein (Partali et al. 1987), presumably because the lutein content of lepidopteran larvae in deciduous woods was higher than that of larvae in coniferous forest. This study was followed almost two decades later by a study of nestling color in Great Tits in which one set of nestlings was supplemented with lutein and zeaxanthin and a control group was not supplemented (Tschirren et al. 2003). The carotenoid supplementation had a significant positive effect on the brightness of juvenile plumage. Fitze et al. (2003a) also studied the coloration of nestling Great Tits. They swapped chicks between nests and showed that, although the environment in which the chicks were reared had the largest effect on yellow plumage coloration, the nest of origin (which could have reflected the effect of genes or maternal investment) also explained a significant portion of the variation in plumage coloration (Fitze et al. 2003a). In a companion study, Fitze et al. (2003b) supplemented nestling Great Tits with carotenoids during the early and late nestling stages and showed that only carotenoid access during the early nestling stage affected coloration. Males provide most food during the early nestling stage, so the color of juvenile plumage is largely the result of paternal provisioning.

The mouth color of nestling birds has been proposed to be due to carotenoids (Saino et al. 2000), but this suggestion has yet to be confirmed biochemically (Chapter 8). Saino et al. (2000) supplemented nestling Barn Swallows (*Hirundo rustica*) with lutein and showed that such carotenoid supplementation caused the mouths of nestlings to get redder. Reddening of chick mouths, however, could have been due to improved heath rather than bolstered carotenoid pigmentation.

Together, these studies clearly indicate that the carotenoid pigments provided by parents can have a large effect on the expression of leg, bill, mouth, and plumage coloration of nestlings. The signal content and function of these juvenile color displays await further study.

Parasites

The Hamilton-Zuk hypothesis proposes that colorful plumages evolved specifically as signals of heritable resistance to parasites (Hamilton and Zuk 1982). In its original conception, this model proposed a single locus that could carry an allele for resistance to one of two key pathogens in the environment. The two pathogens and the two alleles for resistance cycled in prevalence. Only when the genes for ornamentation were expressed with the appropriate genes for disease resistance would individuals have a maximum ornament display. Thus ornament expression signaled genes for resistance to the predominant parasite in the environment. In the 20 years since it was published, this specific model for host-parasite co-evolution and the evolution of ornamental traits has been interpreted as a general hypothesis that ornamental traits, and particularly colorful plumages, signal genetically based resistance to parasites. At times, even the idea that parasites depress the expression of ornament plumage coloration has been called the "Hamilton-Zuk hypothesis," but this is not the novel idea of the paper by Hamilton and Zuk (1982).

Møller (1990) considered the original conception of the Hamilton-Zuk hypothesis and identified three necessary conditions for it to be feasible:

(1) parasites must affect the fitness of hosts; (2) there must be heritable variation in resistance to parasites; and (3) parasites must depresses trait expression. In this chapter, I focus almost exclusively on the third of these necessary conditions—how parasites can affect avian coloration. I review a few studies that indirectly test whether plumage coloration might signal heritable resistance to plumage coloration. No study of colorful plumage has demonstrated or even convincingly tested all three conditions in a single population of birds.

It seems logical that parasites could affect the expression of colorful plumage. If there are energetic costs involved with producing colorful plumage, then energy sacrificed in defense against a parasite infection might reduce the quality of the color display produced. Most tests of the effects of parasites on avian coloration are correlational field studies in which parasite loads, often measured at a time offset from molt and hence color production, are compared to color displays. Such tests provide the weakest support for, and the least convincing falsification of, the effects of parasites on color displays. Much more convincing are infection experiments, either in the lab or the field, but such experiments are technically difficult, and few have been conducted. I review both correlational and experimental studies of the effects of parasites on plumage coloration, but the reader is cautioned to bear in mind the quality of evidence when assessing an effect of parasites on color displays.

Carotenoid Coloration

A variety of parasites has been shown to have a direct negative effect on the expression of carotenoid pigmentation. Coccidia (phylum Apicomplexa; suborder Eimeriorina) were among the first parasites reported to affect the expression of carotenoid-based animal coloration, and they are now the parasite for which we best understand the mechanisms for reduced color display. Coccidia are protozoan parasites that encyst in the lining of the gastrointestinal tract in a wide variety of vertebrate hosts. They are spread when oocysts that are shed in the feces of one animal are ingested by another animal. In most birds, coccidia cause chronic infections that are not debilitating. Under certain circumstances, however, coccidia can cause fatal infections (Giacomo et al. 1997).

The mechanism by which coccidiosis affects expression of carotenoid-based skin and feather coloration apparently lies in the disruption of carotenoid absorption and perhaps also carotenoid transport. When coccidia encyst in the gut lining, they cause hyperplasia, a thickening of the epithelium that inhibits the absorption of carotenoids across the gut lining (Allen 1987b). Hyperplasia and the disruption of absorption has been demonstrated only in poultry and only in response to infection by coccidia in the genus *Eimeria*, but it seems likely that coccidia in the genus *Isospora*, which are common parasites of passerine birds, cause similar effects.

The epithelial tissue of the gut is important not only for the absorption of ingested carotenoid molecules, but also as a site for the production of the high-density lipoproteins that transport carotenoids through the circulatory system (Chapter 5). Hyperplasia caused by coccidiosis inhibits not only carotenoid absorption, but also the production of high-density lipoproteins (Allen 1987a). Thus coccidiosis appears to negatively affect the utilization of dietary carotenodis both directly by inhibiting the absorption of carotenoids across the gut lining and indirectly by slowing the production of carrier proteins, thereby inhibiting the transport of carotenoids to colorful integumentary structures.

Given these mechanisms by which coccidia can disrupt carotenoid utilization, it is not surprising that coccidiosis has been shown to negatively affect the expression of carotenoid coloration. Poultry farmers have long observed that eimerian coccidia diminish the skin color of chickens (Bletner et al. 1966). In the twentieth century, poultry farmers began to supplement the diets of chickens with carotenoids to meet consumer demand for well-pigmented products. The majority of the skin of a chicken is hidden by plumage and cannot serve as an ornamental trait, but leg coloration is bright yellow, exposed, and potentially serves as a signal. Interestingly, in an experimental study, the leg and toe color of chickens was significantly reduced by infection with eimerian coccidia (Marusich et al. 1972).

Coccidia in the genus *Isospora* also appear to have a direct negative effect on carotenoid-based plumage coloration of passerines. In wild-caught House Finches, American Goldfinches, and European Greenfinches, males experimentally infected with isosporan coccidia grew plumage with significantly reduced hue and saturation of carotenoid coloration compared to uninfected males (Brawner et al. 2000; McGraw and Hill 2000; Hõrak et al. 2004; Figure 12.5). There are no published data on the importance of coccidiosis to the carotenoid coloration of wild birds.

The only other endoparasite whose effect on carotenoid-based plumage coloration has been studied in controlled experiments is *Mycoplasma gallicepticum*, a bacterium that causes severe eye and respiratory infection in the House Finch. Brawner et al. (2000) observed that, when mycoplasmosis broke out in a captive flock of House Finches undergoing molt, males that were not infected grew brighter plumage than did infected males. Hill et al. (2004)

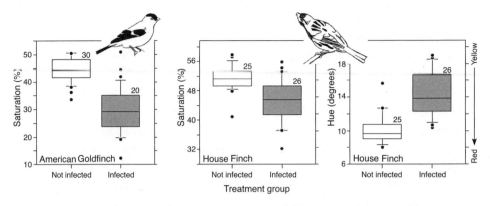

Fig. 12.5. Effects of coccidial infection on the saturation and hue of carotenoid-based plumage coloration in captive male American Goldfinches and House Finches. Coccidia caused male House Finches to grow feathers that were less saturated and less red and male American Goldfinches to grow feathers that were less saturated. Box plots show the tenth, twenty-fifth, fiftieth, seventy-fifth, and ninetieth percentiles as horizontal lines, and solid dots are data for individuals outside this range; sample sizes are shown at the top right of each box. Adapted from McGraw and Hill (2000) and Brawner et al. (2000).

conducted a follow-up study in which they experimentally infected one flock of male House Finches with *M. gallicepticum* and kept a control flock uninfected. Uninfected males grew redder carotenoid-based plumage than did infected males.

In many ways, the demonstration that mycoplasmosis depresses carotenoid plumage coloration is not as compelling as the demonstration that coccidiosis has such an effect. The former pathogen causes severe illness in birds that frequently leads to death. Coccidiosis, in contrast, is a chronic disease that affects the gut lining and for which there are specific mechanisms for reduction of plumage coloration. However, demonstrating that carotenoid ornamentation is sensitive to a variety of pathogens, including those that infect organ systems not in the gastrointestinal tract, shows that carotenoid coloration could signal infection by a wide range of parasites and potentially via multiple physiological mechanisms. Experimental studies on the effects of parasites on carotenoid coloration in fish (Milinski and Bakker 1990; Houde and Torio 1992) corroborate what has been observed in birds.

The studies outlined above show that parasites can affect the expression of carotenoid-based plumage coloration; hence they show the potential for carotenoid coloration to reveal past experience with parasites. The Hamilton-Zuk hypothesis, however, proposes that color displays reliably signal resistance to parasites; that is, the ability to fend off future infections (Hamilton and Zuk 1982). The relationship between carotenoid coloration and ability to resist future infections has recently been tested in several experimental studies. In a study of European Greenfinches, Lindstrom and Lundstrom (2000) found that males with larger patches of yellow plumage coloration on their tails developed less severe viremia and were able to clear the virus faster when they were infected with a Sindbis virus (Togaviridae, *Alphavirus*). Similarly, Hill and Farmer (2005) infected male House Finches of variable plumage hue with *Mycoplasma gallicepticum* and observed a positive relationship between plumage hue and clearance of mycoplasmosis. Dawson and Bortolotti (in press) looked at carotenoid-based yellow skin color in American Kestrels and blood parasites and found that skin color was a poor predictor of current infection but a good predictor of future infection. Rather than studying response to infection directly, Saks et al. (2003) measured both humoral and cell-mediated immune responsiveness in relation to plumage coloration. They found that male European Greenfinches with brighter yellow breast feathers had a stronger humoral immune response, but there was no relationship between cell-mediated response and color. More studies are needed to test whether plumage coloration signals ability to resist specific diseases and ultimately to test whether such disease resistance is heritable, but the growing list of studies linking carotenoid coloration with the ability to fend off future infection supports the idea that a common function of carotenoid coloration in birds is to signal disease resistance.

Although experimental aviary studies provide the best evidence that some parasites directly affect expression of carotenoid-based plumage coloration, field studies of birds also support the idea that parasites negatively affect expression of plumage coloration. As would be expected, however, field studies have yielded much less clear results than captive-animal studies. In the most thorough field study to date, Thompson et al. (1997) analyzed avian pox (*Poxvirus avium*) in relation to plumage coloration in the House Finches. Avian pox is a widespread disease of birds, including House Finches, and can manifest with symptoms as mild as small warts on the feet or as severe as lesions over the feet and eyes that can be fatal (van Riper et al. 2002). Thompson et al. (1997) observed that male House Finches with more severe pox lesions had yellower plumage hue scores than did males with lesser infections. Furthermore, they were able to recapture individuals before and after molt, and they found that males with pox infections going into molt tended to shift in plumage

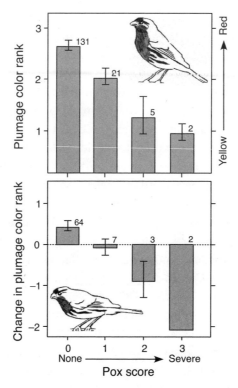

Figure 12.6. Effect of pox virus on the expression of carotenoid-based plumage coloration in male House Finches. Males with more severe pox infections at the time of capture tended to have yellower (as opposed to redder) plumage coloration, and pox infection prior to fall molt predicted change in coloration after molt—males with more severe infections decreased more in plumage coloration. Data shown are mean ± standard error, with sample sizes above each bar. Redrawn from Thompson et al. (1997).

hue from redder to yellower, whereas males going into molt with no pox infection tended to become redder in hue (Figure 12.6).

Most other correlational field studies looking at parasites and carotenoid coloration have focused on hematozoan blood parasites in passerine birds. A key assumption of these studies is that hematozoa have detrimental effects on health, but some authors have been unable to find any negative effects of hematozoa on the health of birds (Bennett et al. 1988; Weatherhead 1990; Weatherhead and Bennett 1992). Others have found a negative effect only for hematozoa in the genus *Trypanosoma* (malaria; Ratti et al. 1993; Dufva 1996). Merino et al. (2000), however, showed experimentally that hematozoan par-

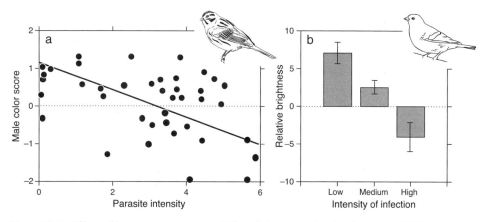

Figure 12.7. Effects of hematozoa on carotenoid-based plumage coloration in two songbird species. (a) There was a significant negative relationship between the carotenoid-based yellow plumage and the number of hematozoan parasites in the blood of male Yellowhammers. Redrawn from Sundberg (1995). (b) Color scores of male European Greenfinches (mean ± standard error) relative to their level of infection with hematozoan parasites. Redrawn from Merilä et al. (1999).

asites have a direct negative affect on individual fitness. They medicated one set of Great Tits, eliminating their hematozoan infections, and left a control group untreated. The birds that had no blood parasites fed offspring at a higher rate.

Given the uncertainty about whether hematozoans negatively impact some birds and the temporal variability of hemotozoan infection, it is not surprising that studies comparing hematozoan infection with ornamental coloration have yielded mixed results. Sundberg (1995) and Merilä et al. (1999) found significant negative relationships between the number of hematozoan parasites and carotenoid-based plumage coloration in the Yellowhammer (*Emberiza citrinella*; Plate 15, Volume 2) and the European Greenfinch, respectively (Figure 12.7). Hõrak et al. (2001) found a similar negative relationship between blood hematozoa and chroma of yellow breast coloration for yearling male and female Great Tits, but they found an opposite, positive, correlation among older males. Hatchwell et al. (2001) found a nonsignificant trend for male Eurasian Blackbirds (*Turdus merula*) infected with *Plasmodium* to have drabber yellow bills than found in uninfected males. Other studies have failed to find the predicted negative relationship between carotenoid coloration and blood parasites in wild birds (Burley et al. 1991; Weatherhead et al. 1993; Seutin 1994; Dufva and Allander 1995). In general, studies have found mixed evidence as to whether hematozoa negatively impact bird health, and perhaps not surprisingly, the results of studies of the effects of hematozoa on plumage color have also been mixed. It may be that the effects of hematozoa vary with season or region, and it seems likely that they have different effects on different species of birds.

Finally, feather mites have been proposed to affect expression of carotenoid-based plumage coloration. In a parallel study to the avian pox study described above, Thompson et al. (1997) looked at the number of mites on the primary feathers of male House Finches in relation to their carotenoid coloration. They found a significant negative relationship between the number of mites and male coloration, and, as with pox, they found that males showed diminished plumage coloration in proportion to the number of mites on their feathers while they were molting. This result was partly corroborated by Harper (1999), who looked at mite loads in four species of passerine birds with carotenoid coloration. He found a significant negative relationship between mite load and coloration in European Greenfinches and Linnets (*Carduelis cannabina*), but no significant relationships between color and mite load in Great Tits and Yellowhammers. Blanco et al. (1999), in contrast, found a tendency for male Linnets with brighter or larger patches of red to have more feather mites. The results of these correlations between feather mites and carotenoid coloration are hard to interpret, however, because it is unclear whether feather mites are parasites or simply commensals living on feather debris, including oil and dander (O'Conner 1992) without negatively affecting their hosts (Blanco et al. 1997, 1999; Dowling et al. 2001). Indeed both Thompson et al. (1997) and Harper (1999) suggested that mites might not be the cause of drab plumage coloration, but rather they might simply be correlated with the true agent causing drab plumage. In support of the idea that mites simply correlate with the parasites that actually depress plumage coloration, Brawner et al. (2000) found that infecting males House Finches with isosporan coccidia increased their mite loads, but he found no relationship between mite loads and coloration independent of coccidiosis.

Even though many biologists have expressed doubt that feather mites can directly depress the expression of plumage coloration, this hypothesis is supported by an experimental study. Figuerola et al. (2003) dusted the feathers of one group of male Serins (*Serinus serinus*), killing all feather mites, and left a control group untreated. The mite-free group grew more brightly colored carotenoid-based yellow plumage coloration than did the group infested with mites. Although this experimental study indicates that a treatment against mites can affect carotenoid-based plumage coloration, there remains no proposed mechanism for how such an ectoparasite, which does not appear to consume the resources of the host, can inhibit the production of carotenoid ornamentation.

To summarize, lab studies clearly demonstrate that parasitic infection during the period of molt and feather pigmentation can directly inhibit the expression of carotenoid coloration. In addition, a few lab studies have shown that carotenoid coloration predicts the ability of an individual to fend off future infection. Observations from field studies have provided much more ambiguous support for the idea that carotenoid color signals parasite resistance, but this ambiguity is to be expected, given the difficulty of controlling confounding variables in the field, and particularly given that most field studies have focused on hematozoan parasites, whose effect on the health of most species of birds is unclear.

Melanin Coloration

Although it is well established that a variety of parasites can suppress expression of carotenoid-based integumentary coloration, experimental evidence suggests that melanin ornaments typically are not directly affected by parasites. Melanins differ fundamentally from carotenoids in that they can be synthesized by birds from basic biological precursors (Chapter 6), although the precursors needed to synthesize melanins may be limiting in some animals (see the section on nutrition below). The idea that carotenoid and melanin ornaments differ in their sensitivity to parasites was first tested in an experiment that exposed male House Finches to either isosporan coccidia or the bacterium *Mycoplasma gallicepticum.* Both parasites had a large effect on carotenoid-based breast coloration but no effect on the melanin coloration of the tail (Hill and Brawner 1998). This experiment suggested that carotenoid coloration is more sensitive to the effects of parasites than is melanin coloration, but the tail coloration of male House Finches is not ornamental (i.e., not densely or boldly pigmented and not sexually dimorphic). To test whether the response to parasites would be different for ornamental melanin coloration, McGraw and Hill (2000) looked at the effect of isosporan coccidia on both the sexually selected carotenoid-based yellow body plumage and the sexually selected (i.e., boldly pigmented and sexually dimorphic) melanin-based black caps of male

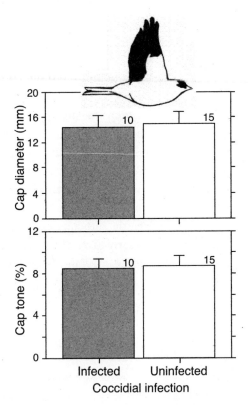

Figure 12.8. Effect (mean ± standard deviation) of coccidial infection on the size and tone of melanin-based black caps in captive male American Goldfinches. Coccidia had no significant effect on either the tone (t-test, $t = 0.70$, $P = 0.49$) or diameter ($t = 0.83$, $P = 0.42$) of caps. Sample sizes are shown at the top right of each bar. Adapted from McGraw and Hill (2000).

American Goldfinches. They found a strong effect of coccidiosis on yellow carotenoid coloration, but none on either the size or blackness of the black caps (Figure 12.8).

The only experimental study to find a direct effect of parasites on melanin coloration tested the effects of the intestinal roundworm (*Ascaridia galli*) on the hue and brightness of hackle feathers of Red Junglefowl (*Gallus gallus*; Zuk et al. 1990). The golden hackle feathers of Red Junglefowl (Plate 26) are pigmented with a mix of eumelanins and phaeomelanins (Hudon 1991), in contrast to the black caps and breast patches of those passerine species most often studied, which have primarily eumelanin pigmentation (Haase et al.

1992; McGraw and Wakamatsu 2004). Zuk et al. (1990) found that male Red Junglefowl that were experimentally infected with roundworms grew paler hackle feathers compared to uninfected (control) males. This is the only experimental study to look at parasite-mediated expression of reddish phaeomelanin pigmentation rather than black eumelanin pigmentation. These observations suggest that more work should be done on the effects of parasites on color displays with a high phaeomelanin content. Korpimaki et al. (1995), however, looked at hematozoa prevalence in relation to the phaeomelanin-containing dorsal coloration of European Kestrels (*Falco tinnunculus*) and found no relationship.

Fitze and Richner (2002) tested the delayed effects of parasites on melanin coloration by modifying the number of fleas (*Ceratophyllus gallinae*) in the nests of Great Tits. They recorded changes in the size and blackness of the melanin breast stripe and the brightness of yellow body plumage in the next year. Parasites had no effect on the carotenoid coloration, but more fleas in the nest had a significant negative effect on the size of the black breast stripe grown in the next molt (Figure 12.9). The parasite manipulations in this study occurred months before the black feathers of the breast patch were grown, so the study demonstrates a delayed effect of parasites on melanin coloration and suggests a different sort of condition dependency for melanin ornaments compared to carotenoid ornaments. More experimental studies of the effects of parasites on melanin coloration conducted during molt are needed.

Few field correlational studies looking at parasites and melanin plumage coloration have been published, which is surprising, given the large number of studies on melanin coloration in relation to dominance and sexual selection. Harper (1999) found a nonsignificant tendency for male Great Tits with wider black breast stripes to have fewer feather mites in their plumage. Interestingly, two of the significant relationships between melanin ornamentation and parasites that have been published involve female coloration. Both male and female Bar-tailed Godwits (*Limosa lapponica*) molt from a drab gray winter plumage to bright rust-colored breeding plumage. The rust coloration has not been studied biochemically, but it is likely that, like other rust colorations of bird plumage, it is melanin-based with a high proportion of phaeomelanins present (Haase et al. 1992; McGraw and Wakamatsu 2004). Piersma et al. (2001) found that the amount of rusty plumage, which is an indicator of the extent of completion of prealternate molt, was significantly negatively related to the number of tapeworms in the guts of female, but not male, godwits. In another study of female coloration, Roulin et al. (2001) found that the number

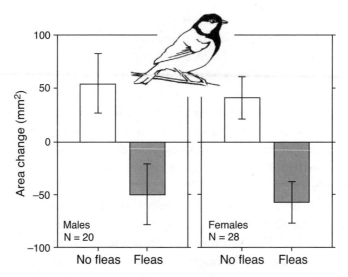

Figure 12.9. Male and female Great Tits infected with fleas showed a decrease in bib area, whereas those dusted with pesticide to remove fleas showed an increase in the area of black. There was no significant effect on carotenoid-based coloration (not shown). Shown are means and standard deviations. Birds were not infected during molt, but several months prior to molt. Redrawn from Fitze and Richner (2002).

of melanin spots on the ventral plumage of female Barn Owls (*Tyto alba*; Plate 29) is negatively correlated to the number of blood-sucking flies on their nestlings (Figure 12.10). The latter study, however, does not provide evidence that parasites directly affect the number of spots on female Barn Owl plumage; indeed in a cross-fostering study, it was shown the breast spottiness is largely under genetic control (Roulin et al. 1998). In a study of Tawny Owls (*Strix aluco*), Galeotti and Sacchi (2003) found that reddish morph birds, presumably with more phaeomelanin pigments (but possibly porphyrin pigments; Chapter 8) in their feathers, were more parasitized than gray morph birds in the same local population. The most likely explanation for this pattern, however, is that red and gray birds occupy different habitats.

One of the oldest hypothesized effects of environment on the expression of plumage color is that feathers get darker in more humid climates (Gloger's rule; Gloger 1833). Zink and Remsen (1986) reported that 94% of the bird species they examined that lived in climates of variable humidity adhered to Gloger's rule. The darkening of feathers with increased humidity has sometimes been expressed as an evolutionary hypothesis—darker plumage evolves

Environmental Regulation of Ornamental Coloration

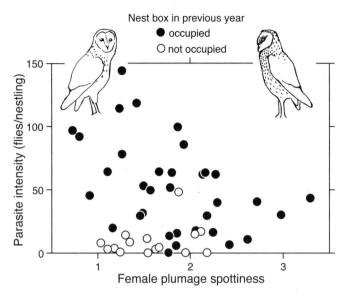

Figure 12.10. Relation between area of melanin breast spots in the plumage of female Barn Owls and number of blood-sucking flies on their nestlings. Female owls with more dark spots tended to have young with fewer parasitic flies. See Figure 11.3 for the definition of spottiness; parasite intensity is measured as the mean number of ectoparasitic flies (*Carnus hemapterus*) per nestling in each brood. Redrawn from Roulin et al. (2001).

in humid environments—but it has also been hypothesized that exposure to humidity during development causes feathers to darken. There is no experimental evidence to support that latter idea, but recently, it was proposed that humid environments promote the growth of bacteria and that greater melanin deposition in feathers is an evolutionary response to increased bacterial activity (Burtt and Ichida 2004; Shawkey and Hill 2004). This idea assumes that melanin in feathers reduces the damaging effects of bacteria, and Goldstein et al. (2004) showed that white chicken feathers are degraded by bacteria faster than black, melanized feathers. Thus the increased melanization of feathers in more humid environments could be an evolutionary response to increased bacterial action.

Clearly there is a need for more direct tests of the effects of parasites during molt on the production of melanin-based plumage coloration. The data at hand suggest that eumelanin-dominated coloration is less affected by parasitism than is carotenoid pigmentation, but this hypothesis needs to be confirmed in more experiments with a greater number of species. The effects of parasites on phaeomelanin-dominated coloration remain largely unstudied. Studies that infect birds with parasites at a time offset from molt do not test the direct effect of parasites on melanin color production (although they can provide insight into other sorts of condition dependencies). The hypothesis that the expression of melanin coloration predicts resistance to future infection also needs to be tested. The signal content of melanin coloration is central to many hypothesis related to sexual selection, and studies of the effects of parasites on melanin coloration should be a priority.

Structural Coloration

In the only experimental study of the effects of parasites on structural coloration, Hill et al. (2005) allowed Wild Turkeys (*Meleagris gallopavo;* Plate 32) to molt on one of three experimental treatments—no coccidia, infected with a single species of coccidia, or infected with multiple species of coccidia. Wild Turkeys have bright iridescent copper/green tail and wing covert feathers that are presented to females during courtship and appear to function in female choice. Males infected with multiple species of coccidia grew feathers with lower brightness and less UV chroma than did males that were infected with one species of coccidia, which in turn grew less brightly colored feathers with less UV chroma than did uninfected males.

There have been a few correlational field studies looking at relationships between parasitic infection and structural coloration. In his analysis of feather mites and plumage coloration, Harper (1999) found a significant negative relationship between mite loads and the brightness of structural blue crown coloration of male Blue Tits (*Parus caeruleus;* Plate 18, Volume 2). Potti and Merino (1996) found no relationship between blood hematozoa and the size of structural white forehead patches in male Collared Flycatchers (*Ficedula albicollis;* Plate 18, Volume 2), but they did find that females with white forehead patches had fewer blood hematozoa than did females lacking white forehead patches. Korpimaki et al. (1995) actually found a positive relationship between blood hematozoa and blue tail coloration in European Kestrels.

Satin Bowerbirds (*Ptilonorhynchus violaceus;* Plate 8) were among the first vertebrates to be studied for a relationship between structural coloration and parasites (Borgia 1986; Borgia and Collis 1989, 1990). Unfortunately, early workers relied on visual assessment of male coloration and could only differentiate between age classes by coloration. As a result, no clear relationships

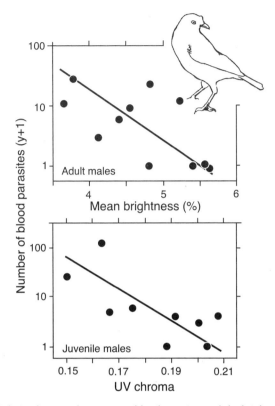

Figure 12.11. Relation between hematozoan blood parasites and the brightness and chroma of purplish structural coloration in male Satin Bowerbirds. Adult males with brighter plumage and juvenile males with more chromatic feather coloration had fewer blood parasites. Note that y-axes are on a log scale, so 1.0 was added to each raw datum for this transformation. Redrawn from Doucet and Montgomerie (2003).

between structural coloration and parasites could be established. In a recent study using a reflectance spectrophotometer and an analysis that included the full range of coloration visible to passerine birds, Doucet and Montgomerie (2003) found significant negative relationships between brightness of the structural blue coloration of male Satin Bowerbirds and abundance of hematozoa in the blood (Figure 12.11).

More carefully controlled experimental investigations of the effects of parasites during molt on structural plumage coloration are needed. One interesting idea that is just beginning to be explored is that feather-degrading bacteria can affect the appearance of structurally based coloration after it is produced (Shawkey and Hill 2004). To test this idea, M. D. Shawkey (unpubl. data) inoculated the blue feathers of Eastern Bluebirds (*Sialia sialis;* Plate 32) with a keratinolytic bacterium. After only 3 days, feathers with bacteria showed an increase in brightness and chroma, but a reduction in ultraviolet (UV) reflectance compared to sterile feathers, and microscopic analysis showed that bacteria had ruptured the cortex of feathers and caused the loss of barbules, which accounted for the increased brightness and chroma. The change in structural coloration of Eastern Bluebird feathers observed in this bacteria-inoculation experiment is similar to the change in coloration across seasons seen in the blue feathers of wild Blue Tits (Ornborg et al. 2002). The importance of bacteria to the structural coloration of feathers in natural environments remains unknown, but there would seem to be a potential for bacteria and perhaps other microbes, such as fungi, to play an important role in altering structural coloration after it has been produced.

Nutritional Condition

Good nutrition is essential for virtually every function of an animal's body, and production of plumage coloration is certainly no exception. Feathers must be built from component molecules that are derived from food. Feather microstructure and hence structural coloration might be directly compromised if inadequate nutrients are ingested. Likewise melanin pigments are made from amino acid precursors, which have to be ingested, and limited intake of the amino acids needed for melanin synthesis might limit the expression of melanin coloration (Jawor and Breitwisch 2003; Chapter 6). Moreover, the enzymatic pathways by which melanins are synthesized require rare mineral elements, such as zinc, copper, and iron, and dietary access to these minerals could limit the expression of melanin coloration (McGraw 2003; Chapter 6). As described in previous sections, carotenoid pigments must be ingested and absorbed as intact molecules, which establishes an inexorable link between pigment access and carotenoid coloration. But carotenoid pigments also have to be absorbed, transported, and deposited in the integument, and, in many cases, metabolically modified. All of these physiological processes require energy, so nutritional condition, independent of access to carotenoid pigments, might also affect the way in which ingested carotenoids are utilized. A growing number of studies has tested the hypothesis that nutritional condition during molt can affect the expression of ornamental coloration.

Carotenoid Coloration

Two experiments have tested the effect of nutritional stress on the expression of carotenoid-based plumage coloration. Hill (2000) maintained four flocks of male House Finches through molt. Two of the four flocks were given ad lib food throughout the molt period; the other two flocks were food-stressed by being subjected to periods of fasting. Male House Finches that were food-stressed during molt grew less brightly colored plumage than did males that were not food stressed, regardless of whether they were fed a pigment precursor or a red pigment that could be deposited directly (Figure 12.12). McGraw et al. (2005) conducted a similar experiment with American Goldfinches and also found that males subjected to food stress during molt grew less chromatic plumage than did unstressed males. A similar effect of nutrition on carotenoid coloration has been observed in fish (Frischknecht 1993). These studies provide experimental evidence that, independent of access to carotenoids, nutritional condition can affect expression of carotenoid coloration. They also suggest that not just carotenoid metabolism but other processes of utilization (absorption, transport, and deposition) require good nutrition.

In the only study of the effects of food restriction on bare-part coloration, Ohlsson et al. (2003) raised Ring-necked Pheasants (*Phasianus colchicus*) on either high- or low-protein diets. Males raised on low-quality diets had less saturated red wattles than did males raised on a higher-quality diet. The presumption in this study was that at least some of the coloration of wattles was due to carotenoids rather than blood, but this hypothesis was not tested. Moreover, the diets differed in both nutritional and carotenoid content, but the latter was not quantified. Bortolotti et al (1996) observed a correlation between levels of blood proteins and skin coloration in American Kestrels, but there was no clear link between blood proteins and nutritional condition, so this study is hard to interpret.

Other evidence that nutritional condition can affect the expression of plumage coloration comes from correlational field studies. Hill and Montgomerie (1994) compared the growth bars of the tail feathers of male House Finches to the brightness of their carotenoid coloration. Growth bars are faint, alternating light and dark bands on feathers, and one dark plus one light bar corresponds to 24 hours of feather growth. By measuring the width of feather growth bars, one can calculate the rate at which feathers were grown during molt. It has been shown in experimental studies that birds with access to better food resources during molt grow feathers at a faster rate than do birds with

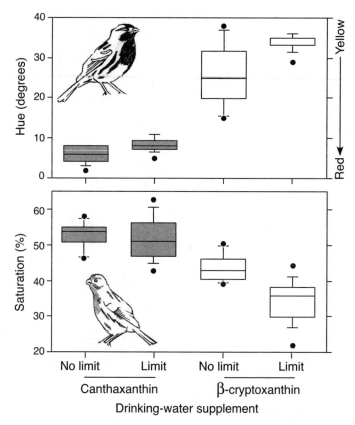

Figure 12.12. Effects of food access on the expression of carotenoid-based plumage coloration in the House Finch. The drinking water for males was supplemented with either the red pigment canthaxanthin (shaded boxes), which can be deposited unchanged into plumage, or with β-cryptoxanthin (white boxes), which is metabolically modified into the red pigment 3-hydroxy-echinenone before being deposited into plumage. In these supplementation groups, males were either subjected to periods of fasting during molt or given unlimited access to food. Regardless of carotenoid supplementation, males that were subjected to fasting grew less colorful plumage than did males with the same carotenoid access that were not subjected to fasting. See Figure 12.5 for an explanation of box plots. Redrawn from original data analyzed in Hill (2000).

access to fewer food resources (Grubb 1989, 1991; Hill and Montgomerie 1994). Male House Finches with brighter plumage had significantly wider growth bars than did males with less colorful feathers. In a study of male Great Tits, Senar et al. (2003) found that carotenoid, but not melanin, coloration was positively related to growth-bar width.

Melanin Coloration

The results of studies of the effects of nutritional condition on melanin coloration have been ambiguous. In the same studies that found a significant effect of nutrition during molt on carotenoid coloration, Hill (2000) and Senar et al. (2003) found no effect of nutritional condition during molt on melanin coloration in House Finches and Great Tits, respectively. Slagsvold and Lifjeld (1992) reported that mass at fledging, which they interpreted as nutritional condition, predicted the blackness of body plumage in subsequent years in male Pied Flycatchers (*Ficedula hypoleuca*). In an aviary study, Veiga and Puerta (1996) showed that the size of melanin ornaments of House Sparrows (*Passer domesticus*; Plate 23) was negatively related to the concentration of blood proteins. It was argued that blood proteins indicated nutritional condition, and that the negative relationship was evidence that badge size was directly affected by nutrition. None of these correlational studies, however, provides a clear test of the effect of nutrition on melanin coloration. In an experimental study, Griffith (2000) manipulated the clutch size of breeding pairs of House Sparrows, causing them to invest more in reproductive effort. After the subsequent fall molt, males that had invested more effort in nesting had smaller black badges. It was argued that energy invested in reproduction reduced the energy available for creation of the black ornament.

In an experimental test of the effect of nutrition on melanin coloration, McGraw et al. (2002) manipulated the food access of molting male Brown-headed Cowbirds (*Molothrus ater*; Plate 7) and House Sparrows. They found no effect of nutritional stress on the coloration of melanin-based brown head feathers of the cowbirds or on the size or blackness of the melanin-pigmented bib of House Sparrows (Figure 12.13), although the same food stress caused a loss of structural coloration in the cowbirds. Buchanan et al. (2001) observed no effect of food quality on the size of melanin badges in House Sparrows.

Two studies have manipulated the amino acid or protein intake of House Sparrows molting in captivity. Poston et al. (in press) manipulated the amount of phenylalanine and tyrosine (PT) in the diets of male sparrows. They found no effect of diet quality on the size of the black bibs of males, but males on the low PT diet grew less black (i.e., brighter) bibs. Similarly, Gonzalez et al. (1999) found no effect of a low-protein diet on badge size and, unlike the Poston study, also failed to find any effect on the blackness of badges.

Although many of the studies have been conducted with House Sparrows, there have now been enough studies of the effect of nutrition on melanin-

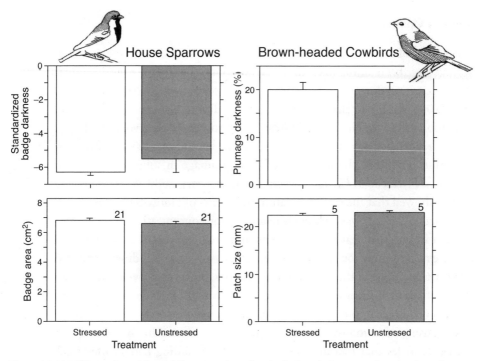

Figure 12.13. Effects of nutrition on the expression of melanin-based plumage coloration in male House Sparrows and Brown-headed Cowbirds. Males were either subjected to periods of fasting during molt ("stressed") or given unlimited access to food ("unstressed"). Food treatment had no significant effect on the size or color quality of melanin ornaments in either species (U-tests, $P > 0.15$). Sample sizes are shown above the bars on the lower graphs. Redrawn from McGraw et al. (2002).

based plumage coloration to tentatively conclude that melanin pigmentation is not very sensitive to food deprivation during molt. It remains possible that depriving birds of certain critical amino acids or minerals may affect melanin pigmentation, as was found by Poston et al. (in press), which might explain some variation in color expression in some species of birds. The size of melanin badges, as opposed to the pigmentation of badges, however, is most frequently the focus of studies of melanin coloration (as well as the target of sexual selection; Chapters 3 and 4, Volume 2), and this trait has never been convincingly demonstrated to vary with diet. From the available studies, it does seem that carotenoid-based color displays are more sensitive to food deprivation than are melanin-based displays.

Structural Coloration

It has been proposed that the nutritional condition of an individual at the time of molt might be reflected in the expression of structural coloration (Fitzpatrick 1998; Andersson 1999; Keyser and Hill 1999; but see Chapter 7). The idea is that for structural coloration to be produced, the fine structure of the feather must be constructed with nanometer-scale precision (Prum 1999; Prum et al. 1999a,b; Shawkey et al. 2003). Any nutritional deficiency would mean less precisely created structures and consequently less brilliant coloration.

This hypothesis has been tested in both correlative and experimental studies with birds. McGraw et al. (2002) conducted an experimental test with Brown-headed Cowbirds, which have iridescent green/black plumage (Plate 7). They divided captive males into two groups; one group was fed ad lib food and the other group was subjected to periods of fasting during molt. Males maintained on ad lib diets grew feathers that were more saturated than were the feathers of males subjected to periods of fasting (Figure 12.14). The treatments had no effect on the body mass of males and no effect on any other component of health or quality that could be measured. Thus even a very minor nutritional stress can cause birds to grow structural coloration with reduced color intensity.

In field tests of the hypothesis that nutritional condition affects structural plumage coloration, Keyser and Hill (1999) and Doucet (2002) looked at the width of the growth bars (which can be used as an index of nutritional condition at the time of molt, see above) in the blue/UV tail feathers of male Blue Grosbeaks (*Passerina caerulea*; Plate 7) and Blue-black Grassquits (*Volatinia jacarina*), respectively. In both studies, there was a significant positive relationship between the hue/chroma/brightness of blue/UV plumage coloration and feather growth rate (Figure 12.15).

The studies conducted to date suggest that the expression of structural coloration may generally be sensitive to nutrition during molt. We need data from more avian species—as well as more experiments on the effects of nutritional deprivation on structural coloration—before more firm conclusions can be drawn. We also need to link specific changes in microstructure to changes in macro-color. This link will lead to a better understanding of the mechanism whereby nutrition can affect structural coloration and ultimately to a better understanding of the circumstances in nature in which nutrition can be expected to play an important role in the expression of structural coloration.

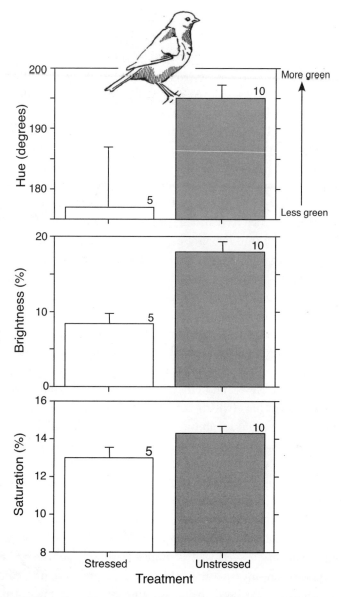

Figure 12.14. Effects of nutrition during molt on the hue, saturation, and brightness of structurally based plumage coloration in male Brown-headed Cowbirds. Males were either subjected to periods of fasting during molt ("stressed") or given unlimited access to food ("unstressed"). Birds that were subjected to fasting grew significantly more yellow-shifted (lower hue values), less saturated, and less bright plumage than did unstressed males. Sample sizes are shown above each bar. Redrawn from McGraw et al. (2002).

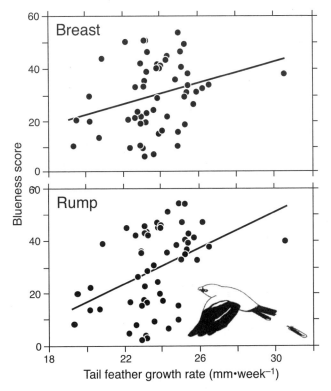

Figure 12.15. Relation between structural plumage coloration and feather growth rate in male Blue Grosbeaks. Males that grew their tail feathers at a faster rate had brighter breast and rump plumage. Feather growth rate is an indicator of nutritional condition. Redrawn from Keyser and Hill (1999).

Social Status

The evidence that avian plumage and bare-part coloration functions in status signaling is reviewed by Senar (Chapter 3, Volume 2). Here I consider the evidence that the social environment in which a bird molts might affect its coloration. Lack of pigments, infection by parasites, and poor nutrition can directly prevent a bird from fully expressing a color display. The social environment may also regulate color production and maintenance through three main pathways: (1) competition over scarce food resources that govern nutritional condition, (2) effects of social status on hormone levels, and (3) effects of group living on parasitism.

Rather than imposing a direct constraint on trait production, the social environment of a bird appears to be most commonly linked to color display via circulating hormones, particularly testosterone. Levels of circulating testosterone affect social status (Wingfield 1985; Wingfield and Moore 1987), and, in at least some species, the expression of ornamental coloration appears to be under direct control of testosterone (Owens and Short 1995; Chapter 10). (The presence or absence of estrogen is more important in other species of birds, but estrogen has been shown to affect the presence of coloration, not the degree of color elaboration). Although there have been many studies that have tested the idea that color displays function as status signals and have addressed the links between dominance, testosterone, and coloration, few studies have tested the idea that social status per se affects the production of color displays. Because so few studies have been conducted, I do not subdivide this section into studies of carotenoid, melanin, and structural coloration.

Zuk and Johnsen (2000) found that the social status of male Red Junglefowl had a large effect on circulating testosterone levels, and circulating testosterone affected the size of red combs. In this study of Red Junglefowl, however, the authors did not look specifically at comb color. In a mammalian study of another red skin ornament (that likely gets its coloration from blood flow), Setchell and Dixson (2001) showed that social status was positively related to testosterone levels and elevated testosterone levels increased the red coloration of the face and genitals of male Mandrills (*Mandrillus sphinx*). This is among the few studies directly linking social status and the expression of ornamental coloration; it would be informative to conduct similar studies linking social status, hormones, and ornaments in birds.

The effect of social status on plumage coloration has been tested in a few bird species. In Superb Fairy-wrens (*Malurus cyaneus*), the speed at which males grow brilliant, structurally based blue coloration in late winter or spring is related to male condition (Mulder and Magrath 1994). Timing of prenuptial molt, in turn, is closely tied to the level of circulating testosterone (Peters et al. 2000). Social status, however, seems to play no role in determining the speed of molt (Mulder and Magrath 1994). In this species, then, the expression of structural coloration is directly tied to hormone levels but is not affected by the social environment of the individual.

In the only experimental test of the role of social status in the expression of plumage coloration, McGraw et al. (2003a) maintained molting male House Sparrows in environments in which they were either dominant or subordinate. They then looked at the size of the melanin-based black badges produced by males. They found that males maintained in a more challenging social environment grew smaller black badges and that, within a group, subordinate males grew smaller patches than did dominant males, regardless of the appearance of the individual before it entered the experiment.

This experimental study shows that there can be a direct link between social status and production of ornamental plumage coloration, but given that only one study has been conducted, it would be premature to attempt to draw conclusions regarding the general importance of social environment on color displays. Because many studies have shown that patches of color, particularly those that are melanin-based, are used as signals of social status (Senar 1999; Chapter 3, Volume 2), it would be surprising if status did not also affect color expression, particularly in melanin-based color displays.

No studies have yet considered how social status might change the acquisition of resources needed for color display and hence the subsequent display of color. Likewise, no study has looked at how status or group living might affect the expression of plumage coloration through changes in exposure to parasites.

Turacins and Psittacofulvins

Few studies have been conducted on the environmental effects on the expression of color displays that result from pigments other than carotenoids or melanins. The brilliant red turacin pigments used by turacos (Family Musophagidae; Plate 12) to color their feathers are constructed around copper atoms (Chapter 8). Copper is rare in the turaco's diet of fruits and leaves, and Moreau (1958) speculated that it could take months for an individual to ingest enough copper to produce bright red plumage coloration. In a calendar distributed at the 2002 Annual Meeting of the American Ornithologists' Union, Sievert Rohwer speculated that only individual turacos in good condition could accrue sufficient copper and molt slowly enough to produce bright red coloration. All of these ideas remain to be tested.

In a recent study of the red belly plumage of Burrowing Parrots (*Cyanoliseus patagonus*; Plate 16, Volume 2), which are presumably pigmented with psittacofulvins—a group of pigments unique to parrots (Chapter 8), Masello and Quillfeldt (2003) found a significant relationship between individual body condition and the size of the red patch. Moreover the patch sizes of individuals were smaller in a drought year, when food supplies were limited, compared to a year with normal rainfall. In general, however, the expression of psittacofulvin pigments of parrots seems very resistant to environmental effects. There is virtually no mention in the zoo or aviculture literature of maintaining coloration in captive parrots, and even mistreated parrots suffering from malnutrition at rehabilitation centers seem fully pigmented (McGraw and Nogare 2005). The effects of environmental conditions on the expression of bright psittacofulvin coloration in parrots need more study. There have been no studies of environmental control of color displays confirmed to result from pterin pigments.

Environmental Toxins

One would expect that pollutants and toxins in the environment would stress birds and lead to the reduced expression of ornamental coloration (Hill 1995). In an experimental study, Bortolotti et al. (2003) found that exposure to PCBs caused a reduction in the brightness of the carotenoid-based yellow coloration of ceres and lores of adult male American Kestrels. McCarty and Second (2000) found a correlation between the reduced expression of structural green coloration and exposure to PCBs in female Tree Swallows (*Tachycineta bicolor*; Plate 27, Volume 2). More research on the effects of pollutants on ornamental coloration is needed.

Summary

Carotenoid-based ornamental coloration can be unequivocally described as condition-dependent. There is a direct link between the ingestion of dietary pigments and the expression of carotenoid-based plumage and integumentary coloration. Even though some authors have proposed that carotenoids are not limiting in natural environments, no field study has yet shown that carotenoids exist in excess in the diets of any population of birds in any natural environment. Several experimental studies have shown that a variety of parasites depresses the expression of carotenoid coloration, and the few studies that have looked at the effect of nutritional condition on carotenoid coloration have shown that poor nutrition also reduces carotenoid ornamentation.

The body of literature on the production of melanin coloration stands in contrast to this carotenoid literature. Melanin coloration does not seem to be condition-dependent in the sense or to the degree that carotenoid coloration is condition dependent. The primary difference between carotenoid and melanin signals is that melanin ornaments do not seem to be constrained by production costs. The same parasites that affect the expression of carotenoid coloration have no effect on melanin ornamentation. No experimental study has yet been published showing a direct effect of parasites on melanin coloration, whereas several studies have shown that parasites do not affect melanin ornamentation. In the few experimental studies reputed to show that parasites directly depress melanin ornamentation, parasitism occurs months before production of the ornament. These studies show that parasites can affect individual condition and that individual condition can then affect the expression of melanin coloration, which is fundamentally different than showing that parasites constrain the production of melanin coloration. Similarly, the same nutritional restrictions that depress the production of carotenoid and structural coloration have no effect on melanin ornamentation.

Rather than being shaped by constraints on production, melanin ornamentation seems to be shaped by the social environment of the individual. Limited evidence suggests that dominant individuals produce larger badges than do subordinate individuals, and this effect of social status does not seem to be a function of access to food or resistance to parasites. The honesty of display of melanin ornaments appears to be enforced via social mediation. Large melanin badges invoke more aggression than do small badges, so only males that can withstand the cost of aggression gain by displaying large melanin badges.

These generalities are drawn from a limited set of studies. With this theoretical framework, however, it should now be possible to design and conduct experimental studies that will either confirm this characterization of caroteniod and melanin ornaments, or that will falsify it and lead to new ideas. The ideal experimental studies must address specific and unique predictions of the various hypotheses and avoid confusing designs in which, for instance, stress is imposed on the bird at a time offset from molt. Perhaps the best studies at this stage would be more experimental studies conducted during molt with species of birds that display both melanin and carotenoid coloration.

Structural coloration remains the least studied of the three main forms of ornamental coloration, but there is currently great interest in the structural coloration of feathers, and new studies are appearing in journals each month. The studies available to date suggest that structural coloration is sensitive to the nutritional condition of the individual during molt. Studies have also been able to link specific elements of microstructure to macro-color display. What is needed now is careful assessment of specifically how nutritional deprivation affects microstructure. Similarly, one experimental test links parasite infection with loss of structural coloration, but specifically how parasites disrupt microstructure needs to be examined.

The idea that color ornaments signal good genes and specifically that expression of coloration indicates ability to resist future infection is among the most popular ideas in evolutionary biology. Despite its popularity, however, it has only been tested very recently and indirectly in experiments assessing whether carotenoid coloration predicts ability to resist or clear infection. In the few studies to date, carotenoid coloration does predict disease resistance. No comparable studies have been conducted with melanin or structural coloration. Moreover, there is at present no way to distinguish between color displays being associated with good phenotypes that promote disease resistance or good genes for disease resistance. With growing ability to genotype birds, however, a true test of the good-genes hypothesis is within grasp; such a test should be a priority for future research.

A nearly universal feature of bright and bold plumage coloration in birds is its function as a signal, conveying specific and often detailed information to a receiver. Understanding the proximate control of such ornamental plumage coloration is key to understanding what information is being conveyed. In turn, knowledge of the information content of color signals will lead to a better picture of how these traits function and by what process they evolved. Much remains to be learned about the environmental regulation of the ornamental color displays of birds, and such studies will feature prominently in expanding our understanding of why and how these colorful displays evolved.

References

Allen, P. C. 1987a. Effect of *Eimeria acervulina* infection on chick (*Gallus domesticus*) high density lipoprotein composition. Comp Biochem Physiol B 87: 313–319.

Allen, P. C. 1987b. Physiological response of chicken gut tissue to coccidial infection: Comparative effects of *Eimeria acervulina* and *Eimeria mitis* on mucosla mass, carotenoid content, and brush border enzyme activity. Poult Sci 66: 1306–1315.

Andersson, M. 1994. Sexual Selection. Princeton, NJ: Princeton University Press.

Andersson, S. 1999. Morphology of UV reflectance in a whistling-thrush: Implications for the study of structural colour signalling in birds. J Avian Biol 30: 193–204.

Beebe, C. W. 1906. The Bird, Its Form and Function. New York: Henry Holt and Company.

Bennett, G. F., J. R. Caines, and M. A. Bishop. 1988. Influence of blood parasites on the body mass of passerine birds. J Wildl Dis 24: 339–343.

Birkhead, T. R., F. Fletcher, and E. J. Pellatt. 1999. Nestling diet, secondary sexual traits and fitness in the Zebra Finch. Proc R Soc Lond B 266: 385–390.

Blanco, G., J. L. Tella, and J. Potti. 1997. Feather mites on group-living Red-billed Choughs: A non-parasitic interaction? J Avian Biol 28: 197–206.

Blanco, G., J. Seoane, and J. de la Puente. 1999. Showiness, non-parasitic symbionts, and nutritional condition in a passerine bird. Ann Zool Fenn 36: 83–91.

Bletner, J. K., R. P. Mitchell, and R. L. Tugwell. 1966. The effect of *Eimeria maxima* on broiler pigmentation. Poult Sci 45: 689–694.

Blount, J. D., D. C. Houston, and A. P. Møller. 2000. Why egg yolk is yellow. Trends Ecol Evol 15: 47–49.

Borgia, G. 1986. Satin Bowerbird parasites: A test of the bright male hypothesis. Behav Ecol Sociobiol 19: 355–358.

Borgia, G., and K. Collis. 1989. Female choice of parasite-free male Satin Bowerbirds and the evolution of bright male plumage. Behav Ecol Sociobiol 25: 445–454.

Borgia, G., and K. Collis. 1990. Parasites and bright male plumage in the Satin Bowerbird (*Ptilonorhynchus violaceus*). Am Zool 30: 279–285.

Bortolotti, G., J. J. Negro, J. L. Tella, T. A. Marchant, and D. M. Bird. 1996. Sexual dichromatism in birds independent of diet, parasites and androgens. Proc R Soc Lond B 263: 1171–1176.

Bortolotti, G. R., K. J. Fernie, and J. E. Smits. 2003. Carotenoid concentration and coloration of American Kestrels (*Falco sparverius*) disrupted by experimental exposure to PCBs. Funct Ecol 17: 651–657.

Brawner, W. R., III, G. E. Hill, and C. A. Sundermann. 2000. Effects of coccidial and mycoplasmal infections on carotenoid-based plumage pigmentation in male House Finches. Auk 117: 952–963.

Bruning, D. 1971. Use of canthaxanthin to maintain the natural colour of captive birds at Bronx Zoo. Int Zoo Yearbook 11: 215–218.

Brush, A. H. 1981. Carotenoids in wild and captive birds. In C. J. Bauernfeind, ed., Carotenoids as Colorants and Vitamin A Precursors, 539–562. New York: Academic Press.

Brush, A. H. 1990. A possible source for the rhodoxanthin in some Cedar Waxwing tails. J Field Ornithol 61: 355.

Buchanan, K. L., M. R. Evans, A. R. Goldsmith, D. M. Bryant, and L. V. Rowe. 2001. Testosterone influences basal metabolic rate in male House Sparrows: A new cost of dominance signaling? Proc R Soc Lond B 268: 1337–1344.

Burley, N., S. C. Tidemann, and K. Halupka. 1991. Bill colour and parasite levels of Zebra Finches. In J. E. Loye and M. Zuk, ed., Bird-Parasite Interactions. Oxford: Oxford University Press.

Burtt, E. H., and J. M. Ichida. 2004. Gloger's rule, feather-degrading bacteria, and color variation among song sparrows. Condor 106: 681–686.

Comben, N. 1976. Note on feeding carotenoid pigments. Int Zoo Yearbk 16: 17–20.

Craig, J. K., and C. J. Foote. 2001. Countergradient variation and secondary sexual color: Phenotypic convergence promotes genetic divergence in carotenoid use between sympatric anadromous and nonanadromous morphs of Sockeye Salmon (*Oncorhynchus nerka*). Evolution 55: 380–391.

Craves, J. A. 1999. White-throated Sparrow with orange lores. Mich Birds Nat Hist 6: 87–88.

Cuervo, J. J., and A. P. Møller. 2001. Components of phenotypic variation in avian ornamental and non-ornamental feathers. Evol Ecol 15: 53–72.

Dale, J. 2000. Ornamental plumage does not signal male quality in Red-billed Queleas. Proc R Soc Lond B 267: 2143–2149.

Dawson, R. D., and G. R. Bortolotti. Carotenoid-dependent coloration of male American Kestrels predicts ability to reduce parasitic infections. Naturwissenschaften (in press).

Doucet, S. M. 2002. Structural plumage coloration, male body size, and condition in the Blue-black Grassquit. Condor 104: 30–38.

Doucet, S. M., and R. Montgomerie. 2003. Multiple sexual ornaments in Satin Bowerbirds: UV plumage and bowers signal different aspects of male quality. Behav Ecol 14: 503–509.

Dowling, D. K., D. S. Richardson, and J. Komdeur. 2001. No effects of a feather mite on body condition, survivorship, or grooming behavior in the Seychelles Warbler, *Acrocephalus sechellensis*. Behav Ecol Sociobiol 50: 257–262.

Dufva, R. 1996. Blood parasites, health, reproductive success, and egg volume in female Great Tits *Parus major*. J Avian Biol 27: 83–87.

Dufva, R., and K. Allander. 1995. Intraspecific variation in plumage coloration reflects immune response in Great Tit (*Parus major*) males. Funct Ecol 9: 785–789.

Endler, J. A. 1980. Natural and sexual selection on color patterns in *Poecilia reticulata*. Evolution 34: 76–91.

Endler, J. A. 1983. Natural and sexual selection on color patterns in poeciliid fishes. Environ Biol Fishes 9: 173–190.

Figuerola, J., J. Domenech, and J. C. Senar. 2003. Plumage colour is related to ectosymbiont load during moult in the serin, *Serinus serinus:* An experimental study. Anim Behav 65: 551–557.

Fitze, P. S., and H. Richner. 2002. Differential effects of a parasite on ornamental structures based on melanins and carotenoids. Behav Ecol 13: 401–407.

Fitze, P. S., M. Kolliker, and H. Richner. 2003a. Effects of common origin and common environment on nestling plumage coloration in the Great Tit (*Parus major*). Evolution 57: 144–150.

Fitze, P. S., B. Tschirren, and H. Richner. 2003b. Carotenoid-based colour expression is determined early in nestling life. Oecologia 137: 148–152.

Fitzpatrick, S. 1998. Colour schemes for birds: Structural coloration and signals of quality in feathers. Ann Zool Fenn 35: 67–77.

Fox, D. L. 1962. Metabolic fractionation, storage and display of carotenoid pigments by flamingoes. Comp Biochem Physiol 6: 1–40.

Frischknecht, M. 1993. The breeding colouration of male Three-sprined Sticklebacks (*Gasterosteus aculeatus*) as an indicator of energy investment in vigour. Evol Ecol 7: 439–450.

Galeotti, P., and R. Sacchi. 2003. Differential parasitaemia in the Tawny Owl (*Strix aluco*): Effects of colour morph and habitat. J Zool (Lond) 261: 91–99.

Giacomo, R., P. Stefania, T. Ennio, V. C. Giorgina, B. Giovanni, and R. Giacomo. 1997. Mortality in Black Siskins (*Carduelis atrata*) with systemic coccidiosis. J Wildl Dis 33: 152–157.

Gloger, C. L. 1833. Das Abändern der Vögel durch Einfluss des Klimas. Breslau, Germany: August Schulz.

Goldstein, G., K. R. Flory, B. A. Browne, S. Majid, J. M. Ichida, and E. H. Burtt, Jr. 2004. Bacterial degradation of black and white feathers. Auk 121: 656–659.

Gonzalez, G., G. Sorci, A. P. Møller, P. Ninni, C. Haussy, and F. de Lope. 1999. Immunocompetence and condition-dependent sexual advertisement in male House Sparrows (*Passer domesticus*). J Anim Ecol 68: 1225–1234.

Goodwin, T. W. 1984. The Biochemistry of Carotenoids, second edition. New York: Chapman and Hall.

Grether, G. F. 2000. Carotenoid limitation and mate preference evolution: A test of the indicator hypothesis in Guppies (*Poecilia reticulata*). Evolution 54: 1721–1724.

Grether, G. F., J. Hudon, and D. F. Millie. 1999. Carotenoid limitation of sexual coloration along an environmental gradient in Guppies. Proc R Soc Lond B 266: 1317–1322.

Griffith, S. C. 2000. A trade-off between reproduction and a condition-dependent sexually selected ornament in the House Sparrow *Passer domesticus*. Proc R Soc Lond B 267: 1115–1119.

Grinnell, J. 1911. The linnet of the Hawaiian Islands: A problem in speciation. Univ Calif Publ Zool 7: 179–195.

Grubb, T. C. 1989. Ptilochronology: Feather growth bars as indicators of nutritional status. Auk 106: 314–320.

Grubb, T. C. 1991. A deficient diet narrows growth bars on induced feathers. Auk 108: 725–727.

Haase, E., S. Ito, A. Sell, and K. Wakamatsu. 1992. Melanin concentrations in feathers from wild and domestic pigeons. J Hered 83: 64–67.

Hamilton, W. D., and M. Zuk. 1982. Heritable true fitness and bright birds: A role for parasites? Science 218: 384–386.

Hardy, L. 2003. The peculiar puzzle of the pink Ring-billed Gulls. Birding 55: 498–504.

Harper, D. G. C. 1999. Feather mites, pectoral muscle condition, wing length and plumage coloration of passerines. Anim Behav 58: 553–562.

Harrison, C. J. O. 1963. Non-melanic, carotenistic, and allied variant plumages in birds. Bull Br Ornithol Club 83: 90–96.

Hatchwell, B. J., M. J. Wood, A. Anwar, D. E. Chamberlain, and C. M. Perrins. 2001. The haematozoan parasites of Common Blackbirds *Turdus merula:* Associations with host condition. Ibis 143: 420–426.

Hill, G. E. 1990. Female House Finches prefer colourful males: Sexual selection for a condition-dependent trait. Anim Behav 40: 563–572.

Hill, G. E. 1992. Proximate basis of variation in carotenoid pigmentation in male House Finches. Auk 109: 1–12.

Hill, G. E. 1993. Geographic variation in the carotenoid plumage pigmentation of male House Finches (*Carpodacus mexicanus*). Biol J Linn Soc 49: 63–86.

Hill, G. E. 1994a. House Finches are what they eat: A reply to Hudon. Auk 111: 221–225.

Hill, G. E. 1994b. Trait elaboration via adaptive mate choice: Sexual conflict in the evolution of signals of male quality. Ethol Ecol Evol 6: 351–370.

Hill, G. E. 1995. Ornamental traits as indicators of environmental health. BioScience 45: 25–31.

Hill, G. E. 2000. Energetic constraints on expression of carotenoid-based plumage coloration. J Avian Biol 31: 559–566.

Hill, G. E. 2002. A Red Bird in a Brown Bag: The Function and Evolution of Ornamental Plumage Coloration in the House Finch. New York: Oxford University Press.

Hill, G. E., and W. B. Brawner, III. 1998. Melanin-based plumage coloration in the House Finch is unaffected by coccidial infection. Proc R Soc Lond B 265: 1105–1109.

Hill, G. E., and K. L. Farmer. 2005. Carotenoid-based plumage coloration predicts resistance to a novel parasite in the House Finch. Naturwissenschaften 92: 30–34.

Hill, G. E., and R. Montgomerie. 1994. Plumage colour signals nutritional condition in the House Finch. Proc R Soc Lond B 258: 47–52.

Hill, G. E., C. Y. Inouye, and R. Montgomerie. 2002. Dietary carotenoids predict plumage coloration in wild House Finches. Proc R Soc Lond B 269: 1119–1124.

Hill, G. E., K. L. Farmer, and M. L. Beck. 2004. The effect of mycoplasmosis on carotenoid plumage coloration in male House Finches. J Exp Biol 207: 2095–2099.

Hill, G. E., S. M. Doucet, and R. Buchholz. 2005. The effect of coccidial infection on iridescent plumage coloration in Wild Turkeys. Anim Behav 69: 387–394.

Hõrak, P., I. Ots, H. Vellau, C. Spottiswoode, and A. P. Møller. 2001. Carotenoid-based plumage coloration reflects hemoparasite infection and local survival in breeding Great Tits. Oecologia 126: 166–173.

Hõrak, P., L. Saks, U. Karu, I. Ots, P. F. Surai, and K. J. McGraw. 2004. How coccidian parasites affect health and appearance of greenfinches. J. Anim Ecol 73: 935–947.

Houde, A. E., and A. J. Torio. 1992. Effect of parasitic infection on male color pattern and female choice in Guppies. Behav Ecol 3: 346–351.

Hudon, J. 1991. Unusual carotenoid use by the Western Tanager (*Piranga ludoviciana*) and its evolutionary implications. Can J Zool 69: 2311–2320.

Hudon, J. 1994. Showiness, carotenoids, and captivity: A comment on Hill (1992). Auk 111: 218–221.

Hudon, J., and A. H. Brush. 1989. Probable dietary basis of a color variant of the Cedar Waxwing. J Field Ornithol 60: 361–368.

Hudon, J., H. Ouellet, E. Benito-Espinal, and A. H. Brush. 1996. Characterization of an orange variant of the Bananaquit (*Coereba flaveola*) on La Desirade, Guadeloupe, French West Indies. Auk 113: 715–718.

Inouye, C. Y., G. E. Hill, R. Montgomerie, and R. D. Stradi. 2001. Carotenoid pigments in male House Finch plumage in relation to age, subspecies, and ornamental coloration. Auk 118: 900–915.

Jawor, J. M., and R. Breitwisch. 2003. Melanin ornaments, honesty, and sexual selection. Auk 120: 249–265.

Keeler, C. A. 1893. Evolution of the Colors of North American Land Birds. San Francisco: California Academy of Sciences.

Keyser, A., and G. E. Hill. 1999. Condition-dependent variation in the blue-ultraviolet coloration of a structurally based plumage ornament. Proc R Soc Lond B 266: 771–778.

Kodric-Brown, A. 1989. Dietary carotenoids and male mating success in the Guppy: An environmental component to female choice. Behav Ecol Sociobiol 25: 393–401.

Korpimaki, E., P. Tolonen, and G. F. Bennet. 1995. Blood parasites, sexual selection and reproductive success of European Kestrels. Ecoscience 2: 335–343.

Koutsos, E. A., A. J. Clifford, C. C. Calvert, and K. C. Klasing. 2003. Maternal carotenoid status modifies incorporation of dietary carotenoids into immune tissues of growing chickens (*Gallus gallus domesticus*). J Nutr 133: 1132–1138.

Kritzler, H. 1943. Carotenoids in the display and eclipse plumage of Bishop Birds. Physiol Zool 16: 241–245.

Krogius, F. V. 1981. The role of resident fish in the reproduction of anadromous Sockeye Salmon, *Oncorhynchus nerka*. J Ichthyol 21: 14–21.

Lindstrom, K., and J. Lundstrom. 2000. Male greenfinches (*Carduelis chloris*) with brighter ornaments have higher virus infection clearance rate. Behav Ecol Sociobiol 48: 44–51.

Linville, S. U., and R. Breitwisch. 1997. Carotenoid availability and plumage coloration in a wild population of Northern Cardinals. Auk 114: 796–800.

Marusich, W. L., E. Schildknecht, E. F. Ogrinz, P. R. Brown, and M. Mitrovic. 1972. Effect of coccidiosis on pigmentation in broilers. Br Poult Sci 13: 577–585.

Masello, J. F., and P. Quillfeldt. 2003. Body size, body condition and ornamental feathers of Burrowing Parrots: variation between years and sexes, assortative mating and influences on breeding success. Emu 103: 149–161.

McCarty, J. P., and A. L. Second. 2000. Possible effects of PCB contamination on female plumage color and reproductive success in Hudson River tree swallows. Auk 117: 987–995.

McGraw, K. J. 2003. Melanins, metals, and mate quality. Oikos 102: 402–406.

McGraw, K. J., and G. E. Hill. 2000. Differential effects of endoparasitism on the expression of carotenoid- and melanin-based ornamental coloration. Proc R Soc Lond B 267: 1525–1531.

McGraw, K. J., and G. E. Hill. 2001. Carotenoid access and intraspecific variation in plumage pigmentation in male American Goldfinches (*Carduelis tristis*) and Northern Cardinals (*Cardinalis cardinalis*). Funct Ecol 15: 732–739.

McGraw, K. J., and M. C. Nogare. 2005. Distribution of unique red feather pigments in parrots. Biol Lett 1: 38–43.

McGraw, K. J., and K. Wakamatsu. 2004. Melanin basis of ornamental feather colors in male Zebra Finches. Condor 106: 686–690.

McGraw, K. J., G. E. Hill, R. Stradi, and R. S. Parker. 2001. The influence of carotenoid acquisition and utilization on the maintenance of species-typical plumage pigmentation in male American Goldfinches (*Carduelis tristis*) and Northern Cardinals (*Cardinalis cardinalis*). Physiol Biochem Zool 74: 843–852.

McGraw, K. J., E. A. Mackillop, J. Dale, and M. E. Hauber. 2002. Different colors reveal different information: How nutritional stress affects the expression of melanin- and structurally based ornamental plumage. J Exp Biol 205: 3747–3755.

McGraw, K. J., J. Dale, and E. A. Mackillop. 2003a. Social environment during molt and the expression of melanin-based plumage pigmentation in male House Sparrows (*Passer domesticus*). Behav Ecol Sociobiol 53: 116–122.

McGraw, K. J., A. J. Gregory, R. S. Parker, and E. Adkins-Regan. 2003b. Diet, plasma carotenoids, and sexual coloration in the Zebra Finch (*Taeniopygia guttata*). Auk 120: 400–410.

McGraw, K. J., G. E. Hill, and R. S. Parker. 2005. The physiological costs of being colorful: Nutritional control of carotenoid utilization in the American Goldfinch (*Carduelis tristis*). Anim Behav 69: 653–660.

McGraw, K. J., E. Adkins-Regan, and R. S. Parker. Maternally derived carotenoid pigments affect offspring survival, sex ratio, and sexual attractiveness in a colorful songbird. Naturwissenschaften (in press).

Merilä, J., B. C. Sheldon, and K. Lindstrom. 1999. Plumage brightness in relation to haematozoan infections in the greenfinch *Carduelis chloris:* Bright males are a good bet. Ecoscience 6: 12–18.

Merino, S., J. Moreno, J. J. Sanz, and E. Arriero. 2000. Are avian blood parasites pathogenic in the wild? A medication experiment in Blue Tits (*Parus caeruleus*). Proc R Soc Lond B 267: 2507–2510.

Milinski, M., and T. C. M. Bakker. 1990. Female sticklebacks use male coloration in mate choice and hence avoid parasitized males. Nature 344: 330–333.

Møller, A. P. 1990. Effects of a haematophagous mite in the Barn Swallow (*Hirundo rustica*): A test of the Hamilton-Zuk hypothesis. Evolution 44: 771–784.

Moreau, R. E. 1958. Some aspects of the Musophagidae. Part 3. Ibis 100: 238–270.

Mulder, R. A., and M. J. L. Magrath. 1994. Timing of prenuptial molt as a sexually selected indicator of male quality in Superb Fairy-wrens (*Malurus cyaneus*). Behav Ecol 5: 393–400.

Mulvihill, R. S., K. C. Parkes, R. C. Leberman, and D. S. Wood. 1992. Evidence supporting a dietary basis for orange-tipped rectrices in the Cedar Waxwing. J Field Ornithol 63: 212–216.

Negro, J. J., and J. Garrido-Fernandez. 2000. Astaxanthin is the major carotenoid in tissues of White Storks (*Ciconia ciconia*) feeding on introduced crayfish (*Procambarus clarkii*). Comp Biochem Physiol B 126: 347–352.

Negro, J. J., J. L. Tella, G. Blanco, M. G. Forero, and J. Garrido-Fernandez. 2000. Diet explains interpopulation variation of plasma carotenoids and skin pigmentation in nestling White Storks. Physiol Biochem Zool 73: 97–101.

Negro, J. J., J. M. Grande, J. L. Tella, J. Garrido, D. Hornero, et al. 2002. An unusual source of essential carotenoids. Nature 416: 807.

Nieboer, E. 1965. Canthaxanthin for Scarlet Ibises *Eudocimus ruber* at Amsterdam Zoo. Int Zoo Yearbook 5: 164–165.

Nilsson, N. A., and G. Andersson. 1967. Food and growth of an allopatric brown trout in northern Sweden. Inst Freshwater Res Reprod 47: 120–125.

O'Conner, B. M. 1992. Evolutionary ecology of astigmatid mites. Annu Rev Entomol 27: 385–409.

Ohlsson, T., H. G. Smith, L. Raberg, and D. Hasselquist. 2003. Effects of nutrition on sexual ornaments and humoral immune responsiveness in adult male pheasants. Ethol Ecol Evol 15: 31–42.

Ornborg, J., S. Andersson, S. Griffith, and B. Sheldon. 2002. Seasonal changes in an ultraviolet structural colour signal in Blue Tits, *Parus caeruleus*. Biol J Linn Soc 76: 237–245.

Owens, I. P. F., and R. V. Short. 1995. Hormonal basis of sexual dimorphism in birds: Implications for new theories of sexual selection. Trends Ecol Evol 10: 44–47.

Partali, V., S. Liaaen-Jensen, T. Slagsvold, and J. T. Lifjeld. 1987. Carotenoids in food chain studies—II. The food chain of *Parus* spp. monitored by carotenoid analyses. Comp Biochem Physiol B 87: 885–888.

Pearson, W. E. 1968. The Nutrition of Fish. Basle, Switzerland: Hoffmann-LaRoche.

Peters, A., L. B. Astheimer, C. R. J. Boland, and A. Cockburn. 2000. Testosterone is involved in acquisition and maintenance of sexually selected male plumage in Superb Fairy-wrens, *Malurus cyaneus*. Behav Ecol Sociobiol 47: 438–445.

Piersma, T., L. Mendes, J. Hennekens, S. Ratiarison, S. Groenewold, and J. Jukema. 2001. Breeding plumage honestly signals likelihood of tapeworm infestation in females of a long-distance migrating shorebird, the Bar-tailed Godwit. Ann Zool Fenn 104: 41–48.

Poston, J. P., D. Hasselquist, I. R. K. Stewart, and D. F. Westneat. Dietary amino acids influence plumage traits and immune responses of male House Sparrows (*Passer domesticus*), but not as expected. Anim Behav (in press).

Potti, J., and S. Merino. 1996. Decreased levels of blood trypanosome infection correlate with female expression of a male secondary sexual trait: Implications for sexual selection. Proc R Soc Lond B 263: 1199–1204.

Prum, R. O. 1999. The anatomy and physics of avian structural colours. In N. Adams and R. Slotow, ed., Proceedings of the 22nd International Ornithological Congress, S29.1: 1633–1653. Durban, South Africa: Bird Life South Africa.

Prum, R. O., R. Torres, S. Williamson, and J. Dyck. 1999a. Two-dimensional Fourier analysis of the spongy medullary keratin of structurally coloured feather barbs. Proc R Soc Lond B 266: 13–22.

Prum, R. O., R. H. Torres, S. Williamson, and J. Dyck. 1999b. Coherent light scattering by blue feather barbs. Nature 396: 28–29.

Ratti, O., R. Dufva, and R. V. Alatalo. 1993. Blood parasites and male fitness in the Pied Flycatcher. Oecologia 96: 410–412.

Reimchen, T. E. 1989. Loss of nutpial color in the Three-spine Stickleback (*Gasterosteus aculeatus*). Evolution 43: 450–460.

Roulin, A., H. Richner, and A. L. Ducrest. 1998. Genetic, environmental, and condition-dependent effects on female and male ornamentation in the Barn Owl *Tyto alba*. Evolution 52: 1451–1460.

Roulin, A., C. Riols, C. Dijkstra, and A. L. Ducrest. 2001. Female plumage spottiness signals parasite resistance in the Barn Owl (*Tyto alba*). Behav Ecol 12: 103–110.

Saino, N., S. Calza, R. Martinelli, F. De Bernardi, P. Ninni, and A. P. Møller. 2000. Better red than dead: Carotenoid-based mouth coloration reveals infection in Barn Swallow nestlings. Proc R Soc Lond B 267: 57–61.

Saks, L., I. Ots, and P. Hõrak. 2003. Carotenoid-based plumage coloration of male greenfinches reflects health and immunocompetence. Oecologia 134: 301–307.

Senar, J. C. 1999. Plumage coloration as a signal of social status. In N. Adams and R. Slotow, ed., Proceedings of the 22nd International Ornithological Congress, S29.1: 1669–1686. Durban, South Africa: Bird Life South Africa.

Senar, J. C., J. Figuerola, and J. Domenech. 2003. Plumage coloration and nutritional condition in the Great Tit *Parus major:* The roles of carotenoids and melanins differ. Naturwissenschaften 90: 234–237.

Serle, W. 1964. A further note on aberrant Yellow Wagtail *Motacilla flava flava*. Bull Br Ornithol Club 84: 40.

Setchell, J. M., and A. F. Dixson. 2001. Changes in the secondary sexual adornments of male Mandrills (*Mandrillus sphinx*) are associated with gain and loss of alpha status. Hormones Behav 39: 177–184.

Seutin, G. 1994. Plumage redness in Redpoll Finches does not reflect hemoparasitic infection. Oikos 70: 280–286.

Shawkey, M. D., and G. E. Hill. 2004. Feathers at a fine scale. Auk 121: 652–655.

Shawkey, M. D., A. M. Estes, L. M. Siefferman, and G. E. Hill. 2003. Nanostructure predicts intraspecific variation in ultraviolet-blue plumage colours. Proc R Soc Lond B 270: 1455–1460.

Siefferman, L., and G. Hill. The signal content and function of ornamental plumage coloration in female Eastern Bluebirds. Evolution (in press).

Slagsvold, T. and J. T. Lifjeld. 1985. Variation in plumage colour of the Great Tit *Parus major* in relation to habitat, season and food. J Zool A 206: 321–328.

Slagsvold, T., and J. T. Lifjeld. 1992. Plumage color is a condition-dependent sexual trait in male Pied Flycatchers. Evolution 46: 825–828.

Steven, D. M. 1947. Carotenoid pigmentation of trout. Nature 160: 540–541.

Stradi, R. 1998. The colour of flight: Carotenoids in bird plumage. Milan: University of Milan Press.

Stradi, R., E. Pini, and G. Celentano. 2001. Carotenoids in bird plumage: The complement of red pigments in the plumage of wild and captive Bullfinch (*Pyrrhula pyrrhula*). Comp Biochem Physiol B 128: 529–535.

Sundberg, J. 1995. Parasites, plumage coloration and reproductive success in the Yellowhammer, *Emberiza citrinella*. Oikos 74: 331–339.

Test, F. H. 1969. Relation of wing and tail color of the woodpeckers *Colaptes auratus* and *C. cafer* to their food. Condor 71: 206–211.

Thompson, C. W., N. Hillgarth, M. Leu, and H. E. McClure. 1997. High parasite load in House Finches (*Carpodacus mexicanus*) is correlated with reduced expression of a sexually selected trait. Am Nat 149: 270–294.

Tschirren, B., P. S. Fitze, and H. Richner. 2003. Proximate mechanisms of variation in the carotenoid-based plumage coloration of nestling Great Tits (*Parus major* L.). J Evol Biol 16: 91–100.

van Riper, C., S. G. van Riper, and W. R. Hansen. 2002. Epizootiology and effect of avian pox on Hawaiian forest birds. Auk 119: 929–942.

Veiga, J. P., and M. Puerta. 1996. Nutritional constraints determine the expression of a sexual trait in the House Sparrow, *Passer domesticus*. Proc R Soc Lond B 263: 229–234.

Villafuerte, R., and J. J. Negro. 1998. Digital imaging for colour measurements in ecological research. Ecol Lett 1: 151–154.

Völker, O. 1934. Die Abhängigkeit der Lipochrombildung bei Vögeln von pflanzlichen Carotinoiden. J Ornithol 82: 439.

Völker, O. 1938. The dependence of lipochrome-formation in birds on plant carotenoids. Proc 8th Int Ornithol Congr 1934: 425–426.

Washington, D., and C. J. O. Harrison. 1969. Abnormal reddish plumage due to "colour feeding" in wild greenfinches. Bird Study 16: 111–114.

Weatherhead, P. J. 1990. Secondary sexual traits, parasites, and polygyny in Red-winged Blackbirds. Behav Ecol 1: 125–130.

Weatherhead, P. J., and G. F. Bennett. 1992. Ecology of parasitism of Brown-headed Cowbirds by haematozoa. Can J Zool 70: 1–7.

Weatherhead, P. J., K. J. Metz, G. F. Bennett, and R. E. Irwin. 1993. Parasite faunas, testosterone and secondary sexual traits in male Red-winged Blackbirds. Behav Ecol Sociobiol 33: 13–23.

Weber, H. 1961. Über die Ursache des Verlustes der roten Federfarbe bei gekäfigten Birkenzeisigen. J Ornithol 102: 158–163.

Wedekind, C., P. Meyer, M. Frischknecht, U. A. Niggli, and H. Pfander. 1998. Different carotenoids and potential information content of red coloration of male Three-spined Stickleback. J Chem Ecol 24: 787–801.

Wingfield, J. C. 1985. Short term changes in plasma levels of hormones during establishment and defense of a breeding territory in male Song Sparrows, *Melospiza melodia*. Hormones Behav 19: 174–187.

Wingfield, J. C., and M. C. Moore. 1987. Hormonal, social, and environmental factors in the reproductive biology of free-living male birds. In D. Crews, ed., Psychobiology of Reproductive Behaviour: An Evolutionary Perspective, 149–175. Englewood Cliffs, NJ: Prentice-Hall.

Witmer, M. C. 1996. Consequences of an alien shrub on the plumage coloration and ecology of Cedar Waxwings. Auk 113: 735–743.

Zahavi, A. 1977. The cost of honesty (further remarks on the handicap principle). J Theor Biol 67: 603–605.

Zahn, S. N., and S. I. Rothstein. 1999. Recent increase in male House Finch plumage variation and its possible relationship to avian pox disease. Auk 116: 35–44.

Zink, R. M., and J. V. J. Remsen. 1986. Evolutionary process and patterns of geographical variation in birds. Curr Ornithol 4: 1–69.

Zuk, M., and T. S. Johnsen. 2000. Social environment and immunity in male Red Junglefowl. Behav Ecol 11: 146–153.

Zuk, M., K. Johnson, R. Thornhill, and J. D. Ligon. 1990. Parasites and male ornaments in free-ranging and captive Red Junglefowl. Behaviour 114: 232–248.

Acknowledgments

For his tremendous help and support in all phases of this project, we give special thanks to Bob Montgomerie. More than anyone else, Bob helped us scale back our preliminary ideas for a review of the colors of all vertebrates to a more focused book on birds. Bob also took charge of the enormous task of redrawing every black and white illustration in this book and overseeing the legends for all figures. John Endler, Alan Brush, Liz Adkins-Regan, and Tim Goldsmith also provided advice or helped review chapters. Two anonymous referees commented on an earlier version of the book. Noah Strycker drew the black and white vignettes that make the graphs and charts of the book much more attractive. Kristen Scott and Meghan Gooldchild helped with the arduous task of converting the authors' original drawings and photocopies into the graphs and drawings for the book; Jason Clarke assisted with revising those figures at the copyediting stage. Members of the Hill lab gave sound advice on the topics to be included in the book. During book preparation, Geoffrey Hill was supported by National Science Foundation grants DEB0077804 and IBN0235778, and Kevin McGraw was supported by the College of Liberal Arts and Sciences and the School of Life Sciences at Arizona State University, Tempe.

Chapter 1

Research at Bristol on vision in birds has been supported by grants from the the National Engineering Research Council, the Biotechnology and Biological Sciences Research Council, the Royal Society, and the Nuffield Foundation to

Julian Partridge, Andy Bennett, and Innes Cuthill. My thanks to Andy, Julian, and all members of the Bristol vision group over the years, notably Sarah Hunt, Nathan Hart, Stuart Church, Sam Maddocks, Sophie Pearn, Verity Greenwood, Emma Smith, and Jennie Evans. Particularly appreciated are Daniel Osorio, Tom Troscianko, and Alejandro Párraga for patiently explaining numerous aspects of color perception to an ignorant behavioral ecologist; Olle Håstad for access to unpublished manuscripts; and John Endler, Kevin McGraw, Tim Goldsmith, and an anonymous referee for comments on the manuscript.

Chapter 2

Many thanks to Geoff Hill for patiently cheering us through this project, to Bob Montgomerie and Jonas Örnborg for excellent comments, and to Leeward Bean at OceanOptics for help with technical questions. Thanks also to Mike Lawes for hosting us at the University of KwaZulu-Natal, South Africa, where most of this project was carried out, and to students back in Göteborg, who were indirectly affected by our absence. Financial support was provided by a Swedish Foundation for International Cooperation in Research and Higher Education (STINT) Swedish–South African exchange program and the Swedish Science Council.

Chapter 3

Thanks to Kristen Scott for searching both the library and the web for relevant material, to Steve Lougheed and Chris Eckert for statistical advice, to Nicole Media for mathematical advice, and to Kevin McGraw, John Endler, and Alexis Chaine for superb, detailed reviews.

Chapter 5

Alan Brush, Geoff Hill, and Paul Nolan provided valuable comments on the manuscript. Funding during manuscript preparation was provided by a U.S. Department of Agriculture grant to Kirk Klasing and by the College of Liberal Arts and Sciences and School of Life Sciences at Arizona State University, Tempe.

Chapter 6

Alan Brush and Geoff Hill provided insightful comments on this chapter. Funding during manuscript preparation was provided by a U.S. Department of Agriculture grant to Kirk Klasing and by the College of Liberal Arts and Sciences and School of Life Sciences at Arizona State University, Tempe.

Chapter 7

Tim Quinn produced many of the excellent electron micrographs used in this research. Rodolfo Torres, Scott Williamson, and Chris Fallen collaborated in the development of the Fourier Tool for Biological Nano-Optics. Funding was provided by a grant from the National Science Foundation (DBI-0078376). Ann Johnson Prum has provided love, support, and patience during endless forays into "Wattle World."

Chapter 8

I am grateful to Geoffrey Hill for his inspirational, field-leading work as a behavioral ecologist, academic advisor, collaborator, and co-editor. Alan Brush also provided valuable comments on the chapter.

Chapter 9

Thanks to Melanie Rathburn for measuring museum specimens; Kristen Scott for analyzing reflectance data; David Hussell for providing winter feathers from Lapland Longspurs; Troy Murphy and Jeroen Reneerkens for information about motmots and preen waxes, respectively; Kevin McGraw and Marc Théry for excellent reviews of an early draft of the chapter; and the Natural Sciences and Engineering Research Council for funding my research on plumage coloration.

Chapter 10

Edward Braun, Brandon Moore, an anonymous reviewer, and particularly Kevin McGraw provided many helpful suggestions that greatly improved this chapter. In addition, I thank David Ligon for stimulating my interest in this topic.

Chapter 11

I thank Ben Sheldon for asistance with the section on quantitative genetics and Kevin McGraw and an anonymous reviewer for helpful comments on the manuscript. My work on the genetics of plumage coloration has been supported by the National Engineering Research Council and the Biotechnology and Biological Sciences Research Council.

Chapter 12

Bob Montgomerie, Gary Bortolotti, Kevin McGraw, and members of the Hill lab read and provided comments on this chapter. While this chapter was being written, I was supported by National Science Foundation grants DEB007804, IBN0235778, and IBN9722971.

Contributors

STAFFAN ANDERSSON
Department of Zoology
Göteborg University
Box 463
SE-405 30 Göteborg
Sweden
staffan.andersson@zool.gu.se

INNES C. CUTHILL
School of Biological Sciences
University of Bristol
Woodland Road
Bristol, BS8 1UG
United Kingdom
I.Cuthill@bristol.ac.uk

GEOFFREY E. HILL
Department of Biological Sciences
331 Funchess Hall
Auburn University
Auburn, Alabama 36849
USA
ghill@acesag.auburn.edu

REBECCA T. KIMBALL
Department of Zoology
University of Florida
Gainesville, Florida 32611
USA
rkimball@zoo.ufl.edu

KEVIN J. MCGRAW
School of Life Sciences
Arizona State University
Tempe, Arizona 85287-4501
USA
Kevin.McGraw@asu.edu

ROBERT MONTGOMERIE
Department of Biology
Queen's University
Kingston, Ontario K7L 3N6
Canada
montgome@biology.queensu.ca

NICHOLAS I. MUNDY
Department of Zoology
University of Cambridge
Downing Street
Cambridge CB2 3EJ
United Kingdom
nim21@cam.ac.uk

MARIA PRAGER
Department of Zoology
Göteborg University
Box 463
SE-405 30 Göteborg
Sweden
maria.prager@zool.gu.se

Contributors

RICHARD O. PRUM
Department of Ecology and Evolutionary Biology, and
Peabody Museum of Natural History
Yale University
New Haven, Connecticut 06520-8105
USA
richard.prum@yale.edu

MARC THÉRY
Muséum National d'Histoire Naturelle
Centre National de Recherche Scientifique
F-91800 Brunoy
France
thery@mnhn.fr

Species Index

Abbott's Starling, 299
Acipenser oxyrhinchus. See Atlantic Sturgeon
Acridotheres cristatellus. See Crested Myna
Acryllium vulturinum. See Vulturine Guineafowl
Aegithalos caudatus. See Long-tailed Tit
Aegithina tiphia. See Common Iora
Aegolius funereus. See Tengmalm's Owl
African Emerald Cuckoo, 297, 328
African Grey Parrot, 21
African Masked-weaver, 186
African Oriole, 212
African Pied Starling, 300, 377
Afropavo congensis. See Congo Peafowl
Agapornis roseicollis. See Peach-faced Lovebird
Agelaius phoeniceus. See Red-winged Blackbird
Agriocharis ocellata. See Ocellated Turkey
Aix galericulata. See Mandarin Duck
Aix sponsa. See Wood Duck
Ajaja ajaja. See Roseate Spoonbill
Alca torda. See Razorbill
Alectoris rufa. See Red-legged Partridge
Amandava amandava. See Red Avadavat
Amandava subflava. See Zebra Waxbill
Amazilia cyanura. See Blue-tailed Hummingbird
Amblyornis inornatus. See Vogelkop Bowerbird
Ambystoma maculatum. See Spotted Salamander
Ambystoma mexicanum. See Mexican Axolotl
American Coot, 183, 203–205, 212, 213
American Crow, 276
American Goldfinch, 109, 113, 115, 116, 183, 188, 197–201, 209, 213, 216, 218–219, 221, 227, 252, 270, 450–452, 509–511, 515, 525–526, 532, 539
American Kestrel, 106, 194, 201, 203, 511, 527, 539, 548
American Redstart, 110, 189, 218

American Robin, 245, 354, 368
Amethyst-throated Sunangel, 297
Ampeliceps coronatus. See Golden-crested Myna
Anas acuta. See Northern Pintail
Anas brasiliensis. See Brazilian Teal
Anas crecca. See Green-winged Teal
Anas discors. See Blue-winged Teal
Anas gibberifrons. See Grey Teal
Anas platyrhynchos. See Mallard
Anas specularis. See Spectacled Duck
Andean Cock-of-the-Rock, 159
Anser anser. See Greylag Goose
Anser caerulescens. See Lesser Snow Goose
Anser rossi. See Ross's Goose
Apaloderma equatoriale. See Bare-cheeked Trogon
Apaloderma narina. See Narina Trogon
Apis mellifera. See Honeybee
Aplonis panayensis. See Asian Glossy Starling
Apple Snail, 214
Aptenodytes forsterii. See Emperor Penguin
Aptenodytes patagonicus. See King Penguin
Aquila chrysaetos. See Golden Eagle
Ara ararauna. See Blue-and-gold Macaw
Ara macao. See Scarlet Macaw
Archilochus alexandri. See Black-chinned Hummingbird
Arctic Skua, 268, 474, 480–482, 495
Ardea herodias. See Great Blue Heron
Ashy Starling, 299
Asian Fairy Bluebird, 299, 331
Asian Glossy Starling, 299
Astrapia rothschildi. See Huon Astrapia
Astrapia splendissima. See Splendid Astrapia
Atlantic Sturgeon, 377
Aythya affinis. See Lesser Scaup

Babbling Starling, 300
Bald Eagle, 374
Bald Starling, 300
Bali Starling, 300
Bananaquit, 186, 268, 269, 472, 474, 480–482, 516
Band-tailed Pigeon, 296
Bannerman's Turaco, 364
Bare-cheeked Trogon, 297
Bare-faced Go-away Bird, 364
Bare-throated Bellbird, 298, 336, 338, 341
Barn Owl, 259–260, 267, 272, 486, 490–491, 534–535
Barn Swallow, 117, 192, 196, 223, 252, 270, 362–363, 523
Bar-tailed Godwit, 533
Basilornis celebensis. See Sulawesi Myna
Baya Weaver, 186, 265, 437
Bearded Vulture, 400–401, 415, 422
Beautiful Rosefinch, 189
Beavan's Bullfinch, 187
Bicolored Antbird, 298, 336
Black Francolin, 296
Black Grouse, 296
Black Guan, 296
Black Lory, 297
Black Partridge, 296
Black Sicklebill, 298
Black Siskin, 188
Black-and-yellow Grosbeak, 187
Black-backed Fruit Dove, 296
Black-bellied Bustard, 368
Black-bellied Glossy Starling, 299
Black-billed Magpie, 299
Black-billed Turaco, 364
Blackcap, 267, 474
Black-capped Chickadee, 115, 120
Black-capped Kingfisher, 331
Black-chinned Hummingbird, 257
Black-headed Bulbul, 299
Black-headed Bunting, 189
Black-headed Gull, 262, 437, 442, 450
Black-hooded Oriole, 185
Blackpoll Warbler, 444
Black-throated Trogon, 297
Blood Pheasant, 365
Blue Coua, 297
Blue Grosbeak, 345, 413, 543, 545
Blue Manakin, 298
Blue Tit, 25–26, 72, 75, 116, 120, 130, 138, 185, 416, 420, 422, 488, 536, 538
Blue Whistling-thrush, 300, 332
Blue-and-gold Macaw, 297
Blue-backed Manakin, 158

Blue-black Grassquit, 345, 543
Blue-chinned Sapphire, 297
Blue-crowned Manakin, 163, 298, 319, 330
Blue-crowned Motmot, 410
Blue-faced Honeyeater, 335
Blue-footed Booby, 296, 335
Blue-necked Tanager, 300
Blue-tailed Hummingbird, 297
Blue-tailed Imperial Pigeon, 296
Bluethroat, 113
Blue-throated Hummingbird, 297
Blue-winged Teal, 436, 451
Bobolink, 450
Bohemian Waxwing, 185
Bombycilla cedrorum. See Cedar Waxwing
Bombycilla garrulus. See Bohemian Waxwing
Bombycilla japonica. See Japanese Waxwing
Bostrychia hagedash. See Hadeda Ibis
Brambling, 187
Brazilian Ruby, 297
Brazilian Teal, 296
Brewer's Blackbird, 374, 377, 437
Bristle-crowned Starling, 300
Bronzed Cowbird, 363–365
Bronze-tailed Glossy Starling, 299
Brown-headed Cowbird, 269, 345–346, 541–542, 544
Bubo virginianus. See Great Horned Owl
Buceros bicornis. See Great Hornbill
Budgerigar, 21, 297, 341, 450, 470, 496
Bufo bufo japonicus. See Japanese Common Toad
Bulwer's Pheasant, 296
Buphagus erythrorhynchus. See Red-billed Oxpecker
Burchell's Glossy Starling, 299
Burrowing Parrot, 382–383, 547
Buteo buteo. See Common Buzzard
Buteo regalis. See Ferruginous Hawk
Butorides striatus. See Striated Heron

Cabot's Tragopan, 296, 301, 336, 341
Calcarius lapponicus. See Lapland Longspur
Calidris alba. See Sanderling
Calidris canutus. See Red Knot
California Quail, 439
Callipepla californica. See California Quail
Callipepla gambelii. See Gambel's Quail
Callipepla squamata. See Scaled Quail
Caloenas nicobarica. See Nicobar Pigeon
Caloperdix oculea. See Ferruginous Partridge
Calyptomena viridis. See Green Broadbill
Campephilus leucopogon. See Cream-backed Woodpecker
Campo Flicker, 185

Species Index

Cape Glossy Starling, 299
Cape Weaver, 186
Capercaillie, 184, 362
Capped Heron, 296
Capuchinbird, 298
Carassius auratus. See Goldfish
Cardinal Quelea, 186
Cardinalis cardinalis. See Northern Cardinal
Carduelis atrata. See Black Siskin
Carduelis cannabina. See Linnet
Carduelis carduelis. See Goldfinch
Carduelis chloris. See Eurasian Greenfinch
Carduelis cucullata. See Red Siskin
Carduelis flammea. See Common Redpoll
Carduelis hornemanni. See Hoary Redpoll
Carduelis lawrencei. See Lawrence's Goldfinch
Carduelis sinica. See Oriental Greenfinch
Carduelis spinoides. See Yellow-breasted Greenfinch
Carduelis spinus. See European Siskin
Carduelis tristis. See American Goldfinch
Carphibis spinicollis. See Straw-necked Ibis
Carpodacus mexicanus. See House Finch
Carpodacus nipalensis. See Dark-breasted Rosefinch
Carpodacus pulcherrimus. See Beautiful Rosefinch
Carpodacus purpureus. See Purple Finch
Carpodacus roseus. See Pallas's Rosefinch
Carpodacus rubicilloides. See Streaked Rosefinch
Carpodacus thura. See White-browed Rosefinch
Carpodacus trifasciatus. See Three-banded Rosefinch
Catharacta skua. See Great Skua
Cedar Waxwing, 185, 192, 208, 516–517
Ceuthmochares aereus. See Yellowbill
Chaffinch, 187
Chalcophaps indica. See Emerald Dove
Chalcopsitta atra. See Black Lory
Chamaea fasciata. See Wrentit
Chamaepetes unicolor. See Black Guan
Channel-billed Toucan, 297, 335
Chen caerulescens. See Snow Goose
Chen canagica. See Emperor Goose
Chestnut Munia, 437
Chestnut-bellied Starling, 300
Chestnut-shouldered Petronia, 450
Chestnut-winged Starling, 300
Chilean Flamingo, 184
Chiroxiphia caudata. See Blue Manakin
Chiroxiphia pareola. See Blue-backed Manakin
Chlamydera maculata. See Spotted Bowerbird
Chloebia gouldiae. See Gouldian Finch
Chlorestes notatus. See Blue-chinned Sapphire
Chlorospingus pileatus. See Sooty-capped Bush Tanager
Chrysococcyx cupreus. See African Emerald Cuckoos

Chrysolampis mosquitos. See Ruby-topaz Hummingbird
Chrysolophus amherstiae. See Lady Amherst Pheasant
Chrysolophus pictus. See Golden Pheasant
Cicinnurus magnificus. See Magnificent Bird-of-parasise
Cicinnurus regius. See King Bird-of-paradise
Ciconia ciconia. See White Stork
Cinnyricinclus femoralis. See Abbott's Starling
Cinnyricinclus leucogaster. See Plum-colored Starling
Cirl Bunting, 201
Citril Finch, 188
Claret-breasted Fruit Dove, 296
Clethrionomys gapperi. See Red-backed Vole
Clytolaema rubicauda. See Brazilian Ruby
Coccothraustes carnipes. See White-winged Grosbeak
Coccothrauses vespertinus. See Evening Grosbeak
Coccycolius iris. See Iris Glossy Starling
Coereba flaveola. See Bananaquit
Colaptes auratus. See Northern Flicker
Colaptes campestris. See Campo Flicker
Colaptes melanochlorus. See Green-barred Woodpecker
Colibri serrirostris. See White-vented Violet-ear
Colinus virginianus. See Northern Bobwhite
Collared Flycatcher, 486, 489, 490, 536
Collared Grosbeak, 187
Collocalia esculenta. See Glossy Swiftlet
Columba fasciata. See Band-tailed Pigeon
Columba livia. See Rock Pigeon
Columba trocaz. See Trocaz Pigeon
Columbina inca. See Inca Dove
Common Bronzewing, 296
Common Buzzard, 471, 474, 475
Common Canary, 107, 178, 181, 188, 204, 212–214, 469, 470, 496, 497, 509, 512
Common Eider, 385
Common Guillemot, 474
Common Lora, 299
Common Moorhen, 183, 204, 205, 209, 212, 213, 221, 450
Common Murre, 21
Common Quail, 14, 436
Common Raven, 244
Common Redpoll, 188, 513–514
Common Rhea, 21
Common Yellowthroat, 189, 209, 214
Congo Peafowl, 296
Copper Sunbird, 300
Copper-tailed Glossy Starling, 299
Coracias benghalensis. See Lilac-breasted Roller
Corapipo gutturalis. See White-throated Manakin
Corvus brachyrhynchos. See American Crow
Corvus corax. See Common Raven

Corythaeola cristata. See Great Blue Turaco
Corythaixoides concolor. See Grey Go-away Bird
Corythaixoides leucogaster. See White-bellied Go-away Bird
Corythaixoides personata. See Bare-faced Go-away Bird
Cosmopsarus regius. See Golden-breasted Starling
Cosmopsarus unicolor. See Ashy Starling
Cotinga cayana. See Spangled Cotinga
Cotinga cotinga. See Purple-breasted Cotinga
Cotinga maynana. See Plum-throated Cotinga
Coturnix coturnix. See Common Quail
Coturnix japonica. See Japanese Quail
Coua caerulea. See Blue Coua
Coua reynaudii. See Red-fronted Coua
Crayfish, 193
Cream-backed Woodpecker, 185
Creatophora cinerea. See Wattled Starling
Crested Fireback Pheasant, 296
Crested Myna, 299
Crested Wood-partridge, 296, 365
Crimson-backed Tanager, 190
Crinifer piscator. See Western Grey Plantain-eater
Crinifer zonurus. See Eastern Grey Plantain-eater
Crotophaga sp., 297
Crow Tit, 369
Cuban Trogon, 297
Cuculus solitarius. See Red-chested Cuckoo
Cyanerpes cyaneus. See Red-legged Honeycreeper
Cyanocorax beechii. See Purplish-backed Jay
Cyanoliseus patagonus. See Burrowing Parrot
Cyanoramphus malherbi. See Orange-fronted Parakeet
Cygnus columbianus. See Tundra Swan
Cygnus olor. See Mute Swan

Danio. See Zebrafish
Dark-breasted Rosefinch, 189
Dendroaspis viridis. See Green Mamba
Dendrocopus major. See Great-spotted Woodpecker
Dendroica coronata. See Yellow-rumped Warbler
Dendroica occidentalis. See Hermit Warbler
Dendroica palmarum. See Palm Warbler
Dendroica petechia. See Yellow Warbler
Dendroica striata. See Blackpoll Warbler
Dendroica townsendii. See Townsend's Warbler
Desert Finch, 189, 208
Diphyllodes magnificus. See Magnificent Bird of Paradise
Dixiphia pipra. See White-crowned Manakin
Dolichonyx oryzivorus. See Bobolink
Domestic Cat, 258
Domestic Chicken, 16, 27, 43, 183, 184, 196, 197, 200–207, 212, 216, 223, 224, 250, 252–254, 256–258, 260, 263–265, 266, 268, 269, 273, 275, 318, 339, 358, 359, 361, 368, 372, 378, 379, 384, 436, 438, 439, 448, 450, 470, 478–480, 484, 492, 522, 525, 535
Downy Woodpecker, 415
Dromaius novahollandiae. See Emu
Dryocopus pileatus. See Pileated Woodpecker
Ducula concinna. See Blue-tailed Imperial Pigeon
Dumetella carolinensis. See Gray Catbird
Dusky Scrubfowl, 296
Dyaphorophyia concreta. See Yellow-bellied Wattle-eye

Eastern Bluebird, 114, 243, 252, 300, 331, 345–346, 538
Eastern Grey Plantain-eater, 364
Eastern Meadowlark, 414
Eastern Rosella, 380
Eastern Screech Owl, 474
Eclectus Parrot, 354, 380
Eclectus roratus. See Eclectus Parrot
Egretta gularis. See Indian Reef Heron
Egyptian Vulture, 184, 194, 519
Elegant Tern, 184, 405
Eleonora's Falcon, 474
Elliot's Pheasant, 296
Elysia chlorotica. See Sea Slug
Emberiza cirlus. See Cirl Bunting
Emberiza citrinella. See Yellowhammer
Emberiza melanocephala. See Black-headed Bunting
Emerald Dove, 296
Emperor Goose, 415
Emperor Penguin, 384
Emu, 296
Enodes erythrophris. See Fiery-browed Myna
Entomyzon cyanotis. See Blue-faced Honeyeater
Epimachus fastuosus. See Black Sicklebill
Erinsceus europaeus. See European Hedgehog
Erithacus rubecula. See Robin
Erythrura psittacea. See Red-headed Parrot Finch
Eskimo Curlew, 416
Eudocimus ruber. See Scarlet Ibis
Eudyptes chrysocome. See Rockhopper Penguin
Eudyptes chrysolophus. See Macaroni Penguin
Eumomota superciliosa. See Turquoise-browed Motmot
Euphagus cyanocephalus. See Brewer's Blackbird
Euphonia affinis. See Scrub Euphonia
Euplectes afer. See Yellow-crowned Bishop
Euplectes capensis. See Yellow Bishop
Euplectes franciscana. See Orange Weaver
Euplectes orix. See Red Bishop
Eurasian Blackbird, 24, 26, 189, 221, 529
Eurasian Bullfinch, 187, 265

Species Index

Eurasian Siskin, 188, 195
European Goldfinch, 188, 209
European Greenfinch, 115, 188, 201, 208, 218, 227, 382, 515, 517, 525, 527, 529–530
European Hedgehog, 358
European Jay, 299
European Kestrel, 533, 536
European Serin, 188, 530
European Starling, 4, 14, 17, 24–26, 119, 128, 300, 324, 326–327, 372, 374, 411, 450, 451
Evening Grosbeak, 187, 208, 214, 417–419, 472

Falco eleonarae. See Eleonora's Falcon
Falco rusticolus. See Gyrfalcon
Falco sparverius. See American Kestrel
Falco tinnunculus. See European Kestrel
Felis domestica. See Domestic Cat
Ferruginous Hawk, 474
Ferruginous Partridge, 296
Ferruginous-backed Antbird, 298
Ficedula albicollis. See Collared Flycatcher
Ficedula hypoleuca. See Pied Flycatcher
Ficedula zanthopygia. See Korean Flycatcher
Fiery-browed Myna, 299
Finch-billed Myna, 300
Fischer's Starling, 300
Fischer's Turaco, 364
Forest Weaver, 186
Fork-tailed Woodnymph, 297
Foudia madagascariensis. See Red Fody
Fox Sparrow, 419
Francolinus francolinus. See Black Francolin
Franklin's Gull, 184, 405
Fringilla coelebs. See Chaffinch
Fringilla montifringilla. See Brambling
Fulicula americana. See American Coot

Galbula ruficauda. See Rufous-tailed Jacamar
Gallinula chloropus. See Common Moorhen
Gallus gallus. See Domestic Chicken, Red Junglefowl
Gambel's Quail, 262, 436
Garrulus glandularis. See European Jay
Gasterosteus aculeatus. See Three-spined Stickleback
Gavia stellata, 415
Geospiza fortis. See Large Ground Finch
Geothylpis trichas. See Common Yellowthroat
Giant Petrel, 474, 484
Glossy Swiftlet, 297
Goldcrest, 186, 210
Golden Bush-robin, 185
Golden Dove, 296
Golden Eagle, 244
Golden Oriole, 185

Golden Pheasant, 296
Golden-breasted Starling, 299
Golden-collared Manakin, 128
Golden-crested Myna, 299
Golden-headed Manakin, 163–164, 190
Golden-winged Grosbeak, 187
Goldfish, 9, 12, 28, 178
Gold-naped Finch, 187
Gold-whiskered Barbet, 297, 330
Gouldian Finch, 186, 437, 472, 473, 474, 497
Gracula religiosa. See Hill Myna
Grafisia torquata. See White-collared Starling
Gray Catbird, 244
Great Blue Hero, 372, 474, 484
Great Blue Turaco, 364
Great Bustard, 360
Great Hornbill, 404, 422
Great Horned Owl, 340, 354, 372
Great Skua, 495
Great Tinamou, 369
Great Tit, 81, 185, 270, 384, 416, 486, 522–523, 529–530, 533–534, 540–541
Greater Blue-eared Glossy Starling, 299
Greater Flamingo, 181, 184, 212
Greater Rhea, 21
Great-spotted Woodpecker, 184
Green Broadbill, 298, 331
Green Mamba, 378
Green Woodhoopoe, 297
Green Woodpecker, 185
Green-barred Woodpecker, 185
Green-winged Teal, 296
Grey Go-away Bird, 364
Grey Partridge, 184
Grey Peacock-pheasant, 296
Grey Teal, 296
Greylag Goose, 184, 205
Grus canadensis. See Sandhill Crane
Guianan Cock-of-the-Rock, 159–163
Guianan Toucanet, 298
Guinea Turaco, 364
Guppy, 195, 196, 517–518
Gymnopithys leucaspis. See Bicolored Antbird
Gypaetus barbatus. See Bearded Vulture
Gyrfalcon, 403

Hadeda Ibis, 296, 324, 326
Haematopus unicolor. See Variable Oystercatcher
Haematospiza sipahi. See Scarlet Finch
Hairy Woodpecker, 184
Hairy-crested Antbird, 298
Halcyon pileata. See Black-capped Kingfisher
Haliaeetus albicilla. See White-tailed Eagle

Haliaeetus leucocephalus. See Bald Eagle
Hartlaub's Turaco, 364
Heliangelus clemenciae. See Blue-throated Hummingbird
Heliangelus strophianus. See Gorgeted Sunangel
Heliangelus viola. See Amethyst-throated Sunangel
Helmeted Guineafowl, 296, 336
Hemiprocne comata. See Whiskered Treeswift
Hepatic Tanager, 190
Hermit Warbler, 494, 497
Herring Gull, 374, 437, 442, 450
Hesperiphona abeillei. See Hooded Grosbeak
Hesperiphona vespertinus. See Evening Grosbeak
Hildebrandt's Starling, 300
Hill Myna, 299
Himalayan Monal, 296
Hirundo rustica. See Barn Swallow
Hoary Redpoll, 188
Hoatzin, 296
Honeybee, 14
Hooded Grosbeak, 187
Horned Guan, 296
House Finch, 104, 109, 113, 189, 191, 192, 195, 196, 201, 204, 205, 208, 210, 211, 216, 217, 219–220, 270, 416, 419, 421, 422, 448, 486, 509–511, 513–516, 518–519, 525–528, 530–531, 539–541
House Sparrow, 106, 243, 258, 261, 262, 265, 266, 269, 270, 411, 414, 417, 418, 422, 423, 437, 447, 448, 450, 451, 486–488, 541–542, 546–547
Huon Astrapia, 298

Icteria virens. See Yellow-breasted Chat
Icterus galbula. See Northern Oriole
Inca Dove, 376
Indian Peafowl, 296, 439
Indian Reef Heron, 474
Indigo Bunting, 300, 330, 437, 450
Irena puella. See Asian Fairy Bluebird
Iris Glossy Starling, 299
Isabelline Shrike, 494
Ithaginis cruentus. See Blood Pheasant
Ivory-breasted Pitta, 298

Jacana spinosa. See Northern Jacana
Jambu Fruit Dove, 296
James's Flamingo, 184
Japanese Common Toad, 377
Japanese Quail, 204, 217, 224–225, 244, 251, 252, 254, 258, 265, 266, 358, 359, 374, 384, 470
Japanese Waxwing, 185
Japanese White-eye, 190

Kalij Pheasant, 296
Kenrick's Starling, 300
Kentucky Warbler, 516
King Bird-of-paradise, 298
King Penguin, 252, 384
Knysna Turaco, 297, 364, 366, 367
Korean Flycatcher, 185

Lady Amherst Pheasant, 436
Lagopus lagopus. See Willow Ptarmigan
Lagopus leucurus. See White-tailed Ptarmigan
Lagopus mutus. See Rock Ptarmigan
Lamprotornis australis. See Burchell's Glossy Starling
Lamprotornis caudatus. See Long-tailed Glossy Starling
Lamprotornis chalcurus. See Bronze-tailed Glossy Starling
Lamprotornis chalybaeus. See Greater Blue-eared Glossy Starling
Lamprotornis chloropterus. See Lesser Blue-eared Glossy Starling
Lamprotornis corruscus. See Black-bellied Glossy Starling
Lamprotornis cupreocauda. See Copper-tailed Glossy Starling
Lamprotornis mevesii. See Meves' Glossy Starling
Lamprotornis nitens. See Cape Glossy Starling
Lamprotornis ornatus. See Principe Glossy Starling
Lamprotornis purpureiceps. See Purple-headed Glossy Starling
Lamprotornis purpureus. See Purple Glossy Starling
Lamprotornis splendidus. See Splendid Glossy Starling
Lanius collurio. See Red-backed Shrike
Lanius isabellinus. See Isabelline Shrike
Lanius senator. See Woodchat Shrike
Lapland Longspur, 413, 423
Large Ground Finch, 486
Larus argentatus. See Herring Gull
Larus atricilla. See Laughing Gull
Larus delawarensis. See Ring-billed Gull
Larus fuscus. See Lesser Black-backed Gull
Larus occidentalis. See Western Gull
Larus pipixcan. See Franklin's Gull
Larus ridibundus. See Black-headed Gull
Laughing Gull, 437, 442, 450
Lawrence's Goldfinch, 211, 411, 412
Leiothrix argentauris. See Silver-eared Mesia
Leiothrix lutea. See Pekin Robin, Red-billed Leiothrix
Lepidothrix coronata. See Blue-crowned Manakin
Lepidothrix nattereri. See Snowy-capped Manakin
Lepidothrix serena. See White-fronted Manakin
Lesser Black-backed Gull, 183, 198, 203–205, 209, 212, 213, 223

Species Index

Lesser Blue-eared Glossy Starling, 299
Lesser Flamingo, 184
Lesser Scaup, 373
Lesser Snow Goose, 267, 268, 472, 474, 475, 480–482
Leucopsar rothschildi. See Bali Starling
Lewis's Woodpecker, 184
Lilac-breasted Roller, 345
Limia perugia, 123
Limosa lapponica. See Bar-tailed Godwit
Linnet, 188, 530
Lissotis melanogaster. See Black-bellied Bustard
Little Shag, 474
Livingstone's Turaco, 364
Lobipes lobatus. See Northern Phalarope
Lonchura atricapilla. See Chestnut Munia
Long-tailed Finch, 120
Long-tailed Glossy Starling, 299
Long-tailed Rosefinch, 188
Long-tailed Skua, 494
Long-tailed Tit, 185
Lophophorus impejanus. See Himalayan Monal
Lophorina superba. See Superb Bird-of-paradise
Lophura bulweri. See Bulwer's Pheasant
Lophura erythropthalmus. See Kalij Pheasant
Lophura ignata. See Crested Fireback Pheasant
Lophura leucomelanos. See White-crested Kalij Pheasant
Loxia curvirostra. See Red Crossbill
Loxia leucoptera. See White-winged Crossbill
Luscinia calliope. See Siberian Rubythroat
Luscinia svecica. See Bluethroat

Macaroni Penguin, 252, 384
Macronectes giganteus. See Giant Petrel
Madagascar Paradise Flycatcher, 299, 336, 471
Madagascar Starling, 300
Magnificent Bird-of-paradise, 158, 298
Magnificent Riflebird, 298, 322
Magpie Starling, 300
Mallard, 120, 184, 245, 252, 263, 265, 267, 366, 401–402, 413, 436, 441, 449
Malurus cyaneus. See Superb Fairy-wren
Malurus leucopterus. See White-winged Fairy-wren
Malurus melanocephalus. See Red-backed Fairy-wren
Manacus vitellinus. See Golden-collared Manakin
Mandarin Duck, 296
Mandrill, 546
Mandrillus sphinx. See Mandrill
Megadyptes antipodes. See Yellow-eyed Penguin
Megalaima chrysopogon. See Gold-whiskered Barbet
Megapodius freycinet. See Dusky Scrubfowl
Melanerpes candidus. See White Woodpecker

Melanerpes lewis. See Lewis's Woodpecker
Melanoperdix nigra. See Black Partridge
Meleagris gallopavo. See Wild Turkey
Melopsittacus undulatus. See Budgerigar
Melospiza melodia. See Song Sparrow
Meves' Glossy Starling, 299
Mexican Axolotl, 377
Mino dumontii. See Yellow-faced Myna
Molothrus aeneus. See Bronzed Cowbird
Molothrus ater. See Brown-headed Cowbird
Molothrus bonariensis. See Shiny Cowbird
Motacilla flava. See Yellow Wagtail
Motmotus mexicanus. See Russet-crowned Motmot
Musophaga johnstoni. See Ruwenzori Turaco
Musophaga porphyreolopha. See Purple-crested Turaco
Musophaga rossae. See Ross's Turaco
Musophaga violacea. See Violet Turaco
Mute Swan, 474
Mycerobas affinis. See Collared Grosbeak
Mycerobas icteroides. See Black-and-yellow Grosbeak
Mycerobas melanozanthos. See Spot-winged Grosbeak
Mycoplasma gallicepticum, 525–527, 531
Myiophonus caeruleus. See Blue Whistling-thrush
Myrmeciza ferruginea. See Ferruginous-backed Antbird

Narina Trogon, 297
Narrow-tailed Starling, 300
Nashville Warbler, 189
Nectarinia coccinigastra. See Superb Sunbird
Nectarinia cuprea. See Copper Sunbird
Nectarinia sperata. See Purple-throated Sundbird
Nelicourvi Weaver, 186
Neochmia ruficauda. See Star Finch
Neocichla gutturalis. See Babbling Starling
Neodrepanis coruscans. See Sunbird Asity
Neodrepanis hypoxantha. See Yellow-bellied Asity
Neophron percnopterus. See Egyptian Vulture
Nesospiza acunhae. See Tristan Bunting
New Zealand Fantail, 474
Nicobar Pigeon, 296
Nine-spined Stickleback, 196
Northern Bobwhite, 439
Northern Cardinal, 104, 105, 109, 116, 189, 192, 208, 214, 252, 509–511, 514–515, 519
Northern Flicker, 106, 185, 204
Northern Jacana, 365
Northern Lapwing, 296
Northern Oriole, 189
Northern Phalarope, 262
Northern Pintail, 441
Numenius borealis. See Eskimo Curlew

Numida meleagris. See Helmeted Guineafowl
Nyctyornis amicta. See Red-bearded Bee-eater

Ocellated Turkey, 296
Oncorhynchus mykiss. See Rainbow Trout
Onychognathus albirostris. See White-billed Starling
Onychognathus blythii. See Somali Starling
Onychognathus frater. See Socotra Starling
Onychognathus fulgidus. See Chestnut-winged Starling
Onychognathus morio. See Red-winged Starling
Onychognathus nabouroup. See Pale-winged Starling
Onychognathus salvadorii. See Bristle-crowned Starling
Onychognathus tenuirostris. See Slender-billed Starling
Onychognathus tristamii. See Tristam's Grackle
Opisthocomus hoazin. See Hoatzin
Oporornis formosus. See Kentucky Warbler
Orange Dove, 296
Orange Bullfinch, 187
Orange Weaver, 263, 264, 437, 450
Orange-fronted Parakeet, 472, 474
Oreophasis derbianus. See Horned Guan
Oriental Greenfinch, 188
Oriolus auratus. See African Oriole
Oriolus oriolus. See Golden Oriole
Oriolus xanthornus. See Black-hooded Oriole
Osprey, 273, 274
Ostrich, 20, 21, 30, 436, 438, 450
Otis asio, 474
Otis tarda. See Great Bustard
Oxyura jamaicensis. See Ruddy Duck

Pale-winged Starling, 300
Pallas's Rosefinch, 189
Palm Warbler, 189
Pandion haliaetus. See Osprey
Paradise Riflebird, 298
Paradise Whydah, 256, 264, 437, 449
Paradisea raggiana. See Raggiana Bird-of-paradise
Paradisea rubra. See Red Bird-of-paradise
Paradoxornis webbiana. See Crow Tit
Parasitic Jaeger. *See* Artic Skua
Parotia sefilata. See Western Parotia
Parus caeruleus. See Blue Tit
Parus major. See Great Tit
Parus spilonotus. See Yellow-cheeked Tit
Passer domesticus. See House Sparrow
Passer montanus. See Tree Sparrow
Passerella iliaca. See Fox Sparrow
Passerina caerulea. See Blue Grosbeak
Passerina cyanea. See Indigo Bunting
Pavo cristatus. See Indian Peafowl
Pavonine Quetzal, 297
Peach-faced Lovebird, 297, 310, 341

Pekin Robin, 16, 127, 186
Perdix perdix. See Grey Partridge
Pericrocotus flammeus, 185
Perissocephalus tricolor. See Capuchinbird
Petronia xanthocollis. See Chestnut-shouldered Petronia
Phainopepla, 276
Phainopepla nitens. See Phainopepla
Phalacrocorax melanoleucos. See Little Shag
Phalaropus lobatus. See Red-necked Phalarope
Phalaropus tricolor. See Wilson's Phalarope
Phaps chalcoptera. See Common Bronzewing
Pharomachrus moccino. See Resplendent Quetzal
Pharomachrus pavoninus. See Pavonine Quetzal
Phasianus colchicus. See Ring-necked Pheasant
Pheucticus ludovicianus. See Rose-breasted Grosbeak
Philepitta castanea. See Velvet Asity
Philomachus pugnax. See Ruff
Phoenicopterus chilensis. See Chilean Flamingo
Phoenicopterus jamesi. See James's Flamingo
Phoenicopterus minor. See Lesser Flamingo
Phoenicopterus ruber. See Greater Flamingo
Phoeniculus purpureus. See Green Woodhoopoe
Pica pica. See Black-billed Magpie
Picoides pubescens. See Downy Woodpecker
Picoides tridactylus. See Three-toed Woodpecker
Picoides villosus. See Hairy Woodpecker
Picus squamatus. See Scaly-bellied Woodpecker
Picus viridis. See Green Woodpecker
Pied Flycatcher, 369–370, 486, 489, 541
Pileated Woodpecker, 185
Pilherodius pileatus. See Capped Heron
Pine Grosbeak, 188, 417–419
Pinicola enucleator. See Pine Grosbeak
Pipilo erythrophthalmus. See Rufous-sided Towhee
Pipra chloromeros. See Round-tailed Manakin
Pipra erythrocephala. See Golden-headed Manakin
Pipra filicauda. See Wire-tailed Manakin
Pipra rubrocapilla. See Red-headed Manakin
Piranga flava. See Hepatic Tanager
Piranga ludoviciana. See Western Tanager
Piranga olivacea. See Scarlet Tanager
Piranga rubra. See Summer Tanager
Pitta maxima. See Ivory-breasted Pitta
Platycercus eximius. See Eastern Rosella
Plectrophenax nivalis. See Snow Bunting
Ploceus bicolor. See Forest Weaver
Ploceus capensis. See Cape Weaver
Ploceus cucullatus. See Village Weaver
Ploceus nelicourvi. See Nelicourvi Weaver
Ploceus philippinus. See Baya Weaver
Ploceus sakalava. See Sakalava Weaver
Ploceus velatus. See African-masked Weaver

Species Index

Plum-colored Starling, 299, 322, 324–325
Plum-throated Cotinga, 298
Poecile atricapilla. See Black-capped Chickadee
Poecilia reticulata. See Guppy
Poephila acuticauda. See Long-tailed Finch
Poeptera kenricki. See Kenrick's Starling
Poeptera lugubris. See Narrow-tailed Starling
Poeptera stuhlmanni. See Stuhlmann's Starling
Polyplectron bicalcaratum. See Grey Peacock-pheasant
Pomacea canaliculata. See Apple Snail
Pompadour Cotinga, 208
Prince Raspoli's Turaco, 364
Principe Glossy Starling, 299
Priotelus temnurus. See Cuban Trogon
Procambarus clarkii. See Crayfish
Procnias nudicollis. See Bare-throated Bellbird
Prosthemadura novaeseelandiae. See Tui
Psittacus erithacus. See African Gray Parrot
Ptilinopus cincta. See Black-backed Fruit Dove
Ptilinopus jambu. See Jambu Fruit Dove
Ptilinopus luteovirens. See Golden Dove
Ptilinopus rivoli. See White-breasted Fruit Pigeon
Ptilinopus superbus. See Superb Fruit Dove
Ptilinopus victor. See Orange Dove
Ptilinopus viridis. See Claret-breasted Fruit Dove
Ptilonorhynchus violaceous. See Satin Bowerbird
Ptiloris paradiseus. See Paradise Riflebird
Ptiloris magnifica. See Magnificent Riflebird
Puffinus pacificus. See Wedge-tailed Shearwater
Pungitius pungitius. See Nine-spined Stickleback
Purple Finch, 413
Purple Glossy Starling, 299
Purple-breasted Cotinga, 298, 330
Purple-crested Turaco, 364
Purple-headed Glossy Starling, 299
Purple-throated Sunbird, 300
Purplish-backed Jay, 299, 331
Pycnonotus atriceps. See Black-headed Bulbul
Pygmy Goose, 385
Pyrrhoplectes epauletta. See Gold-naped Finch
Pyrrhula aurantiaca. See Orange Bullfinch
Pyrrhula erythraca. See Beavan's Bullfinch
Pyrrhula erythrocephala. See Red-headed Bullfinch
Pyrrhula pyrrhula. See Eurasian Bullfinch

Quelea cardinalis. See Cardinal Quelea
Quelea quelea. See Red-billed Quelea

Raggiana Bird-of-paradise, 298, 322
Rainbow Lorikeet, 297
Rainbow Trout, 203
Ramphastos toco. See Toco Toucan
Ramphastos vitellinus. See Channel-billed Toucan
Ramphocelus dimidiatus. See Crimson-backed Tanager
Razorbill, 21
Red Avadavat, 186, 437, 449
Red Bird-of-paradise, 298
Red Bishop, 186
Red Crossbill, 187, 218
Red Fody, 186
Red Junglefowl, 183, 296, 365, 451, 452, 532–533, 546
Red Knot, 400, 401–403, 422
Red Siskin, 188, 469, 497
Red-backed Fairy-wren, 453
Red-backed Shrike, 494
Red-backed Vole, 194
Red-bearded Bee-eater, 331
Red-billed Oxpecker, 299
Red-billed Quelea, 106, 186, 217, 252, 264, 437, 449, 491, 509, 511
Red-chested Cuckoo, 120
Red-crested Turaco, 364
Red-eyed Vireo, 372
Red-fronted Serin, 188
Red-headed Bullfinch, 187
Red-headed Manakin, 190
Red-headed Parrot Finch, 186
Red-legged Honeycreeper, 300, 330
Red-legged Partridge, 106, 201, 203, 511
Red-necked Phalarope, 437, 442
Red-throated Loon, 415
Red-winged Blackbird, 189, 196, 208, 214, 218, 251, 252, 372
Red-winged Starling, 300
Reeve's Pheasant, 436
Regulus regulus. See Goldcrest
Resplendent Quetzal, 297, 322
Rhea americana. See Greater Rhea
Rhegmatorhina melanosticta. See Hairy-crested Antbird
Rhipidura fuliginosa. See New Zealand Fantail
Rhodopechys githaginea. See Trumpeter Finch
Rhodopechys obsoleta. See Desert Finch
Rhodostethia rosea. See Ross's Gull
Rhynchostruthus socotranus. See Golden-winged Grosbeak
Ring-billed Gull, 184, 193–194, 405
Ring-necked Pheasant, 184, 204, 296, 362, 436, 439, 450, 451, 539
Robin, 185
Rock Pigeon, 12, 14, 19, 24, 113, 127, 252, 266, 267, 269, 273, 276–277, 296, 320, 326, 354, 372, 376, 470, 473, 474
Rock Ptarmigan, 318, 400, 403–404, 422
Rockhopper Penguin, 384

Rollolus roulroul. See Crested Wood-partridge
Roseate Spoonbill, 184
Rose-breasted Grosbeak, 189
Ross's Goose, 474
Ross's Gull, 405
Ross's Turaco, 364, 365
Round-tailed Manakin, 190
Ruby-topaz Hummingbird, 297
Ruddy Duck, 296, 335–336
Ruff, 262, 437, 442, 450, 473–475
Rufous-sided Towhee, 444
Rufous-tailed Jacamar, 297
Rupicola peruviana. See Andean Cock-of-the-Rock
Rupicola rupicola. See Guianan Cock-of-the-Rock
Russet-crowned Motmot, 410
Ruwenzori Turaco, 364

Saffron Finch, 189
Sakalava Weaver, 186
Sanderling, 414
Sandhill Crane, 414, 415
Sarcops calvus. See Bald Starling
Saroglossa aurata. See Madagascar Starling
Satin Bowerbird, 345, 406, 407, 536–537
Satyr Tragopan, 296, 336
Scaled Quail, 262, 436
Scaly-bellied Woodpecker, 185
Scarlet Finch, 188
Scarlet Ibis, 184, 200
Scarlet Macaw, 380, 381
Scarlet Minivet, 185, 218
Scarlet Tanager, 190, 218
Schalow's Turaco, 364, 366
Scissirostrum dubium. See Finch-billed Myna
Scrub Euphonia, 300, 321, 322
Sea Slug, 179
Selenidera culik. See Guianan Toucanet
Serinus canaria. See Common Canary
Serinus citrinella. See Citril Finch
Serinus mozambicus. See Yellow-fronted Canary
Serinus pusillus. See Red-fronted Serin
Serinus serinus. See European Serin
Setophaga ruticilla. See American Redstart
Shelley's Starling, 300
Shiny Cowbird, 300
Sialia sialis. See Eastern Bluebird
Siberian Rubythroat, 185
Sicalis flaveola. See Saffron Finch
Silver-eared Mesia, 186
Sindbis virus, 527
Slaty-tailed Trogon, 297, 322
Slender-billed Starling, 300
Snow Bunting, 413, 423

Snow Goose, 415
Snowy-capped Manakin, 298, 319, 330
Socotra Starling, 300
Somali Starling, 300
Somateria mollissima. See Common Eider
Song Sparrow, 273
Song Thrush, 368
Sooty-capped Bush Tanager, 190
South American Tern, 106
Spangled Cotinga, 298
Spectacled Duck, 296
Speculipastor bicolor. See Magpie Starling
Sphyrapicus varius. See Yellow-bellied Sapsucker
Splendid Astrapia, 298
Splendid Glossy Starling, 299
Spotted Bowerbird, 408
Spotted Salamander, 244
Spot-winged Grosbeak, 187
Spreo albicapillus. See White-crowned Starling
Spreo bicolor. See African Pied Starling
Spreo fischeri. See Fischer's Starling
Spreo hildebrandti. See Hildebrandt's Starling
Spreo pulcher. See Chestnut-bellied Starling
Spreo shelleyi. See Shelley's Starling
Spreo superbus. See Superb Starling
Star Finch, 186
Steganura paradisaea. See Paradise Whydah
Stercorarius longicaudus. See Long-tailed Skua
Stercorarius parasiticus. See Arctic Skua
Sterna elegans. See Elegant Tern
Sterna hirundinacea. See South American Tern
Straw-necked Ibis, 296
Streaked Rosefinch, 189
Streptocitta albicollis. See White-necked Myna
Striated Heron, 296
Strix aluco. See Tawny Owl
Struthio camelus. See Ostrich
Stuhlmann's Starling, 300
Sturnella magna. See Eastern Meadowlark
Sturnella neglecta. See Western Meadowlark
Sturnus sinensis. See White-shouldered Starling
Sturnus vulgaris. See European Starling
Sula nebouxii. See Blue-footed Booby
Sulawesi Myna, 299
Sulfur-breasted Bushshrike, 185
Summer Tanager, 190
Sunbird Asity, 298
Superb Bird-of-paradise, 298
Superb Fairy-wren, 437, 443, 446, 453, 454, 546
Superb Fruit Dove, 296
Superb Sunbird, 310
Sylvia atricapilla. See Blackcap
Syrigma sibilatrix. See Whistling Heron

Species Index

Syrmaticus ellioti. *See* Elliot's Pheasant
Syrmaticus reevesi. *See* Reeve's Pheasant

Tachycineta albiventer. *See* White-winged Swallow
Tachycineta bicolor. *See* Tree Swallow
Tadorna tadorna. *See* Common Shelduck
Taeniopygia guttata. *See* Zebra Finch
Tangara cyanocollis. *See* Blue-necked Tanager
Tangara mexicana. *See* Turquoise Tanager
Tarsiger chrysaeus. *See* Golden Bush-robin
Tauraco bannermani. *See* Bannerman's Turaco
Tauraco corythaix. *See* Knysna Turaco
Tauraco erythrolophus. *See* Red-crested Turaco
Tauraco fischeri. *See* Fischer's Turaco
Tauraco hartlaubi. *See* Hartlaub's Turaco
Tauraco leucolophus. *See* White-crested Turaco
Tauraco leucotis. *See* White-cheeked Turaco
Tauraco livingstonii. *See* Livingstone's Turaco
Tauraco macrorhynchus. *See* Yellow-billed Turaco
Tauraco persa. *See* Guinea Turaco
Tauraco ruspolii. *See* Prince Ruspoli's Turaco
Tauraco schalowi. *See* Schalow's Turaco
Tauraco schuettii. *See* Black-billed Turaco
Tawny Owl, 15–16, 360, 534
Telophorus sulfureopectus. *See* Sulfur-breasted Bushshrike
Temminck's Tragopan, 296
Tengmalm's Owl, 15
Tennessee Warbler, 415
Terpsiphone mutata. *See* Madagascar Paradise Flycatcher
Tetrao tetrix. *See* Black Grouse
Tetrao urogallus. *See* Capercaillie
Thalurania furcata. *See* Fork-tailed Woodnymph
Three-banded Rosefinch, 189
Three-spined Stickleback, 196
Three-toed Woodpecker, 184
Tichodroma muraria. *See* Wallcreeper
Tinamus major. *See* Great Tinamou
Toco Toucan, 185, 208, 301, 336, 341
Townsend's Warbler, 494, 497
Tragopan caboti. *See* Cabot's Tragopan
Tragopan melanocephalus. *See* Western Tragopan
Tragopan satyra. *See* Satyr Tragopan
Tragopan temmincki. *See* Temminck's Tragopan
Tree Sparrow, 266
Tree Swallow, 112, 131–132, 453, 458, 548
Trichoglossus haematodus. *See* Rainbow Lorikeet
Tristam's Grackle, 300
Tristan Bunting, 189
Trocaz Pigeon, 297, 320, 327
Trogon massena. *See* Slaty-tailed Trogon
Trogon rufus. *See* Black-throated Trogon
Trogon violaceus. *See* Violaceous Trogon

Trumpeter Finch, 189
Tui, 299
Tundra Swan, 245
Turdus merula. *See* Eurasian Blackbird
Turdus migratorius. *See* American Robin
Turdus philomelos. *See* Song Thrush
Turquoise Tanager, 300
Turquoise-browed Motmot, 297, 314, 410
Tyto alba. *See* Barn Owl

Uragus sibiricus. *See* Long-tailed Rosefinch
Uria aalge. *See* Common Guillemot

Vanellus vanellus. *See* Northern Lapwing
Variable Oystercatcher, 474, 494
Velvet Asity, 298, 310, 334–336, 341
Vermivora peregrina. *See* Tennessee Warbler
Vermivora ruficapilla. *See* Nashville Warbler
Vermivora virginiae. *See* Virginia's Warbler
Village Weaver, 186
Violaceous Trogon, 297
Violet Turaco, 364
Vireo olivaceus. *See* Red-eyed Vireo
Virginia's Warbler, 189
Vogelkop Bowerbird, 408
Volatinia jacarina. *See* Blue-black Grassquit
Vulturine Guineafowl, 296

Wallcreeper, 185
Wattled Starling, 299
Wedge-tailed Shearwater, 24
Western Grey Plantain-eater, 364
Western Gull, 442
Western Meadowlark, 414
Western Parotia, 298
Western Tanager, 190
Western Tragopan, 296
Whiskered Treeswift, 297
Whistling Heron, 296, 336
White Stork, 184, 193, 201, 204, 205, 208, 516
White Woodpecker, 184
White-bellied Go-away Bird, 364
White-billed Starling, 300
White-breasted Fruit Pigeon, 329
White-browed Rosefinch, 189
White-cheeked Turaco, 364
White-collared Starling, 299
White-crested Kalij Pheasant, 296
White-crested Turaco, 364
White-crowned Manakin, 163
White-crowned Sparrow, 474
White-crowned Starling, 300
White-fronted Manakin, 160, 162–163, 319

White-necked Myna, 300
White-shouldered Starling, 385
White-tailed Eagle, 272
White-tailed Ptarmigan, 436, 439
White-throated Manakin, 160, 162–163
White-throated Sparrow, 478, 516
White-vented Violetear, 14
White-winged Crossbill, 187
White-winged Dove, 296
White-winged Fairy-wren, 268, 299, 409–410, 483
White-winged Grosbeak, 187
White-winged Swallow, 299
Wild Turkey, 4, 184, 204, 258, 296, 335, 337, 361, 384–385, 436, 450, 470, 536
Willow Ptarmigan, 262, 272, 436, 439–440
Wilson's Phalarope, 262–263, 437, 442
Wire-tailed Manakin, 120, 163
Wood Duck, 384
Woodchat Shrike, 494
Wrentit, 43

Xipholena punicea. See Pompadour Cotinga

Yellow Bishop, 186
Yellow Wagtail, 186, 515
Yellow Warbler, 189, 209, 214
Yellow-bellied Asity, 298
Yellow-bellied Sapsucker, 184
Yellow-bellied Wattle-eye, 299, 341
Yellowbill, 297
Yellow-billed Turaco, 364
Yellow-breasted Chat, 120, 189, 208, 214, 516
Yellow-breasted Greenfinch, 188
Yellow-cheeked Tit, 185
Yellow-crowned Bishop, 186, 264, 437
Yellow-eyed Penguin, 384
Yellow-faced Myna, 300
Yellow-fronted Canary, 188
Yellowhammer, 189, 529–530
Yellow-rumped Warbler, 189

Zebra Finch, 168, 186, 198, 200–204, 208, 212, 213, 216, 217, 219, 221–223, 245, 251, 252, 259, 267, 300, 450, 470, 488, 496, 509, 518–519, 522
Zebra Waxbill, 186
Zebrafish, 244
Zenaida asciatica. See White-winged Dove
Zonotrichia albicollis. See White-throated Sparrow
Zosterops japonica. See Japanese White-eye

Subject Index

Absorption spectrophotometry
  of melanin pigments, 248
  of psittacofulvins, 381
  of visual pigments, 23
Achromatic brightness
  effect of light environment on, 156, 163–164
  effect of viewing angle on, 53, 70
  in human vision, 6, 8, 19
  perception of, 9–10, 17
  quantification of, 97, 118, 128–129
Achromatic contrast, 128, 129
Adventitious colors, 93, 399, 414–20, 421–423
Albino, 243, 267
Ambient light, 48–49, 66–67, 79, 85, 90, 92–95, 103, 105, 122, 123, 125, 126, 136, 140, 149–158, 160, 163, 165, 170, 316, 323, 338, 345, 347, 408
Androgens
  effects on carotenoid coloration, 446, 448, 452
  effects on melanin coloration, 260–263, 446, 447–448, 450–451
  effects on sexual dichromatism, 433, 434, 436–437, 439–442, 444, 449, 455–457
  effects on structural coloration, 453, 546
Apostatic selection, 476–477
Applied cosmetics, 70, 93, 399, 400–405, 421–423
Assortative mating, 105, 113, 128, 382, 475–476
Avian pox, 527–528, 530

Bare-parts, 55, 90, 92, 93, 95, 103, 107, 115, 137, 179, 181–182, 184, 207–211, 216–217, 224, 226–227, 251, 257, 259, 272, 333–339, 344, 362, 431–432, 446, 449, 470, 489, 507, 539, 545
Bilateral gynandromorphy, 432, 445–446

Bill coloration
  environmental control of, 519
  genetic control of, 489
  hormonal control of, 449–452
  mate choice for, 168
  measuring, 107, 120
  mechanisms of color of, 362
Bonferroni correction, 130, 133–134, 137, 140
Bragg scattering. *See* Coherent scattering
Bragg's Law, 311, 321
Brightness (definition), 102
Bower, 405–408

Camouflage. *See* Crypsis
Canonical variates analysis (CVA), 115
Carotenoid coloration
  concealed with buffy tips, 413
  condition dependence of, 270, 508–531, 538–540
  change over time, 211, 416, 419, 421
  effect of nutrition on, 191–195, 200, 538–540
  effect of parasites on, 524–531
  effect of parental feeding on, 203, 223–224, 521–523
  effect of pigment access on, 508–523
  enhancement of, 515–517
  evolution of, 517
  externally applied, 405,
  genetic control of, 472, 484, 488–489, 491, 493, 494, 497
  hormonal control of, 446
  maternal effects on, 206, 223–224
  in museum specimens, 211
  phylogenetic patterns of, 217–218
  reduction in captivity, 509–515
  role of pigment circulation on, 200–202

Carotenoid coloration (*continued*)
  role of pigment stores on, 204–205
  role of pigment uptake on, 198
  in sexual dichromatism, 201
  signaling immune function, 220–224
Carotenoid limitation hypothesis, 195, 517–521
Carotenoid pigments
  3-hydroxy-echinenone, 511, 513
  absorptance function of, 82
  absorption from food, 196–198, 512, 519, 524–525, 539
  abundance of, 354
  anhydrolutein, 191, 213, 214, 217
  as antioxidants, 203, 206, 220–224, 271, 360
  astaxanthin, 181, 193, 207, 208, 215–216, 218, 362, 514, 516
  in bare-parts, 181, 182, 183, 207–211, 362, 451
  β-carotene, 178–179, 183, 191, 192, 197, 198, 200, 209, 211–213, 215–216, 512–513
  β-cryptoxanthin, 191, 192, 215–216, 511, 513
  binding proteins, 210, 214
  canary xanthophylls, 181, 207–209, 214, 216–219, 515
  canthaxanthin, 183, 193, 198, 200, 207, 209, 215–216, 218, 221, 516, 540
  in chylomicrons, 198–199
  coloration strategies using, 218–220
  colorimetry of, 80–83
  colors due to, 104, 115, 134
  combined with structural coloration, 341–342
  in cone oil droplets, 23
  cost of metabolic conversion of, 511, 513
  deposition of, 512, 515–516, 535, 539
  in diet, 179, 191–196, 200–201, 212, 213, 215, 217, 219, 220, 509–515
  efficiency of use of, 510
  in egg yolk, 522
  esterification, 210–211, 224
  in eyes, 182, 183, 340, 372
  feeding strategies for, 194–196
  in gap-junctional communication, 226
  as immunomodulators, 206, 220–224
  incorporation into tissues, 245
  interactions during digestion, 198
  interactions with keratin, 209
  ketocarotenoids, 179, 181, 207, 208, 217, 218, 497, 513–515
  light absorption by, 179, 209, 248, 249, 385
  lipoprotein binding, 198, 203, 212, 213, 219
  lutein, 80–82, 183, 191, 192, 194, 198, 199, 207, 208, 210, 212–217, 219, 221, 248, 362, 385, 512–515, 522

  metabolism of, 181, 191, 211–218, 496, 511, 512, 539
  in micelles, 198–199
  molecular structure of, 178–179, 211, 220
  in mouth parts, 182, 362–363, 523
  organizational effects of, 203, 206, 223
  as photoprotectants, 210–211, 224–225, 374
  physiological affinities for, 183, 204, 209–214
  in plasma, 183, 191, 199–204, 210, 212, 213, 216, 217, 219, 382
  as pro-oxidants, 225–226
  range of colors, 178–179, 207–208, 433
  reflectance spectra of, 81
  sex ratio manipulation in relation to, 223–224
  stereoisomers of, 179
  storage in internal tissues, 203–207, 212
  transport of, 512, 519, 524–525, 539
  types in birds, 180, 215
  xanthophylls, 179, 197, 198, 200, 207, 209–215, 218, 219, 221
  in young birds, 203, 205, 206, 223, 384
  zeaxanthin, 219, 221, 385, 512–515, 522
Chroma (definition), 101
Chromatic contrast, 128–129, 163–164
Chromatic index, 128
Chromaticity, 7–9, 43–44, 100
CIE XYZ color space, 43–44, 52, 59–60, 97, 99
CIELAB color space, 44, 99, 113, 114, 139
CIELUV color space, 99
CMY(K) color space, 98, 101
Coccidia, 197, 202, 239, 520, 524–526, 530–532, 536
Coherent scattering
  in barbs, 328–333, 341
  in barbules, 319–328
  definition of, 302
  description of, 83, 305, 307–309, 321–327
  Fourier analysis of, 311–315, 324–325, 329, 332–333, 334–335
  iridescence produced by, 315–316, 323–327
  in the iris, 339–340
  methods of analysis, 311–315
  reflectance spectra from, 71
Coincident normal, 53–54, 61–63, 71–73, 86
Collagen, 296–300, 304, 334–339, 342–344
Color charts, 41, 44
Color constancy, 7, 24, 28, 94
Color mixture experiments, 12
Color perception, 3–31, 48, 77, 90, 93–94, 96, 101, 104, 106, 114, 125, 126, 136
Color quantification
  color matching, 43–44, 90, 96, 98, 100, 101, 103, 104, 106, 136, 137

Subject Index

color ranking, 90, 91, 95, 96
comparing colors, 125, 126, 128–130
history of, 43–44
subjectivity, 45–49
using color charts and swatches, 41, 43, 90, 96, 103, 104, 106, 136, 137
using photography, 4, 44, 90, 105–107, 136
using spectrometers, 41, 47–85, 90, 92, 95, 96, 103, 106, 107, 111, 136, 137, 138
of UV, 46–47
Color vision
in birds, 13–30
evolution of, 19–23, 27–29
functions of, 28–30
general principles of, 3–16
in humans, 4–6, 23, 26–27, 46
Colortron™, 107, 109
Comb coloration, 184, 207, 209, 210, 361–362, 365, 449–452, 459, 546
Commission Internationale de l'Eclairage (CIE), 43, 97
Computer software
data analysis, 91, 106, 109, 116, 131, 138–139
graphics, 99, 105, 106
Cone photon capture, 7–9, 24, 28, 79
Cones
avian, 4–5, 16–23
double, 16, 23–26
human, 4–5, 101, 121
sensitivity, 10, 18–19, 21–22, 95, 122, 123, 125, 130, 137, 138, 140
single, 16, 23–26, 92, 101, 125–130, 140
stimulation, 92, 93, 101, 102, 126, 127, 129, 140
Constructive interference. *See* Coherent scattering
Contrast
achromatic definition, 129
chromatic definition, 108, 129
equation for, 108, 129
Cosine receptor, 150
Cosmetic colors, 399, 400–414, 421, 422
Cotingin, 385
Crypsis
effect of light environment on, 145, 155–156, 160, 163–166, 169
effect of soiling on, 399, 403, 414, 422
from melanin pigmentation, 245
Crystal-like nanostructures, 309, 312, 315–316, 321, 323, 327, 344

Dark current, 55, 66–71
Dark signal, 55, 66–71
Diester wax, 401, 402

Digital color meters
description, 90, 106, 107, 109, 137
Colortron™, 107, 109
Digital Swatchbook, 105, 107, 109
Dimensionality of color vision, 7–9, 27–28
Discriminant function analysis (DFA), 115, 120
Disruptive selection, 476
Dominance status, 104, 115, 120, 160, 365, 400, 408, 410–411, 414, 422, 533, 546

Eggshell coloration, 120, 245, 354–355, 358–360, 368–369, 370, 387
Electroretinography, 10
Endocrine disruptors, 453
Energy flux, 50
Estrogens
effects on bare-part coloration, 450–451
effects on carotenoid color, 446
effects on melanin color, 260, 261, 263–264, 266
effects on sexual dichromatism, 433, 434, 436–442, 444, 445, 449, 454–457
Eye coloration, 295, 337, 339–340, 354, 363–364, 370–377, 449

Feather-degrading bacteria, 211, 273, 401, 416, 535, 537–538
Feather mites, 273, 530–531, 533, 536
Feather wear, 70, 92, 113, 272, 345, 399, 401, 410, 411, 413–414, 416, 419, 422, 423
Flavin pigments
chemical methods for extraction, 379
dietary origin of, 379
in egg yolk and albumen, 354, 378–380
fluorescence by, 377–378
light absorption by, 377–379
molecular structure of, 377–378
occurrence in animals, 377–378
photobleaching of, 379
riboflavin, 182, 375, 377–380
role in development, 379
solubility of, 379
Fluorescence
in color displays, 167–168
of flavins, 377, 378
in natal plumage, 354
in penguin plumage, 354, 377
of porphyrins, 355, 360, 368
Follicle-stimulating hormone (FSH)
effects on melanin coloration, 264
effects on sexual dichromatism, 433, 439, 440

Fourier analysis, 311–315, 324–325, 329, 332–333, 334–335
Fourier power spectrum, 302, 312–314, 324–325, 328
Fourier transform, 302, 312–313, 324–325, 329
Fresnel equation, 323, 338

Genetics
  autosomal control, 473
  of bare-part coloration, 489
  captive breeding studies, 469–471
  of carotenoid coloration, 472, 489, 491, 493, 494, 497
  chromosomal bases of color, 477–478, 495
  color mutations, 471
  dominant inheritance, 473, 479–480, 483–484, 494
  epistasis, 473
  heritability of color, 485–491
  heterozygote advantage, 476
  hybrid forms, 471, 493–494
  of melanin coloration, 470, 472, 475, 478–484, 491, 493, 494, 497
  pleiotropy, 482, 487
  plumage polymorphisms, 471–482
  polygenic control, 484–487, 494
  of psittacofulvin coloration, 470–472
  quantitative trait loci (QTL), 492
  single-locus (oligogenic) control, 470–484
  of structural coloration, 470–471
Geographic variation, 104, 268, 273, 276, 519
Gloger's rule, 276–277, 534
Growth bars, 539–540, 543, 545
Guanine
  as a colorant, 374–377
  as a precursor for pterin, 370–374
  as a structural color component, 339–340

Hamilton-Zuk hypothesis, 523–524, 526–527
HCV color space, 97–98
Hematozoa, 528–531, 533, 536–537
Hen fleas, 533–534
High-performance liquid chromatography, 207, 218, 246, 355, 361, 372, 379, 381
Honest signaling
  by carotenoid coloration, 82–83, 177, 195, 224–225, 227, 346, 508–531, 539–540
  by cosmetic coloration, 401–402
  definition of, 507
  by eggshell coloration, 369
  by feather abrasion, 411, 414
  by melanin coloration, 256–257, 259, 261–262, 266, 531–536, 541–542

structural coloration, 29, 344–346, 536–538, 543–545
Hormones
  during development, 432, 447
  during molt, 434, 435, 441, 443, 444, 453, 457
  effects on alternate plumage, 441–444
  effects on coloration via status, 545–546
  effects on eclipse plumage, 441
  effects on intrasexual plumage variation, 446–446
  enzyme conversions, 433–434, 439, 441–443, 446, 458
  phylogenetic links to color, 454–457
  methods of study of, 432–435
  in skin, 439, 442
HSB color space, 96, 97, 99, 101–103, 105, 107, 109, 111–115, 117, 121, 123, 126, 134, 140
HSL color space, 97, 99
HSV color space, 99
Hue (definition), 101
Human color spaces, 7–9, 43–47, 77, 94, 96–99, 102–103, 105, 113–114

Immunocompetence
  handicap hypothesis, 259, 261, 452
  signaled by carotenoid coloration, 206, 220–224
  signaled by melanin coloration, 270
  testosterone and melanin, 259, 261–262
Incident light, 52–54, 66, 84, 85, 92, 93, 109, 121–123, 125, 133
Incoherent scattering, 84, 302, 305–307, 315, 317, 319, 335, 338, 377
Integrating spheres, 63–64, 73, 125
Intensity (definition), 100, 102, 108, 122
Iridescence
  components of, 304, 323–327
  condition dependence of, 344–345
  definition of, 302
  effect of hormones on, 263
  effect of nutrition on, 543
  effect of parasites on, 536
  evolution of, 327–328
  measuring, 43, 113, 128
  production of, 83, 302, 304, 308–310, 315–316, 321, 323–327, 366, 376
  taxonomic distribution, 496
  types of arrays that produce, 320
Iris coloration. *See* Eye coloration
Iron oxides, 400, 401, 415, 423

Keratin, 52, 70, 115, 304, 321, 323–333, 342–343, 399, 416, 419

Subject Index

Lab color system, 97, 99
Laminar nanostructures, 309, 311–312, 315–316, 321, 323, 324–329, 344
Light environment
    effect on colorimetrics, 95, 97, 103, 105, 109, 119, 122–124, 126, 136, 137, 140, 155–158, 161–162, 164, 166
    effect on conspicuousness, 155–156
    effect on perception, 92–94, 136
    effect on tropical lekking birds, 160–165
    of forests, 124, 149–153, 155
    history of the study of, 148–149, 157–160
    irradiance plots of, 124, 153
    measuring, 150–151
    for reflectance spectrometry, 53
    in relation to interspecific plumage variation, 165–167
    as a selective force, 148, 163–167
    spectral analysis of, 156–159
    standardized, 103, 104, 106, 107, 126, 136
    ultraviolet component of, 167
Lipoproteins, 198–203, 213, 219, 226, 525
Luteinizing hormone (LH)
    effects on melanin coloration, 260, 262, 264–265, 449
    effects on sexual dichromatism, 433, 434, 437, 439, 441, 443–445, 449, 451, 455–457, 470, 496
Luv color system, 97, 99

Mate choice, 42, 48, 85, 104, 119, 128, 160, 167–168, 220, 301, 399, 403, 451–453, 475
Maternal effects, 485, 520–521
Mean brightness (definition), 108
Measuring geometry, 53–54, 62–63, 70–73
Melanin coloration
    albinism, 243, 478
    analysis of, 134
    in camouflage, 245
    comparative studies of, 270
    condition-dependence of, 269–270
    effect of metabolic rate on, 262
    effect of parasites on, 531–536
    effect of pigment access on, 508
    effect of social status on, 546–547
    evolution of, 482
    genetic control of, 266–269, 470, 472, 475, 478–484, 487, 491, 493, 494, 497
    hormonal control of, 446–448
    leucism, 471
    melanism, 471, 473, 475, 480, 482, 483, 496
    minerals in synthesis of, 259–260, 272
    organizational effects on, 254, 258, 261
    patterns of, 244–245, 257, 481–484
    polymorphisms, 472
    range of, 244, 249–251
    role of amino acids in, 254, 258
    role of hormonal stimulation, 259–266, 269
    sexual dichromatism from, 257, 262, 263
    tyrosinase in synthesis of, 256, 265, 446
Melanin pigments
    absorptance function of, 82, 83
    abundance of, 177, 178, 243–245, 354
    as antioxidants, 271
    association with minerals, 255, 259, 271–272
    in beaks, 451
    as cation chelators, 271–272
    chemical methods for analysis, 246–248
    co-deposition with carotenoids, 83, 208, 251, 382, 433
    distribution in nature, 244, 250, 270
    effect of concentration on color, 115
    effect of parasites on, 531–536
    enzyme catalysis (tyrosinase), 255–256, 265, 268, 478
    eumelanin, 83, 246–252, 255–257, 263, 269, 305, 319, 472, 473, 478–480, 483, 490, 531–536
    in eyes, 372
    in feathers, 250–252, 438, 441, 442, 478–484, 490
    formation from amino acids, 254–256, 258
    as immunomodulators, 273–275
    light absorption by, 245, 248–250, 275, 355
    masking of other pigments by, 245, 365–366
    melanocortin-1 receptor (*MC1R*), 267–269
    melanophores in poikilotherms, 253
    molecular structure of, 247, 249–250, 271, 275
    in organelles (melanosomes), 253–254, 256
    phaeomelanin, 83, 246–252, 254–257, 263, 264, 266, 269, 320, 472, 478–480, 483, 490, 532–534, 536
    photoprotection by, 251, 275–276
    in pigment cells (melanocytes), 245, 251, 253–254, 267–269, 372, 478
    range of colors from, 244, 249–251, 433
    refractive index of, 305, 323
    in structural colors, 245, 263, 304, 317, 319–321, 324–328 328–329, 333, 338–340, 343–344, 483
    synthesis of (melanogenesis), 251, 253–257, 271, 479, 497
    thermoregulation from, 276–277
    in tissue strengthening, 272–273
    types of, 249–250

Melanocortin-1 receptor gene (*MC1R*), 267–269, 478–484, 495, 496–497
Melanocyte-stimulating hormones (MSH)
  effect on melanin coloration, 260, 268, 269, 479
  effect on sexual dichromatism, 439
Melanosomes, 253–257, 260, 271, 310, 317, 319–321, 323–325, 335, 343–344, 479
Methuen color system, 100, 104
Microspectrophotometry, 10–11
Mie scattering, 305–306, 332–333
Monoester wax, 401, 402
Mouth color
  effect of light environment on, 167
  environmental effects on, 362–363, 523
  measuring, 107
  mechanisms of coloration, 182, 361–363
Multivariate analysis of variance (MANOVA), 115
Multivariate statistics, 103, 115
Munsell color system, 97, 101, 102, 105, 123

Nanostructure
  anatomical classes of, 317–333
  of bare-parts, 333–339
  coherent scattering by, 307–309, 311–316
  combined with pigmentary colors, 341
  composition of, 304
  condition dependency of, 344–346
  definition, 303
  development of, 342–344
  in irides, 339–340
  occurrence in avian integument, 296–300
  pigments in, 316–317
  types of, 309–310
Nutrition, 183, 194, 197, 203, 220, 223, 258, 273, 278, 344, 433, 520–521, 538–545

Ocular filtering, 14, 23, 92, 93, 101, 122–126, 136, 138
Ocular media, 14, 18, 23, 92, 93, 95, 101, 122–126, 138
Oil droplets, 13, 16, 18, 23–26, 28, 30, 92, 93, 125, 128, 152
Opponent processing, 6, 10–11, 27–28
Optical distance, 303, 311, 324
Ostwald color system, 100
Outlier (statistical), 117, 130–133, 135, 140
Overall intensity, 108

Palmer color system, 100
Pantone color system, 100
Parasites
  effects on carotenoid coloration, 194, 197, 202, 221, 524–531

effects of foraging height on, 165
effects of iron oxide on, 401
effects on melanin coloration, 270, 273–274, 531–536
effects on structural coloration, 344, 536–538
measuring effects of, 520
related to bower decoration, 407
Path-length addition, 303, 308, 310, 315, 323, 327
Peak wavelength, 56, 108, 310
Peaky chroma, 108, 114
Phenylalanine, 254, 258, 541
Photon catch, 7–8, 91, 122, 123–128, 129, 130, 136, 138, 140
Photon flux, 50–52, 122, 126
Pigment (definition), 303
Pigments. *See* Carotenoid pigments, Flavin pigments, Melanin pigments, Porphyrin pigments, Psittacofulvin pigments, Pterin pigments, Purine pigments, Undescribed pigments
Polarization sensitivity, 13–14, 19, 155
Porphyrin coloration
  adaptive significance of, 359, 365, 367, 368–369
  effect of pigment access on, 508
  genetic basis of, 359
  physiological factors affecting, 359
  as a signal of diet quality, 367
  as a signal of metabolic state, 365
Porphyrin pigments
  as antioxidants, 360, 365, 369
  association with minerals, 355–356, 361, 366–367
  in bare-parts, 361–363, 365
  bilins, 355–359, 368–369
  bilirubin, 356, 358
  biliverdin, 354–356, 368–369
  chemical methods of extraction of, 357
  co-deposition with melanins, 365–366
  combined with structural coloration, 341
  coproporphyrin, 355–358, 360
  distribution of, 354, 355, 358
  in eggshells, 245, 358–360
  in eyes, 363–364, 372
  in feathers, 245, 354, 360–361, 365–367
  fluorescence from, 355, 368
  heme and hemoglobin, 178, 354–358, 360–363, 365, 368, 369
  insulating properties of, 359, 361
  light absorbance by, 359, 367, 369
  as metal chelators, 367
  metalloporphyrins, 355, 361–367
  molecular structure of, 355–356, 366
  natural porphyrins, 355, 357, 358–361
  photobleaching of, 369
  as prooxidants, 360

Subject Index

protein complexing with, 355
protoporphyrin, 355–358, 360, 361, 368
range of colors from, 245, 355
solubility of, 355, 366
synthesis of, 357–361, 368
turacin, 245, 354, 355, 357, 364–367
turacoverdin, 354, 355, 364–367, 496
uroporphyrin, 355–358, 360, 366
Presenting results, 134–137
Principal components analysis (PCA), 113, 115–121, 123, 140
Psittacofulvin coloration
condition-dependence of, 382
effect of body condition on, 547–548
effect of pigment access on, 508
genetic control of, 470–472
sexual dichromatism, 380
in sexual selection, 382–383
Psittacofulvin pigments
as antioxidants, 383
fluorescence in UV, 380
light absorption by, 380, 381
molecular interaction with keratin, 382
molecular structure of, 381
occurrence in animals, 380, 496
range of colors from, 178, 354, 380, 382
selective incorporation in feathers, 382
solubility of, 380
synthesis of, 381–382
Psychometrics, 44
Pterin coloration
age variability in, 373–374
control mechanisms of, 374
effect of pigment access on, 508
environmental control of, 548
seasonal variation in, 374
sex differences in, 373–374
Pterin pigments
as antioxidants, 374
cell ultrastructure of, 372–373
distribution of, 178, 354
in eyes, 253, 362, 370, 372
fluorescence under UV light, 247, 371, 374
as immunostimulants, 374
light absorption by, 371, 373
molecular structure of, 369–371
occurrence in animals, 370
pterinosomes, 372
range of colors from, 370
solubility, 247, 371
as structural color components, 304, 340, 342
synthesis of, 369–370, 374, 375
variety of forms of, 371–372

Purine coloration
age differences in, 377
sex differences in, 377
Purine pigments
cell ultrastructure, 376
co-deposition with pterins, 377
in eyes, 370, 374–377
guanine, 370, 374–377
hypoxanthine, 375–377
light absorption by, 376–377
molecular structure of, 376
as precursors of pterins, 370, 374
as structural color components, 304, 340

Quantum catch, 91, 93, 125
Quarter-wave stacks. *See* Laminar nanostructures
Quasi-ordered nanostructures, 309–312, 315–316, 321, 328–333, 335

Radiance, 94, 101, 102, 150,
Radiance receptor, 150
Radiant power, 50
Radiometer, 50
Rayleigh scattering, 305–306, 332–335
Reflectance
calculations of, 311
of combined pigmentary and structural colors, 341
definition of, 53–54, 150
infrared, 309, 324–325
measuring, 49–85, 90, 101, 102, 104, 105, 108–114, 117–121, 123, 125, 133, 137–139
of plumage coloration, 43–44, 47–56, 49–85
predicted by Fourier analysis, 314
predicted by thin-film optics, 321–324
Reflectance ratio, 108
Reflectance spectrometer
calibration, 66–67
colorimetrics from, 108
history, 41–42
light sources, 58–60
models, 55–56
operation of, 68–69
overview, 5, 51
reflection probes, 60–62
software, 57–58
Reflectance spectrometry
of carotenoid-based colors, 209
choice of measuring patch, 73–74
discussion of, 50–67, 70–85, 90, 92, 95, 96, 103, 106, 107, 111–114, 136–138
file organization, 75–76
of light environment, 157–159

Reflectance spectrometry (*continued*)
  of melanin-based colors, 247
  overview, 47–49, 111
  procedures, 68–69
  of pterin-based colors, 374
  sampling patches, 74, 107
  to test for incoherent scattering, 334
Reflected radiance, 53
Reflection
  definition, 53
  diffuse, 53
  specular, 53, 71
Refractive index, 303–305, 308–309, 311, 323–324, 338
RGB color space, 98–99, 101, 105, 139
Ridgway color system, 100

Saturation (definition), 101. *See also* Chroma
Segment classification, 91, 113, 121–123, 128, 129, 136, 140
Sensory adaptation, 7
Sensory drive, 148, 157, 160, 164–165, 168, 169
Sexual dichromatism
  in bare-parts, 449
  determined via photon catch, 130
  in plumage, 431, 432, 435–438, 442, 443, 445, 446, 448, 449, 454–457, 476, 483, 495, 496
Signal potential, 92, 94–95, 101, 102, 137
Smithe color system, 100
Social status, 342, 545–548
Soiling of feathers, 399, 402–404, 414–420, 422–423
SoLux® light box, 104
Spatial frequency, 303, 312–314, 325, 328–329, 337
Spectral intensity, 56, 77–78, 81, 84–86, 102, 108
Spectral location, 23, 59, 77–78, 83–84, 101, 108
Spectral power, 51–52, 58–59
Spectral purity, 77–78, 82, 84–86, 101, 108
Spectral radiance, 5, 102
Spectral range, 13, 14, 43, 47, 56
Spectral reflectance, 41–42, 51–52, 218, 385
Spectral saturation, 108, 113
Spectral sensitivity, 4, 11, 16, 18–19, 22–24, 26, 30, 56, 103, 123, 125, 126, 127, 130, 140
Spectral shape, 92, 117–119, 123
Spectralon™, 118
Spectroradiometer, 42, 50–52, 56
Spongy medullary keratin, 83, 310–311, 314, 319, 332–333, 341, 344
Staining of feathers, 399, 400–401, 404, 414–416, 422, 423
Structural coloration
  artificial selection for, 339
  of barbs, 328–333
  of barbules, 319–328, 343
  basal pigmentary layers in, 316–317
  combined with pigmentary coloration, 208, 341–342, 366, 382, 385
  condition dependence of, 344–346
  definition of, 295, 301, 303–304
  description of, 295–346, 496
  development of, 342–344
  effects of nutrition on, 538, 543–545
  effects of parasites on, 536–538
  effect of social status on, 546
  evolution of, 319, 327–328, 346
  of eyes, 339–340, 376
  of feathers, 295–301, 303–337, 341–347, 366, 438, 441
  genetics of, 470–471, 488–490, 497
  glossary of terms, 302–303
  history of the study of, 295
  as an honest signal, 84
  mechanisms of production, 305
  versus pigmentary colors, 301, 303–304
  reflectance from, 83–85, 115
  self assembly of, 342, 344–346
  signal components, 84
  of skin, 333–339, 362
  UV component of, 301
Subtractive color mixing, 80, 83
Surface normal, 52, 54, 62, 63

Testosterone
  effects on bare-part coloration, 450–451
  effects on carotenoid coloration, 448
  effects on melanin coloration, 261–263, 447
  effects on ornamentation, 546
  effects on sexual dichromatism, 433, 434, 442, 443
Thin-film optics, 311, 321, 324
Thyroxin
  effects on melanin coloration, 260, 265–266, 448
  effects on sexual dichromatism, 439
Topography of plumage regions, 76–77
Transmission, 150
Tristimulus color variables, 43, 94, 96–117, 120–121, 133–135, 138, 140
Turacin, 245, 354–355, 364–367, 547
Tyndall scattering, 43, 305–306, 332–335
Tyrosine, 254–258, 541

Ultraviolet vision, 14–16, 19–26, 29–30, 41–42, 337

Subject Index

Undescribed pigments
  in green feathers, 385
  in natal down, 384–385
  in penguin feathers, 384
Univariate statistics, 103, 114
Uropygial gland and oils, 400, 401, 402, 402, 404

Value (definition), 100, 105
Villalobos color system, 100
Visual background, 123, 125, 128–130, 148–149
Visual pigments, 4, 10–11, 14–23, 26, 29, 92, 94–95, 101–102, 126, 128, 137

Wattle coloration, 73, 178, 207, 210, 361–362, 449–452, 539
"Weaver finch test," 264

Websites, 138, 139
White feathers
  coloration affected by carotenoid access, 193, 204
  as displays, 160–164
  evolution of, 319
  genetic control of, 472–475, 478, 480, 487, 489, 491, 494
  hormonal control of, 439–440
  mechanisms of production, 84, 317–319, 376–377
  related to reproductive success, 120
  resistance to degradation, 273–275, 535–536
  soiling of, 403–404, 414–418
  thermoregulation by, 276–277
White standard, 54–56, 63–71, 86, 118, 125